SURFACE SCIENCE
VOL. I

INTERNATIONAL CENTRE FOR THEORETICAL PHYSICS, TRIESTE

SURFACE SCIENCE

LECTURES PRESENTED AT
AN INTERNATIONAL COURSE
AT TRIESTE FROM 16 JANUARY TO 10 APRIL 1974
ORGANIZED BY THE
INTERNATIONAL CENTRE FOR THEORETICAL PHYSICS, TRIESTE

In two volumes

VOL. I

INTERNATIONAL ATOMIC ENERGY AGENCY
VIENNA, 1975

THE INTERNATIONAL CENTRE FOR THEORETICAL PHYSICS (ICTP) in Trieste was established by the International Atomic Energy Agency (IAEA) in 1964 under an agreement with the Italian Government, and with the assistance of the City and University of Trieste.

The IAEA and the United Nations Educational, Scientific and Cultural Organization (UNESCO) subsequently agreed to operate the Centre jointly from 1 January 1970.

Member States of both organizations participate in the work of the Centre, the main purpose of which is to foster, through training and research, the advancement of theoretical physics, with special regard to the needs of developing countries.

SURFACE SCIENCE, IAEA, VIENNA, 1975
STI/PUB/396
ISBN 92-0-130275-4

Printed by the IAEA in Austria
August 1975

FOREWORD

The International Centre for Theoretical Physics, pursuing its objective of research and training with a comprehensive and synoptic coverage in various disciplines, has already presented three courses on solid-state physics: Theory of Condensed Matter (1967); Theory of Imperfect Crystalline Solids (1970); and Electrons in Crystalline Solids (1972). These Proceedings have been published by the International Atomic Energy Agency.

The present Proceedings constitute a fourth international course, held from 16 January to 10 April 1974, devoted to surface science, a new topic, holding the centre of current interest in solid-state physics. The directors of the course were V. Celli and G. Chiarotti (Italy), F. García-Moliner (Spain), S. Lundqvist (Sweden), N.H. March and J.M. Ziman (United Kingdom).

Generous grants from the Swedish International Development Authority and the United Nations Development Programme are gratefully acknowledged.

Abdus Salam

EDITORIAL NOTE

The papers and discussions have been edited by the editorial staff of the International Atomic Energy Agency to the extent considered necessary for the reader's assistance. The views expressed and the general style adopted remain, however, the responsibility of the named authors or participants. In addition, the views are not necessarily those of the governments of the nominating Member States or of the nominating organizations.

Where papers have been incorporated into these Proceedings without resetting by the Agency, this has been done with the knowledge of the authors and their government authorities, and their cooperation is gratefully acknowledged. The Proceedings have been printed by composition typing and photo-offset lithography. Within the limitations imposed by this method, every effort has been made to maintain a high editorial standard, in particular to achieve, wherever practicable, consistency of units and symbols and conformity to the standards recommended by competent international bodies.

The use in these Proceedings of particular designations of countries or territories does not imply any judgement by the publisher, the IAEA, as to the legal status of such countries or territories, of their authorities and institutions or of the delimitation of their boundaries.

The mention of specific companies or of their products or brand names does not imply any endorsement or recommendation on the part of the IAEA.

Authors are themselves responsible for obtaining the necessary permission to reproduce copyright material from other sources.

CONTENTS OF VOLUME 1

Part I. STRUCTURE OF SURFACES

Surface thermodynamics (IAEA-SMR-15/1)	3
F. García-Moliner	
Ideal surface structures (IAEA-SMR-15/4)	77
S. Andersson	
Surface waves (IAEA-SMR-15/5)	113
S. Andersson	
Introduction to the theory of low-energy electron diffraction (IAEA-SMR-15/34)	145
A. Fingerland, M. Tomášek	
Low-energy electron diffraction and surface topography (IAEA-SMR-15/33)	173
G.A. Somorjai	
Nucleation theory and crystal growth (IAEA-SMR-15/11)	265
A.J. Forty	
Brillouin-Wigner theory and some surface problems (IAEA-SMR-15/10)	283
R.O. Jones	
Liquid surfaces (IAEA-SMR-15/9)	291
M.V. Berry	

Part II. PROPERTIES OF SURFACES

Electrons at metal surfaces (IAEA-SMR-15/2)	331
S. Lundqvist	
Electromagnetic surface excitations (IAEA-SMR-15/8)	393
V. Celli	
Surface spectroscopy (IAEA-SMR-15/15)	423
G. Chiarotti	
Electrical transport in the space-charge region (IAEA-SMR-15/31)	447
A. Many	
Faculty	501

Part I
STRUCTURE OF SURFACES

SURFACE THERMODYNAMICS

F. GARCIA-MOLINER
Centro Coordinado del CSIC en la UAM,
Departamento de Física,
Universidad Autónoma de Madrid,
Canto Blanco, Madrid,
Spain

Abstract

SURFACE THERMODYNAMICS.
1. Physical considerations. 2. Basic thermodynamics of a system consisting of two bulk phases with an interface. 3. Solid surfaces: general. 4. Discussion of experimental data on surface tension and related concepts. 5. Adsorption thermodynamics in the Gibbsian scheme. 6. Adsorption on inert solid adsorbents. 7. Systems with electrical charges: chemistry and thermodynamics of imperfect crystals. 8. Thermodynamics of charged surfaces. 9. Simple models of charge transfer chemisorption. 10. Adsorption heat and related concepts. 11. Surface phase transitions.

1. PHYSICAL CONSIDERATIONS (Flood (1967); Defay et al. (1966))

If we start the thermodynamical analysis of surface systems by writing down an appropriate form of the first law, i.e. by making an energy balance, the only novelty concerns the mechanical work terms, because of the extra work involved in changing the interface either in shape or in extent. Vague as this may be, it is clear that exchanges of heat or material contents in which the interface participates are described in terms of parameters like temperature T and chemical potentials μ, just as if we had an equilibrium between two homogeneous bulk phases with no surface effect, while new parameters — other than hydrostatic pressure in the bulk phases — must be introduced to describe the extra 'surface work' terms.

It is clear that these terms will involve geometrical variables and parameters which will have the nature of mechanical forces. Thus we must start with a clear understanding of the conditions required for mechanical equilibrium at an interface. Any model system for which we write down the thermodynamical analysis must be mechanically equivalent to the real system.

Now, if we consider a wall (Fig. 1) and a system of two bulk fluid phases (α and β) and an interface, the forces exerted on the wall will vary between, say, A and B in a complicated fashion. It is conceivable that given sufficient detailed information we could specify the force per unit area of wall for each different position between A and B. The idea is to introduce a geometrical surface which is thought of as a membrane without rigidity, but one that takes work to stretch, i.e. that has a surface tension γ. The model system (Fig. 2) consists then of bulk phases α and β on each side of the 'membrane' plus the said membrane which is called the surface of tension and which is described as pulling the wall with a surface tension γ.

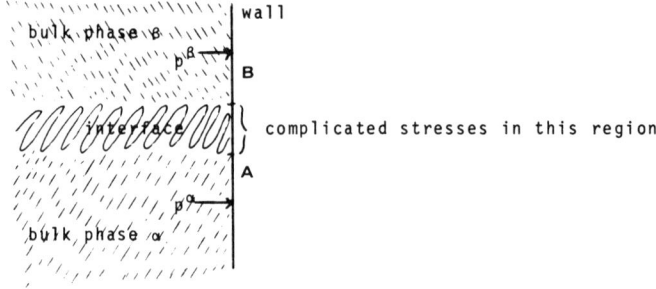

FIG.1. Forces exerted on a wall AB in a system of two bulk fluid phases (α and β) and an interface.

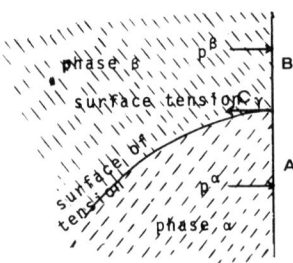

FIG.2. The model system.

Suppose, for example, that the interface has a cylindrical symmetry, so that the surface of tension is cylindrical. We have two unknowns: γ and the location of the intersection of the surface of tension and the wall, somewhere between A and B. And we have two conditions: (i) the total force exerted on AB must be the same, and (ii) the moment of this force about a fixed axis must also be the same (Defay et al. (1966)). With this we can in principle determine the two unknowns so that the model system is mechanically equivalent to the real system. For an increase dA in the area of the surface of tension the surface work term is then γdA, so that the mechanical work involved in an infinitesimal deformation of the system is

$$dW = -p^\alpha dV^\alpha - p^\beta dV^\beta + \gamma dA \qquad (1.1)$$

Suppose we displace the dividing surface. Then $dV^\alpha = -dV^\beta$ and the principle of virtual work yields the equilibrium condition:

$$(p^\alpha - p^\beta) dV^\alpha = \gamma dA \qquad (1.2)$$

If (α) is, for example, a spherical drop of water this leads to the well-known Laplace equation:

$$p^\alpha - p^\beta = \frac{2\gamma}{R} \tag{1.3}$$

(R = radius of the sphere).

Notice that (1.1) is a correct and complete expression for dW because our mechanical model is constructed in terms of the surface of tension. Otherwise we should have to include terms describing the work necessary to change the curvatures of the surface. We could still construct a mechanically equivalent system, but with a different dividing surface we would have

$$dW = -p^\alpha dV^\alpha - p^\beta dV^\beta + \gamma dA + C_1 d\kappa_1 + C_2 d\kappa_2 \tag{1.4}$$

where κ_1 and κ_2 are the principal curvatures (Melrose (1970)).

2. BASIC THERMODYNAMICS OF A SYSTEM CONSISTING OF TWO BULK PHASES WITH AN INTERFACE
(Tolman (1948); Kirkwood and Oppenheim (1961); Flood (1967); Defay et al. (1966))

We shall start out with elementary background and somewhat pedestrian remarks which, nevertheless, will become useful later on.

Recall some basic notions.

Energy balance:

$$dE^\alpha = TdS^\alpha - p^\alpha dV^\alpha + \mu dN^\alpha \tag{2.1}$$

for a homogeneous phase (α) in a system of purely hydrostatic stresses. Otherwise pdV might be replaced by a sum of similar terms, all describing mechanical work. We can think of μdN as short for $\Sigma_i \mu_i dN_i$, if different species (components) are present. Here N = n⁰ of molecules (or atoms, or ions, or electrons).

The usual argument on intensive and extensive variables leads to the integrated relation:

$$E^\alpha = TS^\alpha - p^\alpha V^\alpha + \mu N^\alpha \tag{2.2}$$

This argument is not purely formal. The idealized process in which the system grows, or different parts are added, thus having the same growth or addition of extensive properties — at constant intensive properties — must correspond to a conceivable physical process which is possible for the system under study (see Flood for a careful discussion). Among other considerations it follows that the above equations can only hold in the absence of surface effects, because it would not be the same to have the system grow, say, perpendicular or parallel to the surface, i.e. the thermodynamical analysis of a surface system must include, in the energy balance equation, mechanical work terms in which explicit reference is made to the way in which the system grows relative to the surface.

Now, S is not even an intuitive independent variable, and V is often not a convenient one. Thus one introduces the Gibbs function $G = E - TS + pV = \mu N$, such that, for phase α,

$$dG^\alpha = -S^\alpha dT + V^\alpha dp^\alpha + \mu dN^\alpha \tag{2.3}$$

T and p are convenient intensive properties. But the complete description of the state of the system requires the specification of the material contents, i.e. the numbers N_i^α. We could use the concentrations $n_i = N_i/V$, which are also intensive parameters, and more intuitive than the chemical potentials, and then we should have to specify the size, i.e. V^α. Notice, however, that we can divide all the above equations by V^α; the size does not matter. In fact when we do thermodynamics we assume that we are in the thermodynamic limit and all the fundamental significance remains attached solely to basic equations like energy balance per unit volume.

Thus we have $r+2$ intensive variables (r is the number of components), namely, T, p and the n_i's or, alternatively, T, p and the μ_i's. However, it is impossible to write down an energy balance in terms of differentials of the latter set of variables. If we go on subtracting terms from E we arrive at $E - TS + pV - \mu N = 0$, hence the identity:

$$-S^\alpha dT + V^\alpha dp^\alpha - N^\alpha d\mu = 0 \tag{2.4}$$

This is the Gibbs-Duhem equation. Notice the difference: (2.3) is a formula. From it we can deduce, for example, that $V = +(\partial G/\partial p)_{T,N}$. But (2.4) is an identity which establishes a relationship between the intensive variables. It says, for example, that p cannot be varied quite arbitrarily, or else that its variations may then restrict the possible variations of the other intensive parameters. Thus (2.4) must be taken into account when evaluating all the partial derivatives obtained from the energy balance, i.e. for each relation like (2.4) we have one independent variable less, and if we have ν phases we have $r + 2 - \nu$ independent variables or degrees of freedom (Gibbs' phase rule; we do not treat chemical reactions).

A trivial example is a liquid in equilibrium with its vapour. Here $r + 2 - \nu = 1 + 2 - 2 = 1$. Thus write, for example, dividing (2.4) by V^α for $\alpha = \ell, g$:

$$s^\ell dT + n^\ell d\mu = dp$$
$$s^g dT + n^g d\mu = dp \qquad (2.5)$$

and take p as the one independent variable. This yields dT and $d\mu$ in terms of dp. (For the two phases in equilibrium T, p, μ are equal.) Thus

$$dT = \frac{n^g - n^\ell}{s^\ell n^g - s^g n^\ell} dp \qquad (2.6)$$

which is the Clausius-Clapeyron equation. Considerations of this kind are often important in surface thermodynamics.

Incidentally, the existence of an equation of state for a given phase does not decrease the number of independent thermodynamical variables because this equation does not arise from (macroscopic) thermodynamical equalities or arguments. It is, instead, a relationship which says something about interparticle or intermolecular interactions and therefore must be derived either empirically or from specific models, thereby introducing new parameters. This remark is also relevant to the thermodynamics of adsorption, when one is often interested in an equation of state for the adsorbed substance.

At this stage one introduces into the analysis the surface work terms referred to above and discussed in Section 1. Thus we start from

$$dE = TdS + dW + \mu dN$$

instead of (2.1). We insist now that dW must include the surface work terms, as written in (1.1), i.e. our basic equation is

$$dE = TdS - p^\alpha dV^\alpha - p^\beta dV^\beta + \gamma dA + \mu dN \qquad (2.7)$$

Notice what exactly we are doing here: (i) We introduce a model system, as in Section 1. (ii) This model is mechanically equivalent to the real system. (iii) Equation (2.7) holds under the same assumptions of (1.1). In particular, the dividing surface is the surface of tension. So far the position of the dividing surface is not at all arbitrary. For instance, when curvature effects are studied (Melrose (1970); Defay et al. (1966); Tolman (1948)) one must write down:

$$\gamma dA + C_1 d\kappa_1 + C_2 d\kappa_2 \qquad (2.8)$$

instead of simply γdA. Here κ_1 and κ_2 are the two principal curvatures. We can rewrite

$$\gamma dA + \frac{1}{2}(C_1 + C_2)d(\kappa_1 + \kappa_2) + \frac{1}{2}(C_1 - C_2)d(\kappa_1 - \kappa_2) \qquad (2.9)$$

If the two principal curvatures are equal or sufficiently similar, then the last term can be neglected, but we are still left with the second term. (In the cylindrical case we would have, say, $C_1 d\kappa_1$.) The surface of tension is the particular choice for which this term vanishes. Its intuitive meaning is expressed by Tolman (1948) in these words: "... since change in curvature of the dividing surface would lead to increase in the lamelliform distribution of matter on one side of the surface and decrease on the other, it would seem reasonable to expect that a location for the dividing surface could be found which would make $\frac{1}{2}(C_1 + C_2)$ equal to zero through a balancing of the consequences of such opposing changes in distribution".

The Gibbs method of studying the thermodynamics of the kind of system we are considering is suggested by the notion of the surface of tension. Consider any extensive property — S, N, etc. Down in the bulk of the two phases it has a given uniform density, say s^α. Imagine homogeneous bulk phases α, β supposedly extending right up to the dividing surface with volumes V^α, V^β. They would have total amounts of extensive properties like, say, $S^\alpha = s^\alpha V^\alpha$, etc. This is not of course the actual entropy of all matter on the side α of the dividing surface. Thus $S^\alpha + S^\beta \neq S$, the total entropy of the real system. One then defines the 'surface entropy' S^σ by

$$S = S^\alpha + S^\beta + S^\sigma \qquad (2.10)$$

None of the three quantities of the right-hand side is the actual entropy of any part of the real system, but the scheme is a convenient and correct one to define the surface excess (which can be positive or negative) or surface amount of all extensive properties. The only real division is $V = V^\alpha + V^\beta$.

Now, for each bulk phase we can legitimately write (2.1) and, subtracting $dE^\alpha + dE^\beta$ from (2.7), we obtain the differential equation for the change in surface energy:

$$dE^\sigma = TdS^\sigma + \gamma dA + \Sigma_i \mu_i dN_i^\sigma \qquad (2.11)$$

which leads to the integrated relation:

$$E^\sigma = TS^\sigma + \gamma A + \Sigma_i \mu_i N_i^\sigma \qquad (2.12)$$

From now on we shall restrict our discussion to flat interfaces ($p^\alpha = p^\beta = p$).

Notice that we introduce one more thermodynamical variable, but from (2.11) and (2.12) we have one relation among them:

$$Ad\gamma + S^\sigma dT + \Sigma_i N_i d\mu_i = 0 \tag{2.13}$$

i.e. the Gibbs-Duhem equation for the 'surface phase'. Thus we still have the same number of degrees of freedom as if we had two bulk phases in equilibrium with no surface effects. (For surfaces of arbitrary shape see Defay et al. (1966).) We shall soon come back to this. For the time being let us comment on the thermodynamical meaning of γ as it appears in (2.11 - 13). We introduce the Kramers function:

$$\Omega = F - G \tag{2.14}$$

For a bulk phase $\quad \Omega = -pV$

and

$$-S = \left(\frac{\partial \Omega}{\partial T}\right)_{\mu, V} \quad ; \quad -N = \left(\frac{\partial \Omega}{\partial \mu}\right)_{T, V} \tag{2.15}$$

Of all contributions to the internal energy, Ω measures the part associated with work against external forces. For the surface phase (a term henceforth used but not meant literally) we have

$$\Omega^\sigma = F^\sigma - G^\sigma = \gamma A \tag{2.16}$$

Thus the parameter γ introduced in the analysis is, strictly speaking, the Kramers energy per unit surface area. The characteristic function Ω is particularly convenient to describe the surface phase because it is a thermodynamic potential for (T, μ_i, γ) and (T, μ_i) have common values for two fluid bulk phases in equilibrium through an interface, while the volume terms pV have entirely disappeared from (2.16). Under these conditions, equilibrium at constant (T, μ_i) is described simply by the principle of virtual work and corresponds to Ω^σ = minimum with respect to changes in the surface at constant (T, μ_i) and bulk pressures. The various names and meanings attributed to γ in the literature are the source of considerable confusion, but it should be clear from (2.16) that γ is simply the surface concentration of an unambiguously defined characteristic function or thermodynamical potential. For example, it is not the surface concentration of surface free energy, although it may have the same value under certain circumstances. One is so often misled by obscure presentations and careless wording that it seems worth spelling this out in some detail. Suppose we displace the surface of separation by a normal displacement $\delta \ell$ of all its elements dA (Fig. 3) so that phase α gains a volume:

$$V^\alpha = \int \delta \ell \, dA \equiv \delta V$$

while phase β loses δV. Then the extensive quantities X change by $\pm x \delta V$, where x is the volume density of X in the corresponding homogeneous bulk. Thus the change per unit surface area is $\pm x \delta l$. Therefore, from (2.12), the surface tension changes by

$$\delta \gamma = [(e^\alpha - e^\beta) - T(s^\alpha - s^\beta) - \Sigma_i \mu_i (n_i^\alpha - n_i^\beta)] \delta l \qquad (2.17)$$

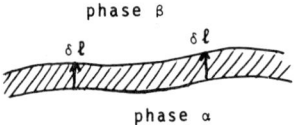

FIG.3. Shaded band shows volume gained by phase α and lost by phase β.

Now, for each bulk phase we have an integrated relation (2.2), divided by the corresponding volume. Thus, in general

$$e^\alpha - Ts^\alpha - \Sigma_i \mu_i n_i^\alpha = p^\alpha \qquad (2.18)$$

and likewise for phase β. Hence

$$\delta \gamma = (p^\alpha - p^\beta) \delta l \qquad (2.19)$$

Notice the particular property of the plane interface. Even starting from the surface of tension — the only case in which we could claim (2.11 - 12) in general — the surface tension depends on the choice of dividing surface, and so of course do the arbitrarily defined surface excesses. However, for a plane interface $p^\alpha = p^\beta$. We thus reach a very important statement: For a plane interface γ is independent of the choice of dividing surface. Now consider the Gibbs-Duhem equation (2.13), also called the Gibbs adsorption equation. Divide by A and define the surface concentrations:

$$s^\sigma = \frac{S^\sigma}{A} \quad ; \quad n_i^\sigma = \frac{N_i^\sigma}{A} \quad ; \quad \text{etc.} \qquad (2.20)$$

Thus

$$d\gamma = -s^\sigma dT - \Sigma_i n_i^\sigma d\mu_i \qquad (2.21)$$

This equation must also be invariant with respect to the choice of dividing surface, otherwise we could start from an initial state with a given γ and then reach by a reversible process another state in which γ would depend on the choice of dividing surface.

Indeed, since each surface excess is defined as the difference between two volume terms whose change upon a displacement of the surface we can easily write down, it is trivial to verify, by using the definitions like (2.10) and the Gibbs-Duhem equations like (2.4), that $d\gamma$ is identically invariant. What this in fact yields is the Gibbs-Duhem equation for the entire system, surface effects included.

Notice that the invariance of (2.21) does not mean that the surface excesses remain invariant — by definition they do not — because the variations dT and $d\mu_i$ are not independent of each other; they are related to dp by (2.4) again. Consider, for example, the one-component case. For isothermal processes, if hydrostatic equilibrium is maintained, μ cannot change, since

$$d\mu = -\frac{s^\beta - s^\alpha}{n^\beta - n^\alpha} dT \qquad (2.22)$$

as is easily obtained from (2.4) applied to α and β with a common pressure. This trivial example shows that the calculation of partial derivatives obtained from equations describing the thermodynamics of the 'surface phase' depends on relations prevailing in the bulk of the homogeneous bulk phases.

We are now prepared to discuss the question previously raised about the meaning of γ. We can write

$$\gamma = f^\sigma - \Sigma_i n_i^\sigma \mu_i \qquad (2.23)$$

which is neither the Helmholtz nor the Gibbs free energy per unit area. However, we have just shown that γ is invariant with respect to the choice of dividing surface. Therefore we are free to make the particular choice for which

$$\Sigma_i n_i^\sigma \mu_i \equiv 0 \qquad (2.24)$$

This, or a similar one for which the surface excess of a chosen component is zero, was already studied by Gibbs, who realized that it is particularly convenient to study adsorption and similar problems. Of course f^σ does change with the choice of the dividing surface and only with the particular choice (2.24) is the excess free energy equal to γ, which remains always invariant.

It follows from the above that the surface work due to a reversible change in area at constant T, μ_i is

$$\left(\frac{\partial \Omega^\sigma}{\partial A}\right)_{T,\mu_i} dA = \gamma \, dA \qquad (2.25)$$

To complete the analysis we must consider a reversible process in which no surface work is involved. It is intuitively obvious that this process does not stretch the surface but it changes the surface tension. In fact, formally, we can write down, for the energy balance in this process,

$$d\omega^\sigma = \left(\frac{\partial \omega^\sigma}{\partial T}\right)_{\mu_i} dT + \sum_i \left(\frac{\partial \omega^\sigma}{\partial \mu_i}\right)_{T,\mu_j \neq \mu_i} d\mu_i \qquad (2.26)$$

But we know

$$\frac{\Omega^\sigma}{A} = \omega^\sigma = \gamma \qquad (2.27)$$

and we have written down $d\gamma$ in (2.21). In other words: (a) The physical meaning of the Gibbs adsorption equation is that it expresses the energy balance for a reversible process in which no surface work is performed. (b) We have obtained, incidentally, the formulae

$$-s^\sigma = \left(\frac{\partial \gamma}{\partial T}\right)_{\mu_i, A} \quad ; \quad -n_i^\sigma = \left(\frac{\partial \gamma}{\partial \mu_i}\right)_{T,\mu_j \neq \mu_i, A} \qquad (2.28)$$

just as in a homogeneous bulk phase.

Let us recapitulate. We have discussed plane fluid interfaces in equilibrium. We have briefly indicated how curvature effects can be included. The extension to situations in which the interface and the two bulk homogeneous phases are not in equilibrium is discussed at length by Defay et al. (1966) and we shall not cover it here. There are two questions of importance for the physics of solid surfaces: (a) the relationship between surface stress and surface tension; and (b) the electrochemical equilibrium between bulk surface and possibly a gas environment. This involves the surface charge, which may be due to intrinsic or extrinsic surface states and leads to the discussion of charge transfer chemisorption on semiconductors, which will be discussed in some detail.

3. SOLID SURFACES: GENERAL

As long as we deal with fluid surfaces it seems that mechanical considerations fit in quite naturally in the thermodynamical scheme. The formula yielding the pressure difference between the inside and the outside of a liquid droplet, for example, can be arrived at from the principle of virtual work, but the argument fits in with the thermodynamical analysis, where it means simply that the surface concentration of the Kramers energy seeks its minimum under conditions of thermal and adsorption equilibrium of the droplet with the environment. A formal thermodynamical analysis corresponds so well to the literal treatment of a high pressure inside a membrane, swelling it up to the point in which this pressure equilibrates the forces of the tension in the membrane plus the outer pressure, that one is forcibly carried almost to the conviction that the model system is more than a formal device. But it is important to remember that the concept of surface tension is arrived at by defining an excess quantity, which implies that, far enough away from the interface, one can define homogeneous concentrations or densities of free energy, matter, etc.

In the case of solid surfaces there are essentially two difficulties. One is that there are mechanical stresses which may become very inhomogeneous and/or anisotropic. For example, the state of stresses is characteristic of each crystal face and this is bound to induce a complicated state of stresses in the interior of the solid. Thus, it is not obvious a priori that, say, a region of homogeneous energy density can be defined. The other difficulty is that the concept of surface tension is isotropic, whereas surface stresses can obviously be anisotropic. Thus we expect some difference between the thermodynamical concept (for which we shall continue to use the term surface tension) and the mechanical concept of surface stress.

Related to the first difficulty is the long-standing one (Herring (1953); Defay et al. (1966)) that in the interior of a solid it is not always possible to define a chemical potential unambiguously. This might be a serious difficulty for small crystals, and might appear in problems of adsorption on very porous specimens in which the interior of the crystallites may be highly strained and diffusion equilibrium with the surface might fail to be achieved. Thus this is not a purely academic question, and points all the more to the desirability of reliable data on more tractable crystalline specimens — fortunately an increasing possibility with modern techniques. With these reservations, however, it would be expected in most cases, for the sort of sizes one is interested in, that it is reasonable to ignore the first difficulty, as suggested by consideration (Defay et al. (1966)) that the elastic energy is only a small part of the total energy of a solid. Although the complicated state of stresses obviously cannot always be reduced to a simple hydrostatic pressure, it is clear that other terms could not amount to a larger order of magnitude than that of the pV term if the crystal is not too small. Consider the work needed to compress isothermally one mole of substance to its actual pressure:

$$W = -\int p\,dV = \int p\left(-\frac{\partial V}{\partial p}\right)_T dp = \kappa \mathscr{V}_m \int_0^p p\,dp = \frac{1}{2}\mathscr{V}_m \kappa p^2 \qquad (3.1)$$

Here κ is the isothermal compressibility and $\mathscr{V}_m$ is some representative value of the molar volume. For typical metals we can take $\mathscr{V}_m \lesssim 10$ cm^3·mole^{-1}. As for κ, we find values ranging from $\sim 0.6 \times 10^{-2}$ cm^2·dyne^{-1} (Fe) to ~ 0.7 (Cs). This yields values of W ranging from $\sim 2.5 \times 10^{-4}$ cal·mole^{-1} to $\sim 2.5 \times 10^{-2}$ cal·mole^{-1} for compressions of up to 100 atm. Now, only the classical Doulong-Petit thermal energy is of order $RT = 5.7$ T cal·mole^{-1}, i.e. $\sim 1.7 \times 10^3$ cal·mole^{-1} at T = 300 K. We shall therefore assume that

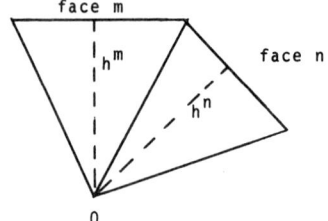

FIG.4. Convex polyhedron corresponding to crystal shape.

except for very small sizes we can define surface excesses predicated on an assumed homogeneity of the bulk.

These considerations allow us to elucidate certain aspects of the purely energetic (or thermodynamical) analysis of solid surfaces. Following Defay et al. (1966), we start by defining the geometrical variables which describe the size and shape of a crystal with plane faces. Take any point 0 in the interior and draw lines perpendicular to the faces. Call h^m, h^n, etc., the distances from 0 to faces m, n, etc. (Fig. 4). We end up with a convex polyhedron corresponding to a certain crystal shape. We can imagine pyramids based on the faces, with a common vertex 0. The volume of the crystal is

$$V^c = \frac{1}{3} \sum_m h^m A^m \tag{3.2}$$

The following relationships are easily established:

$$dV^c = \sum_m A^m dh^m \tag{3.3}$$

This one is obvious. Then, from cross differentials

$$\frac{\partial A^m}{\partial h^n} = \frac{\partial A^n}{\partial h^m} \tag{3.4}$$

from Euler's theorem for homogeneous functions

$$A^m = \frac{1}{2} \sum_m h^n \frac{\partial A^m}{\partial h^n} = \frac{1}{2} \sum_n h^n \frac{\partial A^n}{\partial h^m} \tag{3.5}$$

and finally from (3.2) and (3.3), we have

$$dV^c = \frac{1}{2} \sum_m h^m dA^m \tag{3.6}$$

Defay et al. (1966) give an argument which illustrates the relationship between the thermodynamical and mechanical concepts rather well. Let us develop a somewhat simplified version of this argument. Consider the equilibrium shape of a crystal. (We are not concerned here with kinetic considerations, which are decisive in studying how a crystal grows and which may even imply that a given specimen need not in fact have its true equilibrium shape.) Imagine a virtual displacement of the faces A^a through distances dh^a, carried out reversibly and isothermally and in such a way that the intensive state of the bulk phases does not change. Under these conditions we describe equilibrium by the minimum of the Kramers function of the entire system. Now, for this process

$$d\Omega = \omega^c dV^c + \omega^g dV^g + \sum_m \gamma^m dA^m = 0 \tag{3.7}$$

Here we allow for γ to take different values for different crystal faces. Moreover, by (3.6)

$$dV^c = \frac{1}{2} \sum_m h^m dA^m = -dV^g \tag{3.8}$$

Thus

$$\sum_m \left[\frac{1}{2} (\omega^c - \omega^g) h^m + \gamma^m \right] dA^m = 0 \tag{3.9}$$

One shape which is consistent with these conditions is that for which

$$\frac{\gamma^m}{h^m} = \frac{1}{2} (\omega^g - \omega^c) \quad \text{(all faces m)} \tag{3.10}$$

These are the well-known Wulff's relations. From this some interesting conclusions can be derived. Thus for the gas phase we have

$$\omega^g = \frac{1}{V^g} (F^g - G^g) = -p^g \tag{3.11}$$

while for the crystal we can only write

$$\omega^c = f^c - \sum_i n_i^c \mu_i^c \tag{3.12}$$

We imply that we are now focusing attention on small crystal, otherwise all $h^m \to \infty$ and then (3.10) is identically zero on both sides. Consider the case of one single component. The above relations yield

$$\mu^g = \bar{v}^c \left(\frac{2\gamma^m}{h^m} + p^g + f^c \right) \tag{3.13}$$

where $\bar{v}^c = 1/n^c$. We have obtained a formula for the chemical potential of the vapour in equilibrium with the crystal. All faces are in equilibrium with a vapour with the same T and p^g and therefore with the same μ^g, as is ensured by (3.10). Now, for a very large crystal ($h^m \to \infty$) we have

$$\mu_\infty^g = \bar{v}^c (p_\infty^g + f_\infty^c) \tag{3.14}$$

But $(f^c - f_\infty^c)$ is in practice rather negligible even for linear dimensions of order 10 Å (Defay et al. (1966)). We are also neglecting the compressibility of the solid, which is quite reasonable. Thus

$$\mu^g - \mu_\infty^g = \frac{2\gamma^m}{h^m} \bar{v}^c + (p^g - p_\infty^g) \bar{v}^c \tag{3.15}$$

where in fact the last term is also usually negligible. So far this is a purely thermodynamical formula. Irrespective of the details of a complicated state of stresses in the interior of a small crystal, we can calculate the difference between the chemical potentials of the vapour in equilibrium with a large or small crystal.

Now let us turn the argument round. Suppose we have a given small crystal, but suppose we estimate orders of magnitude by reducing all mechanical effects in the bulk crystal to a uniform hydrostatic pressure. Then we also write:

$$n^c \mu^c - f^c = p^c \tag{3.16}$$

Then (3.16) and (3.13) yield the pressure difference:

$$p^c - p^g = \frac{2\gamma^m}{h^m} \tag{3.17}$$

just as in a small fluid drop. We therefore expect to find very high pressure inside a small crystal. Suppose values of γ of order 10^2-10^3 dyne·cm^{-1} (common values in solids). Take h of order 10 - 100 Å. Then p^c-p^g would range between 10^2 and 10^4 atm. With such pressures, appreciable reduction of the lattice constant would be expected, and this in fact has been observed experimentally by X-ray diffraction with very fine crystalline powders.

Our discussion of the equilibrium shape of a crystal is by no means complete or rigorous (see Herring (1953) for an elegant discussion); we have not actually proved that the shape corresponding to Wulff's relation is the necessary and sufficient condition for the minimum of the Kramers function. But we are not actually interested in discussing this here. It suffices to see how a thermodynamical analysis leads to (3.17) whose nature and consequences strongly suggest that the surface of a solid can be legitimately thought of as being in a state of mechanical tension, just as for small liquid drops.

In relation to (3.15) and (3.17), the reader is referred to Defay et al. (1966), who discuss at length the justification for treating small crystals like liquid droplets. One question which must be clarified here is the following. We have argued in Section 2 that, for two bulk phases in contact through a plane interface, the number of degrees of freedom is equal to the number of components. If the small drop obeyed this rule then the one-component case would have just one degree of freedom, which would imply difficulties in understanding the basis of (3.15) or (3.17). The point is that this rule does not apply for curved interfaces. Chapter VI of Defay et al. (1966) discusses in great detail the problem of variance, i.e. of the number of degrees of freedom for different situations. In the particularly simple case of a small drop the essential point is simply that the very argument leading to Laplace's equation (1.3) precisely establishes that knowledge of the pressure in one bulk phase is not enough to know the pressure in the other phase. We must also know the radius R, i.e. the problem has one more degree of freedom. The vanishing of $p^\alpha - p^\beta$ for a plane interface (R → ∞) just reflects the fact that this extra degree of freedom has disappeared.

For a small crystal this degree of freedom is equivalent to, say, one of the heights h^m. Having fixed this, the others follow from Wulff's relations in the form:

$$\frac{\gamma^1}{h^1} = \frac{\gamma^2}{h^2} = \cdots \tag{3.18}$$

This is the meaning of (3.10) in the problem of the variance of the small crystal system and explains why, for the same temperature, the gas phase can have different chemical potentials depending on whether it is in equilibrium with large or small crystals.

Let us stress that formulae like (1.3) and (3.17) are not purely academic. Consider, for example, small liquid drops. The vapour pressure increases as the size of the drop decreases. For example, for water at 18°C the pressure difference p(liquid) - p(gas) is 1.46 bar (Sanfeld (1971)). Now consider a drop initially in equilibrium with its vapour at a pressure given by (1.3), in which α is liquid and β is vapour. This equilibrium is unstable against fluctuations. Suppose a small amount of liquid evaporates. The size of the drop decreases. Its local vapour pressure then increases and becomes larger than the initial value of p^α, which is the pressure of the surrounding atmosphere. Therefore more liquid tends to evaporate into this atmosphere. This explains why supersaturated vapour can exist. In the case of solid surfaces we have seen that it is a good approximation to treat them as small drops and therefore to replace (3.17) by a formula rather like (1.3), i.e. a precise value of h^m by a mean radius R. Consider then the problem of the solubility of a solid crystal (c) in a liquid solvent (ℓ). We have two components and therefore, as discussed above, precisely three degrees of freedom. This means that R can change — the small crystal particles are dissolving — while T and p^ℓ remain constant — in a liquid solvent at a given temperature and pressure. For an infinitesimal reversible process of this kind we have

$$\delta\left(\frac{2\gamma}{R}\right) = \delta p^c = n^c \delta \mu_1 \tag{3.19}$$

We are calling component 1 the crystal species and we are saying that the liquid species is insoluble in the crystal phase.

Now, the value of μ_1, at equilibrium, is equal to the value of the chemical potential that species 1, the solute, has in the solution. According to standard chemical thermodynamics (Kirkwood and Oppenheim (1961)) this has the expression:

$$\mu_1 = \mu_{st}(T, p^\ell) + kT \ln a^\ell \tag{3.20}$$

Here a^ℓ is the activity of the solute in the solution. All we need to know at this stage is that it is proportional to the concentration of the solute. To be clear, we are referring to molecules — or atoms — and not to moles as amount of matter. This is why we have the Boltzmann constant in (3.20) instead of the gas constant. But the essential point is that μ_{st} depends only on T and p^ℓ. Therefore it remains constant in the process described in (3.19). Thus

$$\delta\left(\frac{2\gamma}{R}\right) = n^c kT \delta \ln a^\ell \tag{3.21}$$

Integrating, at T, p^ℓ constant, from R = 0 to its actual finite value,

$$a^\ell = a_0^\ell \exp\left(\frac{2\gamma}{n^c kTR}\right) \tag{3.22}$$

We have neglected the effect of small changes in n^c for very small values of R.

This says that the concentration of the solute in the solution in equilibrium with a crystal increases — exponentially! — as the linear size of the crystal decreases. In other words, smaller crystals are more soluble, and this is because there is a surface tension. This explains the experimentally observed fact that when crystals of different size are in contact with a liquid solvent the large crystals grow at the expense of the small ones. For a recent and more detailed analysis of some basic problems involved in the thermodynamics of small crystals, see Couchman and Jesser (1973)[1].

Coming back to not too small crystals, it is interesting to analyse the question of anisotropy of the surface tension. We have seen that γ is an acceptable unique parameter for all faces of a not too small crystal (the differences decreasing as the reciprocal of the linear dimensions) from a purely energetic standpoint. We can also draw some interesting and suggestive conclusions which indicate an intimate relationship with a state of mechanical stress. But this does not yet prove that γ can be identified with a stress. In fact, we have no assurance that the surface concentration

---

[1] The simplified argument we have used to obtain (3.17) and (3.22) is actually too crude to distinguish between γ and τ, later on introduced in (3.30). A more careful analysis shows that (3.17) and (3.22) must contain τ instead of γ. See Couchman and Jesser (1973).

of Kramers energy does not depend on the state of strain of the surface and that, therefore, γ itself does not change upon an arbitrary deformation of the surface.

The analysis of this problem has been very clearly formulated by Herring (1953). It runs as follows. Consider (Fig. 5) a plane surface. Define an arbitrary plane normal to the surface. Call $\vec{\pi}$ the vector normal to this plane, and therefore contained in the surface plane. Let F_μ ($\mu = x, y, z$) be the components of the force exerted by the material on one side of the plane $\vec{\pi}$ on the material on the other side of this plane per unit length ℓ. We can write

$$F_\mu = \sum_\nu \tau_{\mu\nu} \pi_\nu \quad (\mu\nu = x, y) \tag{3.23}$$

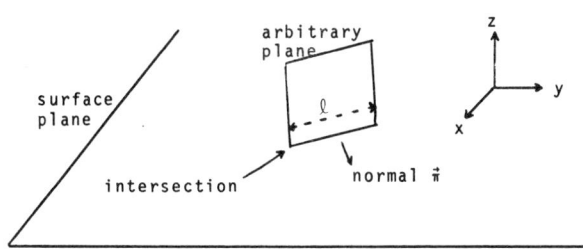

FIG.5. Surface stress acting on a segment of line ℓ in the surface.

Obviously $\tau_{\mu\nu}$ represents just the inhomogeneous part of the stress distribution, i.e. the true surface (mechanical) stress tensor. Arguing as in Section 1, it is clear that the elements $\tau_{\mu\nu}$ have the nature of surface excess quantities and that they do depend in principle on the location of the dividing surface, unlike γ. However, it is reasonable in practice to regard the order of magnitude of γ (or a representative diagonal element $\tau_{\mu\nu}$) as very large compared with the product of the volume stresses and plausible displacements of the dividing surface throughout the physical interface. To this approximation we can regard $\tau_{\mu\nu}$ as practically independent of the choice of the dividing surface.

Let us now imagine a certain surface area A bounded by walls normal to it. Consider an elementary deformation described by the change in the strain tensor, i.e. by $\delta\epsilon_{\mu\nu}$, such that the surface dilatation is

$$\frac{\delta A}{A} = \delta\epsilon_{xx} + \delta\epsilon_{yy} \tag{3.24}$$

We assume the transformation occurs reversibly and in such a way that the intensive state of the bulk phases does not change. The energy change involved in this process can be written in two alternative ways:

(a) From the expression for the mechanical work:

$$A \times \frac{\text{work}}{\text{unit area}} = A \sum_{\mu\nu} \tau_{\mu\nu} \delta\epsilon_{\mu\nu} \tag{3.25}$$

(b) From thermodynamics:

$$\delta\Omega^\sigma = \delta(\omega^\sigma A) = \delta(\gamma A) = \gamma\delta A + A\delta\gamma \tag{3.26}$$

What we are saying here is that the surface concentration of Kramers energy may depend on the state of strain of the surface, i.e.

$$\delta\gamma = \sum_{\mu\nu} \frac{\partial\gamma}{\partial\epsilon_{\mu\nu}} \delta\epsilon_{\mu\nu} \tag{3.27}$$

From (3.24) and (3.27) we can write (3.26) as

$$A \sum_{\mu\nu} \left(\gamma\delta_{\mu\nu} + \frac{\partial\gamma}{\partial\epsilon_{\mu\nu}}\right) \delta\epsilon_{\mu\nu} \tag{3.28}$$

And equating (3.28) and (3.25), since these expressions hold for arbitrary $\delta\epsilon_{\mu\nu}$'s:

$$\tau_{\mu\nu} = \gamma\delta_{\mu\nu} + \frac{\partial\gamma}{\partial\epsilon_{\mu\nu}} \tag{3.29}$$

This formula establishes explicitly the relationship between surface stress and surface tension and shows why in fluids, due to the lack of rigidity, we may expect the tensor $\tau_{\mu\nu}$ to have one single principal value equal to γ.

The anisotropy of the surface stress, apparent in (3.29), does not seem to have been actually verified experimentally in a conclusive manner (Defay et al. (1966)). However, even the isotropic version of (3.29):

$$\tau = \gamma + A \frac{\partial\gamma}{\partial A} \tag{3.30}$$

illustrates very clearly a distinction first made by Gibbs between the energy needed to create new area ($\gamma\delta A$ by definition of γ) and the mechanical work $\tau\delta A$ needed to stretch the area. Equation (3.30) is known as Shuttleworth's equation.

Notice where we stand. The surface entropy, material contents, Helmholtz energy and Gibbs energy are defined in exactly the same way as in the case of a fluid surface. Hence the first equality of (2.16) holds as well, and implies that Ω^σ contains solely the surface mechanical work term. The second equality of (2.16) also holds if we choose to. It is simply a definition of γ as the surface concentration of an unambiguously defined thermodynamical potential. The meaning of γ is further illuminated by considering the concept of reversible cleavage work (Eriksson (1969)).

Imagine that we had a closed system of bulk phases in equilibrium. In a reversible change, at constant T and p, and therefore μ, to a new equilibrium situation, the — necessarily zero — change in the Gibbs function of the entire system:

$$\Delta G = \Delta F - \Delta \Omega = \Delta F + p\Delta V - \Delta \Omega^\sigma = 0$$

would consist of a change in Helmholtz energy plus mechanical work; notice that $p\Delta V$ would be work done by the system. Consider now a crystal, component 1, in equilibrium with a gaseous atmosphere, component 2. Equilibria and reversible processes are supposed to occur at constant T and p. We assume that the gas is practically insoluble in the crystal and the solid component does not evaporate or diffuse appreciably.

In the situation labelled by 0 we ignore all surface effects. Thus

$$G_0 = F_0 + pV_0 \tag{3.31}$$

where V_0 is the total volume of the system — crystal plus gas. We now cleave the crystal reversibly at constant (T, p), allowing for the appearance of surface effects in the freshly formed surface area A. Thus, cleavage leads us to a new situation with the same value of G for the total system, now written as

$$G = F + pV - \Omega^\sigma \tag{3.32}$$

Now, under these conditions, the change ΔF measures the mechanical work of cleavage $(\Delta W)_c$ done on the system. Moreover, if V is sufficiently large, the change ΔV needed to ensure constant pressure is negligible, whence

$$(\Delta W)_c = \Omega^\sigma \tag{3.33}$$

Therefore the meaning of γ, defined as the surface concentration of Kramers energy, is that it is the reversible cleavage work per unit area at constant T and p.

Let us finally complete the thermodynamic analysis of solid surfaces. In the fluid case we started out from the energy balance and thence we obtained an integrated relation making use of a mathematical property, i.e. that E is a homogeneous function, of the first degree, of the extensive variables. With solids we proceed the other way round. Equation (2.12) still holds. We have just discussed its meaning. But it does not follow from anything like (2.11) because γ itself may depend on surface strain. The energy balance is now

$$dE^\sigma = TdS^\sigma + \Sigma_i \mu_i dN_i^\sigma + \tau dA \qquad (3.34)$$

To be more precise we should have used (3.25) or (3.28) to express the surface work, but we have already indicated that the isotropic version, i.e. (3.34), is in practice sufficiently accurate.

The partial derivatives written in (3.29) and (3.30) must of course be understood at constant T and chemical potentials. If we look at the new Gibbs-Duhem equation as an expression for $d\gamma$ in a general reversible process, we must add a term which describes the dependence of γ on A at constant (T, μ_i). It is obvious that (3.34) leads now, together with (2.12), to

$$d\gamma = -s^\sigma dT - \Sigma_i n_i^\sigma d\mu_i + \left(\frac{\partial \gamma}{\partial \epsilon_s}\right)_{T, \mu_i} d\epsilon_s \qquad (3.35)$$

where we have put

$$d\epsilon_s = \frac{dA}{A} \qquad (3.36)$$

or, using (3.30),

$$d\gamma = -s^\sigma dT - \Sigma_i n_i^\sigma d\mu_i + (\tau - \gamma) d\epsilon_s \qquad (3.37)$$

This is Eriksson's (1969) Eq. (24) and constitutes the central result of his analysis. From this and (3.30) one can now rewrite all partial derivatives of physical interest. Some of these are contained in the Appendix to this section.

APPENDIX TO SECTION 3

We derive and collect here a number of formulae applicable to solid surfaces. These were first written by Eriksson (1969). They are reproduced here following the line of Section 3, which is different in notation and in approach. The usefulness of this analysis is apparent in the discussion of Section 4.

First of all, the Shuttleworth equation, to be specific, must be written as

$$\tau = \gamma + \left(\frac{\partial \gamma}{\partial \epsilon_s}\right)_{T, \mu} \qquad (3.A.1)$$

In our notation μ means all μ's constant. An expression of the form $(\partial f/\partial\mu_i)\mu_j$ means differentiation with respect to μ_i while keeping constant all other μ_j ($j \neq i$). Notice that our analysis does not depend on the specific choice of dividing surface made by Eriksson.

The first partial derivatives of γ are, as indicated in Section 4, simply

$$\left(\frac{\partial \gamma}{\partial T}\right)_{\epsilon_s, \mu} = -s^\sigma \quad ; \quad \left(\frac{\partial \gamma}{\partial \mu_i}\right)_{T, \epsilon_s, \mu_j} = -n_i^\sigma \tag{3.A.2}$$

as in a fluid interphase. The new question in a solid surface is whether ϵ_s remains constant or not. We have

$$\left(\frac{\partial \gamma}{\partial T}\right)_\mu = \left(\frac{\partial \gamma}{\partial T}\right)_{\mu, \epsilon_s} + \left(\frac{\partial \gamma}{\partial \epsilon_s}\right)_{\mu, T} \frac{d\epsilon_s}{dT} \tag{3.A.3}$$

i.e. by (3.A.1):

$$\left(\frac{\partial \gamma}{\partial T}\right)_\mu = \left(\frac{\partial \gamma}{\partial T}\right)_{\mu, \epsilon_s} + (\tau - \gamma) \frac{d\epsilon_s}{dT} \tag{3.A.4}$$

As mentioned in Section 4, the last term is usually of order 10^{-3} to 10^{-2} erg·cm^{-2}·deg^{-1}, while s^σ (see (3.A.2)) is of order 10^{-1} erg·cm^{-2}·deg^{-1}. Thus there does not seem to be much difference between (3.A.2) and (3.A.4), although it might just become marginal whether or not the strain dependence of γ has negligible effects for this purpose.

On the other hand,

$$\left(\frac{\partial \gamma}{\partial \mu_i}\right)_{T, \mu_j} = \left(\frac{\partial \gamma}{\partial \mu_i}\right)_{T, \mu_j, \epsilon_s} + \left(\frac{\partial \gamma}{\partial \mu_s}\right)_{T, \mu} \frac{d\epsilon_s}{d\mu_i} \tag{3.A.5}$$

The dependence of A (or ϵ_s) on chemical potentials is usually negligible for solids, so that also

$$\left(\frac{\partial \gamma}{\partial \mu_i}\right)_{T, \mu_j} = -n_i^\sigma \tag{3.A.6}$$

Now for the derivatives of τ. We have already discussed the dependence of τ on T (see (4.13) and (4.16)). Let us study its dependence on ϵ_s and on the μ's. Differentiating (3.A.1) we have

$$\left(\frac{\partial^2 \gamma}{\partial \epsilon_s^2}\right)_{T, \mu} = \left(\frac{\partial \tau}{\partial \epsilon_s}\right)_{T, \mu} - \left(\frac{\partial \gamma}{\partial \epsilon_s}\right)_{T, \mu} \tag{3.A.7}$$

and, by (3.A.1) again,

$$\left(\frac{\partial \tau}{\partial \epsilon_s}\right)_{T, \mu} = \tau - \gamma + \left(\frac{\partial^2 \gamma}{\partial \epsilon_s^2}\right)_{T, \mu} = (\tau - \gamma) + \left(\frac{\partial(\tau - \gamma)}{\partial \epsilon_s}\right)_{T, \mu} \tag{3.A.8}$$

As with many of these formulae, the difference between working in terms of γ or in terms of τ depends on how much γ depends on strain.

Furthermore, differentiating (3.A.1) again:

$$\left(\frac{\partial^2 \gamma}{\partial \epsilon_s \partial \mu_i}\right)_{T, \mu_j} = \left(\frac{\partial \tau}{\partial \mu_i}\right)_{T, \epsilon_s, \mu_j} - \left(\frac{\partial \gamma}{\partial \mu_i}\right)_{T, \epsilon_s, \mu_j} \quad (3.A.9)$$

which, by (3.A.2), yields

$$-\left(\frac{\partial \tau}{\partial \mu_i}\right)_{T, \mu_j, \epsilon_s} = n_i^\sigma + \left(\frac{\partial n_i^\sigma}{\partial \epsilon_s}\right)_{T, \mu} \quad (3.A.10)$$

For the free strain derivative we have

$$\left(\frac{\partial \tau}{\partial \mu_i}\right)_{T, \mu_j} = \left(\frac{\partial \tau}{\partial \mu_i}\right)_{T, \mu_j, \epsilon_s} + \sum_i \left(\frac{\partial \tau}{\partial \epsilon_s}\right)_{T, \mu_i} \left(\frac{\partial \epsilon_s}{\partial \mu_i}\right)_T \quad (3.A.11)$$

but, as already remarked, $(\partial \epsilon_s / \partial \mu_i)_T$ is usually negligible, thus

$$\left(\frac{\partial \tau}{\partial \mu_i}\right)_{T, \mu_j} \approx \left(\frac{\partial \tau}{\partial \mu_i}\right)_{T, \mu_j, \epsilon_s} = (3.A.10) \quad (3.A.12)$$

4. DISCUSSION OF EXPERIMENTAL DATA ON SURFACE TENSION AND RELATED CONCEPTS

A number of different experiments can be and have been performed to measure surface 'tension', or 'stress' or 'energy'. We must now discuss what these experiments actually measure and how we must compare theory with experiment. Roughly, four main types of experiment can be distinguished.

(a) <u>Reversible cleavage work for crystals</u>

We have discussed this in Section 3 and concluded that it measures the surface tension γ. This experimental method is actually used sometimes, although the difficulty seems to be in ensuring reversibility in practice.

(b) <u>Calorimetric experiments on powders</u>

The latent heat of fusion or solution is measured for a powder and for the same mass of a large specimen. The difference is attributed to a difference in surface energy, since both specimens differ only in surface area. In view of the general discussion in Sections 2 and 3 it seems somewhat dangerous to rely too much on these experiments because small crystallites in finely divided powders are probably in a state of internal strain which may invalidate the clear-cut applicability of the concept of surface excess. The latent heats may be affected not only by inhomogeneous strains in the bulk of the small crystallites but also by impurities (another complication), very hard to get rid of in practice. In any case this experiment would not measure surface tension directly.

(c) The 'ripple' or surface wave method

This is a dynamical method proposed and discussed by Vermaak et al. (1968). The essential idea is that the velocity of surface waves is determined by a restoring force which, depending on the experimental situation, will be given by the surface stress or by the surface tension. It is argued that this should make possible an experimental distinction between γ and τ.

The point to notice is that although this was not explicitly invoked in the derivation of (3.29), the strain entering this equation is implicitly assumed to be elastic. Whether the strain is elastic or plastic depends on the rate at which it is produced. If it is sufficiently slow, then the nature of the deformation depends on the material. For liquids the atoms diffuse, re-arrange and re-adapt themselves so that the average density and configuration of the surface atoms does not change. The deformation is then totally plastic, the liquid has no rigidity and $\partial \gamma / \partial \epsilon_{\mu\nu}$ is zero. For solids it depends on the amplitude of the strain, i.e. on whether the flow stress has been reached. This is easily accomplished at high temperatures; the solid is about to melt. On the other hand, for fast deformations the atoms do not have sufficient mobility to follow and the strain is solely elastic.

It follows from the above considerations that the restoring force which determines the velocity of high-frequency surface waves is the surface stress, which could in this way be experimentally measured. For a more elaborate analysis of this point, see Couchman et al. (1972). This, however, seems to be just a conjecture so far. In fact τ has been experimentally measured rather from the lattice contraction of small metallic particles, by an application of (3.17) with τ properly put instead of γ.

(d) The surface tension can therefore be measured in high-temperature plastic flow experiments

The experiment, which is amazingly simple, is actually carried out by hanging weights from wires. This is done near the melting point, for the reasons stated. In fact we might expect the surface stress to be obtained from the velocity of sufficiently slow surface waves, but not by hanging weights from cold wires, because bulk elastic stresses would then mask the measurements. In the experiment the length of the specimen is observed as a function of time for different values of the weight P applied. For small P the wire contracts because γ pulls up. For large P it flows plastically. When the length remains constant there is an equilibrium:

$$P \, d\ell = \gamma \, dA \tag{4.1}$$

between the work that P would do in stretching the length ℓ and the energy γdA it would cost to create new area. If we assume the total volume of the wire remains constant, then, for a cylindrical shape,

$$2\pi \ell r dr + \pi r^2 d\ell = 0 \tag{4.2}$$

while

$$dA = d(2\pi r \ell) = \pi r d\ell \tag{4.3}$$

(4.2) has been used in writing (4.3). Thus, substituting in (4.1),

$$P = \pi r \gamma$$

whence γ is obtained. In fact a correction is needed to take account of extra intergranular surface areas in the wire, but this is easily taken care of.

Other experimental methods are discussed in detail by Benson and Yun (1967) with particular reference to ionic and rare gas crystals, but we are not interested in lengthening this discussion. See also Linford (1973) for further details.

From the point of view of the electron theory of solid surfaces it is important to stress that usually one calculates surface internal energy, and the above discussion must be kept in mind when comparing theory with experiment.

Furthermore, the following equation is usually found in the literature:

$$e^\sigma = \gamma - T \frac{d\gamma}{dT} \tag{4.4}$$

Now, this requires some discussion. For the time being let us study the fluid case. Look at (2.28). The nearest we can get, from first principles, to (4.4) is

$$e^\sigma = \gamma - T\left(\frac{\partial \gamma}{\partial T}\right)_{\mu_i} + \Sigma_i n_i^\sigma \mu_i \tag{4.5}$$

which seems to look rather different. Suppose a one-component system. It has one degree of freedom. Take T as the independent variable. We have

$$\frac{d\gamma}{dT} = \left(\frac{\partial \gamma}{\partial T}\right)_\mu + \left(\frac{\partial \gamma}{\partial \mu}\right)_T \frac{d\mu}{dT} \tag{4.6}$$

Use (2.28):

$$-\left(\frac{\partial \gamma}{\partial T}\right)_\mu = -\frac{d\gamma}{dT} - n^\sigma \frac{d\mu}{dT} \tag{4.7}$$

For an order of magnitude estimate it suffices to treat the gas as ideal. In this case μ has the form:

$$\mu = -kT \ln(\text{const} \times T^{5/2}) \tag{4.8}$$

whence

$$T \frac{d\mu}{dT} = \mu - \frac{5}{2} kT \tag{4.9}$$

Thus

$$e^\sigma = \gamma - T\frac{d\gamma}{dT} + \frac{5}{2} n^\sigma kT \tag{4.10}$$

Now, the value of the last term depends of course on the choice of dividing surface, but let us make the extreme assumption that $n^\sigma \sim 10^{13}$ cm^{-2}. This still yields a term of order 10^{-4} to 10^{-3} erg·cm^{-2} round about room temperature whereas typical values for the first and second term are of order 10^2 erg·cm^{-2}. Thus it is sufficiently accurate just to write (4.4), which means

$$-\frac{d\gamma}{dT} = s^\sigma \tag{4.11}$$

again a sufficiently accurate version of (2.28) in practice. As it stands, (4.4) has a clear interpretation: the energy required to create new area is equal to the work done on the system plus the heat absorbed by the system. Furthermore, since the arrangement of matter should be less orderly near the interface, we would expect the excess surface entropy to be positive and the surface tension to decrease with increasing temperature. This rule is not without exceptions, but by and large it tends to be observed.

When it comes to solid surfaces we have two options. One is to compare with measurements of surface tension. It is clear from the discussion of the reversible cleavage work in Section 3 that we still have the integrated relation (2.12) in terms of γ. It is also clear from (3.35) that

$$s^\sigma = -\left(\frac{\partial \gamma}{\partial T}\right)_{\mu,\epsilon_s} \quad ; \quad n_i^\sigma = -\left(\frac{\partial \gamma}{\partial T}\right)_{T,\epsilon_s,\mu_j} \tag{4.12}$$

(Let us from now on make the convention that μ stands for the entire set of μ's for all components and that μ_j in a term referring specifically to component i means constancy of all μ_j for $j \neq i$.) These are like (2.28), only that ϵ_s = constant must now be specified. Now, in a solid, as emphasized by Eriksson (1969), ϵ_s is generally dependent on T (effect of thermal expansion) and on the μ's (because of changes in surface stresses upon adsorption), but the latter effect is only important for very small crystals. Thus, we have

$$-\left(\frac{\partial \gamma}{\partial T}\right)_\mu = -\left(\frac{\partial \gamma}{\partial T}\right)_{\mu,\epsilon_s} - \left(\frac{\partial \gamma}{\partial \epsilon_s}\right)_{T,\mu} \frac{d\epsilon_s}{dT} \tag{4.13}$$

whence, using the Shuttleworth equation,

$$-\left(\frac{\partial \gamma}{\partial T}\right)_\mu = s^\sigma - (\tau-\gamma)\frac{d\epsilon_s}{dT} \tag{4.14}$$

Typical values for $d\epsilon_s/dT$ are of order 10^{-5} to 10^{-4} deg^{-1}, while $\tau-\gamma$ for, say, a typical dielectric crystal, may be of order 10 to 10^2 erg·cm^{-2}. This is still smaller than s^σ (usually of order 10^{-1} erg·cm^{-2}·deg^{-1}) but already becoming marginal. Still the position seems to be that in terms of γ there is no great novelty and the above discussion still holds essentially unchanged for solid surfaces.

The story may be somewhat different if we look at experimental measurements of τ. Following Eriksson (1969), from the cross derivatives of the Shuttleworth equation we can write

$$\left(\frac{\partial^2 \gamma}{\partial \epsilon_s \partial T}\right)_\mu = \left(\frac{\partial \tau}{\partial T}\right)_{\epsilon_s,\mu} - \left(\frac{\partial \gamma}{\partial T}\right)_{\epsilon_s,\mu} = \left(\frac{\partial \tau}{\partial T}\right)_{\epsilon_s,\mu} + s^\sigma \tag{4.15}$$

which, using (4.12), yields

$$s^\sigma = -\left(\frac{\partial \tau}{\partial T}\right)_{\epsilon_s,\mu} - \left(\frac{\partial s^\sigma}{\partial \epsilon_s}\right)_{T,\mu} \tag{4.16}$$

Hence

$$e^\sigma = \tau - T\left(\frac{\partial \tau}{\partial T}\right)_{\epsilon_s,\mu} + \Sigma_i n_i^\sigma \mu_i - \left[\left(\frac{\partial \gamma}{\partial \epsilon_s}\right)_{T,\mu} + T\left(\frac{\partial s^\sigma}{\partial \epsilon_s}\right)_{T,\mu}\right] \tag{4.17}$$

i.e.

$$e^\sigma + \left(\frac{\partial e^\sigma}{\partial \epsilon_s}\right)_{T,\mu} = \tau - T\left(\frac{\partial \tau}{\partial T}\right)_{\epsilon_s,\mu} + \Sigma_i n_i^\sigma \mu_i + \left(\frac{\partial \Sigma_i n_i^\sigma \mu_i}{\partial \epsilon_s}\right)_{T,\mu} \tag{4.18}$$

Now,

$$\left(\frac{\partial \tau}{\partial T}\right)_{\epsilon_s,\mu} = \left(\frac{\partial \tau}{\partial T}\right)_\mu - \left(\frac{\partial \tau}{\partial \epsilon_s}\right)_{T,\mu} \frac{d\epsilon_s}{dT} \qquad (4.19)$$

Hence

$$e^\sigma + \left(\frac{\partial e^\sigma}{\partial \epsilon_s}\right)_{T,\mu} = \tau - T\left(\frac{\partial \tau}{\partial T}\right)_\mu + \Sigma_i n_i^\sigma \mu_i - \left(\frac{\partial \Sigma_i n_i^\sigma \mu_i}{\partial \epsilon_s}\right)_{T,\mu}$$

$$+ T\left(\frac{\partial \tau}{\partial \epsilon_s}\right)_{T,\mu} \frac{d\epsilon_s}{dT} \qquad (4.20)$$

Also

$$\frac{d\tau}{dT} = \left(\frac{\partial \tau}{\partial T}\right)_\mu + \Sigma_i \left(\frac{\partial \tau}{\partial \mu_i}\right)_{T,\mu_j} \frac{d\mu_j}{dT} \qquad (4.21)$$

Proceeding as in the above discussion in terms of γ, we can approximately, upon insertion of (4.21) in (4.20), cancel the adsorption terms. The last term in (4.20) also seems negligible since $(\partial \tau/\partial \epsilon_s)_{T,\mu}$, from the Shuttleworth equation, is of order $\tau-\gamma$, i.e. of the same order of magnitude as the surface energies usually encountered, while $T(d\epsilon_s/dT)$, at room temperature, is of order 10^{-3} to 10^{-2}. We finally come to the conclusion that, for a solid,

$$e^\sigma + \left(\frac{\partial e^\sigma}{\partial \epsilon_s}\right)_{T,\mu} = \tau - T\frac{d\tau}{dT} \qquad (4.22)$$

is an approximately valid equation. To sum up, on comparing calculations of surface energy with experimental data one must use (4.4) in terms of γ or (4.22) in terms of τ. The meaning of (4.22), which we have here derived through a formal analysis, can be seen by obtaining it from the following quick argument. Take the Shuttleworth equation and write (implying the approximations discussed after (4.4)):

$$\tau - T\frac{d\tau}{dT} = \gamma - T\frac{d\gamma}{dT} + \left(\frac{\partial \gamma}{\partial \epsilon_s}\right)_{T,\mu} - T\frac{d}{dT}\left(\frac{\partial \gamma}{\partial \epsilon_s}\right)_{T,\mu} \qquad (4.23)$$

i.e. using (4.4),

$$e^\sigma = \tau - T\frac{d\tau}{dT} - \frac{\partial}{\partial \epsilon_s}\left(\gamma - T\frac{d\gamma}{dT}\right)_{T,\mu} \qquad (4.24)$$

which, using (4.4) again, is just (4.22). As one would expect, the formal thermodynamical analysis adds nothing new. The only real novelty comes from the strain dependence of γ, which is the feature that characterizes a solid. Let us try to have an idea of the possible magnitude of this term. First we discuss (4.4). Must we take the last term on its right-hand side seriously? We have argued that $d\gamma/dT$ is to a good approximation the surface excess entropy, which is typically of order 10^{-1} erg·cm^{-2}·deg^{-1}. Put $T = 300$ K; this yields $T\, d\gamma/dT \simeq 30$ erg·cm^{-2}. Typical values of γ itself are of order 10^2 erg·cm^{-2}. This means (a) that the entropy term is not negligible and (b) that e^o itself is still of the same order of magnitude as γ. Thus the possible importance of the extra term in (4.22) just depends on how strongly e^o depends on strain. All we can say off-hand is that this is just what makes a solid different from a liquid.

We have laid out here the general principles that must be taken into consideration when comparing theory with experiment. Fortunately, we can convince ourselves that to a good approximation the effect of the adsorption terms is negligible. This allows greater flexibility in the calculation, as we need not take too much care in the choice of dividing surface. There is still the entropy term. This, as we have seen, is not negligible. In the light of these considerations we can see, for example, how the calculations of Lang and Kohn (1970) can relate to experimental data. These authors calculate internal energy and take experimental data of γ (obtained with the hot wire method). These are high-temperature data and they extrapolate them to zero temperature. This extrapolation may be more or less audacious, but the procedure in principle is legitimate. A different consideration is the fact that the crystal lattice is later put in, to the extent that different values are calculated for different lattices. By this time one may begin to wonder whether the features of the solid state are fully taken into account.

Now, let us emphasize that the above discussion aims at interpreting the calculation of surface energy within the general scheme of surface thermodynamics. This is necessary if one wants to compare the results of such a calculation with experimental measurements of the surface tension γ, as is often done by many authors. This leads to a discussion involving the concept of surface excesses and we necessarily run up against the question of choice of dividing surface. It is very important to realize that most of these calculations do not make any reference whatever to the choice of dividing surface. In fact one often does not calculate any surface excess in the Gibbsian sense, but something else. Most of the obscurity prevailing in this field is due to the fact that terms like surface energy, or excess surface energy — where energy may mean any thermodynamic function — are used to calculate something which in principle has nothing to do with the corresponding term in the Gibbsian scheme. Let us try to clear this up.

Suppose we have a system, say a vibrating crystal, or a jellium-electron gas model of a metal, and we ignore surface modes or states. We calculate 'the energy' of the system — which, to begin making things clear, is in thermodynamics the internal energy — and this is a purely bulk property. That is to say, if we study phonons, for example, and we have, say, $3N_b$ bulk degrees of freedom, we have an energy of the form:

$$E = E^b(3N_b) \tag{4.25}$$

A surface is then 'introduced', i.e. according to some chosen model we decide that the system has now $3N_s$ degrees of freedom which go into surface modes. Thus we calculate a new energy:

$$E' = E^b(3N_b - 3N_s) + E^s(3N_s) \tag{4.26}$$

The surface energy is then defined as

$$E_s = E' - E \tag{4.27}$$

Likewise, we may calculate the change $\delta N(E)$ in the density of states of a crystal due to the introduction of a surface. Assuming that we take proper account of sum rules, the surface energy is then defined by

$$E_s = \int dE \; E \; \delta N(E) \tag{4.28}$$

Of course E_s in (4.27) or (4.28) is unambiguously defined and it has nothing to do with any dividing surface. This is why it has nothing to do with the surface energy E^σ of the Gibbsian scheme. This leads to the question: If we calculate surface energy E_s, or surface entropy S_s, etc., in the manner of (4.27) and (4.28), what must we compare it with?

We have seen that the only excess quantity which is invariant in the Gibbsian scheme is Ω^σ, the surface excess of Kramers energy. If we calculated Ω_s in the manner of (4.27) and (4.28) — a calculation which is never done — then we could equate Ω_s to Ω^σ, but only then. On the other hand, we have seen that Ω^σ is what gives the surface tension. If follows that to write, say,

$$(4.28) = \gamma A \tag{4.29}$$

is simply incorrect. This warning seems to be necessary.

The above remarks do not mean that a calculation of E_s is useless. E_s is an invariant property of a surface system and it surely has something to do with some experimental data, but the choice of calculated and measured quantities must be clearly stated and justified, which, as we have just seen, is not altogether a trivial matter. At this stage it is interesting to come back to a discussion of the calorimetric method (b). The experimental method, consisting as it does in global measurements with or without surface, or with different amounts of surface area, is more akin to the philosophy of the calculations leading to formulae like (4.27) and (4.28). So what do these experiments measure?

The answer involves the question of size. On one hand, the small crystalline particles must reach sufficiently small sizes for the method to yield measurable results, otherwise the data reflect a small difference between larger and nearly equal quantities, thus becoming very inaccurate. On the other hand, if the size becomes very small one faces the difficulties pointed out above, in the brief presentation of method (b). Another practical difficulty is that the size of the crystallites, and hence the actual surface area of the sample, must be known with fair accuracy, and this is not too easy. One obvious precaution, as stressed by Benson and Yun (1967), is that the surface effects thus measured should be proportional to the area, and that is not always established. In fact, when it is one often finds that the data do not extrapolate, at zero area, to the correct value for homogeneous bulk, which probably indicates bulk inhomogeneities and/or impurities. All this shows that these experiments are difficult and one must be very careful in making a delicate choice. In practical terms, the right size for crystallites seems to start roundabout the order of magnitude of $1\mu m$. Supposing, with all the above reservations, that one has good data, there are essentially two meaningful variants of the powder experiment, one of which is not really calorimetric.

If the crystalline particles are sufficiently small, then the difference in solubility may be observed and the theory leading to (3.22) used to estimate γ. The experiment tends to work better with ionic crystals, in which case the theory must be retouched to account for the fact that the solid dissociates on solution, but this is easily done. In the example discussed by Defay et al. (1966), the solubility at 25°C of barium sulphate crystals of 'diameter' 10^{-5} cm was found to be 80% larger than that of large crystals, from which one can surmise a value of 1350 erg·cm^{-2} for γ (not really to be taken as very accurate). We emphasize that this experiment measures γ. Moreover, γ is not exactly the surface tension of the free crystal but the interfacial tension between the crystal and the solution. The reason for stressing this point is that the surface tension depends on the state of adsorption. It is conceivable that the effect of adsorption may be negligible in many practical instances, but this requires explicit justification. It is somewhat surprising that one seldom finds in the literature any plausible argument or approximate estimate to justify this point. We have seen an example in the discussion of (4.4), but even then all that the argument proves is that the adsorption terms can be neglected when relating e^σ to γ and its temperature dependence. This is not the same as saying that the value of γ is independent of the state of adsorption. Still, with these considerations in mind, we can say that this experiment measures γ.

The true calorimetric experiments are described in detail by Benson and Yun (1967). We stress two points: (a) these experiments measure global properties, of the kind just discussed, of the entire system as a function of the state of subdivision of the material, i.e. for different values of the surface area; and (b) the quantity actually measured is the heat capacity of the system. The way to proceed in analysing these data is then the following.

We have experimentally C_p^s at different temperatures, i.e. we have a surface quantity of the non-Gibbsian kind. We can construct, according to standard thermodynamical formulae, the various surface thermodynamic functions. The most direct one is simply the entropy:

$$S^s(T) = \int_0^T \frac{C_p^s}{T} dT \qquad (4.30)$$

Likewise we could construct other functions like F^s, etc., depending on what we wish to calculate and compare with experiment. An unusually clear example in this field is the article by Chen et al. (1971). Explicitly prescribed thermodynamic functions are calculated for a slab of a model of NaCl. This includes surface modes (vibrational). The value of the same property, as corresponding to the same degrees of freedom, all bulk, is subtracted, and the results compared with calorimetric data, specifically entropy and heat capacity. The surface tension, very appropriately, is nowhere mentioned.

5. ADSORPTION THERMODYNAMICS IN THE GIBBSIAN SCHEME

Let us now return to (2.21), the equality commonly known as the Gibbs adsorption equation. Suppose, by varying T and/or p, that we achieve changes in the chemical potentials and hence changes in γ. For the time being we shall restrict ourselves again to fluid surfaces. Now, γ can be measured experimentally and what we should like to do is to use the derivatives of (2.28) to obtain the surface amounts of entropy and matter. Let us look into this more closely.

The point is that if we have r components, 1 plane surface and 2 phases, the problem has only r degrees of freedom and we cannot determine all (r+1) independent intensive quantities appearing in (4.12).

We thus seem to run up against two difficulties: (a) We cannot determine all adsorbed amounts; and (b) in any case, it is not clear what the Gibbsian quantities n_i^σ mean in actual physical terms, i.e. it is not obvious how to identify N_i^σ with the number of molecules of species i which have 'actually left a bulk phase and gone into the surface phase'. This seems a real pity. The Gibbsian scheme is so precise and beautifully constructed, but it seems so formal and difficult to interpret in terms of actual quantities obtained out of experimental measurements or model calculations. This point has already been discussed in Section 4.

This objection against the Gibbsian scheme has often been raised, and alternative formulations have been proposed. A number of authors are convinced by the argument that it makes more sense to choose two dividing surfaces rather than just one. Our choice will be to use the Gibbsian scheme throughout. What we have to justify is that the Gibbsian quantities are useful to build up a theory for 'actual' quantities. We must be prepared to relax somewhat formal rigour and rely to some extent on plausible approximations, as indeed is ultimately done when applying any physical theory.

Let us then consider the first difficulty: the problem of variance. We have one unknown too many. But we also have the condition of thermal equilibrium. We can use the Gibbs-Duhem equation (2.4) for the condensed phase (c) and the gas phase (g). Eliminating dp we have

$$(s^c - s^g)dT + \Sigma_i (n_i^c - n_i^g) d\mu_i = 0 \tag{5.1}$$

This allows one of the unknowns to be written in terms of the rest. Call μ_1 the variable to be eliminated, then

$$d\mu_1 = - \frac{s^c - s^g}{n_1^c - n_1^g} dT - \sum_{i \neq 1} \frac{n_i^c - n_i^g}{n_1^c - n_1^g} d\mu_i \tag{5.2}$$

which, inserted in (2.21), yields

$$d\gamma = - s^\sigma_{(1)} dT - \sum_{i \neq 1} n^\sigma_{i(1)} d\mu_i \qquad (5.3)$$

where

$$s^\sigma_{(1)} = s^\sigma - n^\sigma_1 \frac{s^c - s^g}{n^c_1 - n^g_1} \quad ; \quad n^\sigma_{i(1)} = n^\sigma_i - n^\sigma_1 \frac{n^c_i - n^g_i}{n^c_1 - n^g_1} \qquad (5.4)$$

Now we have r independent variations (dT, $d\mu_2, \ldots, d\mu_r$) in (5.3). Obviously the derivatives:

$$\left(\frac{\partial \gamma}{\partial T}\right)_{\mu_2, \ldots, \mu_r} = -s^\sigma_{(1)}$$

$$\left(\frac{\partial \gamma}{\partial \mu_i}\right)_{\mu_j} = -n_{i(1)} \qquad (5.5)$$

are invariant, since $d\gamma$, dT, $d\mu_2, \ldots, d\mu_r$ do not depend on the choice of dividing surface. Indeed it is trivial to verify explicitly that the quantities written in (5.4) are invariant. It is customary to call $n^\sigma_{i(1)}$ the relative adsorption (of component i with respect to component 1) and $s^\sigma_{(1)}$ the relative surface entropy. One could likewise define relative surface amounts of energy or free energy and show that they are invariant.

Suppose then that we make one particular choice: that position $z = z_1$ of the dividing surface for which $n^\sigma_1 = 0$. Since the equalities of (5.4) are invariant we have

$$s^\sigma_{(1)} = (s^\sigma)_{z_1} \quad ; \quad n^\sigma_{i(1)} = (n^\sigma_i)_{z_1} \qquad (5.6)$$

i.e. the relative surface amount of any quantity is precisely equal to the actual surface amount of this quantity for the particular choice for which $n^\sigma_1 = 0$.

If the adsorbed component does not dissolve appreciably in the condensed phase, then we have a scheme in terms of quantities which have a direct experimental meaning. Of course, the adsorbed gases will usually dissolve in the adsorbent liquid. Consider a practical case: component 1 in condensed phase adsorbing component 2 from bulk phase. The concentrations in the gas phase can usually be neglected and (5.4) rewritten:

$$n^\sigma_{2(1)} = n^\sigma_2 - n^\sigma_1 \frac{n^c_2}{n^c_1} \qquad (5.7)$$

This provides an appealing intuitive picture. Suppose the two components are present in the adsorbed surface layer in the same proportions as in the solution, i.e.

$$\frac{n_2^\sigma}{n_1^\sigma} = \frac{n_2^c}{n_1^c}$$

Then the relative adsorption is zero. Thus a positive/negative relative adsorption corresponds to a surface layer which is richer/poorer in component 2 than the bulk solution.

The question is whether the relative adsorption can be measured experimentally. It turns out that it can, to a good approximation, with the microtome method, in which the adsorbed surface layer is swept rapidly with a sharp knife edge and analysed. The agreement between theory and experiment, using the concept of reduced adsorptions and Eqs (5.3) and (5.4), is in fact quite satisfactory. A clear discussion of this point has been given by Defay et al. (chapter 7) and needs no further comment.

6. ADSORPTION ON INERT SOLID ADSORBENTS

In the previous discussion the approach was that we wanted to use (5.3) in order to estimate the adsorption experimentally. Let us now change the viewpoint. We are interested in studying the thermodynamics of the adsorption process itself in a more complete manner and in following the variation of thermodynamic functions of interest as the adsorption phenomenon proceeds. Moreover, we want to resort to some models of adsorption, in which 'the adsorbed amount' is calculated with some theory, and we want to interpret all this in terms of the Gibbsian scheme.

We shall not enter here into discussion of how one produces models based on various types of gas-surface interactions. There are abundant discussions in the literature (Clark (1970); Somorjai (1972); Green (1969)).

Our problem is component 1 adsorbing component 2 from the gas phase. The adsorbate, by definition, means the system formed by matter in the surface phase, and does not include molecules in the gas phase. We shall assume that the adsorbent is non-volatile and also inert, i.e. its thermodynamic state is not appreciably affected by adsorption. We shall make use of the results proved in Section 5 and thus make the choice $n_1^\sigma = 0$. From now on this choice will be implicitly understood without explicit labelling.

We shall develop the analysis for a solid adsorbent. Thus we shall start from (3.27). The basic idea of the Gibbsian scheme can be carried a step further in a way which is very appropriate to study the adsorption process. Suppose the gas pressure is zero. We have the pure solid surface. We write the balance for all thermodynamic quantities and we obtain the surface amounts which we shall label with 0 as superscript. Now consider a finite gas pressure and, therefore, adsorption. We write the Gibbsian balance again and obtain new surface amounts. Given any $X^{\sigma 0}$ which has changed into X^σ, we define the adsorbate amount X_a by

$$X_a \equiv X^\sigma - X^{\sigma 0} \tag{6.1}$$

Only for τ and γ we change the sign in the definition. Thus

$$\gamma_a \equiv \gamma^0 - \gamma \quad ; \quad \tau_a \equiv \tau^0 - \tau \tag{6.2}$$

For a fluid surface this would give the spreading pressure of the adsorbate. For a solid we shall distinguish between the spreading stress (τ_a) and the spreading tension (γ_a), i.e. the adsorbate amount of Kramers energy per unit area. For an inert adsorbent we attribute all the adsorbate amounts to actual gas molecules in the adsorbed layer; the contribution of the solid matter to the adsorbate amounts is in this case zero. This yields a very neat formulation. The Gibbs-Duhem equation for the adsorbate is simply

$$d\gamma_a = s_a dT + (\tau_a - \gamma_a) d\epsilon_s + n_a d\mu \tag{6.3}$$

Since we use the choice $n_1^\sigma = 0$, we omit the label for component 2 ($n_a = n_2^\sigma - n_2^{\sigma 0}$) and μ is of course the chemical potential of the gas. The free energy concentration in the adsorbate changes then by

$$df_a = d(n_a \mu - \gamma_a) = -s_a dT + \mu dn_a + (\gamma_a - \tau_a) d\epsilon_s \tag{6.4}$$

and the total adsorbed free energy by

$$dF_a = d(Af_a) = A df_a + f_a dA = -S_a dT - \tau_a dA + \mu dN_a \tag{6.5}$$

where we have recalled (3.36).

The adsorbed free energy is particularly interesting because when we use models what we have is a partition function for the adsorbed matter, and it is to F_a that this partition function is directly related. Notice that

$$\tau_a = -\left(\frac{\partial F_a}{\partial A}\right)_{T, N_a} \tag{6.6}$$

and

$$\mu = \left(\frac{\partial F_a}{\partial N_a}\right)_{T, A} \tag{6.7}$$

while

$$\gamma_a = \frac{1}{A}(N_a \mu - F_a) \tag{6.8}$$

We have defined a thermodynamical phase, the adsorbate, for which we have defined 'stress' and 'tension'. If we emphasize the features of the solid state we ought to investigate whether τ_a and γ_a are related by the same Shuttleworth equation as τ and γ are for an actual solid surface. This point is considered in a particular example by Eriksson (1969), but otherwise has not apparently been discussed in the literature. Yet the adsorbate, as a thermodynamical phase, has been defined in a purely formal manner, subject to the same objections as are usually raised against the Gibbsian scheme, and the point is a substantial one. Let us prove that τ_a and γ_a satisfy the Shuttleworth equation quite generally. From (6.8):

$$\left(\frac{\partial \gamma_a}{\partial A}\right)_{T,\mu} = -\frac{1}{A^2}(N_a \mu - F_a) + \frac{1}{A}\left[-\left(\frac{\partial F_a}{\partial A}\right)_{T,\mu} + \mu\left(\frac{\partial N_a}{\partial A}\right)_{T,\mu}\right] \tag{6.9}$$

Thus

$$\gamma_a + A\left(\frac{\partial \gamma_a}{\partial A}\right)_{T,\mu} = -\left(\frac{\partial F_a}{\partial A}\right)_{T,\mu} + \left(\frac{\partial N_a}{\partial A}\right)_{T,\mu} \qquad (6.10)$$

Now, from (6.5)

$$d(F_a - \mu N_a) = -S_a dT - N_a d\mu - \tau_a dA \qquad (6.11)$$

Hence

$$\tau_a = -\left(\frac{\partial F_a}{\partial A}\right)_{T,\mu} + \mu\left(\frac{\partial N_a}{\partial A}\right)_{T,\mu} \qquad (6.12)$$

which is an alternative formula for τ_a (compare with (6.6)). Hence, from (6.10) and (6.12),

$$\gamma_a + A\left(\frac{\partial \gamma_a}{\partial A}\right)_{T,\mu} = \tau_a, \quad \text{q.e.d.} \qquad (6.13)$$

Furthermore, if we know F_a (and hence, from (6.7) and (6.8), γ_a) the adsorption equations, instead of (5.5), are, from (6.3),

$$s_a = \left(\frac{\partial \gamma_a}{\partial T}\right)_{A,\mu} \quad ; \quad n_a = \left(\frac{\partial \gamma_a}{\partial \mu}\right)_{T,A} \qquad (6.14)$$

So far the discussion has been purely thermodynamic. It is convenient at this stage to consider briefly the introduction of models in the theory. We shall stress two aspects. One is that the general equations admit practical simplifications when the model is adopted. The other is that, from the model, we calculate the quantities appearing in the thermodynamical relations.

For example, assume mobile, non-localized adsorption. This is often encountered in practice (Clark (1970), chapter V). With all due reservations about sharp classifications, it is what one usually calls physical adsorption, or physisorption. The intuitive image is a kind of two-dimensional liquid condensed on the surface of the solid, the adsorbed molecules being attracted to the neighbourhood of the surface by, say, polarization forces not involving actual chemical bonding of adsorbed molecules to localized entities of the crystal surface. This means that the molecules of the adsorbate can move rather freely throughout the surface without having to break bonds or hop out of localized potential minima or go up and down energy valleys of energy height of at least the order of kT or more. We emphasize that we are discussing solids. However, for the model we are now considering, our assumptions mean that the surface concentrations (e.g. f_a, n_a, etc.) do not depend appreciably on surface strain. Thus, for a process in which the surface is stretched at constant T and μ_a, we have

$$dF_a = f_a dA \quad ; \quad dN_a = n_a dA \qquad (6.15)$$

Hence, from (6.5), for this process

$$dF_a = f_a dA = -\tau_a dA + \mu_a n_a dA \qquad (6.16)$$

or

$$\tau_a = \mu_a n_a - f_a \equiv \gamma_a \qquad (6.17)$$

which is intuitively obvious for the model of non-localized physical adsorption. In other words, the adsorbate has all the properties of a fluid surface phase. Equations (6.3) and (6.4) are considerably simplified. We are back to the simple Gibbsian scheme of Sections 2 and 5. In particular we have the standard Gibbs adsorption equations:

$$s_a = \left(\frac{\partial \gamma_a}{\partial T}\right)_\mu \quad ; \quad n_a = \left(\frac{\partial \gamma_a}{\partial \mu}\right)_T \qquad (6.18)$$

analogous to (5.5). Notice that, as emphasized by Eriksson (1969), these equations hold irrespective of whether or not $d\epsilon_s = 0$. For example, changes in temperature will produce changes in area (thermal expansion of the solid), in spite of which the first equality of (6.18) still holds true because $\tau_a = \gamma_a$ (look at (6.3)).

Now suppose we specify the model for the adsorbate. We then know, in principle, the partition function Q_a of the adsorbed substance. Whence we have

$$F_a = -kT \ln Q_a \qquad (6.19)$$

Thus

$$\mu = -kT \left(\frac{\partial \ln Q_a}{\partial N_a}\right)_{T,A} \qquad (6.20)$$

$$\tau_a = kT \left(\frac{\partial \ln Q_a}{\partial A}\right)_{T,N_a} \qquad (6.21)$$

and

$$\gamma_a = \frac{kT}{A} \left[\ln Q_a - N_a \left(\frac{\partial \ln Q_a}{\partial N_a}\right)_{T,A}\right] \qquad (6.22)$$

These formulae are ready for evaluation. When this is done, two important results are obtained: (a) the equation of state for the adsorbate (this is contained in (6.21) or (6.22)); and (b) the calculation of the adsorption isotherms (this can be obtained from (6.20)). Remember that μ, the chemical potential of the adsorbate, is equal to the chemical potential in the gas phase. This can be written in standard thermodynamics as a function of (p,T). On the other hand, Q_a is a function of (T, A, N_a). Thus we may regard (6.20) as an equation in which μ is given and, solving it for N_a, we can calculate $N_a = N_a(A, p, T)$, i.e. what is usually called the adsorption isotherm. There are other ways of obtaining the formula for N_a, but this is a possible one.

Let us consider a very simple case: an ideal two-dimensional adsorbate. This is really trivial but it is instructive to work it through and illustrate all the above formulae. First with Q_a. For an ideal gas,

$$Q_a = \frac{1}{N_a!} q_a^{N_a} \tag{6.23}$$

where q_a is the molecular partition function,

$$q_a = q_{xy} q_z \exp(-U_0/kT) \tag{6.24}$$

Here U_0 is the depth of the potential well which attracts the gas molecules to the surface, q_z is the vibrational partition function of the molecule trapped in this well, and q_{xy} is the free translational partition function in the (x,y) surface plane. All we need to know is that q_z is some given function of T, with parameters which depend on the particular adsorbate and adsorbent. It is assumed that q_z and U_0 are not appreciably affected by surface strain. For q_{xy} we have simply

$$q_{xy} = \frac{2\pi mkT}{h^2} A \tag{6.25}$$

We shall put

$$q_a = wA \tag{6.26}$$

where w is simply a function of T.

Using Stirling's approximation for $N_a!$,

$$\ln Q_a = N_a \left(1 + \ln \frac{q_a}{N_a}\right) \tag{6.27}$$

Now we evaluate:

Equation (6.20):

$$\mu = -kT \ln\left(\frac{q_a}{N_a}\right) \tag{6.28}$$

The chemical potential is usually written as

$$\mu = \mu_0(T) + kT \ln p \tag{6.29}$$

where $\mu_0(T)$ is the standard chemical potential for an ideal gas phase. Hence the adsorption isotherm:

$$N_a = A\ w(T)\ \exp[\mu_0(T)/kT]p \tag{6.30}$$

Notice: N_a is proportional to p and n_a is independent of A.

Equation (6.21): This yields

$$\tau_a A = N_a kT \tag{6.31}$$

which of course is the ideal equation of state for a two-dimensional gas.

Equation (6.22): Using (6.27), it is trivial to verify that $\gamma_a = \tau_a$, in agreement with (6.17). Evaluating (6.18) we would have the formula for the entropy of the adsorbate.

We are not attempting here a description of the many different cases encountered in adsorption (see Clark (1970) for a rather complete treatment) but it is instructive to consider a few selected examples because the features of the solid state are seldom discussed in the standard literature on adsorption thermodynamics. Let us still assume non-localized mobile adsorption with $q_z \exp(-U_0/kT)$ approximately independent of A, but otherwise not forming an ideal gas. An approximate model is the following: we still write (6.23) and (6.24), but instead of (6.26), q_a is now given by

$$q_a = w(T) A_f e^{B/kT} \tag{6.32}$$

Here $A_f = A - N_a b$ is the free area, b is the area excluded by each adsorbed molecule (because of their mutual interactions) and B comes from an argument in which one inserts a Boltzmann factor $\exp(-V_m/kT)$, where V_m is a mean intermolecular potential which depends on the density (two-dimensional concentration in this case) and on parameters of the molecular interactions. The details are given by Clark (1970), chapter V. The result is that B takes the form:

$$B = \frac{N_a}{A} \bar{\alpha}$$

where $\bar{\alpha}$ depends on the parameters which characterize the molecular interactions. Clearly, this is a van der Waals model in two dimensions.

Now,

$$\ln Q_a = N_a \ln w + N_a - N_a \ln\left(\frac{N_a}{A-N_a b}\right) + \frac{N_a^2 \bar{\alpha}}{AkT} \tag{6.33}$$

The first thing we notice is that the surface concentration of adsorbed free energy is

$$f_a = -n_a kT \ln w - n_a kT + n_a \ln\left(\frac{n_a}{1-bn_a}\right) - \bar{\alpha} n_a^2 \tag{6.34}$$

Thus f_a will be independent of A if n_a is independent of A. We are therefore interested in $(\partial \ln Q_a/\partial N_a)_{A,T}$ which, from (6.33) and (6.20), yields

$$\mu = -kT \ln w + kT \ln\left(\frac{n_a}{1-bn_a}\right) + \frac{kTbn_a}{1-bn_a} - 2\bar{\alpha}n_a \tag{6.35}$$

Writing, as usual, $\mu = \mu(p,T)$, a function we are supposed to know as information we have on the gas phase, (6.35) yields the adsorption isotherm in the form:

$$n_a = \frac{N_a}{A} = f(p,T) \tag{6.36}$$

Therefore n_a is independent of A and, by (6.34), so is f_a. We have then verified the two assumptions (6.15) for non-localized adsorption. The point to stress is that the adsorbate need not form an ideal two-dimensional gas. This is interesting because we can study phase transitions in the adsorbate while still having simple adsorption equations.

The equation of state for the adsorbate is easy to obtain. From (6.33) and (6.21),

$$\tau_a = \frac{n_a kT}{1-bn_a} - \bar{\alpha}n_a^2 \tag{6.37}$$

which can be rewritten in a more familiar form:

$$(\tau_a + \bar{\alpha}n_a^2)(1-bn_a) = n_a kT \tag{6.38}$$

i.e. the two-dimensional van der Waals equation.

We have shown that with these assumptions the two simplifications (6.15) characteristic of non-localized adsorption are satisfied. We also proved on general grounds that this leads to the equality of τ_a and γ_a. Indeed we can directly evaluate (6.8) from (6.35) and (6.34). The result is easily obtained and is identical to (6.37).

This should suffice to give us an idea of what to expect from mobile adsorption. The opposite case is that of strictly localized adsorption. Now the gas molecules are supposed to be attached to fixed entities on the surface. Let us loosely call them adsorption sites, whatever they are (we might be thinking, for example, of actual bonding to certain atoms or ions on the surface). Again, there are many particular cases that might be considered. In relatively easy ones the thermodynamic state of the adsorbate is not appreciably affected by changes in area not involving changes in T and μ (Eriksson (1969)). When this condition is met neither the adsorbed free energy F_a nor the number of adsorbed molecules N_a are changed by a deformation dA at T and μ constant. Hence, by (6.5), τ_a in this case must be zero. This means that the surface stress τ of the entire system (adsorbent, gas and adsorbate) is not affected by adsorption. There is no reason why the spreading tension γ_a should be zero. However, the usual identification of γ_a with the 'spreading pressure' clearly is not suited to emphasize the features of the solid state. Only for mobile adsorption, when $\gamma_a = \tau_a$, both concepts coalesce into a single one, and then the term 'spreading pressure' has an unambiguous meaning. But suppose we wanted to derive the equation of state for the adsorbate by using (6.21). We would simply be verifying that we are in a situation in which Q_a does not depend on A.

For example, consider the following model. We assume a number M of adsorption sites and we assume further that M does not change under a small deformation dA at constant T and μ. If the partition function q_a of <u>one</u> adsorbed molecule on a fixed site is likewise unaffected by this deformation, then the partition function Q_a of N_a adsorbed molecules depends only on N_a, M and q_a. Hence from (6.21), $\tau_a = 0$. The equation of state in this case can only be obtained by using (6.22) to evaluate γ_a, whose differential (6.3) becomes

$$d\gamma_a = s_a dT - \gamma_a d\epsilon_s + n_a d\mu \tag{6.39}$$

Unlike the previous case, it is now important to specify in the derivatives of γ_a whether A remains constant.

In the well-known Langmuir model, the adsorption on different sites is assumed to be independent. Then (Hill (1960), chapter VII),

$$Q_a = \frac{M! \; q_a^{N_a}}{N_a! \, (M-N_a)!} \tag{6.40}$$

The combinatorial factor gives the number of ways in which M sites can be covered by N_a molecules. For this kind of model it is customary to define the degree of coverage:

$$\theta = \frac{N_a}{M} \tag{6.41}$$

which varies between 0 and 1. The calculation of the adsorption isotherm aims at obtaining θ as a function of (p,T).

Now

$$\ln Q_a = M \ln M - N_a \ln N_a - (M-N_a) \ln(M-N_a) + N_a \ln q_a \tag{6.42}$$

Notice that nothing is said about M. For example, the sites might be associated with defects, and M could conceivably depend on T (e.g. thermal creation of point defects) or it might change upon a deformation dA at constant (T, μ). In any case,

$$\left(\frac{\partial \ln Q_a}{\partial N_a}\right)_{T,A} = -\ln N_a + \ln(M-N_a) + \ln q_a \tag{6.43}$$

and this is the derivative we need to evaluate (6.20) and (6.22). Hence the chemical potential:

$$\mu = kT\left[\ln N_a - \ln(M-N_a) - \ln q_a\right] \tag{6.44}$$

and the equation of state:

$$\gamma_a A = -kTMln(1-\theta) \qquad (6.45)$$

Writing μ in the standard way (6.29), we can obtain from (6.44) the well-known Langmuir adsorption isotherm:

$$\theta = \frac{\chi(T)p}{1 + \chi(T)p} \qquad (6.46)$$

where

$$\chi(T) = q_a(T)e^{\frac{\mu^0(T)}{kT}} \qquad (6.47)$$

If N_a and q_a are independent of A, as explained, then by (6.42) F_a is also independent of A and the thermodynamical scheme is further simplified in the manner discussed above. In particular, τ_a is zero, while (6.45) still gives the same equation of state.

We emphasize that this discussion is not a mere repetition of the usual argument. By introducing an ad hoc external parameter (see e.g. Hill (1960), chapter VII) proportional to A, one can use the single concept of 'spreading pressure' and obtain (6.45). However, this argument would be invalid for any model (conceivable for solids) in which M is physically independent of A in the sense just discussed. Our approach, in agreement with Eriksson, circumvents this obstacle in a natural way and is suited to incorporate the features of the solid state.

There are again many possible models of localized adsorption: different gas species, non-uniform surfaces, formation of more than one monolayer, etc. We shall not try to cover them here (see Hill (1960) and especially Clark (1970) for a detailed description) but we shall come back to this question in relation to charge transfer chemisorption on semiconductors and oxides. This will involve the use of the grand canonical formalism. For future reference, let us briefly set it out here in a form valid for an adsorbed phase on a solid surface.

For the time being, we consider one species in the adsorbate, N being a variable number and N_a its thermal equilibrium value. The partition function Q_a will now be explicitly written as $Q_a(N)$, to indicate that there are N molecules in the adsorbate. The grand partition function Ξ_a of the adsorbate is then

$$\Xi_a = \sum_N Q_a(N) \lambda^N \qquad (6.48)$$

where

$$\lambda = e^{\frac{\mu}{kT}} \qquad (6.49)$$

is the absolute activity (Guggenheim (1967)).

General arguments show that, for any thermodynamical system,

$$kT \ln \Xi = TS - E + N\mu = G - F \tag{6.50}$$

If we carry over the Gibbsian process of taking differences — remember the sign convention for γ_a and τ_a — we obtain for the adsorbate

$$kT \ln \Xi_a = \Omega_a = A\gamma_a \tag{6.51}$$

Since, from (6.3), we have

$$d\Omega_a = S_a dT + N_a d\mu + \tau_a dA \tag{6.52}$$

it follows that

$$S_a = k \ln \Xi_a + kT \left(\frac{\partial \ln \Xi_a}{\partial T}\right)_{\mu, A} \tag{6.53}$$

$$N_a = kT \left(\frac{\partial \ln \Xi_a}{\partial \mu}\right)_{T, A} = \lambda \left(\frac{\partial \ln \Xi_a}{\partial \lambda}\right)_{T, A} \tag{6.54}$$

and

$$\tau_a = kT \left(\frac{\partial \ln \Xi_a}{\partial A}\right)_{T, \mu} \tag{6.55}$$

We are all ready to work out the thermodynamics of the adsorbate from the grand canonical partition function. As a trivial example, not entirely a waste of time, let us redo the extreme ideal models of mobile and localized adsorption.

For ideal mobile adsorbate, from (6.23):

$$\Xi_a = \sum_{N=0}^{\infty} \frac{1}{N!} (\lambda q_a)^N = e^{\lambda q_a} \tag{6.56}$$

i.e.

$$A\gamma_a = kT\lambda q_a \tag{6.57}$$

Thus from (6.54)

$$N_a = \lambda q_a \tag{6.58}$$

and from (6.55)

$$\tau_a = kT\left(\frac{\partial(\lambda q_a)}{\partial A}\right)_{T,\mu} = \frac{kT}{kT}\left(\frac{\partial(\gamma_a A)}{\partial A}\right)_{T,\mu} = \gamma_a \qquad (6.59)$$

in agreement with the results obtained from Q_a.

For ideal localized (Langmuir) adsorption, from (6.40):

$$\Xi_a = \sum_{N=0}^{M} \binom{M}{N}(\lambda q_a)^N = (1+\lambda q_a)^M \qquad (6.60)$$

Thus, from (6.54),

$$N_a = M\frac{\lambda q_a}{1+\lambda q_a} \qquad (6.61)$$

identical to (6.46). If M and q_a are independent of A we have again $\tau_a = 0$, while

$$A\gamma_a = M\ln(1+\lambda q_a) \qquad (6.62)$$

From (6.61) we obtain

$$q_a = \frac{\theta}{1-\theta} \qquad (6.63)$$

which used in (6.62) yields (6.45) again. Notice one practical point about the grand canonical approach. The formula for N_a, i.e. the adsorption isotherm, is written down directly, while in the canonical formalism one has to solve for Na_a from the formula (6.20) for μ. Steele (1966) discusses in detail adsorbate equations of state and the use of canonical and grand canonical partition functions.

In simple cases one can write down the formula for γ_a, but life is not always so easy. However, one usually knows the adsorption isotherm, i.e. $n_a(p,T)$, at least empirically. In this case one can resort to the differential equation (6.3). For processes at constant temperature and area we simply have

$$d\gamma_a = n_a d\mu = n_a kT \frac{dp}{p} \qquad (6.64)$$

where (6.29) has been used for μ. Thus, knowing the adsorption isotherm, one can find γ_a at arbitrary coverage (or, alternatively, at arbitrary gas pressure) by integrating:

$$\int d\gamma_a = kT \int \frac{n_a(p,T)}{p} dp \qquad (6.65)$$

This is often a convenient practical procedure. Incidentally, for the Langmuir case it is trivial to verify that using (6.61) in (6.65) we recover (6.45).

7. SYSTEMS WITH ELECTRICAL CHARGES: CHEMISTRY AND THERMODYNAMICS OF IMPERFECT CRYSTALS

Electrical charges play a very important role in surface physics. A semiconductor surface is expected to have a charge due to electrons in surface states. This implies a space charge boundary layer. We also expect to find boundary layers near the surface of an ionic crystal. Electron transfer at the surface takes place in a very important type of adsorption: charge transfer chemisorption. We want to include these facts in the scheme of surface thermodynamics and need to understand how electrical charges — electrons, holes, charged impurities, ions, vacancies — are characterized as chemical entities and then included in the thermodynamical formulation.

The principles of chemical thermodynamics of charged systems are well explained in standard textbooks (see, e.g., Guggenheim (1967) or Kirkwood and Oppenheim (1961)). The chemistry of imperfect crystals is discussed very thoroughly by Kröger (1964). A brief introduction is given by Van Gool (1966). We shall collect here the essential points needed to set out for subsequent discussion. For details and proofs the reader is referred to the above references.

The electrochemical potential: The standard definition as the appropriate derivative of a thermodynamic function is still valid, but its meaning is now more complex. The energy required to introduce dN extra particles includes a 'purely chemical part', which would correspond to a diffusion potential associated with a gradient of particle density, and an electrostatic part, which would correspond to an electrostatic potential associated with a gradient of charge density. Formally we write the energy balance:

$$dE = TdS + dW + \tilde{\mu} dN \qquad (7.1)$$

This defines $\tilde{\mu}$ which we write as

$$\tilde{\mu} = \mu + q\phi \qquad (7.2)$$

We shall call $\tilde{\mu}$ the electrochemical potential and μ the chemical potential. Notice that μ and $q\phi$ have no separate operational meaning. We cannot move the actual particles from one region to another within our physical system without altering simultaneously the chemical composition and the charge density of the two regions affected by this process. Nevertheless the μ part of $\tilde{\mu}$ is still related to the particle density in the standard manner, as μ and n are related for uncharged particles, while ϕ is obtained from the electrostatics of the problem. We can then formally manipulate equalities involving μ or ϕ separately, but in all energy balances they always appear together as $\tilde{\mu}$.

The choice and number of components: Suppose we have a Si crystal. We have one component (Si atoms). Some Si atoms are ionized according to

$$Si^+ + e^- \rightleftarrows Si \qquad ((7.3)$$

where e^- means an electron in the conduction band. Can we regard the mobile electrons as another component? If we do so we ought of course to describe Si^+ as yet another component, on the same footing as e^-. Does this mean that the number of degrees of freedom had increased by two? The answer is no. We have two extra conditions. One is charge neutrality; the other is the equilibrium condition for (7.3).

Suppose we have our Si crystal exposed to a gas of particles P, which can be adsorbed on the Si surface and trap electrons according to

$$P_s + e_s^- \rightleftarrows P_s^- \tag{7.4}$$

We have an extra component: the species P. The variance changes accordingly, as it must. If we wish, we can also introduce yet another component, P_s^-, which of course entails introducing the electrons as a new component. But we have again charge neutrality and an equilibrium condition. Alternatively, this condition alone suffices if the electrons are already independent components in our scheme to describe the solid crystal, in which case the overall charge neutrality is already one of the conditions imposed irrespective of whether or not (7.4) takes place. The number of degrees of freedom again remains the same; we need no extra variables $\tilde{\mu}(e_s^-)$ or $\mu(P_s)$ if the surface phase is in equilibrium with the bulk phases.

It is easy to see that one is always free to choose formally extra components without altering the variance of the problem because of the extra conditions. This is very practical. We are not compelled to use, for example, the temperature and the gas pressure as the two independent variables. We may choose, say, T and free electron density, and we might be interested in doing this because the concentration of charge carriers is what we obtain from the models of the solid crystal which we want to insert into our analysis.

The issues raised in these simple examples are more general. They appear, for example, in the study of point imperfections in compound crystals (e.g. vacancies, interstitials, charged point defects in ionic crystals, etc.). And we always write down the corresponding 'chemical reaction'. For example, in Kröger's (1964) notation the equation

$$A_A \rightarrow V_A + A_S \tag{7.5}$$

describes the formation of a Schottky defect (disregarding charge for the time being). An atom A at an A site (A_A) migrates to the surface. We are left with a vacancy of $A(V_A)$ and an atom A at a surface site (A_S). We regard not only A_A and A_S but also V_A as entities being at particular sites. Kröger (1964) calls these 'structure elements'. Completeness requires that every time a new structure element is added there must necessarily be a change in the number of other structure elements. For instance, if we add an atom A_A to an AB compound crystal, we must say that we have also added a vacancy V_B. Imagine, for example, that the extra atom has come from a gas in equilibrium with the crystal. Then the consistent way to write down the real process is

$$A_g \rightarrow A_A + V_B \tag{7.6}$$

<u>Chemical potentials for separate components:</u> The concept of structure element is useful in a practical sense. We may choose the different structure elements to be the components, or species, of our problem.

There is no difficulty with the variance, as we have seen. However, one point is worth mentioning.

In a problem in which there are different species or components, the definition of the chemical potential for the ith component is

$$\mu_i = \left(\frac{\partial G}{\partial N_i}\right)_{T,p,N_j \neq i} \tag{7.7}$$

Charge is irrelevant. The same definition applies to $\tilde{\mu}_i$ if this component is charged. But if we choose our components to be the structure elements, (7.7) becomes meaningless, because we cannot alter N_i while keeping constant <u>all the other N_j's</u>. This defines a virtual chemical potential (Kröger, 1964). The question is whether it is legitimate to use virtual potentials to describe real processes. The same question arises in electrochemistry (Guggenheim (1967), chapter 7). The answer is yes, provided we are consistent in our description.

This is why we must describe real processes as in (7.6). Here $\mu(Ag)$ is real; it is simply the chemical potential in the gas phase, which we are supposed to know as a function of (p,T), while $\mu(A_A)$ and $\mu(V_B)$ are virtual for the reasons just explained. But we insist that for each one A_A which appears one V_B also appears. The real change in free enthalpy for this process is

$$\Delta G = \mu(A_A) + \mu(V_B) - \mu(Ag) \tag{7.8}$$

Thus, we do not write $\mu(A_A)$ or $\mu(V_B)$ separately — they are virtual and in fact contain an arbitrary additive constant. These two quantities always appear together in our balance. According to (7.8) the combination

$$\mu(A_A) + \mu(V_B) = \mu(Ag) + \Delta G$$

is, of course, real. The undetermined additive constants cancel out.

This question has been studied in detail by Kröger et al. (1959) and we shall not elaborate on it any further. The lesson is that we can do meaningful formal theory using virtual potentials, provided our description is always complete and our balances consistent. We must observe charge neutrality and the conservation of number relation between different structure elements. In essence the justification for using virtual μ's is the same as the justification for using the concepts of μ and $q\phi$ for charged species, where only $\tilde{\mu}$ is real.

<u>Equilibrium condition:</u> The usual analysis proceeds in the following way. Chemical or electrochemical potentials are freely used, whether real or virtual, as explained. One writes down a process of the kind considered in the chemical form:

$$\Sigma_i \nu_i X_i = 0 \tag{7.9}$$

where the ν_i's (some positive, some negative) are the stoichiometric numbers for the reaction. The equilibrium condition is then

$$\Sigma_i \nu_i \mu_i = 0 \tag{7.10}$$

Writing the μ_i's in the standard form in terms of the concentrations, one arrives at the corresponding mass action law in the standard form. This involves the product of concentrations in the form:

$$\Pi_i n_i^{\nu_i} = \text{equilibrium constant} \tag{7.11}$$

if the concentrations are sufficiently dilute. These facts are well known and explained everywhere. We shall come back to this for the case of impurity ionizations in solids when the electron gas is not sufficiently dilute.

With this quick survey of standard theory we can now return to our main business, the study of the surface when electrical charges are involved.

8. THERMODYNAMICS OF CHARGED SURFACES

The concepts summarized in Section 7 are especially useful to discuss semiconductors and ionic crystals. The case of metals lends itself to a discussion exclusively in terms of an electron gas (provided one takes care to immerse it in a background of positive charge), as was done by Lang and Kohn (1970). We discussed the thermodynamics of this case in Section 4. For semiconductors or ionic crystals we resort to a description in terms of neutral and charged structure elements, as explained in Section 7.

Let us summarize some well-known basic facts (Many et al. (1971); Blakemore (1962); Frankl (1967); Kröger (1964)). We contemplate a certain density of surface charge. This means that there is an electrostatic potential difference between surface ($z = 0$) and bulk ($z \to \infty$). We take (Fig.6) $\phi(\infty) = 0$. There is an electrostatic profile $\phi(z)$, which is the solution of the Poisson equation with two boundary conditions: (a) charge neutrality in the bulk; and (b) $\phi(0) = \phi_s$. This yields the corresponding value of the surface charge density — or, vice versa, we take this as given and hence surmise ϕ_s. Charge neutrality is effectively achieved within a distance of the order of the

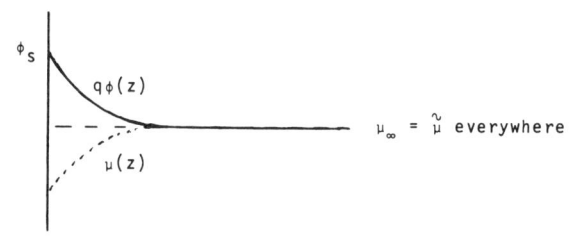

FIG.6. Electrostatic potential difference between surface ($z = 0$) and bulk ($z \to \infty$).

Debye-Huckel length, which characterizes the ability of the crystal to screen out the surface charge. This length gives the effective thickness of the space-charge boundary layer. The space-charge density $\rho(z)$ is due to the fact that charged particles are present in different densities, according to their charge. Since the electrochemical potential $\tilde{\mu}$ must be everywhere constant for equilibrium to exist, by (7.2) μ varies like the dotted line in Fig.6 and the concentration of a charged component varies across the space-charge layer according to some appropriate function of T and $\mu(z)$. Thus we have an inhomogeneous boundary layer.

The surface charge may have different origins. Thus, for example: (i) Clean crystal surfaces may have surface states. Electrons in these states contribute as surface charges. (ii) In a field effect experiment the surface charge is deliberately induced by application of an external voltage. This shifts ϕ_s up and down, thus scanning, for example, different energy levels. (iii) If charge transfer chemisorption takes place, this may change quite appreciably surface charge and potential. In fact, it may be reasonable to attribute all surface charge to charged adsorbed species. This case will be discussed in greater detail. (iv) Ionic crystals offer an amusing example which it is quite instructive to consider. Every crystal contains, at finite temperature, a certain amount of structural disorder, which can be of various types (Kröger (1964), chapter 13). Consider, for example, Schottky defects in an ionic MX compound. We have vacancies V'_M and $V^{\bullet}_X$. Here the prime means a negative charge relative to the perfect crystal. The entity we are removing from a regular lattice site is a positive cation M^+, i.e. a positive charge is missing. Likewise $V^{\bullet}_M$, a negative cation X^- missing, is positively charged relative to the perfect crystal. This is indicated by the dot. Deep in the bulk the concentrations are equal. In Kröger's notation

$$[V'_M]_\infty = [V^{\bullet}_X]_\infty \tag{8.1}$$

Now, these ions migrate to the surface and, because the energy needed to remove an ion from its lattice site to a surface site is different for M^+ or X^-, it turns out that the surface contains different amounts of such ions, and therefore a net surface charge. We have again an inhomogeneous space-charge boundary layer across which the concentrations of the two types of vacancies vary in the form:

$$[V'_M] = [V'_M]_\infty \exp(-q\phi/kT)$$
$$[V^{\bullet}_X] = [V^{\bullet}_X]_\infty \exp(q\phi/kT) \tag{8.2}$$

After a screening length, charge neutrality is practically achieved and we are in the neutral bulk region. At the surface the following reactions are formulated:

$$M_M^x + S_{M,S}^x \to M_{M,S}^+ + V'_M + S_{X,S}^x$$
$$X_X^x + S_{X,S}^x \to X_{X,S}^- + V^{\bullet}_X + S_{M,S}^x \tag{8.3}$$

The first one, for example, means that initially we have an M^+ ion in its normal lattice site — it is a regular structure element and the small cross $(^x)$ indicates that it is neutral relative to the perfect crystal. We also say that we have at the surface an unoccupied site $S_{M,S}^x$, next to an X^- ion. We only introduce one site $S_{M,S}^x$, i.e. <u>one more</u> structure element, when we are going to occupy it with an M^+ ion. To maintain the number ratio between different structure elements we must then say that, when $M_{M,S}^+$ has appeared at the surface, leaving behind a vacancy V_M', a new site $S_{X,S}^x$ has been created. The second of reactions (8.3) is interpreted in a similar way.

The next step is to write down the equilibrium condition (8.9) for these reactions. It is at this stage that we use freely the concepts of virtual chemical and electrochemical potentials. The details are given in Kröger (1964), chapter 19. One writes the chemical potential in the standard form:

$$\mu_i = \mu_i^0 + kT \ln[i] \tag{8.4}$$

where [i] is the fraction of sites occupied by the structure element i. This leads to

$$\mu^0(V_M') + \mu^0(S_{X,S}^x) - \mu^0(S_{M,S}^x) + kT\{\ln[V_M']_\infty$$

$$+ \ln \frac{[S_{X,S}^x]}{[S_{M,S}^x]}\} = -q\phi_s \tag{8.5}$$

and

$$\mu^0(V_X^\cdot) - \mu^0(S_{X,S}^x) + \mu^0(S_{M,S}^x) + kT\{\ln[V_X^\cdot]_\infty$$

$$- \ln \frac{[S_{X,S}^x]}{[S_{M,S}^x]}\} = q\phi_s \tag{8.6}$$

The first three terms of each of these equations give the standard free enthalpy of formation of the corresponding defect, $G(V_M')$ and $G(V_X^\cdot)$, under conditions at which $\phi_s = 0$. At the surface we have

$$[S_{M,S}^x] - [S_{X,S}^x] = [X_{X,S}^-] - [M_{M,S}^+] \equiv \delta \tag{8.7}$$

Here δ measures the surface charge in elementary units per regular surface site and is usually very small, $\delta \ll 1$.
Approximating $\ln(1+\delta) \cong \delta$ and recalling (8.1), one has

$$2q\phi_s = G(V_X^\cdot) - G(V_M') + 2\delta kT \tag{8.8}$$

or, neglecting differences in the vibrational contribution to the entropy term in the free enthalpies,

$$2q\phi_s = H(V_X^{\cdot}) - H(V_M^{'}) + 2\delta kT \tag{8.9}$$

The point of this simple exercise is to demonstrate how one uses models of the solid state. From these models the enthalpies $H(V_X^{\cdot})$ and $H(V_M^{'})$ can be estimated. For ionic crystals it is usually found that their difference is much larger than $2\delta kT$. Thus (8.9) says that the actual amount of surface charge is immaterial. What actually determines the magnitude of the surface potential is the fact that very different energies are needed to create $V_X^{\cdot}$ or $V_M^{'}$. For NaCl, for example, one estimates $\phi_s = +0.28$ volts. Thus we have learnt from simple thermodynamic considerations that the surfaces of ionic crystals are at a different potential from the bulk because of ionic surface charges. Hoyen (1973) gives an extensive discussion of similar ideas applied to surfaces of silver halides, very important for the photographic process.

In semiconductors, on the other hand, the reason we have a surface potential is because of electronic surface charges. Let us investigate how much this can contribute to the surface tension. Here we shall make use of the distinction between global and Gibbsian quantities, established in Section 4. Consider an n-type semiconductor. We assume N_b electrons forming a Maxwell-Boltzmann gas in a box of volume $V = AD$. Here D is the distance between two faces of area A. The free energy is

$$F_b^0 = kT\left[N_b \ln N_b - N_b - \frac{3}{2}N_b \ln w - N_b \ln A - N_b \ln D\right] \tag{8.10}$$

where

$$w = \frac{4\pi mkT}{h^2}$$

Now we allow some of the N_b electrons to go into surface charge. Put

$$N_b = N_b^{'} + N_s \tag{8.11}$$

The $N_b^{'}$ electrons in bulk states have the same total volume available — an important fact to take into account in order to estimate entropy terms correctly — but now a space-charge boundary layer has developed, as in Fig.6. In the new situation the bulk part of the partition function is

$$Q_b^{'} = \frac{1}{N_b^{'}!}(q_v^{'})^{N_b^{'}} \tag{8.12}$$

where $q_v^{'}$ is the partition function of one electron and contains an integral of the form:

$$\int_0^{\text{bulk}} e^{-\beta q\phi(z)}\,dz \quad (\beta = 1/kT) \tag{8.13}$$

We recall the usual definition of surface excess:

$$\Delta n = \int_0^{bulk} [n(z) - n_b] dz = n_b \int_0^{bulk} [e^{-\beta q \phi(z)} - 1] dz \equiv n_b J \quad (8.14)$$

where n_b is the bulk concentration N_b/AD, and the last equality defines the integral J.

Adding and subtracting unity to the integrand in (8.13), we can write

$$q_v' = w^{3/2} AD \left(1 + \frac{2J}{D}\right) \quad (8.15)$$

The thickness of the boundary layer is much smaller than D, hence $2J \ll D$, and we can approximate $\ln(1+2J/D) = 2J/D$. From (8.15) and (8.12) we then obtain the new bulk contribution to the free energy:

$$F_b' = kT \left[N_b' \ln N_b' - N_b' - \frac{3}{2} N_b' \ln w - N_b' \ln A - N_b' \ln D - \frac{2 N_b' J}{D} \right] \quad (8.16)$$

We further assume that the N_s surface electrons form a Maxwell-Boltzmann gas: a point to be justified later. Their partition function is then

$$Q_s' = \frac{1}{N_s!} (q_s)^{N_s} \quad ; \quad q_s = wAe^{-\beta q \phi_s} \quad (8.17)$$

whence their free energy:

$$F_s' = kT [N_s \ln N_s - N_s - N_s \ln w - N_s \ln A + N_s \beta q \phi_s] \quad (8.18)$$

The global change in free energy:

$$F = F_b' + F_s' - F_b^0 \quad (8.19)$$

follows from (8.18), (8.16) and (8.10). In evaluating these terms we use (8.11) to eliminate N_b' everywhere, and also use the fact that $N_s \ll N_b$. Putting $N_s = 2 n_s A$, we find (remember the total area is 2A)

$$\frac{\Delta F}{2A} = kT n_s \ln\left(\frac{n_s w^{1/2}}{n_b}\right) + n_s q \phi_s - \frac{1}{2} kT \Delta n \quad (8.20)$$

If we estimate the global change in free enthalpy, $G_b' - G_s' - G_b^0$, the global difference $\Delta(F-G)$ is, as we saw in Section 4, both a global and a Gibbsian quantity and yields the surface tension γ. From the partition function of the system we can use standard formulae to calculate the electrochemical potential $\tilde{\mu}$. However, in equilibrium $\tilde{\mu}$ is constant everywhere, as indicated in Fig.6. Thus the global change in free enthalpy is simply

$$\Delta G = \tilde{\mu}(N_b' + N_s' - N_b) = 0 \quad (8.21)$$

Therefore the contribution of the space-charge boundary layer to the surface tension is simply

$$\gamma_{b.l.} = \frac{\Delta F}{2A} = (8.20) \tag{8.22}$$

This is very different from a formula repeatedly proposed (Sparnaay (1964, 1969, 1972)) in which only the electrostatic contribution is taken into account. Our formula (8.20) contains three terms. The first is an entropy term and takes account of the redistribution of the electron gas in space. The second is indeed the potential energy of a charge $n_s q$ at a potential ϕ_s. The third is a kinetic energy term. It says that Δn electrons have left states of mean thermal energy $(3/2)kT$ and gone into states of mean thermal energy kT, losing $(1/2)kT$ in kinetic energy each.

Let us see some representative numbers. In a typical situation in n-type Ge or Si at room temperature we may have the Fermi level in the bulk at, say, 10 kT above the intrinsic Fermi level. This corresponds to $\phi_s \cong 10$ kT when there is charge neutrality at the surface. The bulk carrier concentrations are $n_b \cong 5.3 \times 10^{17}$ cm^{-3} for Ge, or $\cong 3.3 \times 10^{14}$ cm^{-3} for Si. A small shift of the surface potential ($\cong 0.1$ kT) produces a surface concentration of electrons of $n_s \cong 10^{12}$ cm^{-2} for Ge, or $\cong 5 \times 10^{10}$ cm^{-2} for Si. These figures can be estimated from tabulated solutions of the Poisson equation for the space-charge boundary layer (Many et al. (1971); Frankl (1967)) and are in agreement with the rough consideration that densities of surface states in the region of charge neutrality are usually round about the order of 10^{13} electrons·cm^{-2}·eV^{-1}; hence a shift in surface potential, with respect to charge neutrality, of order 0.1 kT should fill up levels yielding a value of n_s round about the order of 10^{11} cm^{-2}.

The degeneracy limit in two dimensions can be estimated at about $n_s \cong 10^{12}$ cm^{-2}. Thus it is reasonable, at least for an order of magnitude estimate, to describe the two-dimensional electron gas as a Maxwell-Boltzmann assembly. We are now ready to estimate (8.22). We start from an empirical fact: that surface tensions of solids are usually of the order of a few to several hundred erg·cm^{-2}. Now, with effective masses of order 0.1 m_e, and always in c.g.s. units (erg·cm^{-2}) the first term of $\gamma_{b.l.}$ is of order 10^{-2} for both Ge and Si; the second term is of order 10^{-1} for Ge and 10^{-2} for Si; the third term is an order of magnitude less than the second. A more accurate calculation would undoubtedly change these results, but the orders of magnitude of $\gamma_{b.l.}$ and of γ itself are so wide apart that we can conclude that the contribution of the space-charge boundary layer to the surface tension of semiconductors is completely negligible.

So far, we have considered the pure solid surface. It remains to investigate what happens when charge transfer chemisorption takes place. This will be our task in the following sections.

9. SIMPLE MODELS OF CHARGE TRANSFER CHEMISORPTION

Charge transfer chemisorption is of great interest in physical chemistry of surfaces, and worth considering here because it is a process in which the adsorbent is far from being inert. In fact it suffers important chemical and electrical changes. We shall illustrate the basic problems involved by discussing electrical changes.

In this section we shall loosely use two terms whose meaning should be further elucidated in a more complete discussion of chemisorption as a problem of gas-solid interaction. Here we are simply studying the statistical thermodynamics of the surface phase defined as adsorbate. We are essentially confronted with two issues: (a) We want to study the novelties due to the introduction of electrical charge in the analysis. (b) We want to do this in such a way that we bring into the discussion some standard features of the models of the solid state. Thus, something must be said about the models used to describe chemisorption. We shall use a very simple one, essentially of a Langmuir type, but allowing for lateral interactions in the adsorbate. Thus we shall assume some active sites for adsorption, at which some species, loosely called adatoms, are attached through some sort of binding. The energy levels of these bonds are part of the information needed. In our analysis they appear as parameters, just like the width of a band gap in the solid, for example.

We regard these adatoms as surface impurities, which can act as either donors or acceptors. Again, this involves new binding energies or, in another language, the electro-afinity or the ionization energy of an adatom. It is here that we need the concepts introduced in Sections 7 and 8. Basically, the elements of our model are the following (Fig.7): (i) The solid is characterized by two bands and one gap, where (donor or acceptor) impurity levels may lie, i.e. we are thinking of semiconductors, covalent or ionic, or else of something like non-stoichiometric metal oxides described in semiconductor

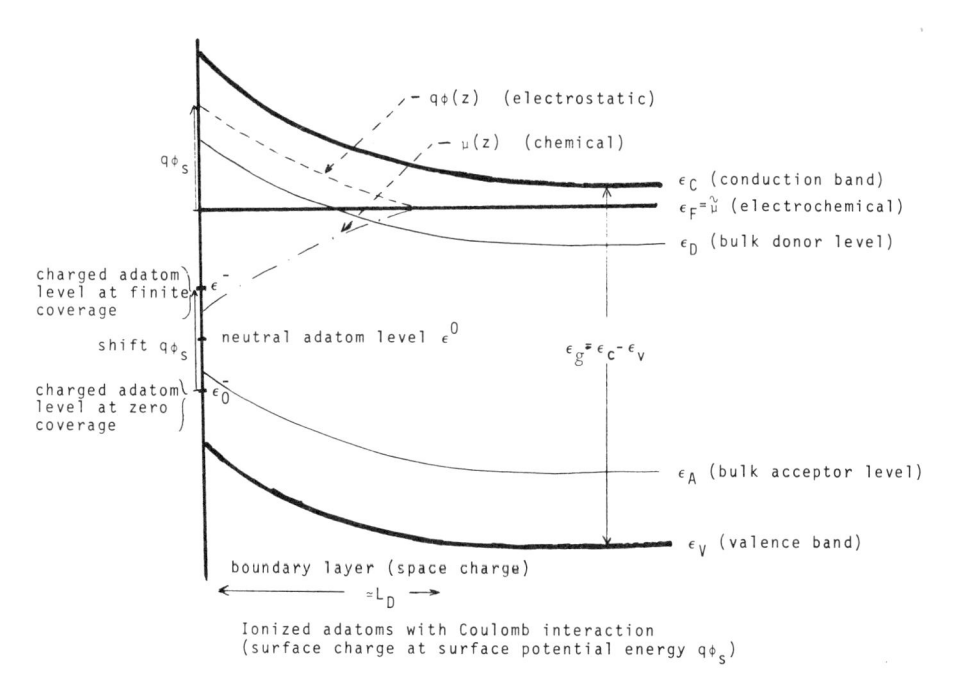

FIG.7. Parameters for charge transfer chemisorption.

language. Incidentally, this is not an altogether trivial question from a
thermodynamic viewpoint. See Kröger (1964) for a very clear discussion.
(ii) Temperature and imperfections determine the bulk position of the Fermi
level, which is now the electrochemical potential. (iii) Adatoms may trap
electrons, either from band or from impurity states. We have an inhomogeneous space-charge boundary layer and a surface charge. (iv) We take
account only of Coulomb interaction between charged adatoms and even this
in a smeared-out way. We simply say that, at finite coverage, the next
surface charge has to sit at the surface potential energy $q\phi_s$ where ϕ_s is due
to the surface charge density, i.e. to existing charged adatoms. For a charged
adatom this means that its energy level $\bar{\epsilon_0}$ changes to $\bar{\epsilon} = \bar{\epsilon_0} + q\phi_s$.

Here it is convenient to use the grand canonical formalism. We can
regard the surface active sites as an open system of traps where different
species can be trapped, namely, adatoms and then perhaps electrons or holes.
There will be some rules. For example, one electron at an active site but
no gas 'atom' (previously adsorbed) is forbidden. These rules must be
properly incorporated in the statistics. It is also conceivable (and sometimes
very likely) that different types of adsorption take place on different types
of active sites. Let M_r ($r = 1,2,\ldots,R$) indicate the number of active sites (traps)
of type r; let z_i be the variable number of particles of species i trapped at
one trap and ν_i be the thermal equilibrium value of z_i. We want to know the
number $M_i(\ldots \nu_i \ldots)$ of traps which, in thermal equilibrium, have precisely ν_1
particles of species 1, ν_2 of species 2, etc. What we need is a suitable
extension of (6.54). The crucial point is the construction of a grand partition
function. We shall make the assumption that the partition function of the
system of active sites can be factorized as the product of the single partition
functions of the different sites. Notice that this does not rule out the possibility of interactions between different adatoms, provided we smear these out
and put the interacting adatoms in the mean potential field of all the others.
Another meaningful point is that we must specify the components and their
surface amounts in our definition of the adsorbate as a thermodynamical
phase. Here chemical changes in the adsorbent — involving atoms or ions —
may introduce significant changes in the analysis. We shall assume that from
this point of view the adsorbent is inert. This would also require a different
formulation if ion diffusion were involved. With our assumptions the choice of
dividing surface as indicated in Section 5 implies that, as adsorption proceeds,
the surface amounts of atoms or ions of the solid adsorbent remain equal to
zero. But electrons appear in the adsorbate as a new component. Their
numbers will be obtained from the numbers $M_r(\ldots \nu_i \ldots)$ in the manner to be
explained. Furthermore we must include in our analysis the fact that the
adsorbate suffers a chemical and an electrical change: both the chemical and
the electrostatic potential of the electrons change across a boundary layer
of thickness $\cong L_D$, typically of order 100 Å.

Now, with the above assumptions the grand partition function of the
adsorbate is

$$\Xi_a = \prod_{r=1}^{R} \left(\sum_{z_1} \ldots \sum_{z_i} \ldots \lambda_1^{z_1} \ldots \lambda_i^{z_i} \ldots q_r(\nu_1, \ldots, \nu_i, \ldots) \right)^{M_r} \quad (9.1)$$

Details are given elsewhere (García-Moliner (1969)). We shall summarize
the main points of the analysis.

The numbers we are interested in are

$$M_r(\nu_1,\ldots,\nu_i,\ldots) = M_r \frac{\lambda_1^{\nu_1}\cdots\lambda_i^{\nu_i}\cdots q_r(\nu_1,\ldots,\nu_i,\ldots)}{\sum_{z_1}\cdots\sum_{z_i}\cdots\lambda_1^{z_1}\cdots\lambda_i^{z_i}\cdots q_r(z_1,\ldots,z_i,\ldots)} \quad (9.2)$$

The ratio of this number to M_r gives the fractional coverage of active sites of type r with ν_1 particles of species 1, etc. The occupation rules are expressed in two ways: (i) every $\sum_{z_i}$ has an upper limit; and (ii) some of the $q_r(\ldots\nu_i\ldots)$ may be forbidden. For example:

(a) Langmuir model: There is only one species, the adatoms, and $z_1 = 0,1$. This yields immediately (6.60) and (6.61).

(b) Acceptor adatoms — one type of active site: There are two species: adatoms (1) and electrons (2). The rules are: $z_1 = 0,1$, $z_2 = 0,1$ and $q_1(0,1) = 0$, i.e. we assume that electrons can only be trapped at a previously adsorbed 'atom'. We now have, omitting the label to indicate the type of active site:

$$M(1,0) = M \frac{\lambda_1 q(1,0)}{q(0,0)+\lambda_1 q(1,0)+\lambda_1\lambda_2 q(1,1)} \quad (9.3)$$

and

$$M(1,1) = M \frac{\lambda_1\lambda_2 q(1,1)}{q(0,0)+\lambda_1 q(1,0)+\lambda_1\lambda_2 q(1,1)} \quad (9.4)$$

Other cases could be treated similarly. For example, multiple ionization, different types of active sites, different types of adatoms, etc. Electronic surface states, labelled by their quantum numbers, could be formally regarded as traps with the rule that they can trap zero adatoms and up to one electron. One would then include two-dimensionally free surface states while doing chemisorption theory, an exercise which might be rather interesting. But let us continue with the simple case.

The notation of the formalism is somewhat abstract. It may be convenient in practice to rewrite the results in a more transparent fashion. For example, following common practice we introduce the coverage degrees:

$$\theta^0 = \frac{M(1,0)}{M}, \qquad \theta^- = \frac{M(1,1)}{M} \quad (9.5)$$

There is an arbitrary energy origin in these formulae. We divide everywhere by $q(0,0)$ and put

$$\frac{q(1,0)}{q(0,0)} = q^0 = \exp\{-\beta f^0\}; \qquad \frac{q(1,1)}{q(0,0)} = q^- = \exp\{-\beta f^-\} \quad (9.6)$$

where f^0 and f^- are, respectively, the free energy of one uncharged or charged adatom. According to our model we have

$$f^- = f_0^- + q\phi_s \tag{9.7}$$

We introduce the dimensionless variable

$$u(z) = \beta q \phi(z) \tag{9.8}$$

and the symbols

$$\zeta^0 = \exp[\beta(\mu_{0g} - f^0)] \quad ; \quad \zeta^- = \exp[\beta(\mu_{0g} + \epsilon_F - f_0^-)] \tag{9.9}$$

and rewrite (9.3) and (9.4) as

$$\theta^0 = \frac{\zeta^0 p}{1 + [\zeta^0 + \zeta^- \exp(-u_s)]p} \quad ; \quad \theta^- = \frac{\zeta^- \exp(-u_s) p}{1 + [\zeta^0 + \zeta^- \exp(-u_s)]p} \tag{9.10}$$

Notice the terms appearing in these formulae: ζ^0 and ζ^- contain information on the gas (μ_{g0}), the solid (ϵ_F) and their interaction (f^0, f_0^-). The formulae allow for the case in which the adatom may have more than one bound state for a trapped electron. We also have the surface barrier u_s. This is where we connect adsorption theory with the description of the surface charge boundary layer.

The idea is elementary: we must set up a Poisson equation with two boundary conditions: (a) charge neutrality as $z \to \infty$ and (b) the surface charge density is precisely $qM\bar{\theta}/A$ or, equivalently, the electrostatic potential at $z = 0$ is the same ϕ_s appearing in (9.10). It is here that we feed into the model all the information we have on the adsorbent, i.e. band gaps, impurity contents, etc. These parameters appear implicitly in ϵ_F and explicitly in the expression of the charge density $\rho(z)$. This may involve additional statistical problems, e.g. multilevel impurity centres, which can also be treated with the same grand canonical formalism. Once all these problems are solved and u_s is fed back into (9.10), the adsorption isotherms have been calculated. Notice that u_s depends on coverage and hence on gas pressure p, so that the total coverage $\theta = \theta^0 + \theta^-$ no longer follows a Langmuir isotherm.

It may be instructive to consider some examples of models proposed in the literature just to see how the notions of chemical thermodynamics are used in problems involving gases and components of the solid adsorbate in the sense of Section 7. One is the model of oxidation of a semiconductor proposed by Sparnaay (1969). This model postulates the following adsorbed species: O_2 molecules; dissociated atoms forming Sc_xO, where Sc stands for a semiconductor atom and x is just a number, say 1 or 2; finally, ionized adatoms in the form Sc_xO^-. We have the following chemical reactions at the surface:

$$O_2(\text{gas}) \rightleftarrows O_2(\text{adsorbed}) \tag{9.11}$$

then the adsorbed O_2 dissociates:

$$O_2 + 2xSc \rightleftarrows 2xScO \qquad (9.12)$$

and then the adsorbed atoms trap electrons:

$$ScxO + e^- \rightleftarrows ScxO^- \qquad (9.13)$$

These are examples of equations like (7.9).

We now define a group of 2xSc atoms as an active site. We define the components:

1	2	3	4	5
O_2	ScxO	ScxO$^-$	e^-	Sc

The occupation rules in this model are that (z_1, z_2, z_3) can only take the values (1,0,0), (0,2,0), (0,1,1) and (0,0,2). From the equilibrium conditions (7.10) applied to (9.11) through (9.13), expressed in terms of absolute activities, we have

$$\lambda_2 = \lambda_5^x \lambda_1^{1/2} \quad ; \quad \lambda_3 = \lambda_5^x \lambda_4 \lambda_1^{1/2} \qquad (9.14)$$

Here λ_1 is λ(gas), given by (6.49) and (6.29); λ_4, for the electron gas, is $\exp(\beta\epsilon_F)$. Thus λ_2 and λ_3 are proportional to $p^{1/2}$. Define

$$\begin{aligned}
\zeta_{100} &= \lambda_0 \exp(-\beta f_{100}) \\
\zeta_{020} &= \lambda_5^{2x} \lambda_0 \exp(-\beta f_{020}) \\
\zeta_{011} &= \lambda_5^{2x} \lambda_0 \lambda_4 \exp(-\beta f_0^-) \\
\zeta_{002} &= \lambda_5^{2x} \lambda_0 \lambda_4^2 \exp(-2\beta f_0^-)
\end{aligned} \qquad (9.15)$$

where $\lambda_0 = \exp \beta\mu_0(\text{gas})$. Application of (9.2) then yields

$$\theta(O_2) = \frac{\zeta_{100} p}{D} \quad ; \quad \theta(O) = \frac{\zeta_{020} p}{D} \quad ;$$

$$\theta(O^-) = \frac{(\zeta_{011} e^{-u_s} + \zeta_{002} e^{-2u_s}) p}{D} \qquad (9.16)$$

where

$$D = 1 + (\zeta_{100} + \zeta_{020} + \zeta_{011} e^{-u_s} + \zeta_{002} e^{-2u_s}) p \qquad (9.17)$$

Another model has been proposed (Ahmad, 1968) for the adsorption of oxygen on TiO$_3$Ba. This involves two types of active sites. Type 1 is oxygen ions, where the adsorption of O$_2$ is assumed to take place, without dissociation or ionization. Type 2 is titanium ions, where the following surface reactions are postulated:

$$O_2(gas) \rightleftarrows 2O(ads) \tag{9.18}$$

and

$$O + e^- \rightleftarrows O^- \tag{9.19}$$

We define components:

1	2	3	4
O$_2$	O	O$^-$	e$^-$

The occupation rules are: for type 1, $(z_1, z_2, z_3) = (1,0,0)$; for type 2, $(z_1, z_2, z_3) = (0,1,0)$ or $(0,0,1)$.

The equilibrium conditions yield

$$\lambda_2 = \lambda_0^{1/2} p^{1/2} \; ; \quad \lambda_3 = \lambda_0^{1/2} \exp(\beta \epsilon_F) p^{1/2} \tag{9.20}$$

The same routine yields isotherms of the form:

$$\theta(O_2) = \frac{\zeta_1 p}{1 + \zeta_1 p} \tag{9.21}$$

and

$$\theta(O) = \frac{\zeta_2^0 p^{1/2}}{1 + (\zeta_2^0 + \zeta_{2,0}^- e^{-u_s}) p^{1/2}}$$

$$\theta(O^-) = \frac{\zeta_{2,0}^- e^{-u_s} p^{1/2}}{1 + (\zeta_2^0 + \zeta_{2,0}^- e^{-u_s}) p^{1/2}} \tag{9.22}$$

The point of these exercises is simply to see how chemical thermodynamics and notions of solid-state physics can be incorporated in the description of charge transfer adsorption at the surface, and to demonstrate how the adsorption isotherms can be obtained without having to postulate them as is sometimes done. The possible merit of the models is another matter, which will not be pursued in this paper. If we are interested in using the known adsorption isotherms together with the (differential) Gibbs adsorption equation, as indicated at the end of Section 6, we must specify the adsorbed components. We recall that, as was proved in Section 5, we can always choose the adsorbed amount of one component equal to zero. Consider, for example, the simple model of charge transfer chemisorption. In this case we have two adsorbed species: neutral adatoms (X) and charged adatoms (X⁻). We are not considering dissociation. If there is dissociation, we simply replace p by $p^{1/2}$ or, in general, by the appropriate fractional power of p, as we have just seen in the last two examples. On the other hand, the surface reaction $X + e^- \rightleftarrows X^-$ tells us that, at constant temperature and area,

$$d\mu(X) = d\mu(X^-) = \frac{dp}{p} \tag{9.23}$$

if, again, (6.29) is used for μ. Thus,

$$d\gamma_a = \frac{M}{A} \theta^0 d\mu(X) + \frac{M}{A} \theta^- d\mu(X^-) = \frac{M}{A} \theta \frac{dp}{p} \tag{9.24}$$

or, by (9.10),

$$d\gamma_a = \frac{M}{A} \frac{[\zeta^0 + \zeta^- \exp(-u_s)]}{1 + [\zeta^0 + \zeta^- \exp(-u_s)]} dp \tag{9.25}$$

At this stage we must remember that u_s is itself a function of p and, to find this function, we must feed into the analysis the Poisson equation which describes the electrical changes in the boundary layer. The solution of this equation relates u_s to θ^-, which is to be eliminated between this relation and (9.10) itself. Then $u_s(p)$, thus obtained, must be used in (9.25) in order to yield the integrand in $\int d\gamma_a$ as a function only of p. Even in simple cases it is practically impossible to do this analytically, but it can be done numerically. At least one has an explicit prescription for using the known adsorption isotherm in a practical way consistent with surface thermodynamics.

10. ADSORPTION HEAT AND RELATED CONCEPTS

Many different quantities can be defined which may legitimately be called adsorption heat and yet they appear to be different. The point is that they are different, as they correspond to different physical concepts. Basically two questions are involved:

(a) The relationship between the different quantities. This depends on how they have been defined.
(b) The relation to what is experimentally given as the adsorption heat. This depends on how the experiment is done.

In both these respects there is variety for all tastes (see e.g. extensive discussions by Hill (1949, 1952), Defay et al. (1966) and Clark (1970), among many others). It seems necessary to say that sometimes there has been positive confusion and misunderstanding on this point, or at least sufficient carelessness to apply uncritically any formula to any experimental data.

To avoid further excessive pollution of the literature on this subject we shall keep our discussion to a bare minimum, just sufficient to enable the reader to go on to more articulate and professional papers.

Following a line of argument by Hill (1949), to begin with one must distinguish between reversible and irreversible processes. Consider an adsorption system in some specified initial and final conditions. Suppose that, given the system in the initial state, more gas is adsorbed by reversible isothermal compression, i.e. by slowly moving a piston. Call q(rev) the corresponding differential heat per gas molecule adsorbed. The entropy change for the system under study is

$$\Delta S = \frac{q(rev)}{T} \tag{10.1}$$

If, on the other hand, given the same initial state, the same final state is reached by simply opening a stopcock and admitting more gas, the process is irreversible. Call q(irrev) the corresponding differential adsorption heat. Since the system goes between the same two states, the differential entropy change ΔS is clearly the same, but now the following inequality holds:

$$\Delta S > \frac{q(irrev)}{T} \tag{10.2}$$

Therefore

$$q(rev) > q(irrev) \tag{10.3}$$

so that the two differential adsorption heats cannot be equal. The relationship between them is derived in the above references, but this is only one among many. One can, using thermodynamic equations — applicable therefore to equilibrium states and reversible processes — define different adsorption heats, differential or integral. Let us simply concentrate on one.

Suppose N_a molecules, or atoms etc., have reversibly disappeared from a gas phase with $\overline{S}_g$ entropy per molecule and appeared in the adsorbate with $\overline{S}_a$ entropy per molecule. The total entropy change is $N_a(\overline{S}_a - \overline{S}_g)$ and the corresponding differential adsorption heat is defined by

$$-\frac{q_{st}}{T} = \left[\frac{\partial N_a(\overline{S}_a - \overline{S}_g)}{\partial N_a}\right]_T \tag{10.4}$$

The reason for the subscript (st) will be explained later.

Now, from (6.3),

$$A d\gamma_a = S_a dT + (\tau_a - \gamma_a) dA + N_a d\mu_a \tag{10.5}$$

We make the reasonable assumption that the area can only change by thermal expansion, excluding processes in which A is deliberately changed by application of mechanical stresses. Then

$$dA = \frac{dA}{dT} dT \tag{10.6}$$

Put

$$\Psi_a = (\tau_a - \gamma_a)\frac{dA}{dT} \tag{10.7}$$

Thus,

$$Ad\gamma_a = (S_a + \Psi_a)dT + N_a d\mu_a \tag{10.8}$$

hence

$$d\mu_a = \bar{A}d\gamma_a - (\bar{S}_a + \bar{\Psi}_a)dT \tag{10.9}$$

For the gas phase

$$d\mu_g = \bar{V}_g dp - \bar{S}_g dT \tag{10.10}$$

and, if equilibrium adsorption prevails, $d\mu_a = d\mu_g$, whence

$$\bar{V}_g dp = (\bar{S}_g - \bar{S}_a - \bar{\Psi}_a)dT + \bar{A}d\gamma_a \tag{10.11}$$

For the adsorbate we can regard the concentration n_a and the temperature as the two independent intensive variables. Hence

$$d\gamma_a = \left(\frac{\partial \gamma_a}{\partial n_a}\right)_T dn_a + \left(\frac{\partial \gamma_a}{\partial T}\right)_{n_a} dT \tag{10.12}$$

and, substituting in (10.11),

$$\bar{V}_g dp = [\bar{S}_g - \bar{S}_a - \bar{\Psi}_a + \bar{A}\left(\frac{\partial \gamma_a}{\partial T}\right)_{n_a}]dT + \bar{A}\left(\frac{\partial \gamma_a}{\partial n_a}\right)_T dn_a \tag{10.13}$$

i.e.

$$\bar{V}_g \left(\frac{\partial p}{\partial T}\right)_{n_a} = \bar{S}_g - \bar{S}_a - \bar{\Psi}_a + \bar{A}\left(\frac{\partial \gamma_a}{\partial T}\right)_{n_a} \tag{10.14}$$

Of course, what we want to do is to extract $\bar{S}_g - \bar{S}_a$ from these equations, which explicitly incorporate the condition of equilibrium and the features of the solid state, in order to use it in Eq.(10.4), which we now write as

$$-\frac{q_{st}}{T} = \bar{S}_a - \bar{S}_g + N_a \left(\frac{\partial \bar{S}_a}{\partial N_a}\right)_T \tag{10.15}$$

Now, from (10.9) and (6.3),

$$d(\mu_a - \bar{A}\gamma_a) = -\gamma_a d\bar{A} - (\bar{S}_a + \bar{\Psi}_a)dT \tag{10.16}$$

Since $\mu_a - \bar{A}\gamma_a$ is a function of state, (10.16) is an exact differential. Also, since $\bar{A} = 1/n_a$, constant $\bar{A}$ means constant n_a. Hence

$$\left(\frac{\partial \gamma_a}{\partial T}\right)_{n_a} = \left[\frac{\partial(\bar{S}_a + \bar{\psi}_a)}{\partial \bar{A}}\right]_T \qquad (10.17)$$

Also,

$$\left[\frac{\partial}{\partial \bar{A}}\right]_T \equiv \frac{1}{A}\left[\frac{\partial}{\partial N_a^{-1}}\right]_T = -\frac{N_a^2}{T}\left[\frac{\partial}{\partial N_a}\right]_T \qquad (10.18)$$

Thus we rewrite (10.17) as

$$\bar{A}\left(\frac{\partial \gamma_a}{\partial T}\right)_{n_a} = -N_a\left[\frac{\partial(\bar{S}_a + \bar{\psi}_a)}{\partial N_a}\right]_T \qquad (10.19)$$

which we can now use in (10.14), obtaining

$$\bar{V}_g\left(\frac{\partial p}{\partial T}\right)_{n_a} = \bar{S}_g - \bar{S}_a - \bar{\psi}_a - N_a\left(\frac{\partial \bar{S}_a}{\partial N_a}\right)_T - N_a\left(\frac{\partial \bar{\psi}_a}{\partial N_a}\right)_T \qquad (10.20)$$

Hence, from (10.20) and (10.15),

$$\frac{q_{st}}{T} = \bar{V}_g\left(\frac{\partial p}{\partial T}\right)_{n_a} + \bar{\psi}_a + N_a\left(\frac{\partial \bar{\psi}_a}{\partial N_a}\right)_T \qquad (10.21)$$

It only remains to transform the last two terms, i.e.

$$\bar{\psi}_a + N_a\left(\frac{\partial \bar{\psi}_a}{\partial N_a}\right)_T = \left(\frac{\partial \psi_a}{\partial N_a}\right)_T = \left\{\frac{\partial}{\partial N_a}\left[(\tau_a - \gamma_a)\frac{dA}{dT}\right]\right\}_T$$

$$= \left[\frac{\partial(\tau_a - \gamma_a)}{\partial n_a}\right]_T \frac{d\epsilon_s}{dT} \qquad (10.22)$$

Hence finally, from (10.22) and (10.21),

$$q_{st} = T\bar{V}_g\left(\frac{\partial p}{\partial T}\right)_{n_a} + T\left[\frac{\partial(\tau_a - \gamma_a)}{\partial n_a}\right]_T \frac{d\epsilon_s}{dT} \qquad (10.23)$$

The last term describes the effects of two features which are characteristic of the solid state: (a) the coefficient of thermal expansion of the solid adsorbent and (b) the fact that, as adsorption proceeds, τ_a may be different from γ_a. This will depend very much on the type of interactions involved in the adsorption process under study. Notice that, as (10.23) stands, nothing restricts it to inert adsorbents. Only the evaluation of the terms will be more involved, which again depends on the models and is no longer a thermodynamical question. It would seem upon superficial inspection that the effects of thermal expansion do not introduce any change when $\tau_a = \gamma_a$. This, however, is not true, and the point merits some discussion.

A formula like (10.23) is useful in practice because the first term on the right-hand side involves pressure, temperature and surface concentration of adsorbate. These are the three experimental variables in adsorption isotherms and so it is practical to have defined a concept of differential adsorption heat which can be obtained directly from the adsorption isotherms. Hence the customary notation q_{st}, which means isosteric differential adsorption heat. But now consider the following argument:

$$\left(\frac{\partial p}{\partial T}\right)_{N_a} = \left(\frac{\partial p}{\partial T}\right)_{N_a, A} + \left(\frac{\partial p}{\partial A}\right)_{N_a, T} \frac{dA}{dT} \tag{10.24}$$

Since absolute determination of the total adsorbent area is not always very easy or very accurate, it is not a waste of time to distinguish between derivatives at constant N_a or at constant n_a. In fact an early controversy arose on the question: which of the two derivatives gives the correct value for q_{st}? According to the foregoing argument there should be no doubt: Eq.(10.23) speaks for itself. Now, looking at (10.24), if N_a and A remain constant so does their ratio n_a. Thus

$$\left(\frac{\partial p}{\partial T}\right)_{n_a} = \left(\frac{\partial p}{\partial T}\right)_{N_a} - \left(\frac{\partial p}{\partial A}\right)_{N_a, T} \frac{dA}{dT} \tag{10.25}$$

Now, at constant T, A remains constant by assumption, and so does N_a in the second term on the right-hand side. Thus

$$\begin{aligned}\left(\frac{\partial p}{\partial T}\right)_{n_a} &= \left(\frac{\partial p}{\partial T}\right)_{N_a} - \frac{1}{N_a}\left(\frac{\partial p}{\partial \overline{A}}\right)_T \frac{dA}{dT} \\ &= \left(\frac{\partial p}{\partial T}\right)_{N_a} - \frac{1}{N_a}\left(\frac{\partial p}{\partial N_a^{-1}}\right)_T \frac{d\epsilon_s}{dT} \\ &= \left(\frac{\partial p}{\partial T}\right)_{N_a} + N_a\left(\frac{\partial p}{\partial N_a}\right)_T \frac{d\epsilon_s}{dT}\end{aligned} \tag{10.26}$$

which means that the two derivatives are only different because of the thermal expansion of the adsorbent.

From (10.26) and (10.23) we have

$$q_{st} = T\bar{V}_g \left(\frac{\partial p}{\partial T}\right)_{N_a} + T\bar{V}_g N_a \left(\frac{\partial p}{\partial N_a}\right)_T \frac{d\epsilon_s}{dT}$$

$$+ T\left[\frac{\partial(\tau_a - \gamma_a)}{\partial n_a}\right]_T \frac{d\epsilon_s}{dT} \tag{10.27}$$

But this formula admits a considerable simplification. Indeed, from (10.13), keeping T constant we have

$$\left(\frac{\partial \gamma_a}{\partial n_a}\right)_T = \bar{V}_g n_a \left(\frac{\partial p}{\partial n_a}\right)_T \qquad (10.28)$$

and, again, since A only depends on T,

$$\left(\frac{\partial \gamma_a}{\partial n_a}\right)_T = \bar{V}_g N_a \left(\frac{\partial p}{\partial N_a}\right)_T \qquad (10.29)$$

which simplifies (10.27) to

$$q_{st} = T\bar{V}_g \left(\frac{\partial p}{\partial T}\right)_{N_a} + T\left(\frac{\partial \tau_a}{\partial n_a}\right)_T \frac{d\epsilon_s}{dT} \qquad (10.30)$$

This agrees with the formula derived by Eriksson (1969) for the perfect gas case:

$$q_{st} = kT^2 \left(\frac{\partial \ln p}{\partial T}\right)_{N_a} + T\left(\frac{\partial \tau_a}{\partial n_a}\right)_T \frac{d\epsilon_s}{dT} \qquad (10.31)$$

The first term on the right-hand side is of the Clausius-Clapeyron type. If the differential adsorption heat were given just by this term, adsorption would be like the condensation of a gas in a liquid phase. This image would seem rather appropriate for fluid adsorbates, i.e. essentially for physical adsorption. But even then there is a difference due to the finite thermal expansion coefficient of the substrate. This typically is in the range of 10^{-5} to 10^{-4} deg^{-1}. The other factor cannot be estimated with such generality. This will depend on the type of interactions involved. For example, for the van der Waals type of adsorbate considered in Section 6, we have

$$\left(\frac{\partial \tau_a}{\partial n_a}\right)_T = \frac{kT}{(1-bn_a)^2} - 2\bar{a}n_a \qquad (10.32)$$

Thus this term depends not only on the interaction parameters but also on the adsorption coverage. Yet when typical values are inserted it seems that the second term on the right-hand side of (10.31) is in practice rather negligible. A similar consideration applies in fact to the difference between the two derivatives expressed in (10.24). On the whole the Clausius-Clapeyron formula for the differential adsorption heat, although far from obvious on general grounds, seems to work quite well in practice. However, the situation may be very different for chemisorption systems, where stronger interactions are involved. This field is by and large rather neglected. It is not always certain that complete reversibility can be achieved. It is difficult to make reliable estimates due to lack of good experimental data. Also, chemisorption is more similar to a chemical reaction than to a condensation and it would be more proper to discuss the concept of chemisorption heat rather in terms of the standard chemical thermodynamics used to study chemical reactions. There is one complication: that the adsorption heat depends on the extent of reaction. I am not aware of any complete discussion of the concept of chemisorption heat in these terms. If this has not been done, it would be worth doing.

Since the adsorption isotherm is a standard experiment, it is interesting to see how much information can be obtained from it. The formula for q_{st} involves a derivative at constant N_a. Let us consider other meaningful derivatives.

One is related to the argument developed in Section 4, which led us to the conclusion that (4.11) is an approximately valid equality. By the same argument, from (6.3) and (10.6) we obtain

$$\frac{d\gamma_a}{dT} = \hat{s}_a + (\tau_a - \gamma_a)\frac{d\epsilon_s}{dT} \tag{10.33}$$

as an approximately valid equality for the adsorbate. But let us consider the actual experiment. It would involve the temperature coefficient of the entire system adsorbent plus adsorbate. Recalling the definitions (6.1) for s_a and (6.2) for γ_a, we have, from (4.11) and (10.33),

$$\frac{d\gamma}{dT} = -s^{\sigma 0} - s_a - (\tau_a - \gamma_a)\frac{d\epsilon_s}{dT} \tag{10.34}$$

Thus, as adsorption proceeds the temperature coefficient of the surface tension changes. Observation of this effect is often taken as indication of adsorbed impurities at the surface. In most applications only the first two terms of the right-hand side of (10.34) are used. In fact, the usual extra term ought to be estimated.

Another derivative, closely related to this term, is obtained from the adsorption isotherm by specifying the condition γ_a = constant. From (10.11), this imposes a relationship between dT and dp, whence

$$\bar{V}_g\left(\frac{\partial p}{\partial T}\right)_{\gamma_a} = S_g - S_a - \frac{\tau_a - \gamma_a}{n_a}\frac{d\epsilon_s}{dT} \tag{10.35}$$

again an equation often used in incomplete form, without the last term.

Before closing this section, one final comment is interesting. Heat is exchanged when molecules are exchanged between gas and adsorbate or between liquid and adsorbate. Heat is also evolved when a solid (either with a clean surface or with a surface containing previously adsorbed molecules) is immersed in a liquid. The heat of immersion in fact is very easy to determine with fair accuracy. It seems somewhat surprising and amusing that such a simple experiment as immersing a solid in a liquid and measuring the heat evolved should furnish such valuable information, but the fact is that it can be easily related to the adsorption heat and it has several advantages. For example, it is a more accurate procedure than using adsorption isotherms at different temperatures, as is needed for the Clausius-Clapeyron type of formula. Above all, it is a method which can be used even if adsorption is irreversible, so that it can serve to obtain chemisorption heats by a procedure which is not only fairly accurate in practice but also legitimate in principle. The details of this method are described by Zettlemoyer and Narayan (1966).

11. SURFACE PHASE TRANSITIONS

Having defined a surface phase in the thermodynamical sense, it is interesting to enquire about the possible phase transitions such a system may have. This is an interesting field of research, whose present state is more promising than mature. It is not unlikely that this field may see active and exciting developments in the near future. Although much remains to be studied, one can try to outline some elementary principles. This is what will be done in this section.

To fix ideas, consider an adsorbed phase. To have a phase transition one needs of course some appropriate kind of interactions between the material particles which form the macroscopic aggregate under study. This is a question of model. One, which is tractable and non-trivial, is based on the idea of a mean field. That is to say, instead of accounting for all the lateral interactions within the adsorbate in detail, each one adatom is put into some sort of average field due to all the others. The situation is similar to the one considered in Section 9 as a model for charge transfer chemisorption, except that there we only took account of Coulomb interactions between charged adatoms. Let us now assume that all adatoms interact in some way. Let us study the following model: localized adsorption on active sites and lateral adsorbate-adsorbate interactions reduced to a mean field. This can be imagined in different ways. For example, suppose there is some interaction two-body potential $U(r)$. Assuming that the surface concentration of adsorbate is on the average uniform, and neglecting the difference between N_a and N_a-1 for $N_a \gg 1$, one adatom sees an average potential energy:

$$\frac{N_a}{A} \int_A U(r) d^2r \equiv \frac{N_a U_m}{A} \tag{11.1}$$

Now, we can look at each one active site as an open system, which can contain either 0 or 1 adatom, in equilibrium with the rest. We can then apply to the mean occupation number of each one site (which is the same as the coverage $\theta = N_a/M$) the formalism of Section 9. From (9.2) we have

$$\theta = \frac{\lambda q_0 \exp(-g\theta)}{1 + \lambda q_0 \exp(-g\theta)} \tag{11.2}$$

where q_0 is the canonical partition function of one adatom in the absence of interaction, $g = \beta U_m M/A$ and λ is the absolute activity of the gas. Hence

$$\beta\mu = \ln\lambda = \ln\left[\frac{\theta \exp(g\theta)}{(1-\theta)q_0}\right] \tag{11.3}$$

Now, at constant T and A we have (6.64), i.e.

$$d\left(\frac{A\gamma_a}{MkT}\right) = \theta\, d(\ln\lambda) \tag{11.4}$$

which, from (11.2), is

$$d\left(\frac{A\gamma_a}{MkT}\right) = \left(\frac{1}{1-\theta} + g\theta\right) d\theta \tag{11.5}$$

Since $\gamma_a = 0$ when $\theta = 0$, integration of (11.5) yields

$$\frac{A\gamma_a}{MkT} = -\ln(1-\theta) + \frac{1}{2} g\theta^2 \tag{11.6}$$

Compare with (6.45) for the Langmuir case, with no interaction between adatoms. The same formulae are obtained for a slightly different model, rather more like a Bragg-Williams model in problems of lattice statistics. Let c be the average number of nearest neighbours that each active site has. In a lattice this would be the co-ordination number. Suppose we only take account of nearest-neighbour interactions and let U_i be this interaction. One adatom interacts with an average number $c\theta$ of nearest-neighbour adatoms. This yields the same formulae (11.2)-(11.6), only that g is then equal to $\beta c U_i$. There is one difference. Knowing μ from (11.3) and γ_a from (11.6), we can obtain F_a from (6.8) and τ_a from (6.6). Notice that the latter involves partial differentiation at constant T and N_a. This leaves us with

$$\tau_a = \frac{1}{2} MkT\theta^2 \left(\frac{\partial g}{\partial A}\right)_{T,N_a} \tag{11.7}$$

For the Bragg-Williams type of model τ_a is zero, while with $g = \beta U_m M/A$ we have

$$\tau_a = -\frac{1}{2} U_m \left(\frac{M}{A}\right)^2 \theta^2 \tag{11.8}$$

Thus the two models look the same but have in fact a non-trivial physical difference.

The key term in this analysis is $\frac{1}{2} g\theta^2$ in (11.6). It is this term which causes the isotherm to exhibit a loop characteristic of a phase transition. The analysis is rather standard and essentially follows the lines of well-known arguments (e.g. Hill (1960) chapter XIV). In our notation it is convenient to define $\Theta^{-1} = \theta M/A$ and to plot γ_a versus Θ (Fig.8). For a loop to appear it turns out that U_m must be negative, i.e. the net interaction must be attractive. The critical temperature is given by

$$kT_c = \frac{M|U_m|}{4A} \tag{11.9}$$

so that $T < T_c$ means that, at the given temperature T, the interaction must not only be attractive but also sufficiently strong that $M|U_m| > 4AkT$.

Now, at constant T and A we have, from (6.64),

$$d\mu = \Theta d\gamma_a \tag{11.10}$$

Thus on moving from A to B in an isotherm at $T < T_c$ (Fig.8), the change in chemical potential is given by the dotted area. On moving beyond A,B, etc., μ changes qualitatively as shown in Fig.9. The stable phase is the one with the smaller chemical potential and equilibrium between two phases occurs when the two shaded areas in Fig.8 are equal. This model is of course greatly oversimplified and even somewhat incorrect, as is well known, but it does exhibit the features essential to understand phase transitions in an adsorbed system.

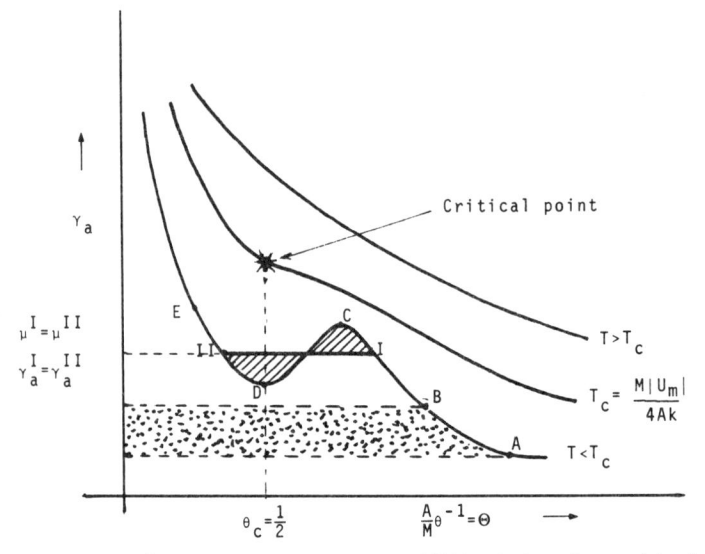

FIG.8. The isotherms of Eq.(11.6) for attractive interaction exhibiting the loop characteristic of a phase transition for temperatures below T_c.

Rather than delve into more realistic and complicated models, which is not really a thermodynamical question, it is interesting to investigate another point. We have approached the surface phase as a system in itself, thermodynamically defined according to some scheme. Consequently we refer to the γ_a versus θ (or Θ) curve as an isotherm in the sense of the equation of state. Here γ_a and Θ play the role of pressure and molar volume of a bulk phase. But it is also necessary to change the viewpoint and consider the state of affairs at the surface as imposed by the equilibrium with the surrounding atmosphere. The chemical potential μ is not something which changes because we vary Θ. Although this is useful in the formal argument to characterize a phase transition, the real fact is that we control temperature and pressure in the atmosphere, and it is in terms of these variables that we want to formulate the thermodynamics of the phase transition. That is to say, we take the view that we change the gas pressure and this decides the value μ must take. The question then is whether values of (p,T) are possible for which two different surface phases have the same μ. From a look at Fig.8, they also have the same γ_a. Thus, if we impose simultaneously the conditions

$$d\gamma_a^I = d\gamma_a^{II} \; ; \qquad d\mu^I = d\mu^{II} \qquad (11.11)$$

this will yield the differential equation of the phase equilibrium curve in the (p,T) diagram. Thus, from (10.8),

$$d\gamma_a^I = (s_a^I + \psi_a^I)dT + n_a^I d\mu = (s_a^{II} + \psi_a^{II})dT + n_a^{II} d\mu \qquad (11.12)$$

SURFACE THERMODYNAMICS

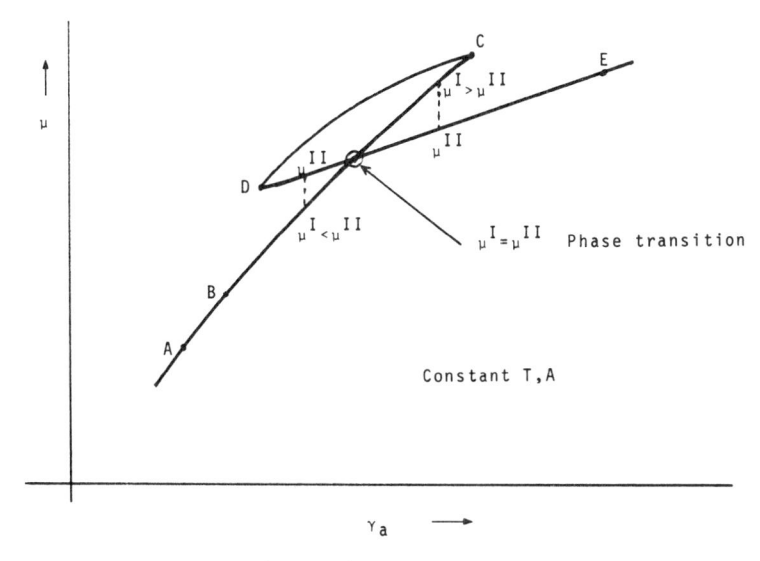

FIG.9. Qualitative variation of μ as a function of γ_a for T and A constant. The points A, B, ..., E correspond to those indicated in Fig.8.

where, assuming ideal gas behaviour, we put

$$d\mu = kTd\ln p - \bar{S}_g dT \qquad (11.13)$$

Thus, since $\gamma_a^I = \gamma_a^{II}$ at equilibrium between the two surface phases, the condition is

$$(s_a^I + \tau_a^I \frac{d\epsilon_s}{dT} - n_a^I)dT + n_a^I kTd(\ln p) = \text{same}^{II} \qquad (11.14)$$

Let us write $\Delta n_a = n_a^I - n_a^{II}$, etc.; then we have

$$kT \frac{d\ln p}{dT} = \bar{S}_g - \frac{\Delta s_a}{\Delta n_a} - \frac{\Delta \tau_a}{\Delta n_a} \frac{d\epsilon_s}{dT} \qquad (11.15)$$

The first term on the right-hand side is just a supposedly known property of the gas. The others must be evaluated from the model for the surface phase. For example, n_a^I and n_a^{II} could correspond to the two values of coverage θ associated with points I and II in Fig.8. It is here that the equality or otherwise of τ_a^I and τ_a^{II} becomes relevant. This will decide whether the last term survives or not in (11.15). Some plausible figures are quoted by Eriksson (1969) for the system $N_i(110)-H_2$. With $\Delta\tau_a = -400$ dyne/cm and $\Delta n_a = 10^{-9}$ mole/cm^2 one finds, at T=400 K,

$$T \frac{\Delta \tau_a}{\Delta n_a} \frac{d\epsilon_s}{dT} = 0.1 \text{ kcal/mole} \qquad (11.16)$$

The experimental value of $RT^2 \, d\ln p/dT$ seems to be 28 kcal/mole. Thus, for this system it seems that the last term is negligible. However, a precise evaluation is, as usual, difficult in general, due to the customary lack of data.

Finally, the analysis leading up to (11.15) is also applicable to the clean surface-vacuum interface. In this case the particular choice can still be made in which the surface amount of the solid component is zero, as we showed on general grounds. Since there is no gas component to be adsorbed, in this case $\Delta n_a = 0$ and (11.15) is reduced to

$$-\Delta s^\sigma + \Delta \tau \frac{d\epsilon_s}{dT} = 0 \qquad (11.17)$$

The analysis just outlined assumes that there is a change of entropy, and hence a latent heat, associated with the transition. Including the finite change in τ_a, this would be an example of a first-order phase transition, in Ehrenfest's sense. This classification is no longer in fashion. Current theories of phase transitions and critical phenomena go a different way (Stanley (1971), Green (1971)). While it is true that Ehrenfest's classification has its shortcomings, it can nevertheless be a useful phenomenological scheme, provided it is used carefully (Pippard (1957), Callen (1960)). At the very least it emphasizes the fact that transitions can be abrupt (first order) or smooth (higher order). If the changes in entropy etc. are zero, then one must do the analysis again. It would not be difficult to derive the appropriate formulae for an adsorbed phase. In this case the quantities whose discontinuities are related to dp/dT are the derivatives of γ_a with respect to p and T. However, at the present time such formulae are not very relevant, due to lack of sufficient experimental data.

In this model of an adsorbed phase we have assumed that there is some interaction. The model is thus really phenomenological. If we had derived the interaction and calculated its parameters from first principles, we would have developed a complete theory. It would not have been a very accurate one, but would have contained all the elements needed for a complete dynamical theory, from the explanation of the mechanism which causes the transition to the dynamical description of the transition itself, so as to be able to evaluate all the terms appearing in the thermodynamical formulae.

If our current knowledge of adsorbed phases is still far from attaining this ideal with a reasonable degree of reliability, the situation is much more primitive as regards clean surfaces. This is the problem referred to as surface reconstruction or as formation of a superstructure, the term 'surface relaxation' being also often employed, thus adding to the usual semantic confusion ever present in scientific writing. The meaning of these terms is explained in a paper by Blandin et al. (1973). From a survey of current literature it seems that at the very most we are beginning to understand what may cause an instability of a surface structure, and even in this there is no unanimous agreement.

One type of explanation is based on a description of electron orbitals near surface atoms. Haneman (1961) proposes to explain one of the observed reconstructions of the clean 111-Si surface as a consequence of orbital hybridization. While for bulk atoms the valence orbitals are s-p, surface hybridization results, according to Haneman, in orbitals which have predominant s-like or p-like character, alternately. In this description the structure, in the sense of a geometrical array of atoms, need not change to explain observed LEED patterns. Another line of approach, more in the nature of a dynamical theory of a phase transition, is based on the idea of soft modes (Samuelsen et al. (1971), Cochran (1973)). Briefly, a mode with zero frequency means a permanent displacement of some atoms according to a certain law which depends on the wave vector for which the mode 'goes soft'. In other words, atoms tend to re-arrange themselves, forming another structure. Blandin et al. (1973) have discussed some general features of the problem from the point of view of lattice dynamics. The real task consists in finding the kind of interactions which can cause the appropriate mode to go soft. Models of this kind have been proposed by Trullinger and Cunningham (1973) and by Tosatti and Anderson (1974), who also propose alternative mechanisms. Blandin (1973) has attempted to develop for soft surface modes the connection with Landau's phenomenological theory. Such a connection has been known for a long time in the case of soft bulk modes. It is also appropriate to remember that Landau's theory has its own non-trivial limitations.

Summing up what is known about surface phase transitions, we can say that plausible models are beginning to emerge which may explain why a given surface structure may be unstable and exhibit a tendency towards another structure. But nothing has yet been done which even attempts to describe the dynamics of the phase transition in a reasonably complete way. All that is known so far is that different mechanisms could plausibly be responsible for the fact of the change of structure, but none of these models has yet been studied in sufficient detail to find out what description of the phenomenon is contained therein. A direct question would be, for instance: "Does this particular model predict a transition of first or second order?" Experimental evidence is also rather too meagre, so that this is really an interesting open field.

ACKNOWLEDGEMENT

During the preparation of these notes I have often benefitted from constructive criticism through discussions with G. Navascués.

BIBLIOGRAPHY

The subject matter of Sections 1-4 is very well covered in the article "Surface thermodynamics of solids" (Linford (1973)), whose article happens to be the closest thing, in contents and approach, to the first part of my paper, but it is considerably more detailed. It contains, for example, the most complete discussion I have been able to find of the different methods for measuring γ and related concepts. For those particularly interested in this area within the thermodynamics of solid surfaces, Linford's article is especially recommended as supplementary to these notes. The rest of the references listed below include also some articles or books of a general nature, but not with the same degree of coincidence, in contents and approach, with any part of my paper.

Confucius said (more or less, I suppose): "... If the designations are not accurate, language will not be clear ... . Hence when Great Man has given something a name it may with all certainty be expressed in language; when he expresses it, it may with certainty be set in operation. In regard to this language, Great Man is never careless in any respect." It would be sensible, no doubt, to avoid the use of the term 'surface tension' for γ, and call it 'specific surface work', as recommended by Linford. But Confucius also said: "I wasn't born knowing what I teach you. Being fond of the past, I sought it through diligence". I came across Linford's article after these notes had been written, and it did not seem worth while making all the material changes in the text. Although, of course, the term may be misleading, I hope I have warned sufficiently against this possible misunderstanding. For to quote Confucius once more: "I do not instruct the uninterested; I do not help those who fail to try."

AHMAD, L., Phys. Stat. Solidi 29 (1968) 29.

BENSON, G.C., YUN, K.S., "Surface energy and surface tension of crystalline solids", The Solid-Gas Interface 1, (FLOOD, E.A., Ed.), M. Dekker, New York (1967) 203.

BLAKEMORE, J.S., Semiconductor Statistics, Pergamon, Oxford (1962).

BLANDIN, A., Remarks on phase transitions of surfaces, Phys. Lett. A 45 (1973) 275.

BLANDIN, A., CASTIEL, D., DOBRZYNSKI, L., Examples of surface instabilities and superstructures, Solid State Commun. 13 (1973) 1175.

BROWN, R.C., MARCH, N.H., Solid and liquid surfaces, Rep. Progr. Phys. to be published.

CALLEN, H.B., Thermodynamics, Wiley, New York (1960).

CHEN, T.S., ALLDREDGE, G.P., DE WETTE, F.W., ALLEN, R.E., Surface thermodynamic functions for NaCl, J. Chem. Phys. 55 (1971) 3121.

CLARK, A., The Theory of Adsorption and Catalysis, Academic Press, New York (1970).

COCHRAN, W., The Dynamics of Atoms in Crystals, Arnold, London (1973).

COUCHMAN, P., JESSER, W.A., On the thermodynamics of surfaces, Surf. Sci. 34 (1973) 212.

COUCHMAN, P., JESSER, W.A., KUHLMAN-WILSDORF, D., HIRTH, J.P., On the concepts of surface stress and surface strain, Surf. Sci. 33 (1972) 429.

CRAIG, R.A., Dynamic contribution to the surface energy of simple metals, Phys. Rev. B6 (1972) 1134.

DEFAY, R., PRIGOGINE, I., BELLEMANS, A., EVERETT, D.H., Surface Tension and Adsorption, Longman, London (1966).

ERIKSSON, J.C., Thermodynamics of surface phase system — V: Contribution to the thermodynamics of the solid-gas interface, Surf. Sci. 14 (1969) 221.

FLOOD, E.A., "The Gibbs and Polanyi thermodynamic description of adsorption", The Solid-Gas Interface 1 (FLOOD, E.A., Ed.), M. Dekker, New York (1967) 11.

FRANKL, D.R., Electrical Properties of Semiconductor Surfaces, Pergamon, Oxford (1967).

GARCIA-MOLINER, F., "The band picture in the electronic theories of chemisorption in semiconductors", Catalysis Reviews 2, (HEINEMAN, H., Ed.), M. Dekker, New York (1969) 1.

GREEN, M.S. (Ed.), Critical Phenomena (Proc. Int. School of Physics Enrico Fermi, Course 51, 1970), Academic Press, New York (1971).

GREEN, M., (Ed.), Solid State Surface Science 1, 2, M. Dekker, New York (1969, 1973).

GUGGENHEIM, E.A., Thermodynamics, North-Holland, Amsterdam (1967), Section 1.53 onwards.

HANEMAN, D., Surface structures and properties of diamond-structure semiconductors, Phys. Rev. 121 (1961) 1093.

HANSEN, R.S., On basic concepts in surface thermodynamics, J. Phys. Chem. 55 (1951 a) 1195.

HANSEN, R.S., Calculation of heats of adsorption from adsorption isosteres, J. Phys. Chem. 54 (1951 b) 411.

HARASIMA, A., "Molecular theory of surface tension", Advances in Chemical Physics 1 (PRIGOGINE, I., Ed.), Interscience, New York (1958) 203.

HERRING, C., "The use of classical macroscopic concepts in surface-energy problems", Structure and Properties of Solid Surfaces (GOMER, R., SMITH, C.S., Eds), University of Chicago Press (1953) 5-81.

HILL, T.L., Statistical mechanics of adsorption — V: Thermodynamics and heat of adsorption, J. Chem. Phys. 17 (1949) 520. (Shows how to break down experimental and theoretical free energies of adsorbate molecules into internal energies, heat contents and entropies.)

HILL, T.L., Statistical mechanics of adsorption — IX: Adsorption thermodynamics and solution thermodynamics, J. Chem. Phys. 18 (1950) 246.

HILL, T.L., Statistical mechanics of adsorption — X: Thermodynamics of adsorption on an elastic adsorbent, J. Chem. Phys. 18 (1950) 791.

HILL, T.L., "Theory of physical adsorption", Advances in Catalysis 4 (FRANKENBURG, W.G., Ed.), Academic Press, New York (1952) 211.

HILL, T.L., An Introduction to Statistical Thermodynamics, Addison-Wesley, Reading, Mass. (1960).

HOYEN, A.H., Jr., Physics and chemistry of silver halide surfaces space charge effects, Photogr. Sci. Eng. 17 (1973) 188.

KIRKWOOD, J.G., OPPENHEIM, I., Chemical Thermodynamics, McGraw-Hill, New York (1961).

KRÖGER, F.A., The Chemistry of Imperfect Crystals, North-Holland, Amsterdam (1964).

KRÖGER, F.A., STIELTJES, F.H., VINK, H.J., Thermodynamics and formulation of reactions involving imperfections in solids, Philos. Res. Rep. 14 (1959) 557.

LANDAU, L.D., LIFSHITZ, E.M., Statistical Physics, Ch. XV, Pergamon, Oxford (1958).

LANG, N.D., KOHN, W., Theory of metal surfaces: charge density and surface energy, Phys. Rev. B 12 (1970) 4555.

LINFORD, R.G., "Surface thermodynamics of solids", Solid State Surface Science 2 (GREEN, M., Ed.), M. Dekker, New York (1973).

MANY, A., GOLDSTEIN, Y., GROVER, N.B., Semiconductor Surfaces, North-Holland, Amsterdam (1971).

MELROSE, J.C., "Thermodynamics of surface phenomena", International Conference on Thermodynamics (Proc. Conf. Cardiff, 1970), Butterworth, London (1970) 273. Also published in Pure Appl. Chem. 22 3-4 (1970) 215.

PACE, E.L., "Adsorption thermodynamics and experimental measurement", The Solid-Gas Interface 1 (FLOOD, E.A., Ed.), M. Dekker, New York (1967) 105.

PIPPARD, A.B., Elements of Classical Thermodynamics, Cambridge University Press, (1957).

SAMUELSEN, E.J., ANDERSEN, E., FEDER, J., (Eds), Structural Phase Transitions and Soft Modes, Universitetsforlaget, Oslo (1971).

SANFELD, A., "Thermodynamics of surfaces", Ch. 2C of Physical Chemistry, An Advanced Treatise. Vol.1: Thermodynamics (EYRING, H., HENDERSON, D., JOST, W., Eds), Academic Press, New York (1971).

SOMORJAI, G.A., Principles of Surface Chemistry, Prentice Hall, New Jersey (1972).

SOMORJAI, G.A., LEED and Auger spectroscopy studies of the structure of adsorbed gases on solid surfaces, Surf. Sci. 34 (1973) 156.

SPARNAAY, M.J., On the electrostatic contribution to the interfacial tension of semiconductor/gas and semiconductor/electrolyte interfaces, Surf. Sci. 1 (1964) 213.

SPARNAAY, M.J., The free energy of double layer system with an emphasis on semiconductor surfaces and on membrane system, Surface Sci. 13 (1969) 222.

SPARNAAY, M.J., The Electrical Double Layer, Pergamon, Oxford (1972).

STANLEY, H.E., Introduction to Phase Transitions and Critical Phenomena, Clarendon Press, Oxford (1971).

STEELE, W.A., "Adsorbate equation of state", The Solid-Gas Interface 1 (FLOOD, E.A., Ed.), M. Dekker, New York (1966) 307.

TOLMAN, R.C., Consideration of the Gibbs theory of surface tension, J. Chem. Phys. 16 (1948) 758.

TOSATTI, E., ANDERSON, P.W., "Charge and spin density waves on semiconductor surfaces", in Proc. 2nd Int. Conf. on Solid Surfaces, Kyoto, March 1974.

TRULLINGER, S.E., CUNNINGHAM, S.L., Soft-mode theory of surface reconstruction, Phys. Rev. Lett. 30 (1973) 913.

VAN GOOL, W., Principles of Defect Chemistry of Crystalline Solids, Academic Press, New York (1966).

VERMAAK, J.S., MAYS, C.W., KUHLMAN-WILSDORF, D., On surface stress and surface tension — I: Theoretical considerations, Surf. Sci. 12 (1968) 128.

ZETTLEMOYER, A.C., NARAYAN, K.S., "Heats of immersion and the vapour solid interface", The Solid-Gas Interface 1 (FLOOD, E.A., Ed.), M. Dekker, New York (1966) 145.

IDEAL SURFACE STRUCTURES

S. ANDERSSON
Department of Physics,
Chalmers University of Technology,
Göteborg, Sweden

Abstract

IDEAL SURFACE STRUCTURES.
 1. Diperiodic structures. 2. Electron diffraction from diperiodic structures. 3. Structure of clean ordered surfaces.

1. DIPERIODIC STRUCTURES

1.1. Surface region

A set of conventions for surface crystallography has been presented by Wood [1] and will be summarized here. Figure 1 shows in a schematic way a layer of adsorbed material on a substrate and the suggested nomenclature. The true substrate or bulk material is periodic in the direction normal to the surface (z direction). The surface structure may be related to that of a parallel planar section of the substrate either by small displacements of atoms or by 'reconstructive displacements'. This region is called the selvedge. The substrate surface is marked by the disappearance of this periodicity (which in general occurs gradually).

The adsorption of atoms of different type will change the structure of the selvedge. If the adsorbed materials constitute a thick film it may be periodic in the z direction and have a selvedge of its own.

It is important to have relations between the substrate structure and a superposed related structure, the surface structure, which is not necessarily just one plane of atoms but, rather, is periodic in two dimensions, i.e. diperiodic.

It was noted by Wood that terms like 'superstructure' and 'superlattice' for the structure and lattice of the superposed structure should be avoided.

1.2. The five nets

In a diperiodic crystal surface the equivalent points form a net and the area units are unit meshes. The five nets or plane lattices are shown in Fig. 2. They correspond to the fourteen Bravais lattices for a triperiodic crystal (see Bürger [2], Ch.7). These nets derive from mutual restrictions among the invariance operations, i.e.

(i) Translations T

$$T = n_1 \vec{a}_1 + n_2 \vec{a}_2 \qquad (1.1)$$

where $\vec{a}_1$ and $\vec{a}_2$ are the translation vectors along the side of the unit mesh.

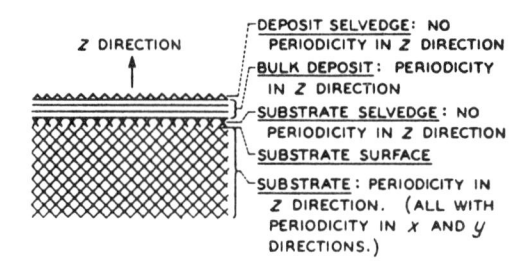

FIG. 1. The surface region (after Wood [1]).

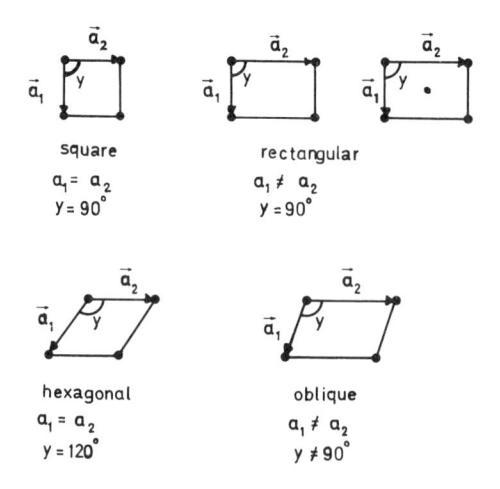

FIG. 2. The five diperiodic surface nets.

(ii) Rotation by an angle

$$R = \frac{2\pi}{n}; \quad n = 1, 2, 3, 4 \text{ and } 6 \tag{1.2}$$

(iii) Mirror reflections

The five nets (or plane lattices) describe just the symmetry of a plane point lattice. Each lattice point may consist of an arrangement of atoms (a basis) which itself is invariant under certain symmetry operations. These are the 10 crystallographic point groups in a plane. They are shown in Fig. 3 and are five axial groups consistent with the permissible rotations (ii) and five combining the rotations (ii) and the mirror reflections (iii). The five nets are of course consistent with the 10 plane point groups. Combining the five nets and the 10 plane point groups yields the 17 strictly two-dimensional plane groups. An example of these plane groups repeating

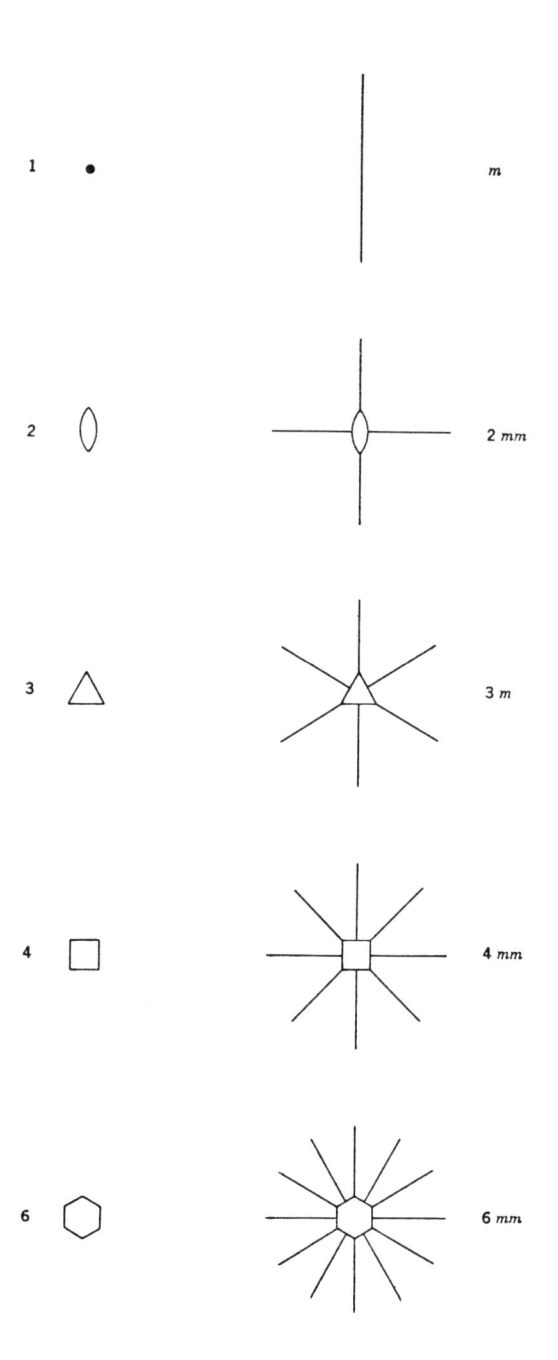

FIG.3. The ten plane point groups.

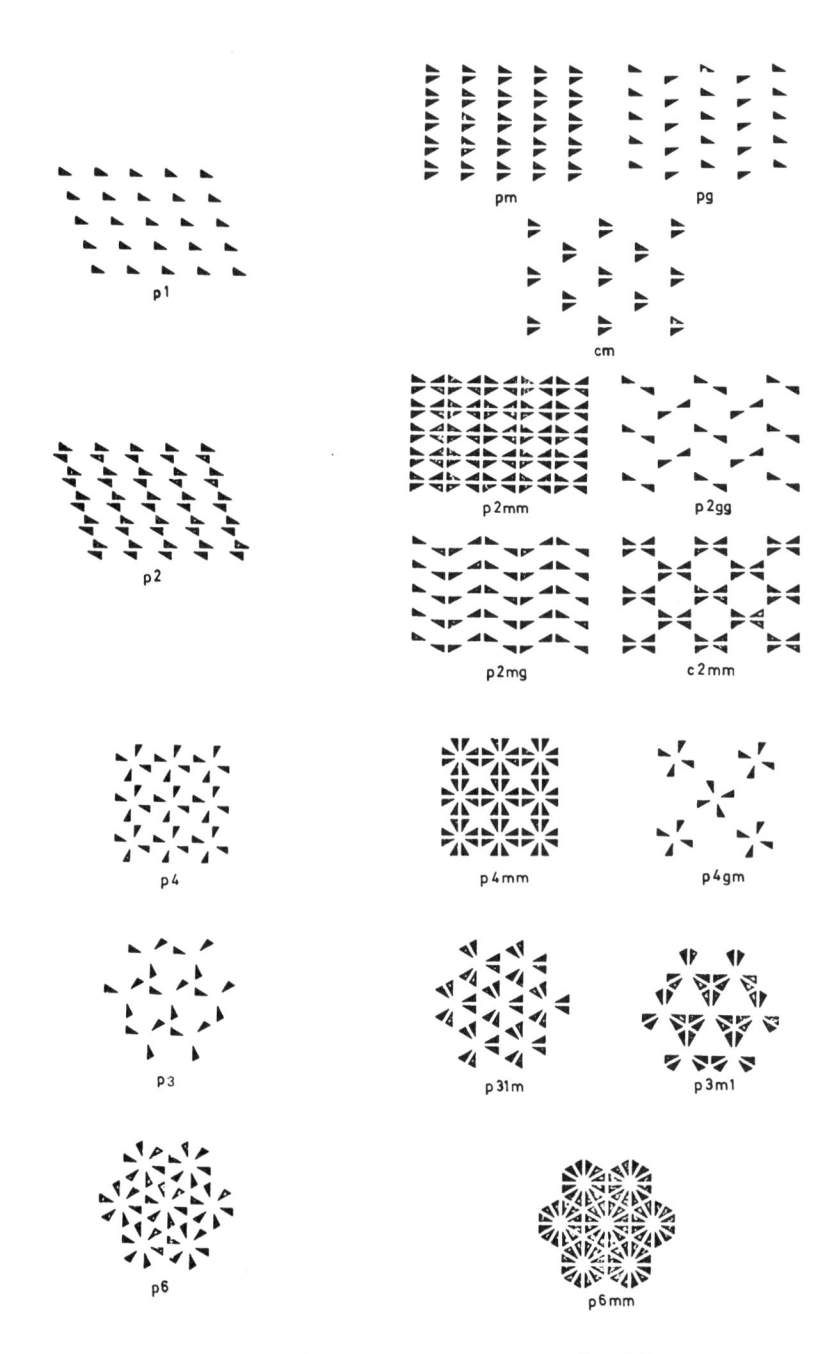

FIG. 4. The seventeen plane groups (after Bürger [2]).

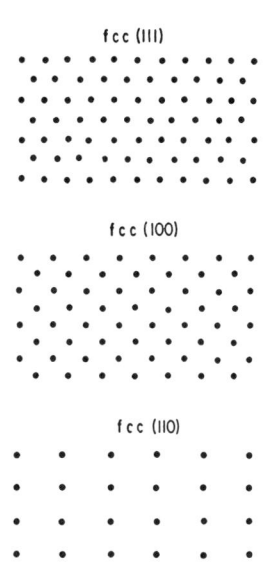

FIG. 5. Symmetry of low-index (h, k, ℓ) planes of the fcc Bravais lattice.

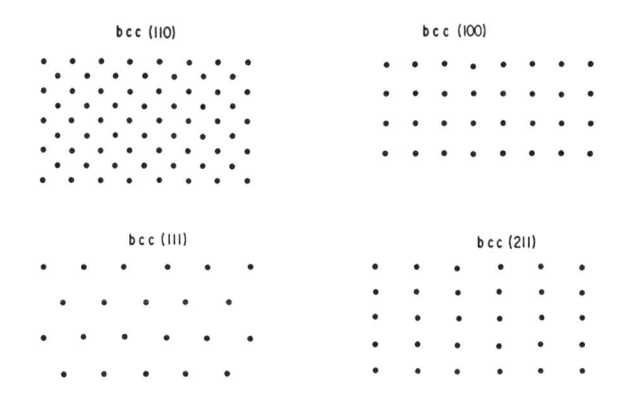

FIG. 6. Symmetry of low-index (h, k, ℓ) planes of the bcc Bravais lattice.

an asymmetrical triangle is shown in Fig. 4 (from Bürger). It was pointed out by Wood that another plane group (the 80 diperiodic groups in three dimensions) would be as useful when denoting diperiodic surface structures. We may anticipate the possible occurrence of the surface structures described by the five nets. Figures 5 and 6 show the symmetries of low-index (h,k,ℓ) planes of the face- and body-centred cubic lattices and we notice square, rectangular and hexagonal nets.

1.3. Superposition of nets

When denoting the structure of re-arranged surfaces and ordered adsorbate layers it is usually required that one expresses the translations $\vec{a}_j$ of one net of area A (e.g. the adsorbate layer net) to the translations $\vec{b}_j$ of another net of area B (e.g. the ideal substrate surface net). The transformation between the two nets may be described by the matrix:

$$\overleftrightarrow{M} = \begin{bmatrix} m_{11} & m_{12} \\ m_{21} & m_{22} \end{bmatrix} \tag{1.3}$$

i.e. the translation vectors are related:

$$\vec{a}_1 = m_{11}\vec{b}_1 + m_{12}\vec{b}_2$$
$$\vec{a}_2 = m_{21}\vec{b}_1 + m_{22}\vec{b}_2 \tag{1.4}$$

or short:

$$\vec{a} = \overleftrightarrow{M} \cdot \vec{b} \tag{1.5}$$

Since the areas are given by

$$A = |\vec{a}_1 \times \vec{a}_2|$$
$$B = |\vec{b}_1 \times \vec{b}_2| \tag{1.6}$$

$$A = |\vec{a}_1 \times \vec{a}_2| = |(m_{11}\vec{b}_1 + m_{12}\vec{b}_2) \times (m_{21}\vec{b}_1 + m_{22}\vec{b}_2)|$$
$$= |m_{11}m_{22} \cdot \vec{b}_1 \times \vec{b}_2 + m_{12}m_{21} \cdot \vec{b}_2 \times \vec{b}_1|$$
$$= B \det \overleftrightarrow{M} \tag{1.7}$$

The values of $\det \overleftrightarrow{M}$ define the character of the superposition of the two nets $\vec{a}$ and $\vec{b}$.

$\det \overleftrightarrow{M}$ integer: $\vec{a}$ and $\vec{b}$ are simply related and the superposition is <u>simple</u>;

$\det \overleftrightarrow{M}$ rational fraction: $\vec{a}$ and $\vec{b}$ are rationally related and the superposition is <u>coincidence-site</u>;

$\det \overleftrightarrow{M}$ irrational fraction: $\vec{a}$ and $\vec{b}$ are irrationally related and the superposition is <u>incoherent</u>.

The superpositioned nets (but for the <u>incoherent</u>) can be denoted by one net $\overleftrightarrow{C}$ such that

$$\overleftrightarrow{C} = \overleftrightarrow{M} \cdot \vec{a} = \overleftrightarrow{P} \cdot \vec{b} \tag{1.8}$$

It is usually suitable to choose the net $\overleftrightarrow{C}$ with the smallest unit mesh. Thus $\det \overleftrightarrow{M}$ and $\det \overleftrightarrow{P}$ must be integers with no common factor.

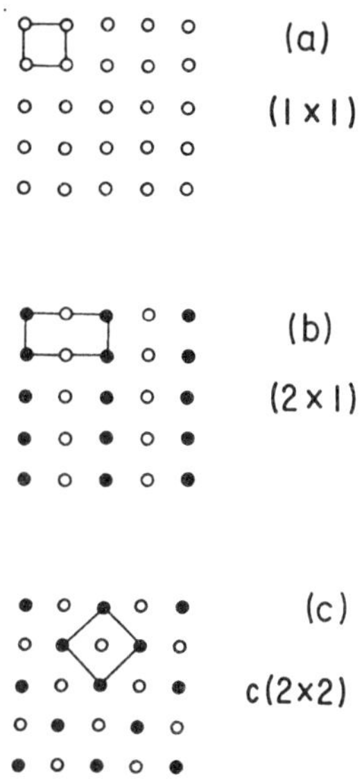

IDEAL SURFACE STRUCTURES

FIG. 7. Examples of simple superpositions on a square symmetric substrate.

1.4. Surface structure notation

There are several different ways of denoting surface structures in the literature. One obvious notation is in terms of the matrices described in Section 1.3 [3].

(i) If $\overleftrightarrow{b}$ is the net of a reference surface structure the superposition is denoted by the matrix $\overleftrightarrow{P}$ connecting the nets. Figure 7 gives a few illustrative examples of the use of this notation for a square symmetric substrate. Figure 7(a) shows a case where the translation periodicity is unaltered and thus

$$\overleftrightarrow{C} = \overleftrightarrow{P} \cdot \overleftrightarrow{b}; \qquad \overleftrightarrow{P} = \begin{bmatrix} 1 & 0 \\ 0 & 1 \end{bmatrix}$$

In Fig. 7(b) the superposed net has in one direction a translation vector that is twice that of the reference net while the other vector is unaltered. There are obviously two possible orientations. So

$$\overleftrightarrow{P}_1 = \begin{bmatrix} 2 & 0 \\ 0 & 1 \end{bmatrix} \quad \text{or} \quad \overleftrightarrow{P}_2 = \begin{bmatrix} 1 & 0 \\ 0 & 2 \end{bmatrix}$$

Figure 7(c) shows a situation where the superposed net is rotated by 45° and the modulus of the translation vector is $\sqrt{2}$ times that of the reference net.

The notation is

$$\overleftrightarrow{P} = \begin{bmatrix} 1 & -1 \\ 1 & 1 \end{bmatrix}$$

and the areas

$$C = A = B \det \overleftrightarrow{P} = 2B$$

(ii) The Wood [1] notation has been more frequently used. The superposition $\overleftrightarrow{C}$ and reference net $\overleftrightarrow{b}$ are related by the quotient of the lengths of the translation vectors and a rotation R:

$$(a_1/b_1 \times a_2/b_2) R$$

This notation can only be used if the unit meshes $\overleftrightarrow{a}$ and $\overleftrightarrow{b}$ have the same included angles.

The examples of Fig. 5 will be denoted:

(a) (1×1) or $p(1 \times 1)$ where p denotes primitive
(b) (2×1) or $p(2 \times 1)$
(c) $(\sqrt{2} \times \sqrt{2}) 45°$

(iii) In experimental literature a notation based on the choice of non-primitive unit meshes is commonly used. Returning to Fig. 7, the notation will be:

(a) $p(1 \times 1)$
(b) $p(2 \times 1)$
(c) $c(2 \times 2)$ where c denotes centred

1.5. The reciprocal net

An important item when describing physical phenomena related to periodic objects is the construction of a related reciprocal space. This is borne out by the fact that periodic functions are expressed as Fourier series in such a space. Here we shall be particularly concerned with reciprocal space when describing diffraction from diperiodic objects.

The reciprocal net unit vectors $\vec{a}_1^*, \vec{a}_2^*$ are related to the direct net unit vectors $\vec{a}_1, \vec{a}_2$ by

$$\vec{a}_i^* \cdot \vec{a}_j = 2\pi \delta ij \tag{1.9}$$

or in matrix form:

$$\overleftrightarrow{a}^* \widetilde{\overleftrightarrow{a}} = 2\pi \overleftrightarrow{I}$$

where $\overleftrightarrow{I}$ is the 2×2 unit matrix.

An arbitrary reciprocal net vector is given by

$$\vec{g}hk = h\vec{a}_1^* + k\vec{a}_2^* \tag{1.10}$$

Previously we had

$$\overleftrightarrow{a} = \overleftrightarrow{M} \cdot \overleftrightarrow{b}$$

for the superposition of two nets. We now prove that for the reciprocal nets we have

$$\overleftrightarrow{a}^* = \widetilde{\overleftrightarrow{M}}^{-1} \overleftrightarrow{b}^* \qquad (1.11)$$

Postmultiply both sides by $\widetilde{\overleftrightarrow{a}}$ and use $\widetilde{\overleftrightarrow{a}} = \widetilde{\overleftrightarrow{b}}\, \widetilde{\overleftrightarrow{M}}$:

$$\overleftrightarrow{a}^* \widetilde{\overleftrightarrow{a}} = \widetilde{\overleftrightarrow{M}}^{-1} \overleftrightarrow{b}^* \widetilde{\overleftrightarrow{a}}$$

Left-hand side:

$$\overleftrightarrow{a}^* \widetilde{\overleftrightarrow{a}} = 2\pi \overleftrightarrow{I}$$

Right-hand side:

$$\widetilde{\overleftrightarrow{M}}^{-1} \overleftrightarrow{b}^* \widetilde{\overleftrightarrow{a}} = \widetilde{\overleftrightarrow{M}}^{-1} \overleftrightarrow{b}^* \widetilde{\overleftrightarrow{b}} \cdot \widetilde{\overleftrightarrow{M}} = \widetilde{\overleftrightarrow{M}}^{-1} 2\pi \overleftrightarrow{I}\, \widetilde{\overleftrightarrow{M}} = 2\pi \overleftrightarrow{I}$$

The reciprocal net has some elementary properties:

(i) Each vector of the reciprocal net is normal to a set of net rows of the direct net.

Let $\vec{g}$ be a reciprocal net vector and $\vec{\ell} = n_1 \vec{a}_1 + n_2 \vec{a}_2$ a direct net vector; then

$$\vec{g} \cdot \vec{\ell} = 2\pi(hn_1 + kn_2) = 2\pi N \qquad (1.12)$$

Thus the projection of $\vec{\ell}$ along $\vec{g}$ has the length:

$$\frac{2\pi N}{|\vec{g}|}$$

but we could instead choose $\vec{\ell}'$:

$$n_1' = n_1 + mk, \qquad n_2' = n_2 - mh,$$
$$\vec{g} \cdot \vec{\ell} = \vec{g} \cdot \vec{\ell}' = 2\pi N \qquad (1.13)$$

All these $\vec{\ell}$'s lie on the line normal to $\vec{g}$. In fact,

$$\vec{g} \cdot \vec{\ell} - 2\pi N = 0$$

is the equation of the line.

(ii) $|\vec{g}|$ is inversely proportional to the spacing of the net rows normal to $\vec{g}$ provided the components h, k of $\vec{g}$ have no common factor. Consider

$$d = \frac{2\pi N}{|\vec{g}|} \qquad (1.14)$$

Since h, k have no common factor we can always find $\vec{\ell}''$ such that

$$d'' = \frac{2\pi(N+1)}{|\vec{g}|}$$

i.e. the spacing between the rows is just

$$d_{(hk)} = d'' - d = \frac{2\pi}{|\vec{g}|}$$

(iii) The area of the reciprocal unit mesh is inversely proportional to the area of the direct mesh:

$$A = |\vec{a}_1 \times \vec{a}_2| = |\vec{a}_1| \cdot |\vec{a}_2| \cdot \sin\gamma \qquad (1.15)$$

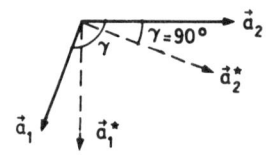

$$A^* = |\vec{a}_1^* \times \vec{a}_2^*| = |\vec{a}_1^*| \cdot |\vec{a}_2^*| \cdot \sin(180-\gamma)$$

from

$$\vec{a}_i^* \cdot \vec{a}_j = 2\pi ij$$

$$|\vec{a}_1^*| \cdot |\vec{a}_1| \cdot \cos(\gamma - 90) = 2\pi$$

$$|\vec{a}_2^*| \cdot |\vec{a}_2| \cdot \cos(\gamma - 90) = 2$$

$$A^* = \frac{4\pi^2 \sin\gamma}{|\vec{a}_1| \sin\gamma \; |\vec{a}_2| \; \sin\gamma} = \frac{4\pi^2}{A} \qquad (1.16)$$

It is obvious that a particular set of net rows can be denoted by the appropriate low-index reciprocal vector. This notation has a close resemblance to the Miller indices in crystallography:

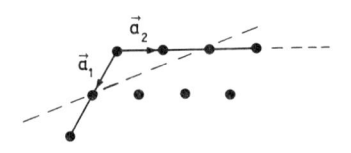

The Miller indices are determined by finding the intercepts with the two unit mesh axes, taking the reciprocals of these numbers and reducing to the smallest integers with the same ratio. The notation is (h,k).

We can construct this from the equation for the net row:

$$\vec{g} \cdot \vec{\ell} = 2\pi N$$

The intercepts are obtained for:

$$\vec{\ell}_1' = n_1' \vec{a}_1, \qquad \vec{\ell}_2' = n_2' \vec{a}_2$$

$$2\pi h n_1' = 2\pi N, \qquad n_1' = \frac{N}{h}$$

$$2\pi k n_2' = 2\pi N, \qquad n_2' = \frac{N}{k} \tag{1.17}$$

The reciprocals are:

$$\frac{1}{n_1'} = \frac{h}{N}, \qquad \frac{1}{n_2'} = \frac{k}{N} \tag{1.18}$$

and reduction to the smallest integer with the same ratio gives the Miller indices (h, k) provided that h and k have no common factor.

1.6. Inter-row spacing for the five nets

As we have seen, the inter-row spacings are important when we want to describe the properties of a net, and an expression for this spacing for the five nets can easily be derived:

Oblique:

$$\frac{1}{d_{(hk)}^2} = \frac{h^2}{a^2 \sin^2 \gamma} + \frac{k^2}{b^2 \sin^2 \gamma} - \frac{2hk \cos \gamma}{ab \sin^2 \gamma}$$

Rectangular p and c:

$$\frac{1}{d_{(hk)}^2} = \left(\frac{h}{a}\right)^2 + \left(\frac{k}{b}\right)^2$$

Hexagonal:

$$\frac{1}{d_{(hk)}^2} = \frac{4}{3} \frac{h^2 + hk + k^2}{a^2}$$

Square:

$$\frac{1}{d_{(hk)}^2} = \frac{h^2 + k^2}{a^2}$$

We shall prove the hexagonal case as an example. From the foregoing we have:

$$d_{(hk)} = \frac{2\pi}{|\vec{g}|}$$

$$|\vec{g}|^2 = (h\,\vec{a}_1^* + k\,\vec{a}_2^*) \cdot (h\,\vec{a}_1^* + k\,\vec{a}_2^*)$$

$$= h^2 |\vec{a}_1^*|^2 + 2hk\,\vec{a}_1^* \cdot \vec{a}_2^* + k^2 |\vec{a}_2^*|^2$$

$$|\vec{a}_1^*| = \frac{2\pi}{a\cos(120°-90°)} = \frac{2}{\sqrt{3}}\frac{2\pi}{a}$$

$$\vec{a}_1^* \cdot \vec{a}_2^* = \frac{16\pi^2 \cos 60}{3a^2} = \frac{8\pi^2}{3a^2}$$

$$\frac{1}{d_{(hk)}^2} = \frac{1}{4\pi^2}\left(h^2 \frac{4}{3}\frac{4\pi^2}{a^2} + hk\,\frac{16\pi^2}{3a^2} + k^2\,\frac{4}{3}\frac{4\pi^2}{a^2}\right) = \frac{4}{3}\left(\frac{h^2+hk+k^2}{a^2}\right)$$

2. ELECTRON DIFFRACTION FROM DIPERIODIC STRUCTURES

2.1. Methods

The obvious means of investigating the relative positions of atoms in a periodic arrangement is diffraction. Depending on the properties of the periodic object, different methods are employed. Electron diffraction has been found to be a powerful tool in the case of surface structures. Two different techniques are employed: low-energy electron diffraction (LEED) and reflection high-energy electron diffraction (RHEED). As is obvious from the nomenclature, different electron energies E are employed. In LEED, typically, $E < 1$ keV; in RHEED, E is in the range 10 - 100 keV. The diffraction geometry is also different as can be seen in Fig. 8. Both methods rely on the observation of electrons that are elastically diffracted at the surface. In LEED, normal or nearly normal incidence of the primary electron beam onto the surface is usually employed, while in RHEED the beam makes a small glancing angle of a few degrees with the surface plane. Figure 9 shows the principle of the LEED technique together with the respective diffraction observation from a Cu(100) surface (square symmetric). We shall return later to the interpretation of the pattern. The diffracted electrons are 'backreflected' from the specimen and the main contribution comes from the outermost atomic planes. In LEED this is due to the small

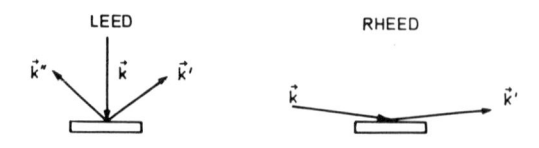

FIG. 8. Diffraction geometry of LEED and RHEED (schematic).

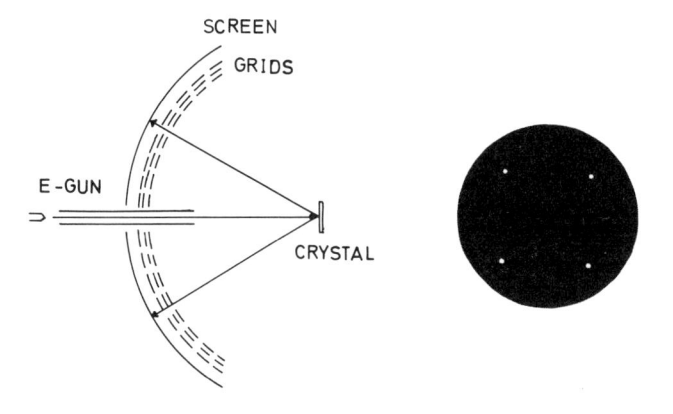

FIG.9. Schematic LEED display equipment and a diffraction pattern from a Cu(100) surface (50 eV).

mean free path of the electrons (typically 5-10 Å) and in RHEED the shallow angle with the surface plane limits the penetration normal to the surface. More thorough presentations of both LEED [4, 5] and RHEED [6] are available in recent surveys. Most of the present knowledge about the structure of clean surfaces and adsorbed layers have been obtained by the LEED technique and we shall be concerned with this method.

2.2. Diffraction conditions

In a LEED experiment a mono-energetic beam of electrons is incident on a crystal surface. The experiment is characterized by the diffraction parameters which together specify the energy E of the primary electron beam and the direction of this beam relative to the crystal axes. The directions are the angle of incidence θ relative to the surface normal and the azimuthal angle φ relative to a unit vector $\vec{a}$ parallel to the crystal surface (Fig.10). The energy is related to $\vec{k}$ (the propagation vector) by

$$E = \frac{\hbar^2}{2m} k^2 \tag{2.1}$$

$$k = \frac{2\pi}{\lambda} \tag{2.2}$$

and thus

$$\lambda = \frac{12.26}{E^{\frac{1}{2}}} \quad (\lambda \text{ in Å}, \; E \text{ in eV}) \tag{2.3}$$

$\vec{k}$ is conveniently expressed in terms of its projection $\vec{k}_\parallel$ and k_z onto the surface and the surface normal, respectively:

$$\vec{k} = (\vec{k}_\parallel, \; k_z) \tag{2.4}$$

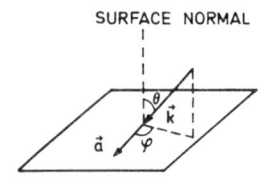

FIG.10. Diffraction parameters.

By using θ, φ and $\vec{a}$ we get

$$k_z = k\cos\theta \tag{2.5}$$

$$\vec{k}\cdot\vec{a} = k_\parallel \cos\varphi \tag{2.6}$$

The observed quantities are the back-diffracted beams, their directions and intensities. The wave vector $\vec{k}'$ of a diffracted beam may, as above, be given in terms of angles θ', φ'. The diffraction process is elastic, i.e. energy and the surface component of momenta are conserved. These conditions can be visualized by the Ewald sphere construction in the reciprocal space. The construction is shown in Fig.11. A set of reciprocal net normals are drawn through the points $\vec{g}_{hk}$ of the reciprocal net in the direction normal to the net (which is also normal to the direct net since $\vec{a}_i$ and $\vec{a}_i^*$ are coplanar). Let the propagation vector $\vec{k}$ terminate at the origin of the reciprocal net and let $\vec{k}$ be the radius vector of the Ewald sphere. The intersection of the net normals with the sphere gives the directions of forward and backward diffracted beams such that

$$k' = k \tag{2.7}$$

i.e. $E' = E$ (energy is conserved)

$$\vec{k}'_\parallel = \vec{k}_\parallel + \vec{g}_{hk} \tag{2.8}$$

i.e. the surface component of momentum is conserved.
The propagation vectors are thus given by

$$\vec{k}^\pm = (\vec{k}_\parallel + \vec{g}_{hk}, \pm k_z) \tag{2.9}$$

$$k_z = [k^2 - |\vec{k}_\parallel + \vec{g}_{hk}|^2]^{\frac{1}{2}} \qquad (k_z \text{ real}) \tag{2.10}$$

where the back-diffracted beams $\vec{k}^-$ are observed in the LEED experiment. These beams will generate a diffraction pattern which reveals the reciprocal unit mesh. From this we can deduce the symmetry and size of the direct unit mesh. The kinds and relative positions of atoms in the unit mesh have to be found from the diffraction intensities.

$$k' = k$$
$$\vec{k}'_\| = \vec{k}_\| + \vec{g}_\|$$

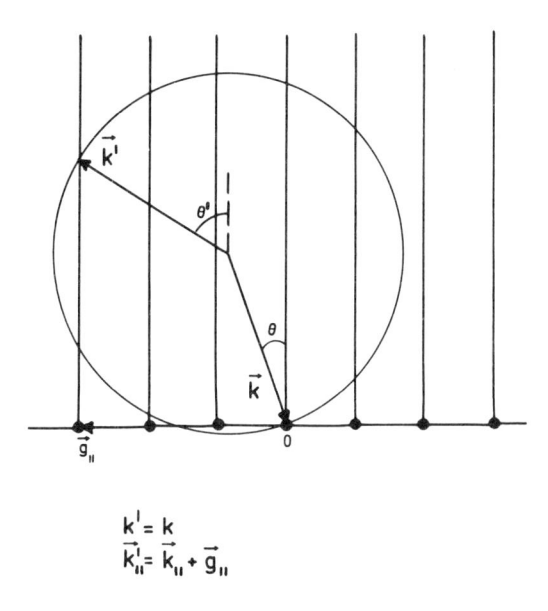

FIG. 11. Ewald sphere construction for a two-dimensional reciprocal net.

2.3. Diffraction intensities and the interference function

The Ewald sphere construction in the reciprocal net is a purely mathematical construction and cannot be realized in nature. The requirements would be a perfectly periodic in two dimensions, infinitely extended, object and a primary electron beam that is a coherent pure plane wave. In reality the object is finite and the electron beam is not a plane wave and it has limited coherence. It is certainly useful to have a more realistic physical picture of the diffraction process.

Let us consider the distribution of the diffracted low-energy electrons and how it is related to the surface structure. Several different approaches to this problem have been presented during recent years and these have been reviewed in a condensed manner in Ref. [7].

At this stage, a scattering treatment is presented that complements the Ewald sphere construction and also gives us a general feeling for the diffraction intensities. When treating low-energy electron diffraction some important factors must be kept in mind:

(a) Although the incident electron wave is rapidly attenuated, several layers contribute to the diffracted intensity.
(b) Atomic scattering cross-sections are generally large, so the effective wave amplitude incident on one particular atom is related to the scattering from all the other atoms.
(c) Lateral periodicity is apt to be persistent even if there is no periodicity in the z direction. Thus summation of amplitudes over layers is a natural approach.

Let $\psi(\vec{r}')$ be the wave function of the electron involved in the scattering process in the scatterer and $V(\vec{r}')$ be the scattering potential. The scattering amplitude of the scattering volume Ω at large distance is then given by

$$A = -\frac{1}{4\pi} \int_\Omega e^{-i\vec{k}'\cdot\vec{r}'} V(\vec{r}') \psi(\vec{r}') d\vec{r}' \qquad (2.11)$$

The lateral periodicity is given by the unit mesh vectors $\vec{a}_1$, $\vec{a}_2$ and thus for any point $\vec{r}' = \vec{r}_0' + n_1\vec{a}_1 + n_2\vec{a}_2$

$$V(\vec{r}') = V(\vec{r}_0') \qquad (2.12)$$

The wave function at equivalent points of the net differs by only a phase factor. For an incident plane wave $\psi = e^{i\vec{k}\cdot\vec{r}}$ we have

$$\psi(\vec{r}') = \exp[i\vec{k}(n_1\vec{a}_1 + n_2\vec{a}_2)] \cdot \psi(\vec{r}_0') \qquad (2.13)$$

Making use of (2.12) and (2.13), we can break up the integration (2.11) over the whole scattering volume into a summation over the net and integration over the unit cell:

$$A = \sum_{n_1 n_2} \exp[-i(\vec{k}'-\vec{k})(n_1\vec{a}_1+n_2\vec{a}_2)] \times \left(-\frac{1}{4\pi}\int_{\Omega_0} e^{-i\vec{k}'\cdot\vec{r}_0} V(\vec{r}_0') \psi(\vec{r}_0') d\vec{r}_0\right)$$

$$= S(\vec{k}',\vec{k}) \cdot F(\vec{k}',\vec{k}) \qquad (2.14)$$

The intensity distribution of the scattered wave is proportional to $|A|^2$:

$$I(\vec{k}',\vec{k}) = |S|^2 |F|^2 \qquad (2.15)$$

$|S|^2$ is called the interference function, written $\mathscr{I}$, and determines the direction of the diffracted waves. The dynamical structure amplitude F determines their relative intensities. The integration in $F(\vec{k}',\vec{k})$ extends over all points within the 'unit cell'. This is a column of layers normal to the surface with an area $|\vec{a}_1 \times \vec{a}_2|$ and a height $z \cdot a_3$ determined by the penetration depth of the wave field.

Performing the summation over n_1, n_2 for a perfect net of dimensions $N_1\vec{a}_1 \times N_2\vec{a}_2$ yields

$$\mathscr{I} = \frac{\sin^2(\tfrac{1}{2}N_1\delta_1)}{\sin^2(\tfrac{1}{2}\delta_1)} \cdot \frac{\sin^2(\tfrac{1}{2}N_2\delta_2)}{\sin^2(\tfrac{1}{2}\delta_2)} \qquad (2.16)$$

$$\delta_1 = (\vec{k}-\vec{k}')\cdot\vec{a}_1 = (\vec{k}_\parallel - \vec{k}_\parallel')\cdot\vec{a}_1$$
$$\delta_2 = (\vec{k}-\vec{k}')\cdot\vec{a}_2 = (\vec{k}_\parallel - \vec{k}_\parallel')\cdot\vec{a}_2 \qquad (2.17)$$

$\mathscr{I}$ yields the symmetry, spacings and beam profiles in the diffraction pattern. $\mathscr{I}$ has maxima of height $N_1^2 \times N_2^2$ whenever

$$\delta_i = 2\pi h_i \quad (h_i \text{ integer}) \qquad (2.18)$$

This statement is equivalent to the Laue conditions for diffraction from a two-dimensional grating.

The first zero in the interference function may be taken as a measure of the beam width. Consider

$$\sin(\tfrac{1}{2} N \delta^0) = 0$$

$$\tfrac{1}{2} N \delta^0 = \tfrac{1}{2} N \, 2\pi h + \pi \tag{2.19}$$

The first zero from the maximum at $\delta = 2\pi h$; thus

$$\Delta \delta = \delta^0 - \delta = (\vec{k}'_{\|} - \vec{k}'^{\,0}_{\|}) \, \vec{a} = \frac{2\pi}{N}$$

For simplicity consider $\vec{k}'_{\|}$ to be parallel to $\vec{a}$, and $\vec{k}$ to be normal to the surface:

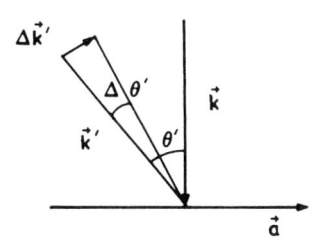

$$\Delta k' = k' \Delta \theta$$

$$\Delta k'_{\|} = \Delta k' \cos \theta = k' \Delta \theta \cos \theta$$

Furthermore,

$$\vec{k}'_{\|} \cdot \vec{a} = k' \sin \theta' a = 2\pi h$$

This gives

$$\Delta \delta = k' \Delta \theta' \cos \theta' a = \frac{2\pi h}{\sin \theta'} \cos \theta' \Delta \theta' = \frac{2\pi}{N}$$

Thus

$$\Delta \theta' = \frac{\tan \theta'}{hN} \tag{2.20}$$

is a measure of the angular broadening of the diffraction beam due to limited periodicity in the surface net.

The structure amplitude $F(\vec{k}', \vec{k})$ carries information about the relative position of atoms within the 'unit cell' Ω_0. If we divide Ω_0 into Wigner-Zeitz cells Ω_ν centred at the equilibrium positions $\vec{r}_\nu$ of the atoms, $F(\vec{k}', \vec{k})$ can be written as

$$F(\vec{k}', \vec{k}) = -\frac{1}{4\pi} \sum_\nu e^{-i\vec{k}' \cdot \vec{r}_\nu} \left(\int_{\Omega_\nu} e^{-i\vec{k}' \vec{r}'_0} \cdot V(\vec{r}_\nu + \vec{r}'_0) \psi(\vec{r}_\nu + \vec{r}'_0) \, d\vec{r}'_0 \right) \tag{2.21}$$

which is similar to the usual kinematical expression for the structure amplitude:

$$F(\vec{k}', \vec{k}) = \sum_\nu f_\nu \exp[-i(\vec{k}' - \vec{k}) \cdot \vec{r}_\nu] \tag{2.22}$$

Accordingly we can expect any periodicity normal to the surface to show up as a third Laue condition:

$$\delta_3 = (\vec{k} - \vec{k}') \cdot \vec{a}_3 = (k_z - k'_z) \cdot a_3 = 2\pi h_3 \tag{2.23}$$

$\vec{a}_3$ representing the periodicity in the z direction. The multiple scattering contained in the integral will complicate this simple relationship.

There is no Fourier transform of $F(\vec{k}', \vec{k})$ as in the X-ray case and a structure analysis will involve the calculation of $F(\vec{k}', \vec{k})$ for a number of trial structures. This is a difficult task and can only be done for simple surface structures. Accordingly most surface structure models have been inferred tentatively from the interference function and other available physical and chemical information.

2.4. Coherence

According to (2.20) the width of the diffraction beams provides information about the dimension $N_1 \times N_2$ of the coherent scattering domains. The instrumental resolution, however, sets a lower limit to the area over which the primary wave field is coherent, the coherence zone. The character of the primary wave is determined by the electron source. For an extended, thermally heated source as used in LEED optics, the coherence zone will be limited by time incoherence due to the thermal energy spread ΔE and by space incoherence which derives from divergence of the electron beam as measured by the half-angle β_s subtended at the crystal surface by the electron source [8]. Let the electron propagation vector $\vec{k}$ be normal to the crystal surface.

$E = \dfrac{\hbar^2}{2m} k^2$ gives the spread;

$$\Delta k = \dfrac{2m}{\hbar^2} \dfrac{\Delta E}{2k} = k \dfrac{\Delta E}{E} \quad (\text{along } \vec{k}) \tag{2.24}$$

Due to the divergence there is a spread in the surface component of $\vec{k}$, $\approx 2k\beta_s$. Combining the uncertainties in quadrature gives

$$\Delta k_\parallel \approx 2[(k\beta_s)^2 + (\Delta k \beta_s)^2]^{\frac{1}{2}}$$

$$= 2\beta_s \dfrac{2\pi}{\lambda} \left[1 + \left(\dfrac{\Delta E}{2E}\right)^2\right]^{\frac{1}{2}} \tag{2.25}$$

Let the coherence criterion be given by

$$\Delta x \cdot \Delta k_\parallel = 2\pi$$

where Δx is the coherence zone diameter,

$$\Delta x \approx \frac{\lambda}{2\beta_s [1+(\Delta E/2E)^2]^{\frac{1}{2}}} \tag{2.26}$$

No area larger than $\approx (\Delta x)^2$ can contribute coherently to the diffraction pattern.

We may estimate Δx for ordinary LEED optics. Assume that $\Delta E \approx (3/2)kT$, i.e. the average velocity of the thermally emitted electrons, and putting $\beta_s \approx 0.005$ (for a source radius 0.5 mm at 100 mm distance from the crystal surface). $\Delta E \approx 0.2$ eV for a low-temperature source at 800°C. It is obvious from (2.26) that the time-incoherence due to ΔE will only be of importance at the lowest energies. At $E = 150$ eV, $\lambda \approx 1$ Å gives

$$\Delta x \approx 100 \text{ Å}$$

The perfectly periodic areas on clean, well prepared crystal surfaces certainly considerably exceed $(\Delta x)^2$ and the diffraction beam profile is thus determined by the coherence of the incident electron wave. Park [9] has investigated the profiles of the beams diffracted from a clean Pd(110) surface at a number of electron energies and at 85° angle of diffraction (Fig. 12). At this angle the diameter of the primary beam (≈ 1 mm) does not influence the measured beam broadening too much. The coherence diameter is derived from the estimated number N of coherently contributing scatters as measured by the half-angular broadening according to (2.20). Δx is apparently dependent on λ and at $\lambda = 1$ Å ($E = 150$ eV), $\Delta x \approx 100$ Å as estimated above.

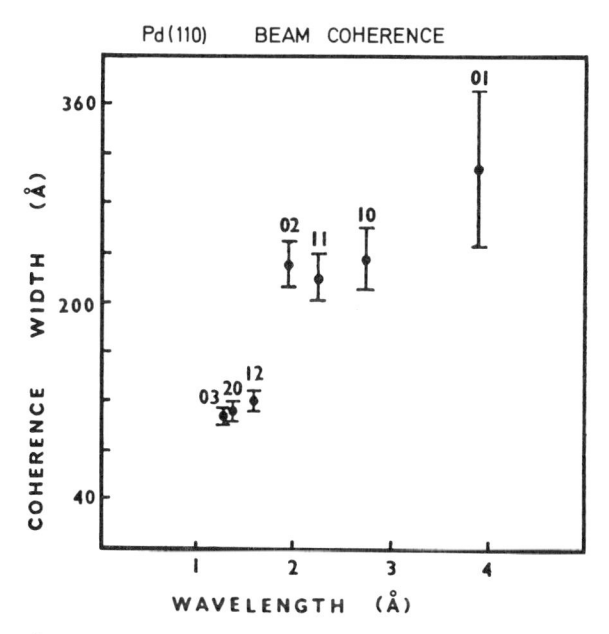

FIG. 12. Beam coherence versus electron wavelength λ as determined from diffraction beam broadening (Pd(110), after Park [9]).

2.5. Surface disorder

There are of course situations applying to clean single-crystal surfaces when limited order in the surface net gives rise to beam broadening. Obviously the surface order can be made to deteriorate by, e.g., noble-gas ion bombardment provided that the specimen is kept at a temperature low enough to prevent the introduced disorder from annealing. In this way disorder may accumulate until the diffraction pattern is completely obliterated (e.g. Si, Ge kept at room temperature). The disorder can usually be leaked out by careful annealing (suitable temperatures for Si and Ge are 700°C and 300°C respectively) and the surface becomes perfect over regions wide enough compared to $(\Delta x)^2$.

Another kind of disorder may occur when clean single-crystal surfaces are produced by cleavage. Usually such surfaces are perfectly periodic and flat over large areas but <u>cleavage steps</u> may appear in a very large number breaking long-range order in the surface direction perpendicular to the step direction. High step densities can also be introduced by deliberately cutting the crystal face at a slightly wrong angle.

In the case of adsorbed layers on single-crystal surfaces, a number of different mechanisms may give rise to beam broadening, e.g. effects of domain size, different kinds of one-dimensional and two-dimensional disorder (out-of-phase domains). Such effects have been discussed in the literature [10, 11].

Figure 13 illustrates disordering of the Ni(100) surface by noble-gas ion bombardment (200 eV argon ions, 1 A/cm^2, 2 hours). The observed beam broadening reveals that the size of the ordered domains has been reduced below the size of the coherence zone $(\Delta x)^2$. We may estimate the size of the domains from

$$\Delta \theta = \frac{\tan \theta}{h \cdot N}$$

$\Delta \theta = 0.04$ radians, $h = 1$, $\theta = 34°$

which gives $N = 18$, i.e.

$Na = 18$, $2.50 \text{ Å} = 45 \text{ Å}$

Compare $\Delta x = 150$ Å at $\lambda = 1.5$ Å from Fig. 12. The disordered surface is easily annealed.

3. STRUCTURE OF CLEAN ORDERED SURFACES

3.1. Preparation of clean surfaces

An important step in a surface structure investigation is the preparation and cleaning of the single-crystal specimen. In the LEED experiment suitable specimens should have an area considerably larger than the area of the electron beam. The bulk material should be of high purity since

Ni(100)
diffraction
pattern

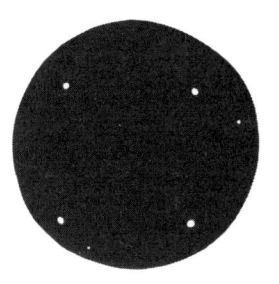

ordered

disordered

FIG. 13. Diffraction patterns from an ordered and a disordered Ni(100) surface (LEED).

contaminants often tend to segregate at the surface during heating of the specimen. Some commonly used preparation and cleaning techniques are listed as follows:

(a) The specimen may be prepared by cleavage in situ at UHV conditions (10^{-10} Torr). The surfaces obtained in this way are usually free of contamination and of very high structural perfection. The technique has been applied to a large number of compounds (semiconductors and insulators) and some single-element materials, e.g.

    Metals: Be, Zn [12]
    Semimetals: Bi, Sb [13]
    Single-element semiconductors: Si, Ge [14], Te [15]

(b) Epitaxial growth from the vapour phase on a single crystalline substrate can yield excellent specimens. The substrate surface may be prepared by cleavage (e.g. alkali halides, mica). Contamination may be a problem.

(c) The specimens are often prepared outside the vacuum system by <u>cutting and polishing</u>. The orientation is obtained by X-ray back reflection to within 0.1°-1°. The specific cutting and polishing technique used depends on the material. Softer metals like Al, Cu and Ni are preferably cut by spark erosion or chemical means while hard materials may be cut by a diamond saw. Mechanical and electrolytic (or chemical) polishing is applied to obtain an undistorted flat surface. These specimens have to be cleaned in the diffraction system. Commonly used methods are:

>Heating
>Ion bombardment, annealing
>Chemical reaction

The appropriate technique depends again on the material. Wolfram surfaces, for instance, may be cleaned by heating alone since segregated carbon can be flashed off. Carbon segregated on, for example, Ni surfaces has to be removed by ion bombardment or chemical reaction.

The definition of clean surface conditions (unless cleavage is used) is an important problem. A combination of the diffraction observations and a sensitive spectroscopic technique (e.g. grazing angle Auger electron spectroscopy) will certainly give the best diagnostics of surface order and purity. Actually the diffraction intensities are very sensitive to small amounts of ordered or disordered foreign species on the surface but a relevant initial definition is difficult to obtain by other means than those discussed.

3.2. Diffraction patterns and surface structure models for clean surfaces

In this section the surface structure models for some metal and semiconductor surfaces are discussed. A great number of materials have been investigated (see the bibliography of Ref. [16]) and a few illustrative examples are presented here.

(a) <u>Metals</u>

The main aim has in many cases been to produce suitable substrates for adsorption studies.

In some cases a sequence of low-index surfaces have been investigated like (111), (100) and (110) of fcc metals (e.g. Cu, Ni, Pt) and (110), (100), (111) and (112) of bcc metals (e.g. W).

Some materials have been investigated with particular emphasis on obtaining diffraction intensities useful for structure analysis.

<u>For most clean metal surfaces the lateral periodicity has been found to be identical to that of a parallel plane in the bulk.</u> This conclusion is in general based on the observation that the diffraction pattern does not contain any other diffraction spots than those expected from the ideal periodicity. Figure 14 shows diffraction patterns from the clean low-index (111) and (100) surfaces of Cu and we recognize hexagonal and square symmetries as expected. The clean low-index (110), (100) and (211) surfaces of W yield the patterns of Fig. 15, and they are also characteristic of ideal lateral

LEED

diffraction pattern
net

Cu(111)

Cu(100)

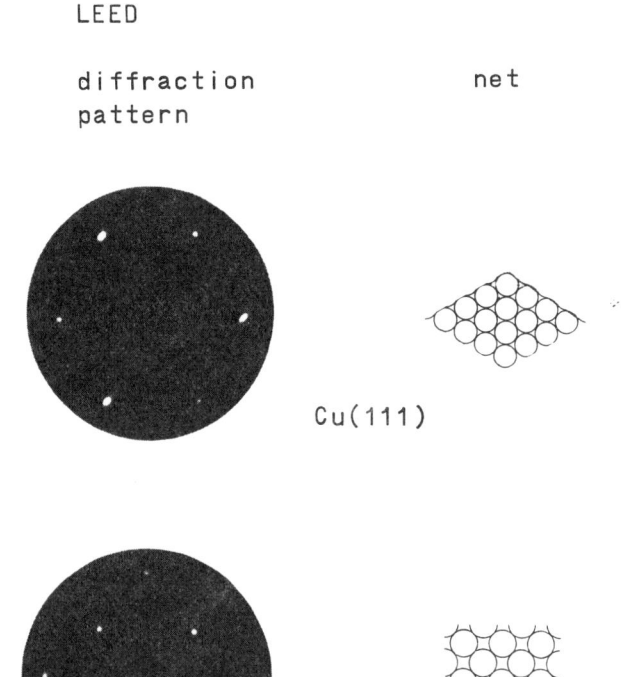

FIG. 14. Diffraction patterns and direct net models for Cu(111) and Cu(100) surfaces (LEED).

periodicity. It is of course possible to determine the net dimensions from the diffraction pattern as long as the spot positions on the photographs have been calibrated in terms of diffraction angle θ'. From (2.8) and Fig.11 we have (at normal incidence)

$$\vec{k}'_{\parallel} = \vec{g}_{hk}$$

or

$$\frac{2\pi}{\lambda} \sin \theta' = \frac{2\pi}{d_{(hk)}}$$

$$d_{(hk)} \sin \theta' = \lambda$$

(3.1)

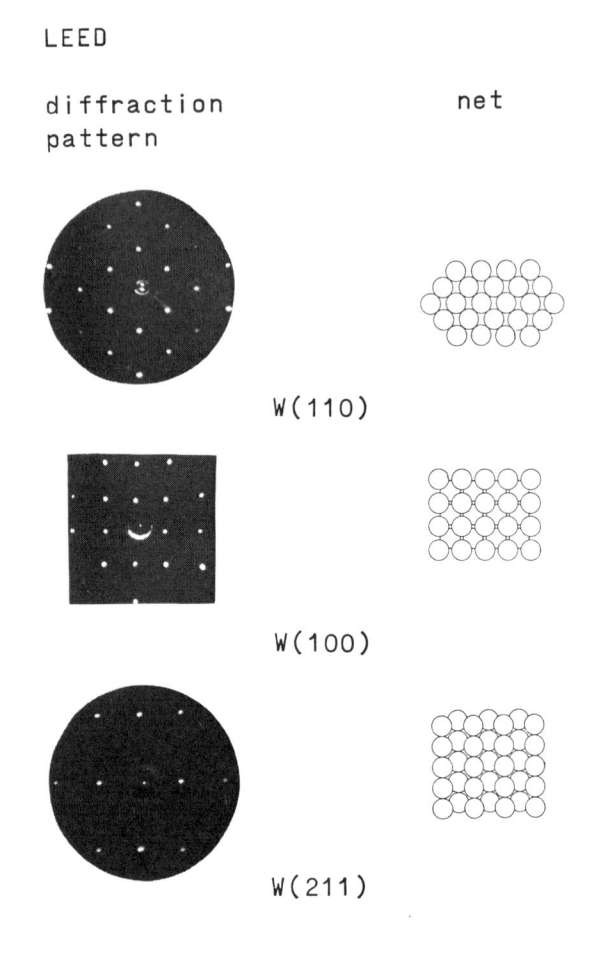

FIG. 15. Diffraction patterns and direct net models for W(110), W(100) and W(211) surfaces (LEED).

This is called the plane grating formula. Making use of (2.3) we get

$$E = V - V_c = \frac{150.4}{d_{(hk)}^2} \frac{1}{\sin^2\theta}$$

where V is the accelerating potential and V_c is the contact potential between cathode and crystal. Figure 16 shows the experimental relation V versus $1/\sin^2\theta$ for Ni(100). The observed spacing is (2.46 ± 0.03) Å to be compared with the X-ray value 2.49 Å. The accuracy is typically 1% provided stray electric and magnetic fields are well cancelled.

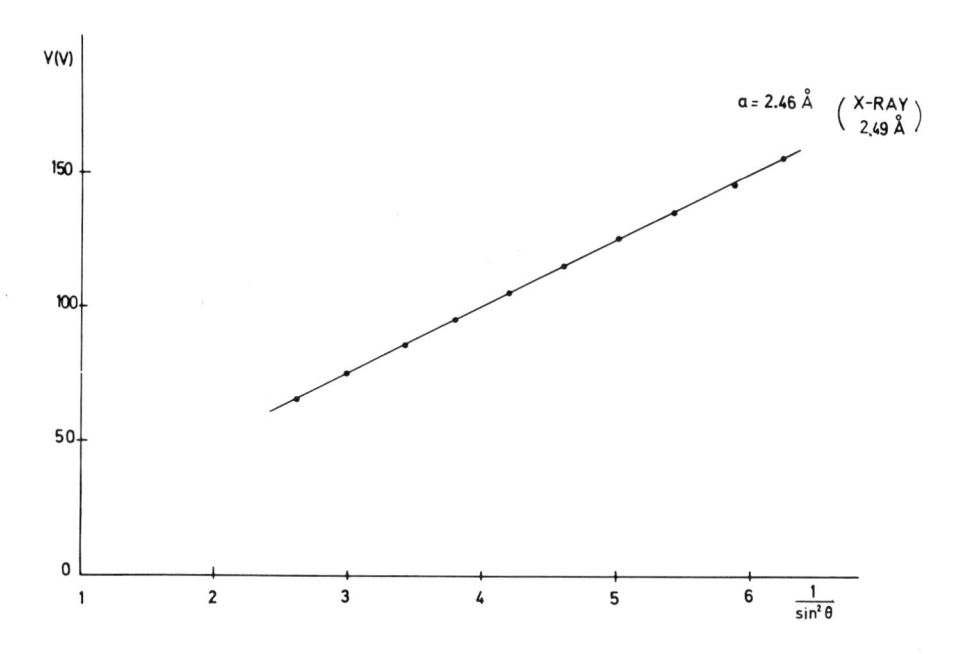

FIG. 16. Surface net spacing determination for Ni(100). Electron potential V versus $1/\sin^2\theta$. (θ is diffraction angle.)

Clean surfaces of Pt, Au, Ir and Pd have been thoroughly investigated since persistent diffraction features indicating surface re-arrangement are observed. The (100) faces of Pt [17] Au [18, 19] and Ir [20] yield diffraction patterns indicating a (5 × 1) unit mesh. Such a pattern from Au(100) is shown in Fig. 17 together with a model structure assuming a hexagonal arrangement of Au atoms superpositioned coincidently on the underlying (100) substrate. It should be pointed out that these structures have been suggested as due to contamination (O_2) [21].

So far we have only considered the two-dimensional diffraction conditions. In Section 2.3 we noticed that any periodicity normal to the surface should show up in the diffraction intensity. The specularly reflected (00) beam (see Fig. 11) ought to be particularly sensitive to such a periodicity. From (2.23) we get

$$2 \frac{2\pi}{\lambda} \cos\theta \, a_3 = 2\pi h_3$$
$$2a_3 \cos\theta = h_3 \lambda \tag{3.2}$$

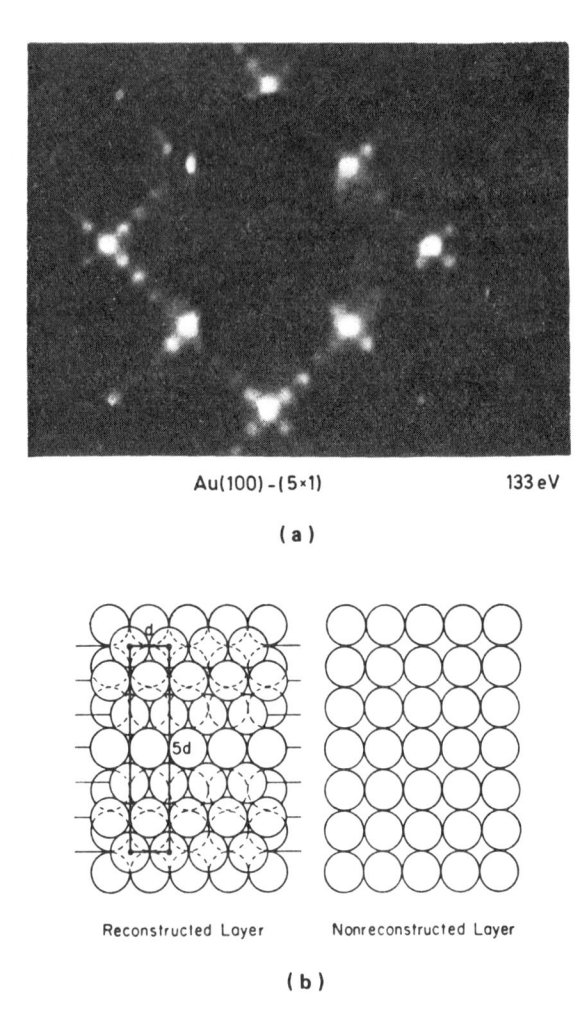

FIG.17. (a) Diffraction pattern from Au(100) - (5×1); (b) Model structure of coincidental hexagonal Au layer superpositioned on the underlying (100) substrate (after Palmberg and Rhodin [19]).

which is simply the Bragg law. Rewriting (3.2) we get

$$E_0 = h_3^2 \frac{150,4}{4a_3^2 \cos^2\theta} - U_0 \qquad (3.3)$$

where U_0 is the inner potential and E_0 is the electron energy relative to the vacuum level. The specular beam intensity I_{00} from Ni(100) and Cu(100) [22] is shown in Fig.18. The data contain a number of peaks; the bulk

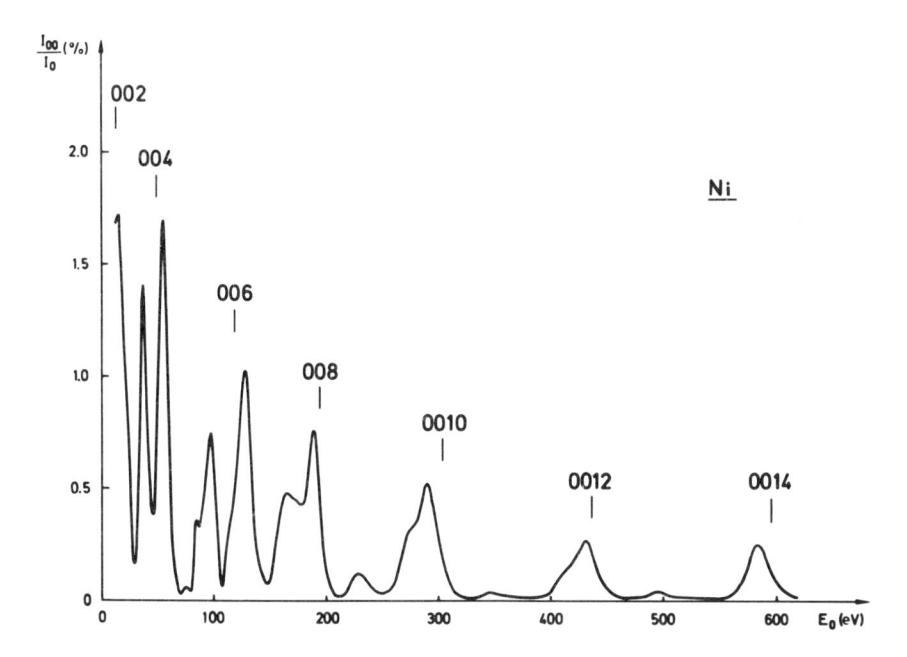

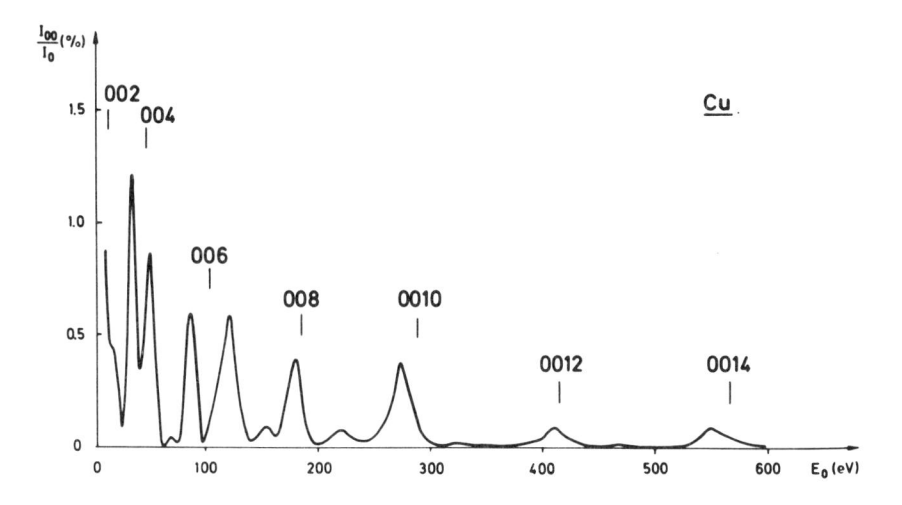

FIG. 18. Specular beam intensity I_{00} from Ni(100) and Cu(100) versus electron energy, E_0 (angle of incidence $\theta \approx 3°$). The bars denote the kinematical Bragg conditions 001. (From Ref. [22].)

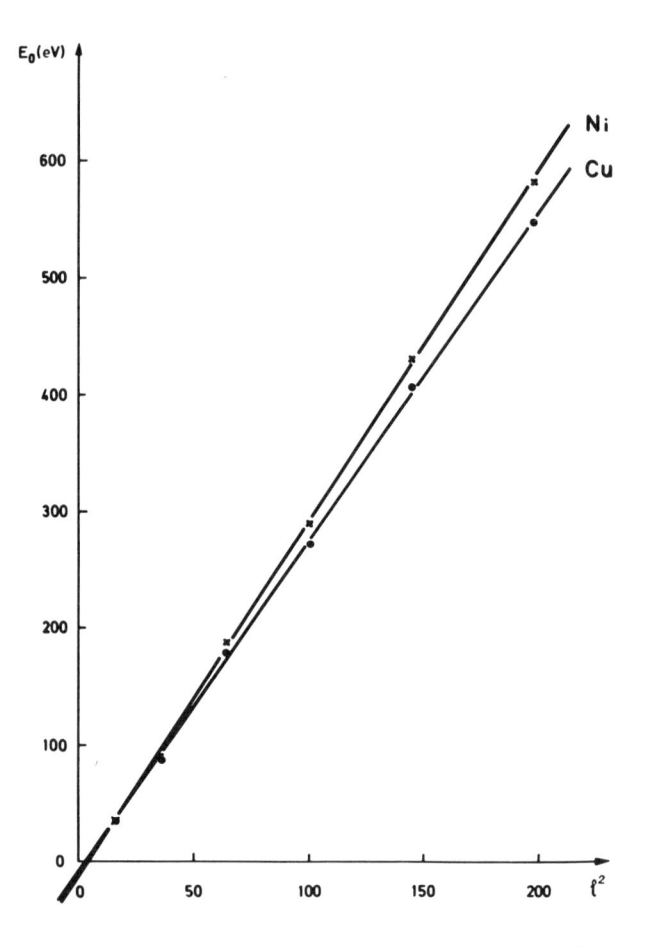

FIG. 19. Position in energy of peaks in I_{00} from Cu(100) and Ni(100) plotted versus ℓ^2. The solid lines are least square fits to the experimental points. (After Ref. [22].)

Bragg conditions are denoted by bars. Fitting these experimental peak positions (Fig. 19) to (3.3) yields a_3 and U_0:

$a_3 = 3.62$ Å , $U_0 = -11.5$ eV for Cu(100)

$a_3 = 3.51$ Å , $U_0 = -13.8$ eV for Ni(100)

The error in a_3 due to scattering in peak positions is about 1%. The X-ray values are 3.61 Å and 3.52 Å for Cu and Ni, respectively, which tells us that there is no appreciable distortion in the plane spacing. Similar results have been obtained for other metals.

A more complete diffraction intensity analysis has been carried out for some low-index surfaces of Al, Be, Zn, Cu, Ni, Ag and W. The conventional way of doing this analysis is to calculate the diffraction intensities of some low-index diffraction beams for some range of diffraction parameters (e.g. normal incidence, $0 < E < 200$ eV). The atomic scattering and the penetration of the wave field have to be calculated in a physically correct way. The adjustable parameters are the atomic positions, and usually one calculates for the rigid bulk position of all atoms except for the first layer which one allows to shift around the ideal bulk position. In general, it has been found that the bulk positions yield very good agreement with experimental data and that a deviation of more than 5% in the first layer spacing gives a less good agreement (Al seems to be an exception). Figure 20 illustrates this kind of comparision for Cu(100). It is obvious that peak positions as well as relative peak heights are well reproduced. The calculation [23] assumes bulk atomic positions.

(b) Semiconductors

Low-index surfaces of semiconductors and semimetals have frequently been found to yield diffraction patterns different from what is expected for a parallel plane in the bulk. Tentative surface structure models have been inferred from the diffraction patterns and from bonding considerations. These materials are characterized by directed bonds, and the change in co-ordination number at the surface may leave unfilled electron orbitals. Saturation of these orbitals is believed to occur by more or less dramatic atomic re-arrangements in the surface. Unfortunately, there has not yet been any reliable diffraction intensity analysis that could support the proposed models.

The freshly room-temperature cleaved (111) surfaces (the cleavage plane) of these materials yield rather simple diffraction patterns characterized by 'half-order spots' (see Ref. [14]). The pattern from Si(111) together with an index scheme is shown in Fig.21. The pattern characterizes a (2×1) unit mesh, referring to the (1×1) hexagonal unit mesh of the ideal layer. Lander et al. [14] used a rectangular mesh and denoted the structure $(1 \times \sqrt{3})$. In the (111) surface of the diamond lattice there will be one 'dangling' bond per surface atom (see Fig.21) and the saturation of these bonds via double bonds between neighbours yields the surface structure model in Fig.21. This model gives a (2×1) unit mesh which is compatible with the observed diffraction pattern. Upon heat treatment, these surface structures transform irreversibly:

Si(111) (2×1) → Si(111) (7×7)

Ge(111) (2×1) → Ge(111) (2×8)

The transition temperatures are about 700°C for Si and 300°C for Ge. These are also the observed structures when Si(111) and Ge(111) surfaces are cleaned by ion-bombardment annealing cycles. Occasionally other structures have been observed upon heat treatment and some of them have conclusively been shown to be related to contamination (e.g. Ni). The diffraction pattern from the Si(111) (7×7) surface structure is shown in Fig.22. These structures are probably characteristic for clean surface conditions although doubts have been raised. Several different structure models have been proposed to explain the unit mesh sizes (in particular for Si(111) (7×7)).

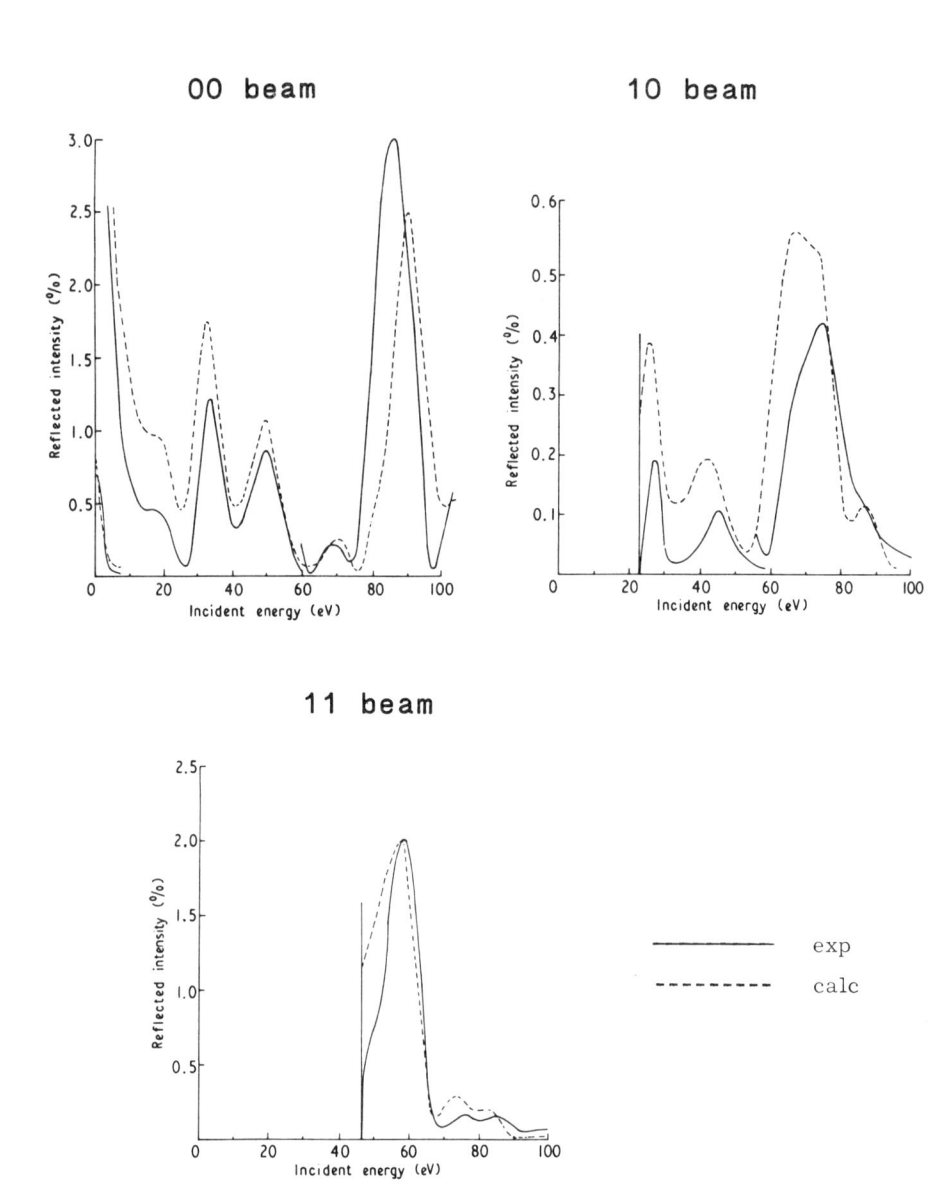

FIG. 20. Comparison of experimental and theoretical LEED beam intensities for Cu(100) (after Pendry [23]).

IDEAL SURFACE STRUCTURES 107

Si(111) - (2×1)
diffraction
pattern

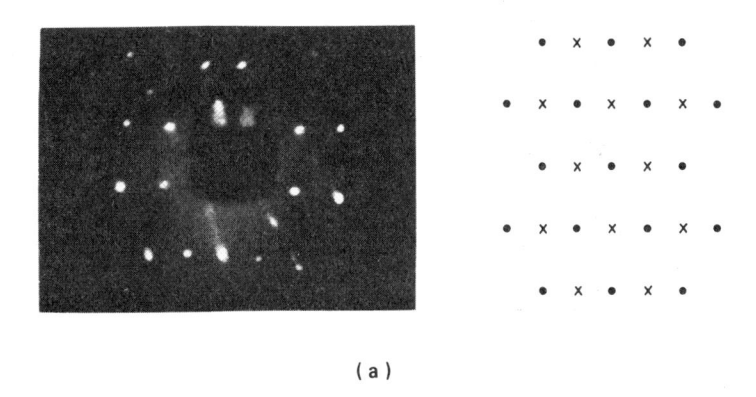

(a)

model

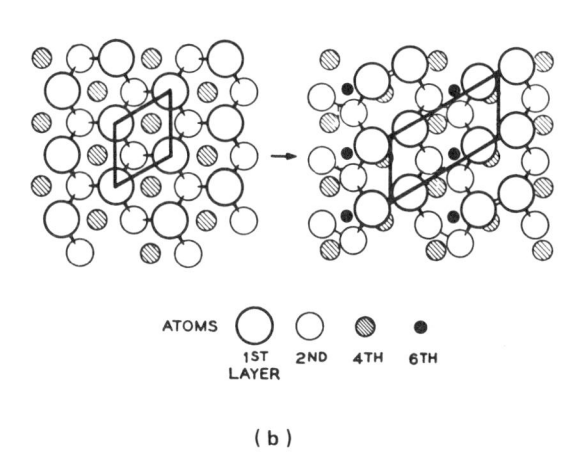

ATOMS ○ ○ ◍ ●
       1ST 2ND 4TH 6TH
       LAYER

(b)

FIG. 21. (a) Diffraction pattern and index scheme for cleaved Si(111); (b) Model structure for Si(111) - (2×1) (after Lander et al. [14]).

Si(111) - (7×7)
diffraction
pattern

model

(a)

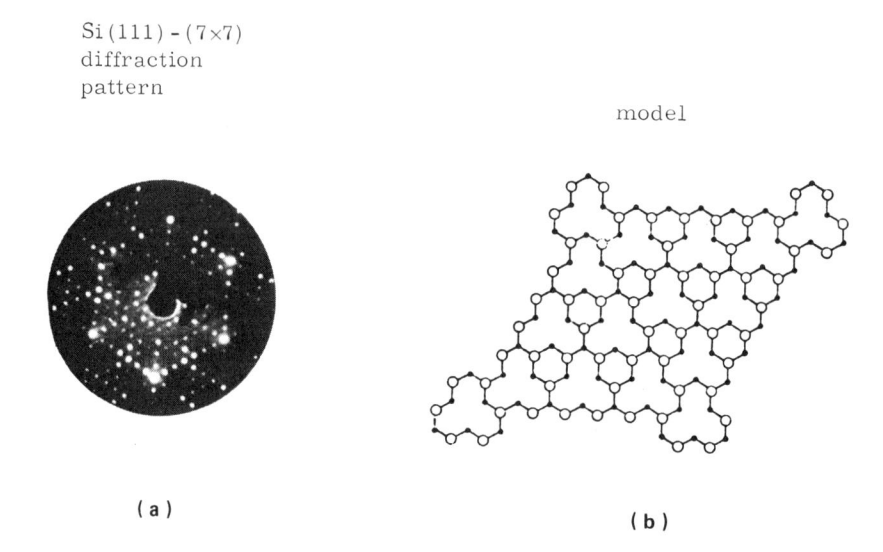

(b)

FIG. 22. (a) Diffraction pattern from the Si(111) - (7×7) surface structure (LEED); (b) Warped-benzene ring model structure for Si(111) - (7×7) (after Lander [4]).

Si(100) - (4×2)
diffraction
pattern

model

(a)

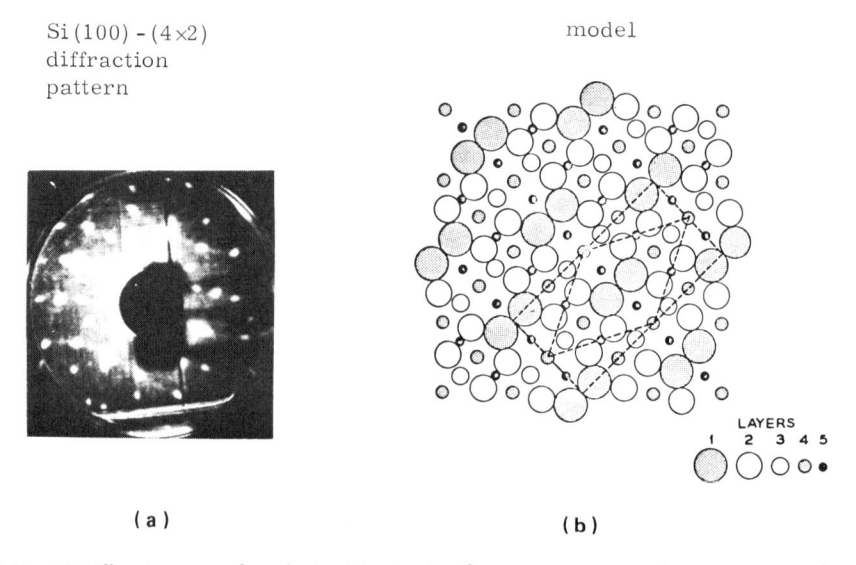

(b)

FIG. 23. (a) Diffraction pattern from the Si(100) - (2×4) surface structure (LEED); (b) Model structure for Si(100) - (2×4) (after Lander [4]).

Te(10$\bar{1}$0)
diffraction
pattern

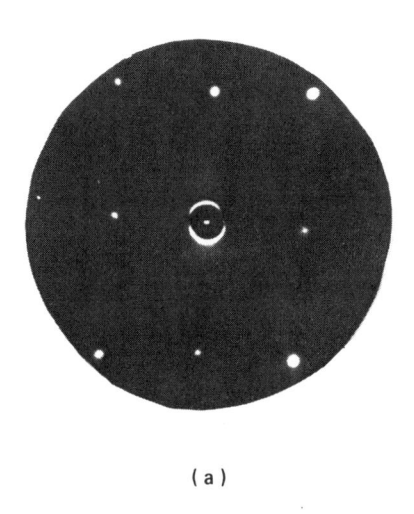

(a)

model

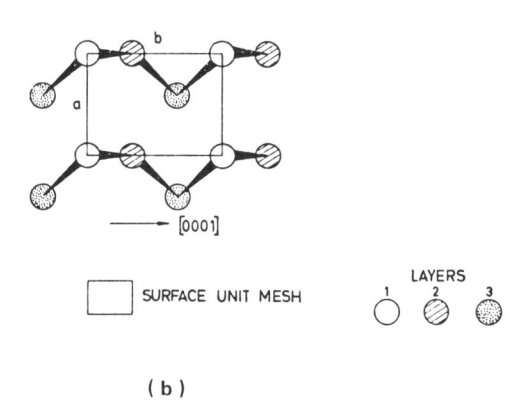

[0001]

SURFACE UNIT MESH

LAYERS
1　2　3

(b)

FIG. 24. (a) Diffraction pattern from Te(10$\bar{1}$0) (LEED). (b) Te(10$\bar{1}$0) model structure (bulk plane).

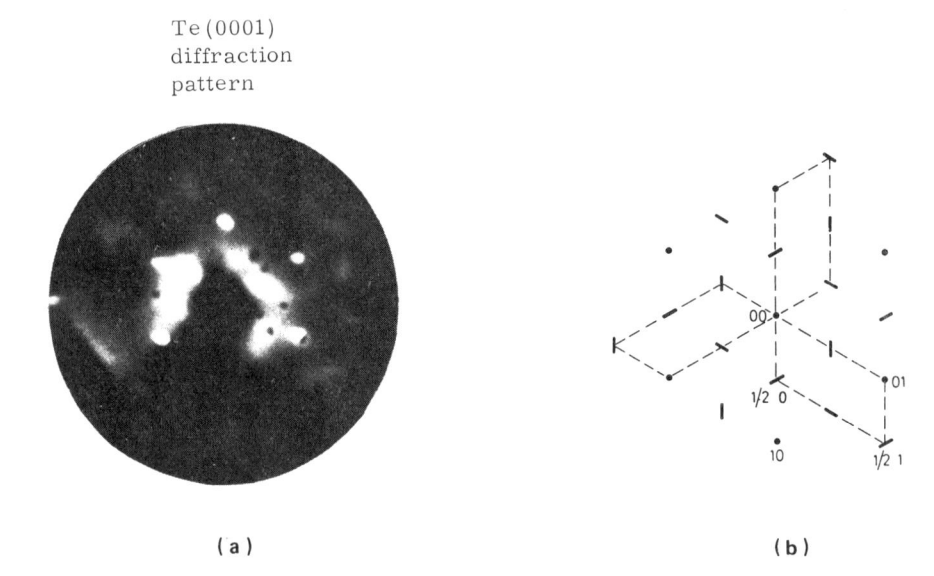

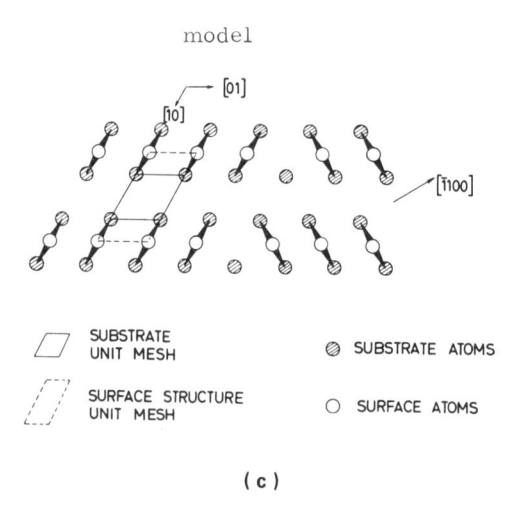

FIG. 25. (a) Diffraction pattern from Te(0001) (LEED). (b) Index scheme showing the elongated half-orders. The three 120° rotated reciprocal unit meshes are sketched. (c) Model structure for Te(0001) - (2×1).

Lander et al. [10] have proposed a 'warped-benzene-like' ring as a structural unit to explain the stable (111) surface structures of Si, Ge and C (diamond). Their model is shown in Fig. 22. Such a structure involves considerable atomic re-arrangement. The observed transition temperatures could be compatible with the activation energies for such re-arrangements. In fact, fast annealing of lattice disorder introduced by ion bombardment occurs at about the temperatures stated.

The (100) faces of Si and Ge cannot be produced by cleavage but have to be cleaned by ion bombardment and annealing and can therefore be expected to correspond to a relaxed atomic arrangement (like the annealed (111) planes). The diffraction pattern from Si(100) which is shown in Fig. 23 is characterized by $\frac{1}{4}$-order spots. Ordering and symmetry suggests a (2×4) direct unit mesh [24] (provided the surface is clean). The same result is obtained for Ge [25]. Thus tentative structure notations are Si(100) (2×4) and Ge(100) (2×4). Bonding considerations have lead to the model shown in Fig. 23.

Te has a hexagonal lattice constructed from helical chains. The cleavage Te($10\bar{1}0$) surface produces a diffraction pattern (Fig. 24) compatible with ideal termination of the bulk crystal (see Ref. [15]). This is expected since no covalent bonds are 'broken'. The hexagonal Te(0001) surface is cleaned by means of ion bombardment. The diffraction pattern (Fig. 25) reveals the direct unit mesh (2×1). There is one 'broken' covalent bond per surface atom in the Te(0001) face; a similar double bonding model to what was put forward for both Si(111) and Ge(111) will explain the Te(0001) (2×1) surface structure.

A great number of compound semiconductors such as III-V compounds (GaAs, GaSb, InAs, InSb) and II-VI compounds (PbTe, PbSe, CdS) have also been investigated. In particular, the cleavage surfaces of the III-V compounds [26] and II-VI compounds exhibit diffraction patterns compatible with the lateral periodicity of a corresponding bulk plane.

REFERENCES

[1] WOOD, E.A., J. Appl. Phys. 35 (1964) 1306.
[2] BÜRGER, M.J., Elementary Crystallography, Wiley, New York (1963).
[3] PARK, R.L., MADDEN, H.H., Surf. Sci. 8 (1967) 426.
[4] LANDER, J.J., in Progress in Solid State Chemistry 2, Pergamon, New York (1965) 26.
[5] BAUER, E., in Techniques for Metals Research 2, Wiley, New York (1969) Pt. 2, Ch. 16.
[6] BAUER, E., ibid, Ch. 15.
[7] SOMORJAI, G.A., FARRELL, H.H., Adv. Chem. Phys. 20 (1971) 215.
[8] HEIDENREICH, R.D., Fundamentals of Transmission Electron Microscopy, Interscience, New York (1964) 104.
[9] PARK, R.L., J. Appl. Phys. 37 (1966) 295.
[10] LANDER, J.J., Surf. Sci. 1 (1964) 125.
[11] ESTRUP, P.J., McRAE, E.G., Surf. Sci. 25 (1971) 1.
[12] BAKER, J.M., Ph.D. Thesis, Cornell University (1970).
[13] JONA, F., Surf. Sci. 8 (1967) 57.
[14] LANDER, J.J., GOBELI, G.W., MORRISON, J., J. Appl. Phys. 34 (1963) 2298.
[15] ANDERSSON, S., ANDERSSON, D., MARKLUND, I., Surf. Sci. 12 (1968) 284.
[16] SOMORJAI, G.A., Surf. Sci. 34 (1973) 156.
[17] LYON, H.B., Jr., SOMORJAI, G.A., J. Chem. Phys. 46 (1967) 2539.
[18] FEDAK, D.G., GJOSTEIN, N.A., Surf. Sci. 8 (1967) 77.

[19] PALMBERG, P.W., RHODIN, T.N., Phys. Rev. 161 (1967) 586.
[20] GRANT, J.T., Surf. Sci. 18 (1969) 288.
[21] GRANT, J.T., HAAS, T.W., Surf. Sci. 18 (1969) 457.
[22] ANDERSSON, S., KASEMO, B., Surf. Sci. 25 (1971) 273.
[23] PENDRY, J.B., J. Phys. C 4 (1971) 3095.
[24] LANDER, J.J., MORRISON, J., J. Appl. Phys. 33 (1962) 2089.
[25] LANDER, J.J., MORRISON, J., J. Appl. Phys. 34 (1963) 1403.
[26] MacRAE, A.U., GOBELI, G.W., J. Appl. Phys. 35 (1964) 1629.

SURFACE WAVES

S. ANDERSSON
Department of Physics,
Chalmers University of Technology,
Göteborg, Sweden

Abstract

SURFACE WAVES.
1. Surface waves of a semi-infinite elastic medium. 2. Long-wavelength optical surface modes in ionic crystals. 3. Surface phonons.

INTRODUCTION

In this paper we shall consider surface waves, i.e. excitations of the surface system, characterized by a wave vector $\vec{q}_\parallel$ for propagation parallel to the surface. The envelope of the amplitude of the excitation decreases with increasing depth into the crystal. In crystals of finite thickness, mixed surface modes consisting of both oscillatory and decaying waves may exist within the energy range of the corresponding bulk modes. Otherwise in the simplest cases the energy of the surface excitations lies outside or within 'gaps' in the bulk bands. There are several types of excitation obeying these properties, e.g. surface lattice waves, surface electron states, surface plasmons, etc. We shall exclusively consider surface lattice waves, i.e. vibrations of the surface atoms round their equilibrium positions.

Three types of surface lattice waves may be distinguished:
(a) Long-wavelength elastic surface waves;
(b) Long-wavelength optical surface waves in ionic crystals; and
(c) Short-wavelength acoustic and optical surface waves controlled by the short-range force constants. Short-wavelength surface waves depend upon microscopic features such a crystallographic orientation of the surface and changes in the short-range interactions at and near the surface.

1. SURFACE WAVES OF A SEMI-INFINITE ELASTIC MEDIUM

Theoretical work on surface waves in solids extends back to the work of Lord Rayleigh [1] who studied an isotropic elastic medium with a planar free surface. The elastic surface waves he found, Rayleigh waves, propagate in the surface and decay exponentially into the solid. The vibrational amplitude is in the sagittal plane (Fig.1) defined by the propagation vector $\vec{q}_\parallel$ and the normal to the surface.

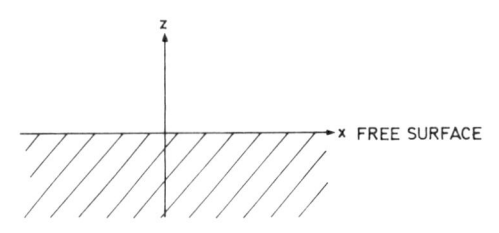

FIG. 1. Sagittal plane.

We shall need some general laws and concepts from the theory of elasticity of isotropic media [2]. The general equation of motion is

$$\ddot{u} - c^2 \nabla^2 u = 0 \tag{1.1}$$

u is any component of the vectors $\vec{u}_\ell$ and $\vec{u}_t$ for longitudinal and transverse displacements. The corresponding velocities are c_ℓ and c_t.

The following conditions are obeyed by $\vec{u}_t$ and $\vec{u}_\ell$:

$$\text{div } \vec{u}_t = 0 \tag{1.2}$$

$$\text{curl } \vec{u}_\ell = 0 \tag{1.3}$$

If $\vec{P}$ is the external force per unit area, we have for the stress tensor σ_{ik}

$$\sigma_{ik} \cdot n_k = P_i \tag{1.4}$$

where n_k is a component of a unit vector normal to the surface. At a free surface we thus have the boundary conditions:

$$\sigma_{xz} = \sigma_{yz} = \sigma_{zz} = 0 \tag{1.5}$$

i.e. the normal components of the stress vanish. The strain tensor u_{ik} is given by

$$u_{ik} = \tfrac{1}{2}\left(\frac{\partial u_i}{\partial x_k} + \frac{\partial u_k}{\partial x_i} \right) \tag{1.6}$$

In particular we have

$$\left.\begin{aligned}
u_{xz} &= \frac{1+\sigma}{E}\, \sigma_{xz} = 0 \\
u_{yz} &= \frac{1+\sigma}{E}\, \sigma_{yz} = 0 \\
\sigma_{zz} &= \frac{E}{(1+\sigma)(1-2\sigma)}\left[(1-\sigma)u_{zz} + \sigma(u_{xx}+u_{yy}) \right] = 0
\end{aligned}\right\} \tag{1.7}$$

where σ is Poisson's ratio and E is Young's modulus.

For the Rayleigh wave we now assume a plane wave solution to (1.1), propagating parallel to the surface in the x direction and at a frequency ω.

$$u = \exp[i(q_{\|} \cdot x - \omega t)] \cdot f(z) \tag{1.8}$$

Thus

$$\nabla^2 u = \frac{d^2 u}{dx^2} + \frac{d^2 u}{dz^2} = \exp[i(q_{\|} \cdot x - \omega t)] \left(-q_{\|}^2 f(z) + \frac{d^2 f}{dz^2} \right)$$

$$\ddot{u} = -\omega^2 \exp[i(q_{\|} \cdot x - \omega t)] f(z)$$

and we get the equation for $f(z)$:

$$\frac{d^2 f}{dz^2} - \left(q_{\|}^2 - \frac{\omega^2}{c^2} \right) f = 0 \tag{1.9}$$

If

$$q_{\|}^2 - \frac{\omega^2}{c^2} < 0 \tag{1.10}$$

we get a periodic function, i.e. we obtain an ordinary plane wave undamped inside the solid:

$$f(z) = A_1 e^{i\mathcal{H}z} + A_2 e^{-i\mathcal{H}z} \tag{1.11}$$

for

$$q_{\|}^2 - \frac{\omega^2}{c^2} > 0 \tag{1.12}$$

$$f(z) = B e^{\mathcal{H}z} \tag{1.13}$$

Actually we have two solutions for $\mathcal{H}$:

$$\mathcal{H} = \pm \left[q_{\|}^2 - \frac{\omega^2}{c^2} \right]^{\frac{1}{2}} \tag{1.14}$$

but the minus sign would correspond to an unlimited increase in displacement for $z \to \infty$. Thus

$$u = B \exp[i(q_{\|} x - \omega t)] e^{\mathcal{H}z} \tag{1.15}$$

i.e. a wave that propagates along the surface and whose amplitude decays exponentially into the solid. The damping is determined by the quantity $\mathcal{H}$.

The true displacement $\vec{u}$, which is a linear combination of $\vec{u}_t$ and $\vec{u}_\ell$, is determined from the boundary conditions.

From (1.6) and (1.7) we have

$$u_{yz} = \tfrac{1}{2}\left(\frac{\partial u_y}{\partial z} + \frac{\partial u_z}{\partial y}\right) = \tfrac{1}{2}\frac{\partial u_y}{\partial z} = 0 \tag{1.16}$$

because all quantities are independent of y. From (1.15) we then get

$$u_y = 0 \tag{1.17}$$

The displacement vector $\vec{u}$ is thus confined to the sagittal plane.
Considering the components of $\vec{u}$, we have from (1.2) and (1.3)

$$\frac{\partial u_{tx}}{\partial x} + \frac{\partial u_{tz}}{\partial z} = 0$$

$$\frac{\partial u_{\ell x}}{\partial z} - \frac{\partial u_{\ell z}}{\partial x} = 0$$

Consider u_{tx} and u_{tz}. From (1.15) and $\mathcal{H}_t = \left[q_{\|}^2 - \frac{\omega^2}{c_t^2}\right]^{\frac{1}{2}}$ we get

$$iq_{\|}\, u_{tx} + \mathcal{H}_t\, u_{tz} = 0$$

and hence

$$u_{tx} = a\,\mathcal{H}_t\, \exp(iq_{\|}x + \mathcal{H}_t z - i\omega t) \tag{1.18}$$

$$u_{tz} = -a\,iq_{\|}\, \exp(iq_{\|}x + \mathcal{H}_t z - i\omega t) \tag{1.19}$$

where a is a constant.
Similarly for $u_{\ell x}$ and $u_{\ell z}$:

$$iq_{\|}\, u_{\ell z} - \mathcal{H}_\ell\, u_{\ell x} = 0$$

$$\mathcal{H}_\ell = \left[q_{\|}^2 - \frac{\omega^2}{c_\ell^2}\right]^{\frac{1}{2}}$$

This gives

$$u_{\ell x} = bq_{\|}\, \exp(iq_{\|}x + \mathcal{H}_\ell z - i\omega t) \tag{1.20}$$

$$u_{\ell z} = -bi\,\mathcal{H}_\ell\, \exp(iq_{\|}x + \mathcal{H}_\ell z - i\omega t) \tag{1.21}$$

We can rewrite the first and third condition of (1.7) (utilizing $c_t/c_\ell = [(1-2\sigma)/2(1-\sigma)]^{1/2}$) as:

$$\frac{\partial u_x}{\partial z} + \frac{\partial u_z}{\partial x} = 0 \tag{1.22}$$

$$c_\ell^2 \frac{\partial u_z}{\partial z} + (c_\ell^2 - 2c_t^2) \frac{\partial u_x}{\partial x} = 0 \tag{1.23}$$

Substituting u_x and u_z as

$$u_x = u_{\ell x} + u_{tx}$$

$$u_z = u_{\ell z} + u_{tz}$$

gives

$$a(q_\parallel^2 + \mathcal{H}_t^2) + 2bq_\parallel \mathcal{H}_\ell = 0 \tag{1.24}$$

$$2ac_t^2 \mathcal{H}_t q_\parallel + b[c_\ell^2(\mathcal{H}_\ell^2 - q_\parallel^2) + 2c_t^2 q_\parallel^2] = 0 \tag{1.25}$$

Divide the last equation by c_t^2 and substitute

$$\mathcal{H}_\ell^2 - q_\parallel^2 = -\frac{\omega^2}{c_\ell^2} = -(q_\parallel^2 - \mathcal{H}_t^2)\frac{c_t^2}{c_\ell^2} \tag{1.26}$$

This gives

$$2a\mathcal{H}_t q_\parallel + b(k^2 + \mathcal{H}_t^2) = 0 \tag{1.27}$$

Combining (1.24) and (1.27) and substituting $\mathcal{H}_t^2 = q_\parallel^2 - \omega^2/c_t^2$, $\mathcal{H}_\ell^2 = q_\parallel^2 - \omega^2/c_\ell^2$,

$$\left(2q_\parallel^2 - \frac{\omega^2}{c_t^2}\right) = 16q_\parallel^4 \left(q_\parallel^2 - \frac{\omega^2}{c_t^2}\right)\left(q_\parallel^2 - \frac{\omega^2}{c_\ell^2}\right) \tag{1.28}$$

This equation yields the dispersion relation, i.e. the relation between ω and $q_\parallel$. Put

$$\omega = \xi \cdot c_t \cdot q_\parallel \tag{1.29}$$

into (1.28):

$$\xi^6 - 8\xi^4 + 8\xi^2\left(3 - 2\frac{c_t^2}{c_\ell^2}\right) - 16\left(1 - \frac{c_t^2}{c_\ell^2}\right) = 0 \tag{1.30}$$

It is obvious that ξ only depends on c_t/c_ℓ, which is a constant for a given isotropic material and just depends on Poisson's ratio σ. Equation (1.29) tells us that the Rayleigh wave has a linear dispersion relation just like an elastic wave in the volume (it belongs to the acoustic branch of the phonon spectrum). The proportionality constant is the velocity of propagation:

$$c_R = \xi \cdot c_t$$

$$\omega = c_R \cdot q_\parallel \qquad (1.31)$$

Since

$$q_\parallel^2 - \frac{\omega^2}{c^2} > 0$$

we have

$$\frac{\omega^2}{c_R^2} - \frac{\omega^2}{c^2} > 0$$

and hence

$$c_R < c \qquad (1.32)$$

It is obvious that $0 < \xi < 1$. Equation (1.30) has only one root that satisfies this condition for any given value of c_t/c_ℓ. The ratio c_t/c_ℓ varies from $1/\sqrt{2}$ to 0 for real materials ($0 < \sigma < \frac{1}{2}$) and ξ is then in the range 0.87-0.95. We have seen that surface waves exist for an isotropic solid. Anisotropic solids of different symmetries have also been found to carry surface waves [3(a),4] although not necessarily in all directions.

Experimentally the surface waves can be excited by converting volume waves into surface waves by means of, for example, a quartz transducer. Figure 2 shows a wedge transducer. A longitudinal wave is transformed into a surface wave. The angle α is dimensioned so that the projection of the wave vector of the longitudinal wave on the surface is equal to the wave vector of the surface wave, i.e.

$$q_\parallel = k_\ell \cos(90-\alpha)$$

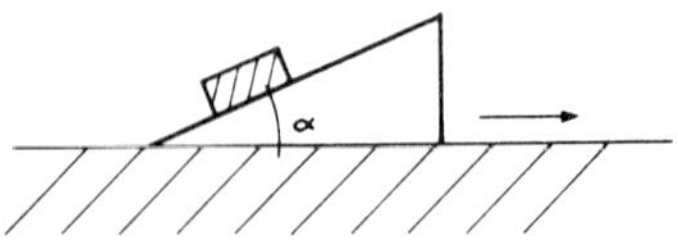

FIG. 2. Wedge transducer.

Piezoelectric materials are of particular practical importance since the surface waves can be generated electrically via an interdigital transducer. Such devices are used, for example, as signal delay lines and they are characterized by low damping compared to similar delay lines utilizing volume waves. In piezoelectric materials the elastic waves are usually accompanied by piezoelectric surface waves. Consider surface waves propagating in the x direction of a semi-infinite piezoelectric crystal with z direction normal to the surface. The solution to the stress equation of motion and Maxwell's wave equation can be expressed by [5]

$$U_j = A_j \sum_n^3 a_{jn} \exp(q_\| \Omega_n z) \exp[i(q_\| x - \omega t)] \qquad (j = x,y,z)$$

$$E_j = B_j \sum_n^3 b_{jn} \exp(q_\| \Omega_n z) \exp[i(q_\| x - \omega t)] \qquad (n = 1,2,3)$$

where U_j are the particle displacements, E_j are the electric fields, $q_\|$ is the wave vector, Ω_n are the dimensionless decay constants, and A_j, B_j, a_{jn} and b_{jn} are amplitude constants. For the case of a free surface the continuity of tangential electrical field and normal electric displacement must be satisfied at the surface:

$$E_x(z=0+) = E_x(z=0-)$$
$$E_y(z=0+) = E_y(z=0-)$$
$$D_z(z=0+) = \epsilon_0 E_z(z=0-)$$

Furthermore, the normal stresses vanish at the surface:

$$\sigma_{xz} = \sigma_{yz} = \sigma_{zz} = 0$$

Under these boundary conditions, surface waves are found to exist on the basal plane of hexagonal crystals. As an example, Table I shows velocity, decay constant, amplitude constant, as calculated for surface waves in CdS ($k \perp$ c axis) [6]. The surface waves are pure Rayleigh waves with real decay constants and the displacements are confined to the sagittal plane.

Figure 3 shows the displacements and piezoelectric fields versus depth for surface elastic waves on CdS [5]. At a frequency of 8 MHz, the wavelength is $1.73 \times 10^3 / 8 \times 10^6$ m = 218 μm. The displacements and field decay roughly to zero within this distance (calculated). Figure 4 shows an experiment that proves the change in velocity when the waves pass a region of thin metallic coating ($\approx$ 1000 Å). Since there is a velocity change (due to changes in the electric boundary conditions), reflection will occur and can be detected in a Bragg reflection arrangement, $2d \sin\theta = \lambda$ [6].

TABLE I. CALCULATED VELOCITY,
DECAY CONSTANT AND AMPLITUDE
CONSTANT FOR SURFACE WAVES IN CdS
($c \perp k$) (from Ref.[6]).

	Free surface	Metallized surface
V (10^5 cm/sec)	1.731	1.730
Ω_1	1.6549	1.6556
Ω_2	0.1939	0.1996
Ω_3	0.7474	0.7469
A_x	A_0[a]	A_0
a_{x1}	1.0	1.0
a_{x2}	−0.5014	−0.5846
a_{x3}	0.5229	0.7047
A_z	$-i0.3793 A_0$	$-i0.3792 A_0$
a_{z1}	1.0	1.0
a_{z2}	−7.644	−8.662
a_{z3}	1.640	2.215
B_x (10^{10} V/m)	$-0.8516 k A_0$[b]	$-0.851 k A_0$
b_{x1}	1.0	1.0
b_{x2}	1.073	1.230
b_{x3}	−1.652	−2.230
B_z (10^{10} V/m)	$i1.409 k A_0$	$i1.409 k A_0$
b_{z1}	1.0	1.0
b_{z2}	0.1255	0.1483
b_{z3}	−0.7465	−1.005

[a] A_0 is an arbitrary constant with dimension of particle displacement.
[b] k is a wavenumber. The same unit must be used for the wavelength ($2\pi/k$) and A_0, so that kA_0 is a dimensionless constant. Displacement and electric field normal to sagittal plane are zero, i.e., $A_y = B_y = 0$.

2. LONG-WAVELENGTH OPTICAL SURFACE MODES IN IONIC CRYSTALS

Long-wavelength optical surface modes in ionic crystals involve electromagnetic fields and must satisfy Maxwell's equations. The optical surface modes have been discussed by, e.g., Fuchs and Kliewer [7], in a continuum model. Microscopic models for slabs consisting of a small number of atomic layers have been discussed by a number of authors.

A brief presentation of the continuum models will serve to illustrate many of the qualitative features. Cubic diatomic polar crystals of the NaCl, CsCl or ZnS structure are all characterized by two long-wavelength optical phonon frequencies ω_{TO} and ω_{LO} of transverse and longitudinal vibrations.

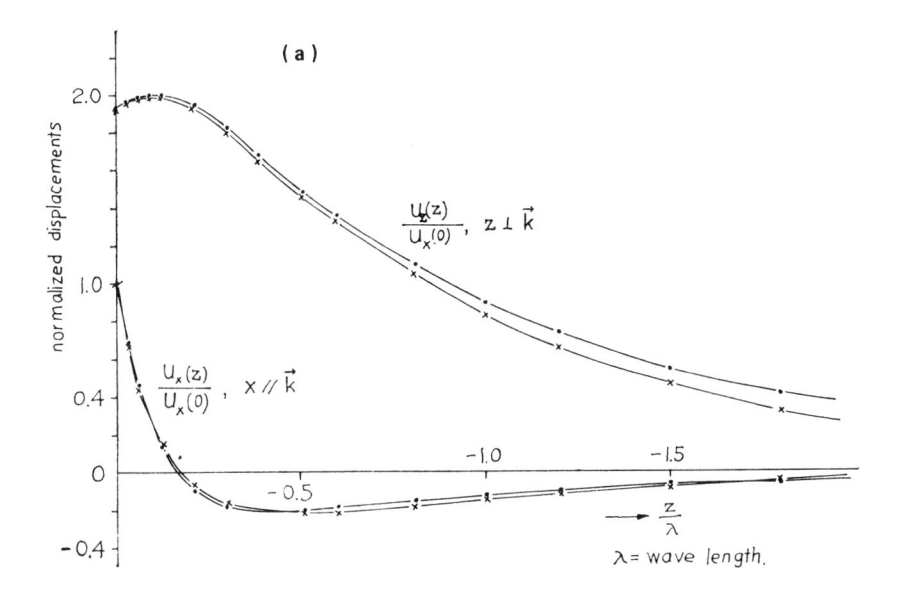

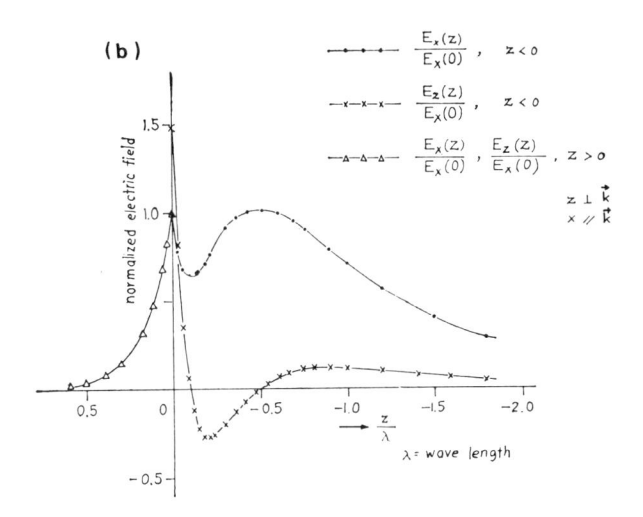

FIG. 3. (a) Normalized particle displacement versus depth for surface wave propagating on the basal plane of CdS. – • – • – : based on Ref. [3(b)]; – x – x – : based on Section III of Ref. [5].
(b) Normalized piezoelectric field versus depth for surface wave propagating on the basal plane of CdS. (From Tseng and White [5].)

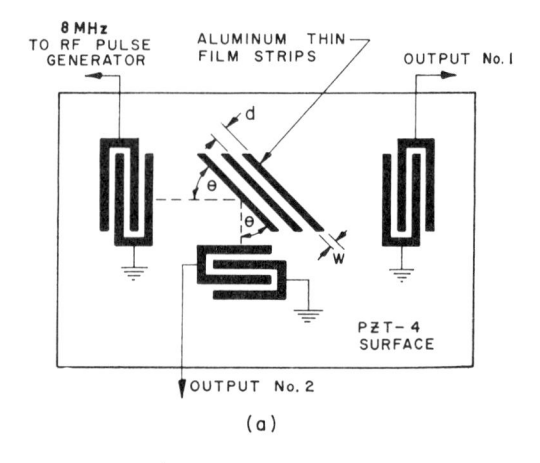

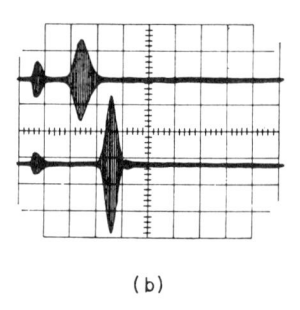

FIG. 4. (a) Arrangement for Bragg reflection of surface waves by metallic thin-film grating. (b) Oscilloscope traces for reflection and transmitted waves. Upper trace for reflected wave received at transducer 2, lower trace for transmitted wave received at transducer 1; vertical scale 0.1 V/cm, horizontal scale 5 μs/cm. (From Tseng [6].)

From the Lyddane-Sachs-Teller relation we know the splitting of these modes:

$$\left(\frac{\omega_{LO}}{\omega_{TO}}\right)^2 = \frac{\epsilon_0}{\epsilon_\infty} \tag{2.1}$$

where ϵ_0 and ϵ_∞ are the static and the high-frequency dielectric constants, respectively. The frequency-dependent dielectric function has the form (see e.g. Kittel [8]):

$$\epsilon(\omega) = \epsilon_\infty + \frac{\epsilon_0 - \epsilon_\infty}{1 - \omega^2/\omega_{TO}^2} \tag{2.2}$$

Obviously $\epsilon(\omega) = 0$ at ω_{LO}.

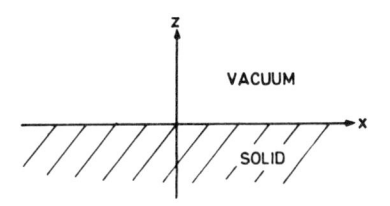

FIG. 5. Semi-infinite crystal bounded by vacuum ($\epsilon_V = 1$).

The condition for the so-called Fuchs-Kliewer surface phonon may be derived as follows. Consider (Fig.5) a semi-infinite crystal bounded by vacuum ($\epsilon_V = 1$).

Since there are no singular charges on the surface we have for the potential ϕ

$$\nabla^2 \phi = 0 \tag{2.3}$$

The electric field obeys

$$\vec{E} = -\nabla \phi \tag{2.4}$$

The harmonic functions from which $\vec{E}$ has to be generated must have the form:

$$\phi(+) = A e^{-kz} e^{ikx} \quad z > 0$$
$$\phi(-) = B e^{kz} e^{ikx} \quad z < 0 \tag{2.5}$$

At the boundary $z = 0$, the tangential components of the electric field $\vec{E}$ and the normal component of the displacement field $\vec{D}$ have to be continuous:

$$E_x(\pm) = -\frac{\partial \phi(\pm)}{\partial x} = -ik\phi \quad z > 0, z < 0 \tag{2.6}$$

$$E_z(+) = -\frac{\partial \phi(+)}{\partial z} = k\phi \quad z > 0$$
$$E_z(-) = -\frac{\partial \phi(-)}{\partial z} = -k\phi \quad z < 0 \tag{2.7}$$

$$D_z(+) = \epsilon_V E_z(+) \quad z > 0$$
$$D_z(-) = \epsilon E_z(-) \quad z < 0 \tag{2.8}$$

Thus we require

$$E_x(+) = E_x(-) \quad , \quad D_z(+) = D_z(-) \tag{2.9}$$

which give the relations

$$-ikA = -ikB \quad , \quad \epsilon_V kA = -\epsilon kB$$

and combining these we get

$$\epsilon = -\epsilon_V = -1 \tag{2.10}$$

Together with $\epsilon(\omega)$ from Eq.(2.2) we get

$$\left(\frac{\omega_s}{\omega_{TO}}\right)^2 = \frac{\epsilon_0 + 1}{\epsilon_\infty + 1} \tag{2.11}$$

for the frequency of the F-K optical surface phonon. Notice that this relation is similar to (2.1).

These surface phonons have been observed in both optical and electron reflection experiments and examples of both techniques will be presented.

Optical experiments on polar semiconductors, alkali halides etc. have been performed by the attenuated total reflection method (ATR). The optical arrangement is shown in Fig.6. The technique utilizes the coupling between the evanescent wave at the totally reflecting prism surface and the surface wave in the sample. The sample and prism have to be brought within a distance of the order of the decay length of the evanescent wave. The frequency and wave vector have to be matched according to the dispersion relation for the surface wave. It is important that the k vector k_s of the evanescent wave is increased compared to the wave vector $k_0 = \omega/c$ in vacuum:

$$k_s = n_p(\omega/c)\sin\alpha$$

where

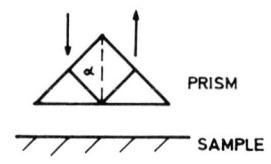

$$1/n_p < \sin\alpha < 1$$

FIG.6. Schematic optical arrangement for attenuated total reflection.

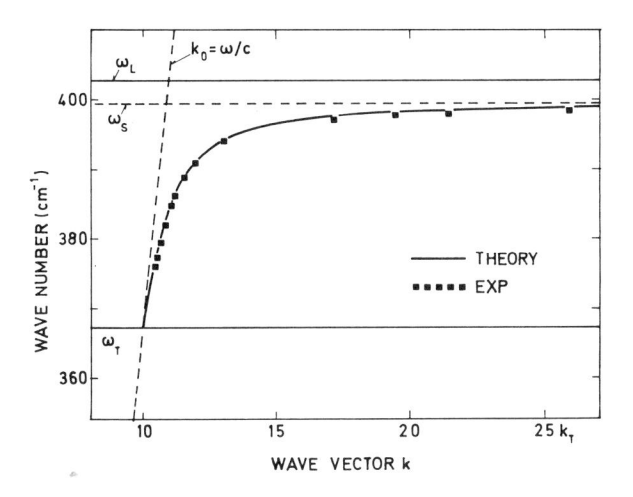

FIG. 7. Dispersion of surface polaritons in GaP (from Ref. [9]).

Thus the range of k_s is

$$\frac{\omega}{c} < k_s < n_p \frac{\omega}{c}$$

Experimentally one observes the excitation as a reduction of the internally reflected intensity which is lowered around the resonance frequency. Figure 7 shows the dispersion relation for the surface excitation in GaP [9]. The solid curves are dispersion relations for the excitation. In the long-wavelength region the surface eigenstate is a mixed type of a photon and a surface phonon. These modes are called surface polaritons in analogy with bulk polaritons. The F-K surface phonons have been observed by Ibach [10] in inelastic scattering of low-energy electrons on ZnO. The electron spectrometer is shown in Fig. 8. The monochromator and the analyser are cylinder condensors of 119° deflection angle and have an optimal resolution of ≈10 meV which is required in phonon scattering experiments. An experimental energy loss spectrum from the ZnO(1$\bar{1}$00) surface is shown in Fig. 9. Notice that there are single and double losses observable and also a gain. The quantum energy is 68.8 ± 0.5 meV, while 69 meV is calculated from the energy loss function $-\text{Im} 1/(\epsilon(\omega) + 1)$ (which has a strong maximum when $\text{Re}\,\epsilon(\omega) = -1$) using $\epsilon(\omega)$ determined from infra-red measurements on ZnO. The infra-red active optical bulk modes $\hbar\omega^{TO}$ and $\hbar\omega^{LO}$ occur at the energies 50.3 meV and 73.1 meV. The half-width (Fig. 10) of the one-phonon loss is about 1.7 meV wider than the half-width of the elastic peak. The increase in half-width is due to damping, which is larger than one finds from calculated $-\text{Im} 1/(\epsilon(\omega) + 1)$ and may be due to surface contamination.

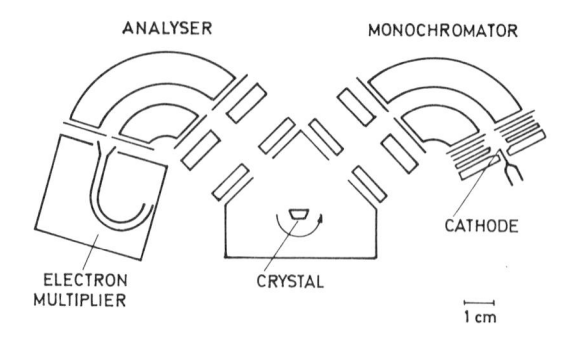

FIG. 8. Electron spectrometer with scattering angle fixed at 90° (from Ref. [11]).

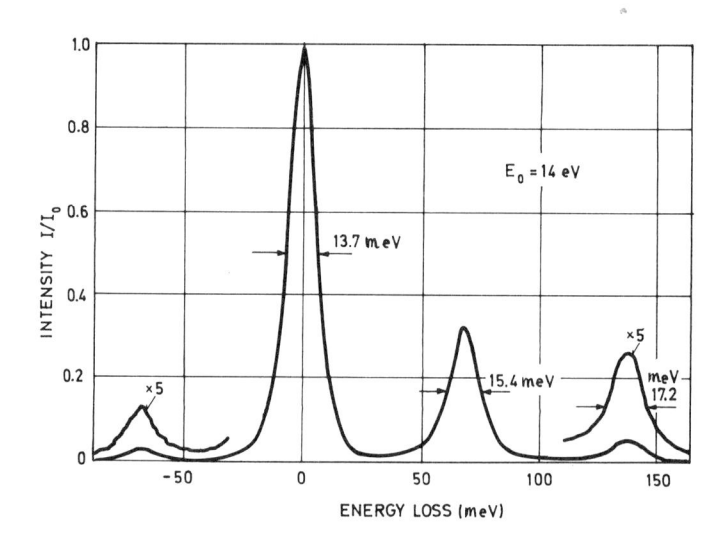

FIG. 9. Energy loss spectrum of 14 eV electrons specular-reflected from a (1̄100) surface of ZnO cleaved in ultrahigh vacuum. Angle of incidence is 45°. (From Ref. [11].)

At low temperatures the surface phonon is in its ground state and hence no energy gain is observed. At elevated temperatures the ratio of the one-phonon gain intensity I_{-1} and the one-phonon loss intensity I_{+1} is given by the Boltzmann factor:

$$\frac{I_{-1}}{I_{+1}} = \exp\left(-\frac{\hbar\omega_s}{k_B T}\right)$$

This ratio is also shown in Fig. 10 for $\hbar\omega_s = 68.8$ meV.

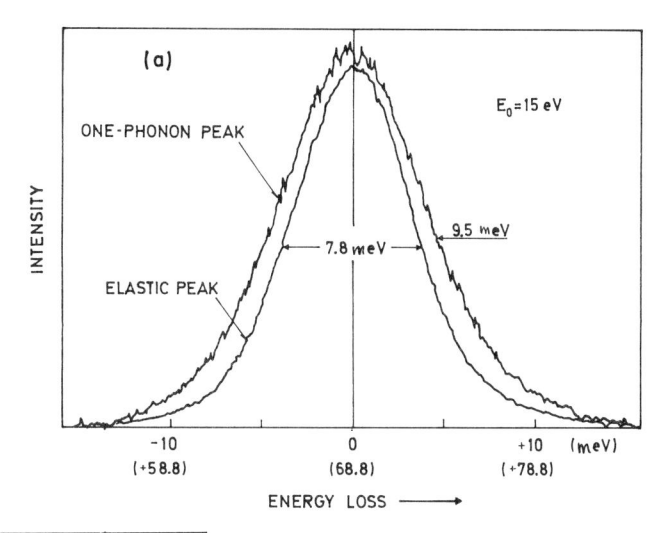

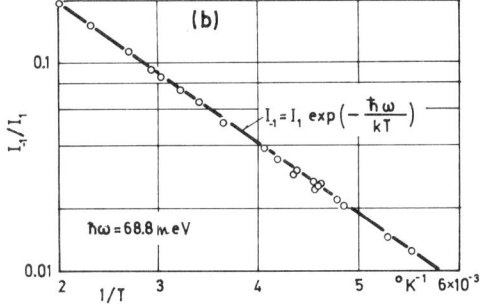

FIG.10. (a) ZnO spectrum with higher resolution. The greater half-width of the phonon loss is due to phonon damping. An excitation of the LO bulk is not observed. (b) Intensity ratio of one-phonon energy gain I_{-1} to one-phonon energy loss I_{+1} versus temperature. (c) Intensity versus angle of crystal rotation for 15 eV electrons at 45° angle of incidence. (From Ref.[11]).

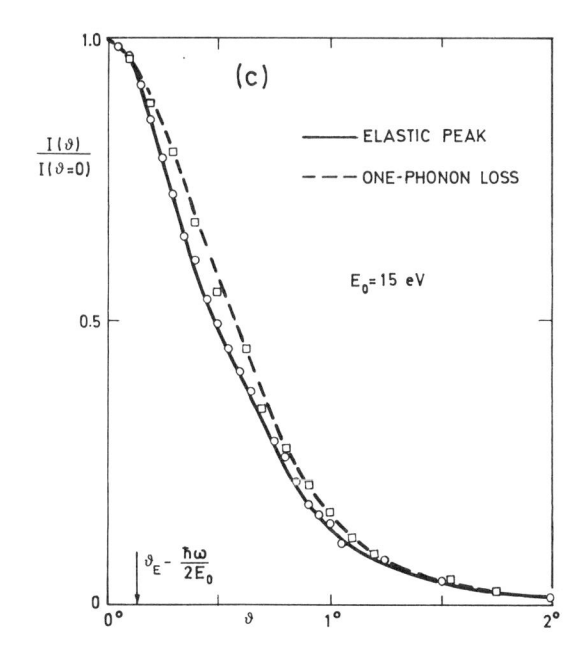

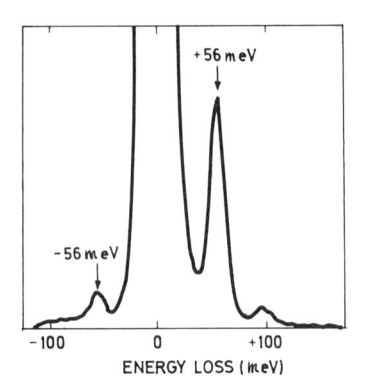

FIG.11. Energy loss spectrum of a cleaved (111) silicon surface in specular reflection (recorder trace). Impact energy is 3.6 eV. The [112] direction is normal to the plane of incidence. (From Ref.[11].)

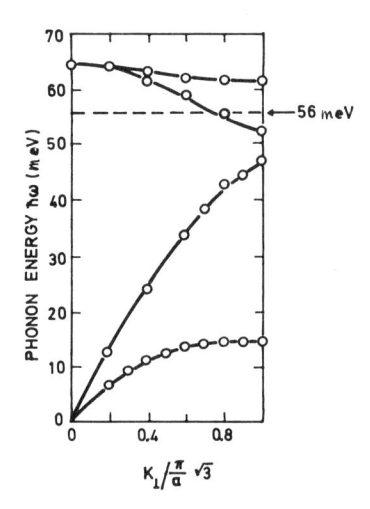

FIG.12. Phonon dispersion relations of bulk phonons with a wave vector perpendicular to a (111) surface (see Ref.[11]). Surface phonons are indicated by an arrow at 56 meV.

The excitation mechanism is dipole scattering, i.e. the electrons interact with the extended dipole field related to the surface vibrations. This means that the loss is just observable close to specular electron reflection (Fig.10). Another possible excitation mechanism is impact scattering, which means that the electrons are inelastically scattered from the vibrating localized atomic potential. Figure 10 shows that the intensity of the one-phonon loss peak follows the specular elastic peak closely in angular distribution. Thus only low-q phonons are excited. The phonon wavelength is estimated to be about 20a where a is the lattice constant.

Ibach [11] has also observed a phonon loss at 56 meV in electron scattering on the Si(111) surface (Figs 11,12). The nature of this phonon is not so evident. The excitation is believed to be a surface phonon since excitation of the LO bulk phonon would give a broad loss band from 57 to 65 meV (see Fig.11).

3. SURFACE PHONONS

So far we have only discussed continuum models. Numerous theoretical investigations of surface phonon spectra and associated thermodynamical functions are available, however, where discrete dynamical models are used. Most calculations are based on atomic force constant models. The simplest case that illustrates the use of these models is the finite linear chain. When the linear chain is used to illustrate bulk lattice waves, cyclic boundary conditions are used, i.e. the translational periodicity is expressed via periodic functions (Bloch waves). Let us first consider the infinite diatomic chain (Fig.13). The equations of motion are:

$$M\ddot{u}_{2j} = -\alpha\left[(u_{2j-1} - u_{2j}) - (u_{2j+1} - u_{2j})\right] \qquad (3.1)$$

$$m\ddot{u}_{2j-1} = -\alpha\left[(u_{2j-1} - u_{2j-2}) - (u_{2j} - u_{2j-1})\right] \qquad (3.2)$$

where u_{2j-1} and u_{2j} specify the light and heavy atoms m and M, respectively, and α is the force constant. The displacements obey the Bloch condition (infinite chain):

$$u_{2j} = U_1 e^{iq\ell} \qquad (3.3)$$

$$u_{2j-1} = U_2 e^{iq(\ell-a)} \qquad (3.4)$$

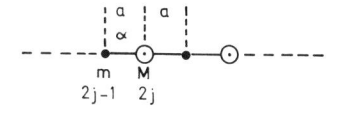

FIG.13. The infinite diatomic chain.

Combined with the equations of motion (3.1),(3.2) we get

$$M\ddot{U}_1 = -\alpha\left[2U_1 - U_2(e^{-iqa} + e^{iqa})\right]$$

$$m\ddot{U}_2 = -\alpha\left[2U_2 - U_1(e^{-iqa} + e^{iqa})\right]$$

The time dependence is given by $e^{i\omega t}$ and the solution for the angular frequency ω is found from the secular determinant:

$$\begin{vmatrix} 2\alpha - M\omega^2 & -2\alpha \cos qa \\ -2\alpha \cos qa & 2\alpha - m\omega^2 \end{vmatrix} = 0 \tag{3.5}$$

which has the roots

$$\omega_{\pm}^2 = \alpha \left(\frac{1}{m} + \frac{1}{M}\right) \pm \sqrt{\left(\frac{1}{M} + \frac{1}{m}\right)^2 - \frac{4\sin^2 qa}{mM}} \tag{3.6}$$

For small q we find

$$\omega_{-} \approx \sqrt{\frac{2\alpha}{m+M}} \; aq \tag{3.7}$$

which is called the acoustic mode. Notice that the Rayleigh wave had this kind of linear dispersion:

$$\omega_{+} \approx \sqrt{2\alpha \left(\frac{1}{m} + \frac{1}{M}\right)} \tag{3.8}$$

which is called the optical mode.

The actual dispersion in the Brillouin zone is sketched in Fig.14.

3.1. Finite diatomic chain

Let us now return to our surface problem. To obtain a free surface it is necessary to set to zero a certain number of the interatomic forces linking the atoms of the cyclic crystal. The creation of a surface in this manner results in a lower 'stiffness' of the lattice and therefore a lowering of some or all of the normal mode frequencies. Another consequence of the creation of a surface is that certain of the normal modes of vibration may become localized surface waves. Consider a chain with free ends constituted of 2N atoms of light masses m and heavy masses M. The force constant is α and only nearest-neighbour interactions are considered. We will follow a treatment due to Wallis [12]. The equations of motion can be written:

$$\left.\begin{aligned} m\ddot{u}_1 &= \alpha(u_2 - u_1) \\ M\ddot{u}_{2j} &= \alpha(u_{2j+1} + u_{2j-1} - 2u_{2j}) \quad & 1 \leq j \leq N-1 \\ m\ddot{u}_{2j-1} &= \alpha(u_{2j} + u_{2j-2} - 2u_{2j-1}) \quad & 2 \leq j \leq N \\ M\ddot{u}_{2N} &= \alpha(u_{2N-1} - u_{2N}) \end{aligned}\right\} \tag{3.9}$$

2j-1 and 2j specify the light and heavy atoms m and M, respectively.

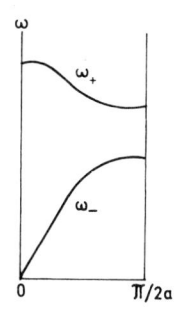

FIG.14. Vibration frequencies of diatomic chain in the Brillouin zone.

We may write the displacements associated with a normal mode of circular frequency ω for $q = 0$ as

$$u_{2j-1}(t) = U_{2j-1} e^{i\omega t} \qquad u_{2j}(t) = U_{2j} e^{i\omega t} \qquad (3.10)$$

where U_{2j-1} and U_{2j} are amplitudes independent of the time t. The equations of motion are thus transformed into a system of linear homogeneous algebraic equations. Let $g_m = \alpha/m$ and $g_M = \alpha/M$. The secular determinant can be written as

$$D_{2N}(u,v) = \begin{vmatrix} u+1 & 1 & & & & & \\ 1 & v & 1 & & & & \\ & 1 & u & \ddots & 1 & & \\ & & 1 & \ddots & v & 1 & \\ & & & 1 & u & 1 & \\ & & & & 1 & v+1 \end{vmatrix}_{2N} = 0$$

(3.11)

$$u = \frac{\omega^2}{g_m} - 2 \quad , \quad v = \frac{\omega^2}{g_M} - 2 \qquad (3.12)$$

The determinant D_{2N} can be evaluated if one introduces θ such that

$$(uv)^{1/2} = -2\cos\theta \qquad (3.13)$$

Thus

$$D_{2N} = (uv + u + v)\sin 2N\theta / \sin 2\theta \qquad (3.14)$$

The solutions to the secular equation can then be obtained as solutions to the equations:

$$\sin 2N\theta / \sin 2\theta = 0 \tag{3.15}$$

$$uv + u + v = 0 \tag{3.16}$$

Substituting (3.12) into (3.13) gives the relationship of ω and θ:

$$\omega_\pm^2 = \sigma \{1 \pm (\cos^2\theta + \epsilon^2 \sin^2\theta)^{\frac{1}{2}}\} \tag{3.17}$$

$$\sigma = g_m + g_M$$

$$\epsilon = \frac{M-m}{M+m}$$

The solutions of (3.15), (3.16) which yield independent values of ω are

$$\theta = \frac{n\pi}{2N} \qquad 1 \leq n \leq N-1 \tag{3.18}$$

Equations (3.17), (3.18) yield 2(N-1) frequency values. The solutions of (3.16) are

$$\omega^2 = 0 \tag{3.19}$$

$$\omega^2 = \sigma \tag{3.20}$$

The zero frequency corresponds to translation of the lattice. The frequency $\omega = \sqrt{\sigma}$ lies in the region between the acoustical and optical branches (see Fig.15). The corresponding value of θ is complex and is given by

$$\theta = \frac{\pi}{2} + i \sinh^{-1}\left(\frac{(M-m)^2}{4mM}\right)^{1/2} \tag{3.21}$$

(thus the mode is dashed in Fig.15).
The range of frequencies for the acoustical branch is ($\theta = 0 \to \pi/2$)

$$0 \leq \omega \leq \sqrt{\frac{2\alpha}{M}} \tag{3.22}$$

while that of the optical branch is

$$\sqrt{\frac{2\alpha}{m}} \leq \omega \leq \sqrt{\frac{2\alpha(m+M)}{mM}} \tag{3.23}$$

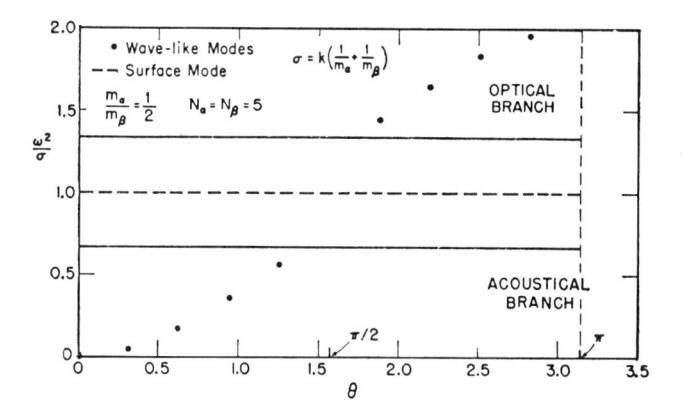

FIG.15. The square of the frequency as a function of θ when the end atoms are not alike (from Ref.[12]).

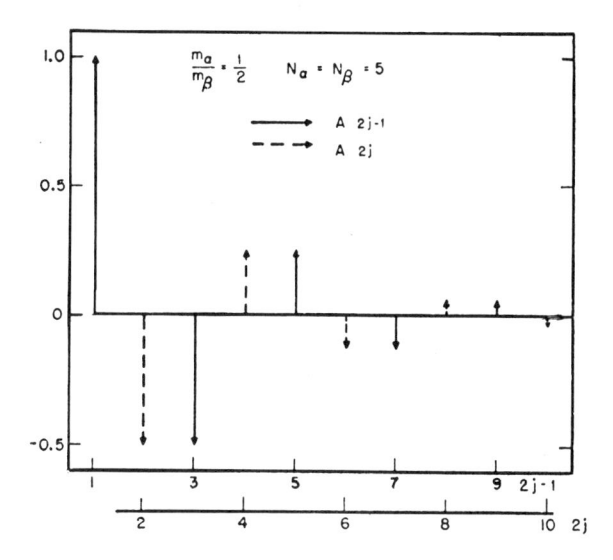

FIG.16. The maximum atomic displacement as a function of position in the lattice for the surface mode when the end atoms are not alike (from Ref.[12]).

The remaining normal mode has a frequency

$$\omega = \sqrt{\frac{\alpha(m+M)}{mM}} \tag{3.24}$$

which is in the gap.

This last mode can be considered as a surface mode, i.e. the maximum displacements of the atoms decrease roughly exponentially from the end having the lighter atom as shown in Fig.16 (notice that this is a graph; the actual displacements are parallel to the chain). When the masses are equal, the surface mode passes into a wave-like mode. Surface modes analogous to the mode just discussed should occur even in three-dimensional diatomic crystals. If the atoms of different masses have opposite electric charges a net dipole moment localized near the surface may be set up during the vibrations of the atoms in an excited surface mode. Such a mode can be expected to interact with, for example, electromagnetic radiation in the infra-red region.

3.2. Noble gas crystal slabs

So far we have discussed the lattice vibrations in terms of the force constant model with only nearest-neighbour interactions. Allen and De Wette [13] have used a potential formalism. The Lennard-Jones (LJ) potential they used describes well the pair interaction in noble gas crystals. The LJ potential makes it possible to perform first-principle calculations of surface properties without assuming, for example, that the force constants in the surface and bulk are the same!

They considered thin slab-shaped fcc crystals, the faces of which were (111), (100) and (110) planes, respectively. The number of atoms per plane is $N = N_1 \times N_2$, the number of planes is N_3 and each plane is labelled by an integer m. The atomic positions in a plane are specified by ℓ_1, ℓ_2 and the position of each plane by ℓ_3. Define the distance between the equilibrium positions of the atoms:

$$\vec{r}_0^{\ell\ell'} = \vec{\ell}(\ell_1, \ell_2, \ell_3) - \vec{\ell}'(\ell_1', \ell_2', \ell_3') \tag{3.25}$$

3.3. Static displacements

Since the model allows the calculation of the equilibrium plane positions, we shall look at the static displacements just for curiosity. These are calculated by minimizing the total energy of the crystal slab with respect to the position of the atomic planes. The static displacements δ_m are defined as:

$$d(1+\delta_1) \quad \text{———————} \quad m = 1$$
$$d(1+\delta_2) \quad \text{———————} \quad m = 2$$
$$\text{———————} \quad \ell_3 = z = 0$$

$$\delta_m = \frac{(z_m - z_{m+1})}{d} - 1$$

where d is the bulk interplanar spacing. The total static energy of the crystal is

$$\Phi = \tfrac{1}{2} N_1 \times N_2 \sum_{\ell_3 \ell_3'} \sum_{\ell_1 \ell_2} 4\epsilon \left[\left(\frac{\sigma}{r_0^{\ell\ell'}} \right)^{12} - \left(\frac{\sigma}{r_0^{\ell\ell'}} \right)^6 \right] \tag{3.26}$$

where ϵ and σ are the parameters of the LJ(6,12) potential:

$$\phi = 4\epsilon\left[\left(\frac{\sigma}{r}\right)^{12} - \left(\frac{\sigma}{r}\right)^{6}\right] \tag{3.27}$$

Φ is minimized with respect to δ_m:

$$\frac{\partial \Phi}{\partial \delta_m} = 0 \qquad\qquad m = 1, 2, \ldots \tag{3.28}$$

Numerically δ_m was calculated iteratively assuming $\delta_{m'} = 0$ for $m' > m$ in the first iteration. Slabs from 11 to 51 planes were calculated for and it was found that the thickness had only a small effect on the results for $N_3 > 30$. Figure 17 shows δ_m for (111), (100) and (110) surfaces. The δ_m's fall off roughly as

$$\delta_m \propto \frac{1}{m^3}$$

(except for δ_1 in the (100) and (111) case).

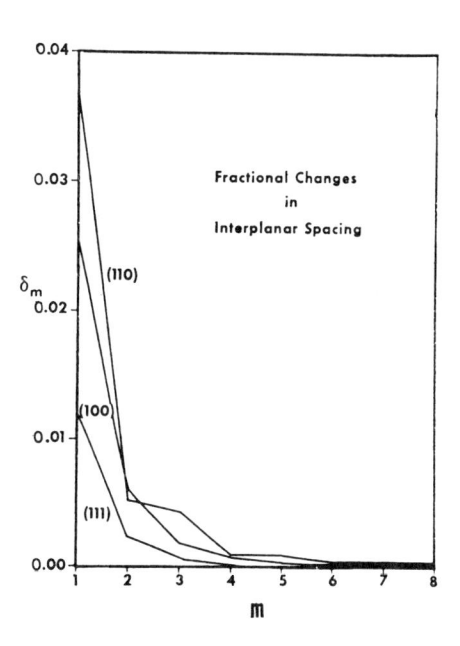

FIG. 17. Static displacements δ_m for (100), (111) and (110) surfaces expressed as fractional changes in the interplanar spacing (from Ref. [13]).

3.4. Dynamical calculations

When the atoms are vibrating, the vibrations obey two-dimensional periodic boundary conditions with respect to translations parallel to the surface. Let $u_\alpha(\vec{\ell})$ be the α component (α = x, y, z) of the displacement of the $\vec{\ell}^{th}$ atom from its mean position. The mean position is $x_0^\ell y_0^\ell z_0^\ell$ and the normal mode solutions have the form (notice two-dimensional periodicity):

$$u_\alpha(\vec{\ell}) = \frac{1}{\sqrt{M}} \xi_\alpha(\ell_3) e^{i\omega t} \exp[-i(q_x x_0^\ell + q_y y_0^\ell)] \quad (3.29)$$

where M is the atomic mass and $\xi_\alpha(\ell_3)$ are the components of the eigenvectors of the dynamical matrix. The eigenvalue equation is (see Ziman [15], p.30)

$$\sum_{\ell_3' \beta} D_{\alpha\beta}(\ell_3, \ell_3') \xi_\beta(\ell_3'; p) = \omega_p^2 \xi_\alpha(\ell_3; p) \quad (3.30)$$

where p = 1, 2,..., $3N_3$ labels the branches of ω for a given q = (q_x, q_y). The dynamical matrix is defined by $(\ell = (\ell_1, \ell_2); r_0^\ell = (x_0^\ell, y_0^\ell))$

$$D_{\alpha\beta}(\ell_3 \ell_3') = \frac{1}{M} \sum_{\ell'} \Phi_{\alpha\beta}(\vec{\ell}, \vec{\ell}') \exp[iq(r_0^\ell - r_0^{\ell'})] \quad (3.31)$$

with

$$\Phi_{\alpha\beta}(\vec{\ell}, \vec{\ell}') = \left(\frac{\partial^2 \Phi}{\partial u_\alpha(\vec{\ell}) \partial u_\beta(\vec{\ell}')}\right)_0 \quad (3.32)$$

The force constants $\Phi_{\alpha\beta}(\vec{\ell}, \vec{\ell}')$ are to be evaluated at the true mean positions of the atoms, which are indicated by the subscript 0 (thus including the static displacement). Notice the summation over ℓ', i.e. over planes parallel to the surface. The allowed q values may be taken to lie in the first two-dimensional Brillouin zone for a primitive net associated with such planes (Fig.18). The general solution for $u_\alpha(\vec{\ell})$ is a superposition of normal mode solutions:

$$u_\alpha(\vec{\ell}) = \sqrt{\frac{1}{NM}} \sum_{q,p} Q(qp) \xi_\alpha(\ell_3, qp) \exp(-iqr_0^\ell) \quad (3.33)$$

The specific heat at constant volume c_v is obtained from the frequencies $\omega_p(q)$ alone (Ziman [15], p.42):

$$c_v = \frac{1}{v} \frac{\partial \overline{E}}{\partial T}$$

$$\overline{E} = \sum_{qp} \exp\left(\frac{\hbar \omega_p(q)}{k_B T}\right) - 1$$

and

$$c_v = k_B \sum_{qp} \frac{x^2 e^x}{e^x - 1} \tag{3.34}$$

where $x = \hbar\omega_p(q)/k_B T$ and k_B is the Boltzmann constant. Using this expression, Allen and De Wette calculated the surface specific heat c_v^s in the following way: c_v is assumed to be broken up into a bulk contribution proportional to the total number of atoms NN_3 and a surface contribution proportional to the number of surface atoms $2N$. Thus

$$c_v = NN_3 c_v^b + 2N c_v^s \tag{3.35}$$

Using (3.35) and the calculated values of c_v for two different thicknesses, N_3 and N_3', yield c_v^s. Figure 19 shows c_v^s from two calculations. c_v^s was found to be almost independent of thickness and to show an expected T^2 behaviour at very low temperatures. Equation (3.34) is usually calculated

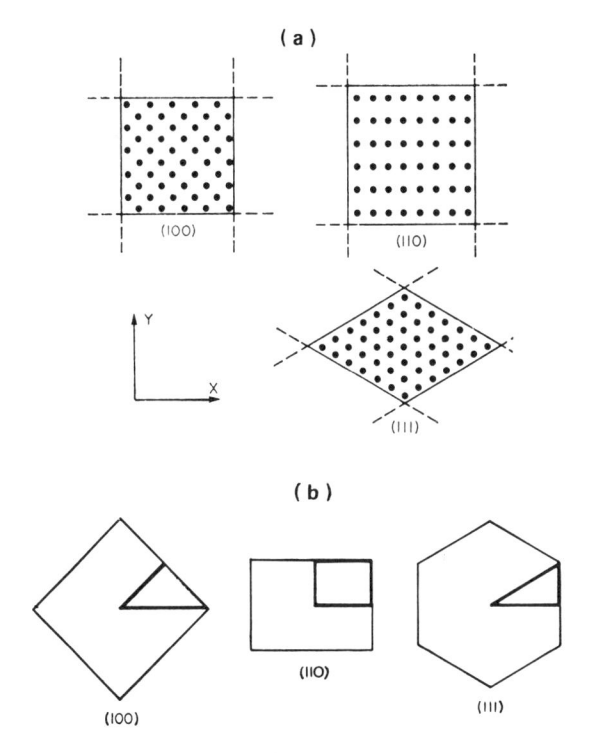

FIG. 18. (a) Structure of (100), (110) and (111) surfaces in an fcc crystal. Each dot represents an atom. Solid lines indicate the boundaries in the xy plane of the slab-shaped models used in the molecular dynamics calculations. (From Ref.[14].)
(b) Two-dimensional Brillouin zones associated with (100), (110) and (111) surfaces for fcc lattices (from Ref.[13]).

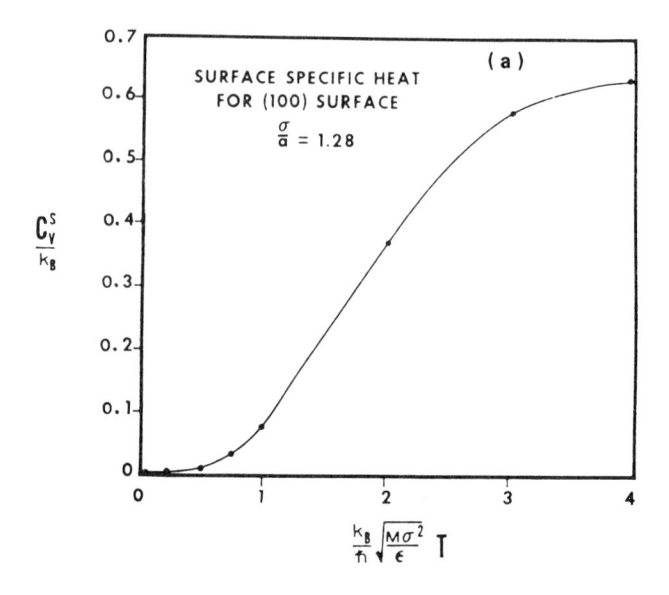

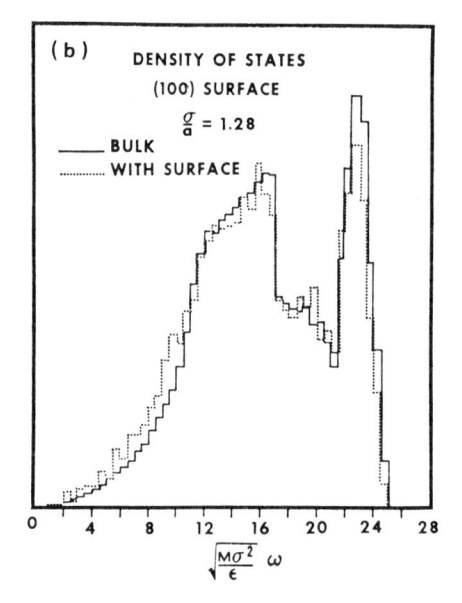

FIG. 19. (a) Surface specific heat C_V^S for (100) surface at $\sigma/a = 1.28$. The results for $N_3 = 11$, $N_3 = 21$ and for $N_3 = 21$, $N_3' = 31$ coincide to within the thickness of the line drawn.
(b) Density of vibrational states at $\sigma/a = 1.28$ in the bulk and for a slab-shaped crystal 21 layers thick with a (100) surface. (From Ref.[13].)

within the Debye approximation with a parabolic lattice spectrum. In this case the calculations are based on a lattice spectrum of the kind shown in Fig.19. In Fig.19 is also noticed the downward shift of some of the vibrational frequencies when a surface is present which is caused by missing and weakened interactions. Not only has the density of states been calculated but also the frequency versus wave-vector relations along symmetry directions [16]. An example is shown in Fig.20. There are three bulk bands corresponding roughly to one longitudinal and two transverse modes. These bands show gaps in some part of the Brillouin zone, and surface modes are found in the gaps. Notice the mode s_1 which comes out below the bulk bands at the centre of the zone with roughly linear dispersion. This is in accordance with our derivation in Section 1 where we found

$$\omega_s = c_R \cdot q_\parallel < \omega_b = c \cdot q_\parallel$$

since $c_R < c$. Using the general solution for $u_\alpha(\vec{\ell})$, one can obtain an expression for the mean-square amplitude $\langle u_\alpha^2(\ell_3) \rangle$, which is defined as the thermal averages of $u_\alpha^2(\vec{\ell})$, in terms of the eigenvalues ω^2 and eigenvectors $\xi_\alpha(\ell_3)$ of the dynamical matrix. The result is (see Ziman [15], p.60):

$$\langle u_\alpha^2(\ell_3) \rangle = \frac{\hbar}{2NM} \sum_{qp} \frac{|\xi_\alpha(\ell_3 q,p)|^2}{\omega_p(q)} \frac{e^x+1}{e^x-1} \tag{3.36}$$

where $x = \hbar \omega_p(q)/k_B T$.

The mean-square amplitudes $\langle u_\alpha^2(\ell_3) \rangle$ at T = 0 K for (111), (100) and (110) surfaces are shown in Fig.21. Notice that all directions x, y, z give different results in the (110) case because of the rectangular symmetry of the (110) plane. If we look particularly at the surface plane and normalize to the bulk value for $\langle u_\alpha^2 \rangle$, i.e. the quantity:

$$\frac{\langle u_\alpha^2 \rangle_{\text{surface}}}{\langle u_\alpha^2 \rangle_{\text{bulk}}}$$

we notice that this ratio increases and levels off with increasing temperature [17] (see Fig.22; θ_D is the Debye temperature). This can be explained as follows. At T = 0 K the summand in (3.36) $\propto |\xi_\alpha(q,p)|^2$ surface/ω and in the high-temperature limit $\propto |\xi_\alpha(q,p)|^2$ surface/ω^2. Therefore as the temperature increases, the lower frequencies are weighted more heavily, and in particular the surface modes have low frequencies. Thus the ratio in (3.42) increases with temperature!

The mean-square amplitudes are ingredients in the Debye-Waller factor for X-ray and electron diffraction (see Ziman [15], p.60). The diffraction intensities are in the kinematical limit reduced by the factor e^{-2W} where W is given by

$$W = \tfrac{1}{2} \sum_q |\vec{K} \cdot \vec{U}_q|^2 = \frac{8\pi^2}{\lambda^2} \cdot \cos^2\varphi \cdot \langle u_{\alpha,\vec{K}}^2 \rangle \tag{3.37}$$

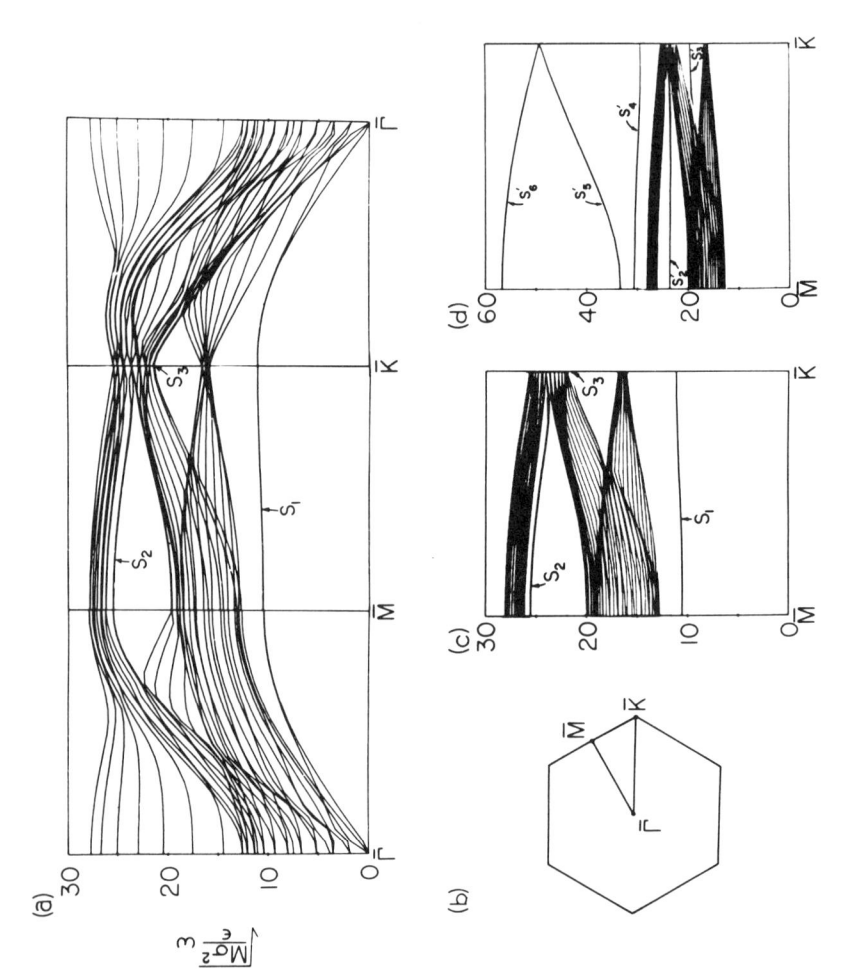

FIG. 20. (a) Frequency versus wave-vector curves along the symmetry directions $\overline{\Gamma M}$ and $\overline{\Gamma K}$, and along the Brillouin zone edge $\overline{MK}$, for an 11-layer monatomic crystal with two (111) surfaces. The surface modes are labelled by S_1, S_2 and S_3 with S_1 corresponding to the Rayleigh mode. M is the atomic mass, ϵ and σ are the potential parameters, and ω is the frequency. (b) Diagram showing the Brillouin zone and the symmetry points $\overline{\Gamma}$, $\overline{M}$ and $\overline{K}$. (c) Frequency versus wave-vector curves along the Brillouin zone edge for a crystal 21 layers thick. (d) Frequency versus wave-vector curves along the Brillouin zone edge for a 21-layer crystal whose outermost layers consist of light adsorbed atoms. The potential parameters of substrate and adsorbed layer are taken to be the same, and the mass ratio is 1:5. The curves of (a), (c) and (d) were computer generated, and no attempt was made to connect the curves properly when crossovers between modes occur. (From Ref. [16].)

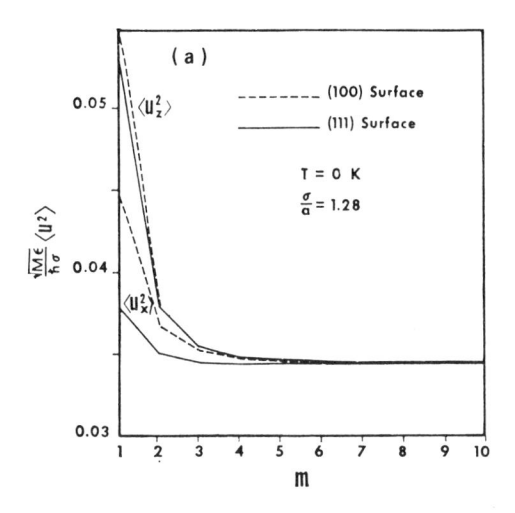

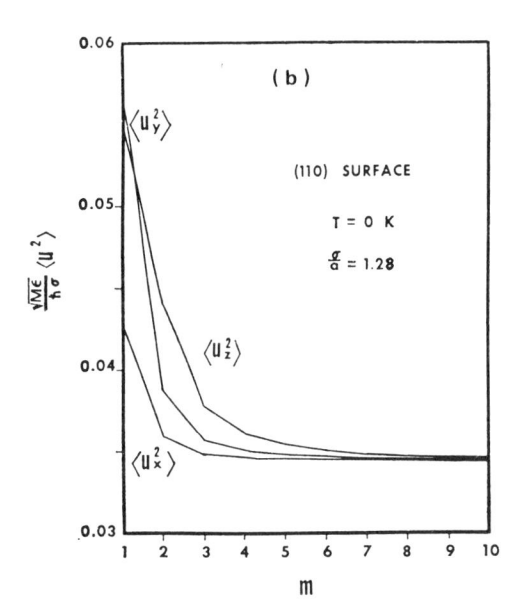

FIG. 21. (a) Mean-square amplitudes $\langle u_x^2 \rangle$ and $\langle u_z^2 \rangle$ for (100) and (111) surfaces at T = 0 K and σ/a = 1.28. (b) Mean-square amplitudes $\langle u_x^2 \rangle$, $\langle u_y^2 \rangle$ and $\langle u_z^2 \rangle$ for (110) surface at T = 0 K and σ/a = 1.28. (From Ref. [13].)

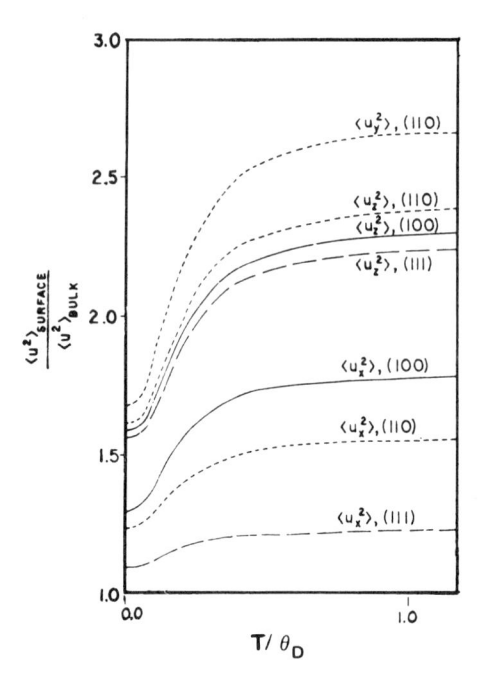

FIG. 22. Temperature dependence of $\langle u_\alpha^2 \rangle_{surface} / \langle u_\alpha^2 \rangle_{bulk}$ for (100), (111) and (110) surfaces (from Ref.[17]).

$\vec{K} = \vec{k} - \vec{k}'$ and φ is the angle of incidence on the particular Bragg plane. Notice that $\langle u_{\alpha,\vec{K}}^2 \rangle$ is along $\vec{K} = \vec{k} - \vec{k}'$. For comparison it is useful to use

$$\langle u_\alpha^2 \rangle = \frac{3\hbar^2 T}{M k_B \theta_D^2} \tag{3.38}$$

which results from the Debye continuum theory for an isotropic elastic medium at high temperature. Thus the experimental data may be expressed in terms of an effective Debye temperature related to the combination of Eqs (3.37) and (3.38). Results obtained from LEED experiments on Ag(111) [18] and Ni(100) [19] are shown in Figs 23 and 24. Notice the increase in θ_D as a function of electron energy. This is because at high energies the scattering from the surface layer is relatively weaker due to its larger vibrational amplitudes (it is not due to any drastic change in the electron penetration!). At high energies the experimental θ_D's approach the bulk value. If $\vec{K}$ is perpendicular to the surface and the electron penetration is known, one can obtain the ratios $\langle u_z^2 \rangle_{surface} / \langle u_z^2 \rangle_{bulk}$. For Ni(100) this figure is 2.1 ± 0.2 around θ_D (400°C). The bulk θ_D values are 427 K and 228 K for Ni and Ag, respectively (from specific heat measurements).

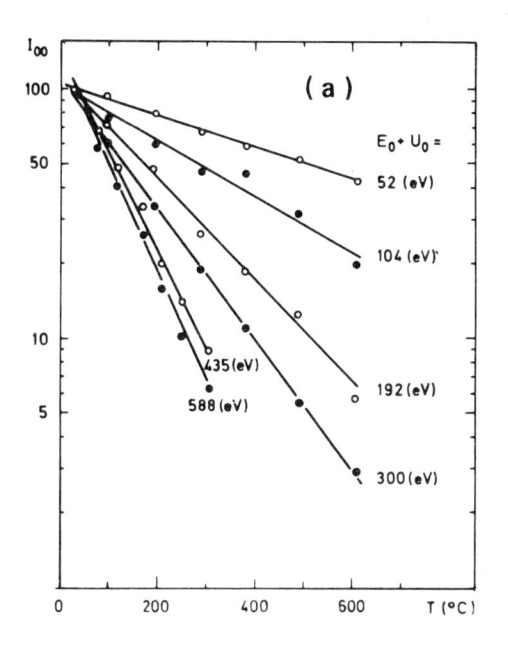

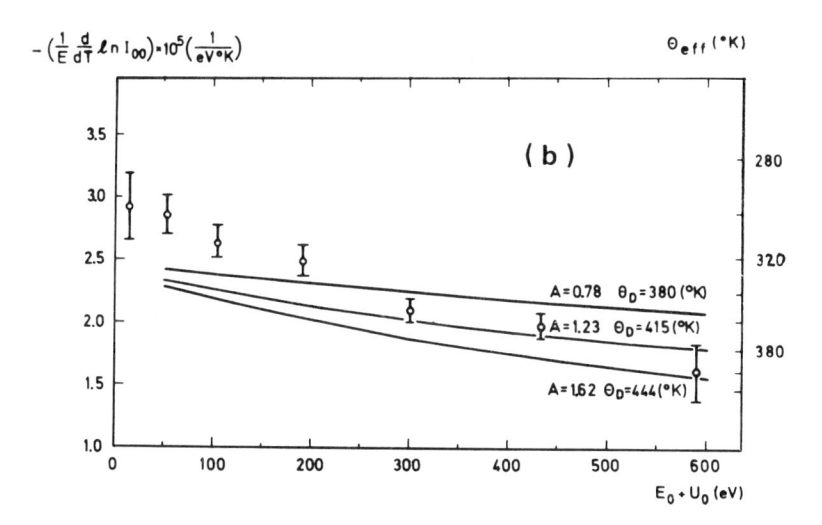

FIG. 23. (a) Specular reflectivity from the Ni(001) surfaces versus incident electron energy at different temperatures. Angle of incidence ≈4°. (b) Experimental values of $1/E$, d/dT ($\ln I_{00}$) (circles) for the integer order peaks 002-0014 versus electron energy $E = E_0 + U_0$. The solid curves are due to calculations described in the text of Ref.[19]. (From Ref.[19].)

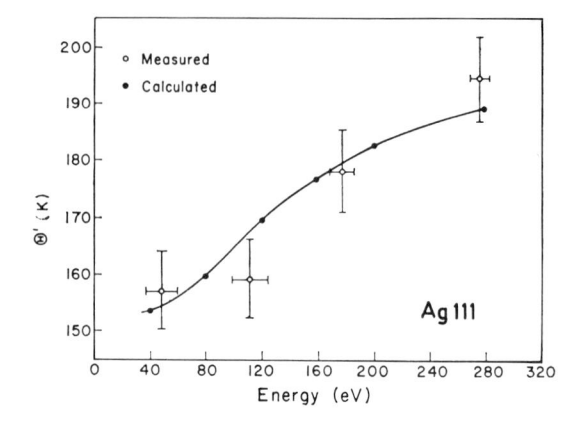

FIG. 24. Derived effective Debye temperatures versus energy for Ag(111) (from Ref. [18]).

REFERENCES

[1] LORD RAYLEIGH, Proc. London Math. Soc. 17 (1885) 4.
[2] LANDAU, L. D., LIFSHITS, E. M., Theory of Elasticity, Pergamon, Oxford (1959).
[3] (a) STONELEY, R., Proc. R. Soc. (London) Ser. A. 232 (1955) 447.
    (b) STONELEY, R., Geophys. Suppl. to Monthly Notes, R. Astron. Soc. 5 (1949) 343.
[4] SYNGE, I. L., J. Math. Phys. 35 (1967) 323.
[5] TSENG, C.C., WHITE, R.M., J. Appl. Phys. 38 (1967) 4274.
[6] TSENG, C.C., J. Appl. Phys. 38 (1967) 4281.
[7] FUCHS, R., KLIEWER, K. L., Phys. Rev. 140A (1965) 2076.
[8] KITTEL, C., Introduction to Solid State Physics, Wiley, New York (1966).
[9] MARSHALL, N., FISCHER, B., Phys. Rev. Lett. 28 (1972) 811.
[10] IBACH, H., Phys. Rev. Lett. 24 (1970) 1416; 27 (1971) 253.
[11] IBACH, H., J. Vac. Sci. Technol. 9 (1972) 713.
[12] WALLIS, R. F., Phys. Rev. 105 (1957) 540.
[13] ALLEN, R. E., DE WETTE, F. W., Phys. Rev. 179 (1969) 873.
[14] ALLEN, R. E., DE WETTE, F. W., RAHMAN, A., Phys. Rev. 179 (1969) 887.
[15] ZIMAN, J. M., Principles of the Theory of Solids, Cambridge University Press (1965).
[16] ALLEN, R. E., ALLDREDGE, G. P., DE WETTE, F. W., Phys. Rev. Lett. 23 (1969) 1285.
[17] ALLEN, R. E., DE WETTE, F. W., Phys. Rev. 188 (1969) 1320.
[18] JONES, E. R., McKINNEY, J. T., WEBB, M. B., Phys. Rev. 151 (1966) 476.
[19] ANDERSSON, S., KASEMO, B., Solid State Commun. 8 (1970) 1885.

IAEA-SMR-15/34

INTRODUCTION TO THE THEORY OF LOW-ENERGY ELECTRON DIFFRACTION

A. FINGERLAND, M. TOMÁŠEK
J. Heyrovský Institute of Physical Chemistry
  and Electrochemistry,
Czechoslovak Academy of Sciences,
Prague,
Czechoslovakia

Abstract

INTRODUCTION TO THE THEORY OF LOW-ENERGY ELECTRON DIFFRACTION.
An elementary introduction to the basic principles of the theory of low-energy electron diffraction is presented. General scattering theory is used to classify the hitherto known approaches to the problem.

1. INTRODUCTION

The object of this work is to offer the reader both an elementary survey of the present state of the theory of low-energy electron diffraction (LEED) and an introduction which should make it easier to read original papers.

Today LEED theory is scattered among hundreds of papers, the first being undoubtedly that by Bethe [1] in 1928, published one year after the discovery by Davisson and Germer [2]. Although it is always the same problem which has to be solved, many of the subsequent papers show little resemblance to the original work of Bethe, so that it is often difficult to find one's way through some of them. Naturally, besides the original papers a number of reviews have appeared, several being quoted in the References (see e.g. Refs [3-9]). Most authors, however, treat LEED from their own point of view, making no attempt to present a unifying discussion of the various theoretical approaches currently used. For this reason a certain classification of the known approaches may prove useful. Such a classification can be done by means of a fairly general formulation which, in our opinion, the scattering theory offers.

2. SELECTION OF THE THEORETICAL MODEL ON THE BASIS OF EXPERIMENT

Apart from various experimental modifications, the typical LEED experiment appears in principle as indicated in Fig. 1. A beam of electrons impinges on a crystal and is scattered in various directions. The scattered electrons are detected after elimination of those which lost part of their energy through inelastic collision with the crystal. Our aim is to obtain information on the crystal from the distribution of the scattered electrons. Consequently, the experiment is here very similar to the diffraction experiment in X-ray, neutron or the usual electron diffraction. However, let us consider some of the experimental facts in more detail.

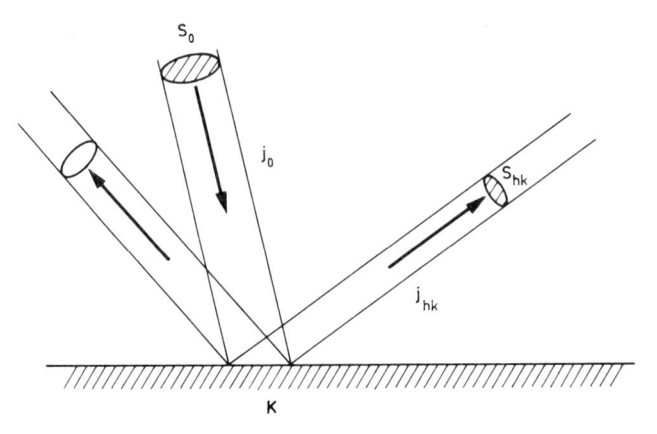

FIG.1. Arrangement of the diffraction experiment. j_0 is the current of incident electrons, j_{hk} the diffracted current, S_0, S_{hk} cross-sections of corresponding beams, and K is the crystal.

The incident beam is a beam of electrons with a relatively well defined energy [10]. Authors give $\delta V \sim 0.2$ eV, which in the usual energy range of 20 - 200 eV means an energy uncertainty better than 1%. In our model, therefore, we shall consider the electrons to be monochromatic. Since the electrons leaving the electron gun are thermal and hence incoherent [11] and the electron optics ensures a relatively well defined direction (the distance of the sample from the source is 10 cm, the cross-section of the beam ~ 1 mm^2), it is therefore reasonable to describe the incident electron as a monochromatic plane wave.

A large part of experiments is carried out by measuring the intensity of the particular diffracted beam as a function of the primary energy. It appears that the width δE of the measured profiles varies in the range of one to tens of eV. On the basis of a consideration similar to that in Ref.[12] we can conclude that the penetration depth of electrons into the crystal is very small, representing practically several atomic layers.

From the small penetration depth it follows that, on one hand, interesting information about the situation on the surface of the crystal can be expected from the experiment and, on the other, the greatest difficulty will be met in choosing the model to describe the crystal surface (whether 'clean' or not). Finally, let us pay attention to the detection of diffracted electrons. The separation of the inelastically reflected electrons by the retarding field is not complete, so that we cannot strictly speak about the detection of elastically scattered electrons only. This concerns primarily electrons which lost part of their energy in collisions with phonons. It is therefore necessary to include lattice vibrations in the same way as in X-ray diffraction. The resulting experiment thus carries information on the time-averaged atom. The calculation of the so-called Debye-Waller temperature factor has naturally to take into account the surface of the crystal, as can be found, for example, in Ref.[13]. The result of the measurement can be expressed as the convolution of the 'actual' distribution and the instrumental distortion. These questions are examined in detail in Ref.[14];

they will not be treated here. Summarizing, it appears that our theoretical model corresponds to a quasi-elastic scattering of a monochromatic plane wave on a time-averaged absorbing crystal with a surface.

3. GENERAL FORMULATION OF THE PROBLEM: OPTICAL POTENTIAL AND ONE-ELECTRON APPROXIMATION

The Hamiltonian of our problem reads:

$$H = H_e + H_k + H_{ee} + H_{ej}$$

where H_e is the operator of the (kinetic energy of the) incident electron, H_k that of the crystal, H_{ee} the operator of the electron-electron interaction and H_{ej} describes the interaction of the electron with the nuclei. Here we have already the averaged term in which the interaction with phonons is considered in this way. The first two terms will be denoted by H_0 because they represent the energy of the electron-crystal system without mutual interaction. The remaining terms will be denoted by H'. Our task is to solve the equation:

$$H|\psi\rangle \equiv (H_0 + H')|\psi\rangle = E|\psi\rangle$$

Before the collision, the system is described by the equation:

$$H_0|\Phi\rangle = E|\Phi\rangle$$

which is solved by the so-called unperturbed states $|\Phi_a\rangle$, corresponding to the energies E_a. Since we assume that the energy of the system is not changed by the collision, we shall seek the solution of the following equation:

$$(H_0 + H')|\psi_a\rangle = E_a|\psi_a\rangle \tag{1}$$

This represents a complicated many-body problem which we shall try to avoid by a further simplification analogous to that in Ref.[15]. The initial state can be characterized by a free electron (plane wave) described by the wave vector $\vec{k}_a$ with the energy k_a^2 and by the ground state of the crystal, which we denote $|0\rangle$ with the energy $E_{k,0}$. Consequently

$$|\Phi_a\rangle = |k_a\rangle|0\rangle \tag{2}$$

$$E_a = k_a^2 + E_{k,0}$$

If $|n\rangle$ is a general eigenstate of the crystal, we obtain

$$H_k|n\rangle = E_{k,n}|n\rangle$$

Now we define the one-particle states:

$$\langle n|\psi_a\rangle = |\psi_n\rangle; \quad \langle n|\Phi_a\rangle = |k_a\rangle\langle n|0\rangle = \delta_{n,0}|k_a\rangle$$

and the one-particle operators:

$$\langle n|A|m\rangle = A_{n,m}$$

When multiplied by $\langle n|$ from the left, Eq.(1) gives

$$H_e|\psi_n\rangle + E_{k,n}|\psi_n\rangle + \sum_m H'_{n,m}|\psi_m\rangle = E_a|\psi_n\rangle \quad (3)$$

For $n = 0$ we obtain, using Eq.(2),

$$H_e|\psi_0\rangle + \sum_m H'_{0,m}|\psi_m\rangle = k_a^2|\psi_0\rangle$$

If we define

$$V|\psi_0\rangle = \sum_m H'_{0,m}|\psi_m\rangle = H'_{0,0}|\psi_0\rangle + \sum_{m\neq 0} H'_{0,m}|\psi_m\rangle \quad (4)$$

we obtain a one-particle Schrödinger equation:

$$(H_e + V)|\psi_0\rangle = k_a^2|\psi_0\rangle$$

Explicit expression for the 'optical potential' V can be obtained as follows. From Eq.(3) for $n = 0$ we obtain

$$(H_e + E_{k,n} - E_a)|\psi_n\rangle = -\sum_m H'_{n,m}|\psi_m\rangle$$

Putting

$$(E_a - E_{k,n} - H_e)^{-1} = G_n$$

one has

$$|\psi_n\rangle = G_n \sum_m H'_{n,m}|\psi_m\rangle = G_n\left(H'_{n,0}|\psi_n\rangle + \sum_{m=0} H'_{n,m}|\psi_m\rangle\right)$$

By substituting successively for $|\psi_m\rangle$ into the right-hand side of this equation we arrive at a formal series:

$$|\psi_n\rangle = \left[G_n H'_{n,0} + G_n \sum_{m \neq 0} H'_{n,m} G_m H'_{m,0} + G_n \sum_{m \neq 0} H'_{n,m} G_m \sum_{p \neq 0} H'_{m,p} G_p H'_{p,0} + \ldots \right] |\psi_0\rangle$$

By inserting this result into (4), i.e. into the definition of V, we get

$$V = H'_{0,0} + \sum_{n \neq 0} H'_{0,n} G_n H'_{n,0} + \sum_{n \neq 0} H'_{0,n} G_0 \sum_{m \neq 0} H'_{n,m} G_m H'_{m,0} + \ldots \tag{5}$$

Yoshioka [15] uses the expansion to the second order so that his equation (see Eq. (11), p. 619 of Ref. [15]) reads:

$$(H_e + H'_{0,0} - k_a^2)|\psi_0\rangle = \sum_{n \neq 0} H'_{0,n} G_n H'_{n,0} |\psi_0\rangle = A|\psi_0\rangle$$

The first term in expansion (5) of the potential V, i.e. $H'_{0,0}$, is nothing more than the averaged potential of the crystal as seen by the incident electron. From the definition of V it is obvious that it depends on the energy of the incident electron, because the operators G_n depend on it. Furthermore, it must contain the absorption as it depends on the excited states of the crystal which arise at the cost of electrons disappearing from the elastic channel. It is therefore natural to assume that the potential V can be approximated by a generally complex energy-dependent pseudopotential. Most authors go so far that they consider the imaginary part of V even constant (for the first time probably in Ref. [16]), although the full complex potential is explicitly derived in Ref. [15]. In the present paper we shall not concern ourselves with the important and complicated question of the construction of the potential. The main result of our discussion is that we have succeeded in converting the many-particle problem into a single-particle one in which the optical potential V includes the influence of inelastic processes upon the elastic scattering. This way, however, remains impracticable and further simplifications are therefore necessary. Nevertheless its mere existence justifies our further treatment, namely work on the one-electron theory.

4. FORMAL SCATTERING THEORY: BORN EXPANSION AND MULTIPLE SCATTERING

The equations:

$$(H_e + V)|\psi\rangle = E|\psi\rangle$$

where

$$H_e |\vec{k}_i\rangle = k^2 |\vec{k}_i\rangle \qquad (k^2 = E)$$

can be written in the integral form (Lippman-Schwinger equation):

$$|\psi^+\rangle = |\vec{k}_i\rangle + G^+ V |\psi^+\rangle \qquad (6)$$

where

$$G^+ = (k^2 - H_e + i\epsilon)^{-1} \qquad (7)$$

Repeated substitution of $|\psi^+\rangle$ into the right-hand side of (6) leads to the series:

$$|\psi^+\rangle = |\vec{k}_i\rangle + G^+ V |\vec{k}_i\rangle + G^+ V\, G^+ V |\vec{k}_i\rangle + \ldots \qquad (8)$$

Expansions of this type are called Born series (see, e.g., Ref.[17]) and the corresponding terms are the first, second, etc., Born approximations.

If we denote

$$\begin{aligned} T &= V + VG^+V + VG^+VG^+V + \ldots \\ &= V + VG^+T \end{aligned} \qquad (9)$$

Eq. (8) can be written in an abbreviated form:

$$|\psi^+\rangle = |\vec{k}_i\rangle + G^+ T |\vec{k}_i\rangle \qquad (10)$$

Eqs (6) and (10) in the co-ordinate representation read:

$$\langle r | \psi^+\rangle = \psi^+(\vec{r}) = e^{i\vec{k}_i\vec{r}} + \int G^+(\vec{r},\vec{r}') V(\vec{r}') \psi^+(\vec{r}') dv'$$

or

$$\psi^+(\vec{r}) = e^{i\vec{k}_i\vec{r}} + \iint G^+(\vec{r},\vec{r}') e^{i\vec{k}'\vec{r}'} T_{\vec{k}',\vec{k}_i} dv' d^3k'$$

By using the asymptotic expression in the last equation:

$$G^+(\vec{r},\vec{r}') = \frac{e^{ikr}}{r} e^{-i\vec{k}_f\vec{r}'}$$

we obtain

$$\psi^+(\vec{r}) \approx e^{i\vec{k}_i\vec{r}} + \frac{e^{ikr}}{r} T_{\vec{k}_f,\vec{k}_i}$$

It is seen that the solution contains both the incident plane wave and the scattered wave, whose amplitude,

$$T_{\vec{k}_f, \vec{k}_i} = \langle \vec{k}_f | T | \vec{k}_i \rangle$$

in the direction k_f depends on the direction (k_i) of the incident beam. The relative intensities are then given by $|T_{fi}|^2$ where we take the matrix element between the initial ($\vec{k}_i$) and the final ($\vec{k}_f$) electron states.

Let us now assume that the potential V can be composed from the contributions of the individual parts, i.e. that it is given by the relation

$$V = \sum_\alpha v_\alpha$$

By substituting it into (9) we get

$$T = \sum_\alpha T_\alpha$$

where

$$T_\alpha = v_\alpha + v_\alpha G^+ T$$

Putting the former relation for T into the latter, we obtain

$$(1 - v_\alpha G^+) T_\alpha = v_\alpha + v_\alpha G^+ \sum_{\beta \neq \alpha} T_\alpha$$

If we further introduce the notation

$$t_\alpha = (1 - v_\alpha G^+)^{-1} v_\alpha$$

or

$$t_\alpha = v_\alpha + v_\alpha G^+ t_\alpha$$

we see that the problem of calculating the intensity consists in solving the following system of equations:

$$t_\alpha = v_\alpha + v_\alpha G^+ t_\alpha = v_\alpha + v_\alpha G^+ v_\alpha + \ldots \tag{11a}$$

$$T_\alpha = t_\alpha + t_\alpha G^+ \sum_{\beta \neq \alpha} T_\beta = t_\alpha + t_\alpha G^+ \sum_{\beta \neq \alpha} t_\beta + \ldots \tag{11b}$$

$$T = \sum_\alpha T_\alpha = \sum_\alpha t_\alpha + \sum_\alpha t_\alpha G^+ \sum_{\beta \neq \alpha} t_\beta + \ldots \tag{11c}$$

$$I = |T_{fi}|^2 \tag{11d}$$

Now let us assume that the scattering centres are translationally equivalent, i.e.

$$v_\alpha(\vec{r}) = v(\vec{r} - \vec{R}_\alpha)$$

Then we can write

$$\begin{aligned}\langle \vec{k}' | v_\alpha | \vec{k}'' \rangle &= \langle R_\alpha | \vec{k}'' - \vec{k}' \rangle \langle \vec{k}' | v | \vec{k}'' \rangle \\ &= \exp\left[i(\vec{k}'' - \vec{k}')\vec{R}\right] \langle \vec{k}' | v | \vec{k}'' \rangle\end{aligned} \tag{12}$$

From Eq. (11a) we obtain

$$\langle \vec{k}' | t_\alpha | \vec{k}'' \rangle = \langle R_\alpha | \vec{k}'' - \vec{k}' \rangle \langle \vec{k}' | t | \vec{k}'' \rangle \tag{13}$$

where

$$\langle \vec{k}' | t | \vec{k}'' \rangle = \langle \vec{k}' | v | \vec{k}'' \rangle + \int \langle \vec{k}' | v | \vec{k} \rangle G^+(k) \langle \vec{k} | t | \vec{k}'' \rangle d^3k$$

From Eq. (11b) it follows that

$$\begin{aligned}\langle \vec{k}' | T_\alpha | \vec{k}'' \rangle &= \langle \vec{R}_\alpha | \vec{k}'' - \vec{k}' \rangle \langle \vec{k}' | t | \vec{k}'' \rangle \\ &+ \int \langle \vec{R}_\alpha | \vec{k} - \vec{k}' \rangle \langle \vec{k}' | t | \vec{k} \rangle G^+(k) \sum_{\beta \neq \alpha} \langle \vec{k} | T_\beta | \vec{k}'' \rangle d^3k \\ &= \langle \vec{R}_\alpha | \vec{k}'' - \vec{k}' \rangle \Big\{ \langle \vec{k}' | t | \vec{k}'' \rangle \\ &+ \int \langle \vec{R}_\alpha | \vec{k} - \vec{k}'' \rangle \langle \vec{k}' | t | \vec{k} \rangle G^+(k) \sum_{\beta \neq \alpha} \langle \vec{k} | T_\beta | \vec{k}'' \rangle d^3k \Big\}\end{aligned}$$

We seek the solution in the form

$$\langle \vec{k}' | T_\alpha | \vec{k}'' \rangle = \langle \vec{R}_\alpha | \vec{k}'' - \vec{k}' \rangle \langle \vec{k}' | T^{(\alpha)} | \vec{k}'' \rangle \tag{14}$$

where

$$\langle \vec{k}' | T^{(\alpha)} | \vec{k}'' \rangle = \langle \vec{k}' | t | \vec{k}'' \rangle$$

$$+ \int \langle \vec{k}' | t | \vec{k} \rangle G^+(k) \sum_{\beta \neq \alpha} \langle \vec{R}_\alpha - \vec{R}_\beta | k - k'' \rangle \langle \vec{k} | T^{(\beta)} | \vec{k}'' \rangle d^3k$$

If we denote

$$G^{\alpha,\beta}(\vec{k}, \vec{k}'') = G^+(k) \langle \vec{R}_\alpha - \vec{R}_\beta | \vec{k} - \vec{k}'' \rangle \tag{15}$$

we can write

$$\langle \vec{k}' | T^{(\alpha)} | \vec{k}'' \rangle = \langle \vec{k}' | t | \vec{k}'' \rangle + \int \langle \vec{k}' | t | \vec{k} \rangle \sum_{\beta \neq \alpha} G^{\alpha,\beta}(\vec{k}, \vec{k}'') \langle \vec{k} | T^{(\beta)} | \vec{k}'' \rangle d^3k$$

which, substituted into Eq. (11c), gives

$$\langle \vec{k}' | T | \vec{k}'' \rangle = \sum_\alpha \langle \vec{k}' | T_\alpha | \vec{k}'' \rangle = \sum_\alpha \langle \vec{R}_\alpha | \vec{k}'' - \vec{k}' \rangle \langle \vec{k}' | T^{(\alpha)} | \vec{k}'' \rangle \tag{16}$$

Before proceeding, let us try to find the physical meaning of some partial results of this section. If we substitute for T into Eq. (10) only the first Born approximation (V only) the result is the same as if we put $|\vec{k}_i\rangle$ instead of $|\psi^+\rangle$ in the right-hand side of (16). This means that we treat the scattered wave as the scattering of the incident plane wave on the potential only and neglect any interaction between the scattered and the incident waves. By combining all multiple scattering on a single centre into (11a) we achieve the situation where only scattering on different successive 'modified' centres is left. Equation (11b) represents all the multiple scatterings which have ended on the centre α, and naturally the result (11c) is the mere sum of all these contributions.

The translational equivalence of the centres manifests itself in the fact that the scatterings from individual (12) (as well as modified (13)) centres differ by the phase term only. However, a different situation is met with T_α, which represents the scattering ending on the centre α (Eq. (14)). Spatial distribution of other centres around the α-th centre is always decisive here and therefore in finite systems T_α will contain, besides the phase term, something specific depending on the surroundings. This is reflected in the very form of the quantities $G^{\alpha,\beta}(\vec{k}, \vec{k}'')$, Eq. (15). It is to be expected that the symmetry of the surroundings will manifest itself to a large extent at this point. The symmetry is discussed in the next section.

5. TRANSLATIONAL SYMMETRY: EWALD CONSTRUCTION

So far, we have tacitly assumed that we have somehow determined the model potential. Now we shall be concerned with one of its specific properties, the symmetry. We shall assume that the crystal consists of two parts which join on a plane parallel to a certain crystallographic plane. One part, involving most of the crystal, exhibits a structure identical with that of the crystal bulk and this part is called the substrate. The other part, called the selvedge [18], has an unknown structure. When the selvedge is also crystalline, i.e. has an ordered structure, it must coincide with the symmetry of the substrate. Consequently, the entire crystal has a translational symmetry, which is a subgroup of the translational symmetry of this crystallographic plane. (More complicated cases [19] could be considered, but this one must suffice.)

Let us therefore consider the case where the potential V(r) has a two-dimensional periodicity parallel to the surface. This means that the potential can be divided into translationally equivalent parts v_α so that $v_\alpha(\vec{r}) = v(\vec{r} - \vec{R}_\alpha)$, where $\vec{R}_\alpha$ are translational vectors of the two-dimensional net. Then the surroundings of each potential are the same and obviously $\langle \vec{k}_f | T^{(\alpha)} | \vec{k}_i \rangle$ from (14) is equal to $\langle \vec{k}_f | T^{(0)} | \vec{k}_i \rangle$, i.e. independent of α; at the same time, the sum over α (16) includes the entire net.

By writing this down we obtain

$$\langle \vec{k}_f | T | \vec{k}_i \rangle = \sum_{\text{net}} \langle \vec{k}_f | T_\alpha | \vec{k}_i \rangle$$

$$= \langle \vec{k}_f | T^{(0)} | \vec{k}_i \rangle \sum_{\text{net}} \exp[i(\vec{k}_i - \vec{k}_f)R_\alpha]$$

Let us introduce a co-ordinate system such that R_α lies in the (x, y) plane; then

$$(\vec{k}_i - \vec{k}_f)\vec{R}_\alpha = (\vec{k}_{i,p} - \vec{k}_{f,p})\vec{R}_\alpha$$

where $\vec{k}, p$ denotes the component parallel to the surface. The sum over the lattice differs from zero only when

$$\vec{k}_{f,p} = \vec{k}_{i,p} + \vec{g}_{hk} \qquad (17)$$

$\vec{g}_{hk}$ is the vector of the net reciprocal to the net $\{\vec{R}_\alpha\}$. This result is important because it shows that the directions in which diffracted electrons can be expected depend only on the translational symmetry of the crystal with the surface. From the geometry of the diffraction pattern it is therefore possible to determine the respective translation group. The change of the pattern upon adsorption of a foreign layer on a clean surface enables the change of this translation group to be ascertained. The intensity, however, is exclusively the matter of scattering on the whole set of scattering centres $\langle \vec{k}_f | T^0 | \vec{k}_i \rangle$ and thus holds information about the arrangement of the atoms.

Elastic scattering means that the energy of the electron is conserved, i.e. $k_i^2 = k_f^2$. Together with (17), this enables a geometrical construction known as the Ewald construction to be carried out (Fig. 2). Furthermore, from this clearly follows the possibility of denoting the individual diffracted

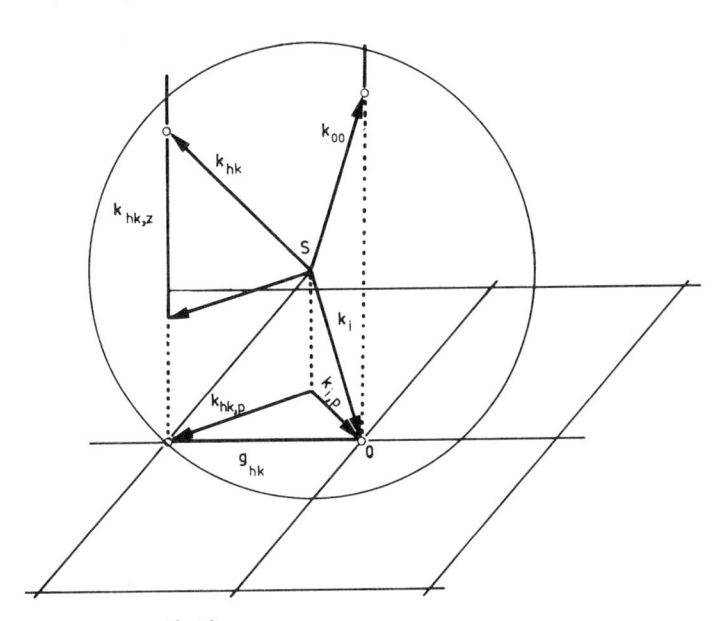

FIG.2. Ewald construction. $\vec{k}_i$, $\vec{k}_{hk}$: wave vectors of the incident and diffracted wave, respectively; $k_i^2 = k_{hk}^2$ defines the reflection sphere; $\vec{k}_{i,p} + \vec{g}_{hk} = \vec{k}_{hk,p}$ the law of conservation of the momentum component parallel to the surface. k_{00} is the specular beam.

beams by pairs of indices (h, k) giving the reciprocal vector $\vec{g}_{hk}$ of (17), so that the resulting solution can be written as a superposition of plane waves:

$$A_{hk} \exp(i\vec{k}_{hk} \cdot \vec{r})$$

where

$$\vec{k}_{hk,p} = \vec{k}_{i,p} + \vec{g}_{hk}$$

and $k_{hk,z}$ are determined from the condition:

$$k_{hk,z}^2 = E - (\vec{k}_{i,p} + \vec{g}_{hk})^2$$

and from the requirement that the plane waves should propagate from the surface into the vacuum (opposite sign to that of $k_{i,z}$), or be damped in the vacuum ($k_{hk,z} = -i\sqrt{-k_{hk,z}^2}$). An important role is played by the beam with the index (0,0), which is always present. This is the so-called specular beam (see Fig.2).

The symmetry of the diffraction pattern will be given by the symmetry common to the reciprocal net and the vector $\vec{k}_{i,p}$. If $\vec{k}_{i,p} = 0$ holds, the symmetry is obviously the largest. For $\vec{k}_{i,p} \neq 0$ either the identity or the plane of symmetry remains. This is probably the main reason why theoreticians mainly treat 'normal incidence'.

TABLE I. CLASSIFICATION OF DIFFERENT THEORIES BY MEANS OF THE T MATRIX

t_α \ T	Simple scattering	Multiple scattering
First Born approximation	(a)	(c)
Higher Born approximations	(b)	(d)

(a) .... kinematical theory
(b,c,d) .... dynamical theory
(a,b) .... (pseudo)kinematical theory
(c,d) .... multiple scattering

t_α is defined by Eq.(11a) and T by Eq.(11c).

6. CLASSIFICATION OF INDIVIDUAL LEED THEORIES

One of the ways to classify the LEED theories is to start with the Born series (9) for the T matrix. This method is widely used and leads to a division of the theories into two main groups:

A) The kinematical theory in this sense is equivalent to the first Born approximation
B) The dynamical theories are all the other theories

If we examine expansion (9) with $V = \sum v_\alpha$ substituted in it, the kinematical theory appears as a theory completely neglecting the multiple scattering. It is naturally possible to make the classification in an analogous way using expansion (11c). In this expansion, the individual terms again represent single, double and multiple scattering events with already renormalized 'atomic' potential (11a), i.e. with the inclusion of multiple scattering on individual 'atoms'. From this point of view it is therefore possible to divide the theories into:

A') The (pseudo)kinematical theories equivalent in this sense to simple scattering
B') The multiple scattering theories or dynamical theories in a restricted sense

These are schematically indicated in Table I.
In further discussions we shall follow the second classification because it reflects better the atomistic nature of the crystal and separates the scattering on the 'atom' from the scattering on a set of scattering centres, i.e. the physical part from the structurally crystallographic part.

7. (PSEUDO)KINEMATICAL THEORY

From the preceding explanations it follows that Eqs (11) now reduce to

$$t_\alpha = v_\alpha + v_\alpha G^+ t_\alpha$$

$$(T_\alpha = t_\alpha)$$

$$T = \sum t_\alpha$$

$$I = |T_{fi}|^2$$

where the subscript α describes the individual 'atoms'.

We shall use the method described in Ref.[20]. We shall divide the whole crystal into subplanes as follows. We start with an arbitrary atom. From translational symmetry we know that a whole net of atoms exists equivalent to the initial one. These atoms form the subplane in question. In each subplane we select one atom as the origin, from which we can determine the position of each atom:

(a) By giving the lattice vector $\vec{R}$ in the subplane in which the atom is situated; and
(b) By giving the order n of this subplane, i.e. the position of its origin $\vec{d}_n$.

Hence

$$\vec{R}_\alpha = \vec{R} + \vec{d}_n$$

Since each subplane contains only one type of atom, we write (13) as:

$$\langle \vec{k}_f | t_\alpha | \vec{k}_i \rangle = \langle \vec{R}_\alpha | \vec{k}_i - \vec{k}_f \rangle \langle \vec{k}_f | t^{(n)} | \vec{k}_i \rangle$$

$$= \exp[i(\vec{R} + \vec{d}_n)(\vec{k}_i - \vec{k}_f)] \langle \vec{k}_f | t^{(n)} | \vec{k}_i \rangle$$

Since $T_\alpha = t_\alpha$, we have from (11c)

$$\langle \vec{k}_f | T | \vec{k}_i \rangle = \sum_{n,R} \exp[i(\vec{R} + \vec{d}_n)(\vec{k}_i - \vec{k}_f)] \langle \vec{k}_f | t^{(n)} | \vec{k}_i \rangle$$

$$= \sum_R \exp[i\vec{R}(\vec{k}_i - \vec{k}_f)] \sum_n \exp[i\vec{d}_n(\vec{k}_i - \vec{k}_f)] \langle \vec{k}_f | t^{(n)} | \vec{k}_i \rangle$$

The first sum includes the whole translational net and therefore differs from zero only for $\vec{k}_{i,p} - \vec{k}_{f,p} = \vec{g}_{hk}$, as shown already. The second sum is the structure factor:

$$F'_{hk} = \sum_n \exp[i\vec{d}_n(\vec{k}_i - \vec{k}_f)] \langle \vec{k}_f | t^{(n)} | \vec{k}_i \rangle \qquad (18)$$

and represents the total single scattering from all representants of individual subplanes, i.e. from a certain 'elementary cell'. The reflections (hk) are obviously the only ones obtained. On the other hand, the form of F'_{hk} implies that the intensity of these reflections depends also on the arrangement of atoms in the elementary cell and that, for some $\vec{k}_i$ and $\vec{k}_f$, this intensity can become zero. Let us analyse at least two cases:

(a) <u>A crystal with a 'clean' surface</u>: We assume that starting from the surface, the crystal structure repeats periodically (i.e. the selvedge is missing). Let us denote $\langle \vec{k}_f | t^{(n)} | \vec{k}_i \rangle$ from the first period as $\langle \vec{k}_f | t^{(s)} | \vec{k}_i \rangle$ (s = 1, 2, ..., ℓ) and put $\vec{d}_n = \vec{d}^{(s)}$. Since $n = q \cdot \ell + s$, we can write $\vec{d}_n = \vec{d}^{(s)} + \vec{c} \cdot q$, where $\vec{c}$ is the translation directed (not necessarily perpendicularly) from the surface into the crystal. Consequently, the structure factor F'_{hk} can be written in the form:

$$F'_{hk} = \sum_{q=1}^{\infty} \exp[i\vec{c}q(\vec{k}_i - \vec{k}_f)] \sum_{s=1}^{\ell} \exp[i\vec{d}^{(s)}(\vec{k}_i - \vec{k}_f)] \langle \vec{k}_f | t^{(s)} | \vec{k}_i \rangle$$

$$= \sum_{q=1}^{\infty} \exp[i\vec{c}q(\vec{k}_i - \vec{k}_f)] F_{hk}$$

where we have defined

$$F_{hk} = \sum_{s=1}^{\ell} \exp[i\vec{d}^{(s)}(\vec{k}_i - \vec{k}_f)] \langle \vec{k}_f | t^{(s)} | \vec{k}_i \rangle \qquad (18')$$

F_{hk} now corresponds to the arrangement of the atoms in the unit mesh which includes only one period.

The sum

$$\sum_{q=1}^{\infty} \exp[i\vec{c}q(\vec{k}_i - \vec{k}_f)]$$

has non-zero values for $\vec{c}(\vec{k}_i - \vec{k}_f) = 2\pi m$ (m integer) only. We put

$$\vec{c} = \vec{c}_p + \vec{c}_z$$

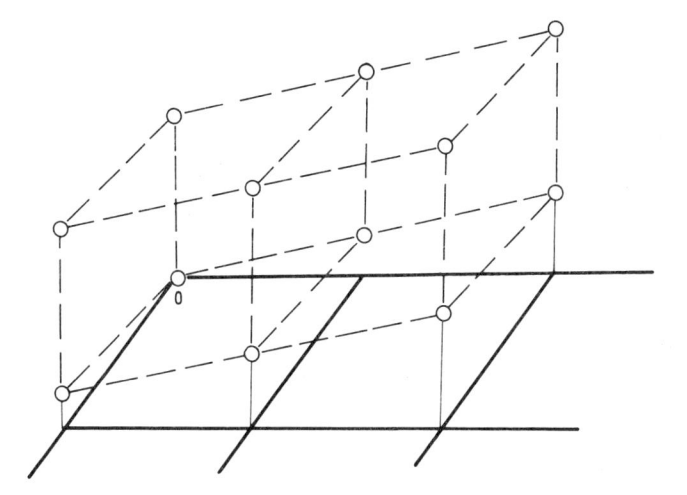

FIG.3. Relation between the two-dimensional reciprocal net and the three-dimensional reciprocal lattice. The two-dimensional net (solid lines) is parallel to the surface of the crystal and is a projection of the three-dimensional lattice (dashed lines) in the direction normal to the surface.

Then

$$\vec{c}(\vec{k}_i - \vec{k}_f) = \vec{c}_p \vec{g}_{hk} + c_z(k_{i,z} - k_{f,z}) = 2\pi m$$

whence follows the third diffraction condition:

$$c_z(k_{i,z} - k_{f,z}) = 2\pi m - \vec{c}_p \vec{g}_{hk}$$

In the Ewald construction this represents certain points on the 'reciprocal rods' erected in the points of the reciprocal lattice perpendicularly to the surface. These points altogether form a normal three-dimensional reciprocal lattice of the crystal. Our two-dimensional reciprocal net mentioned earlier is the projection of the three-dimensional lattice in the direction of the normal-to-the-surface plane. If the crystal has no lattice translation perpendicular to the surface, the situation looks as indicated in Fig.3.

(b) <u>A crystal with an adsorbed monomolecular layer</u>: In this case we assume for simplicity that in the $\vec{a}$ direction the adsorbed layer has double periodicity with respect to the substrate and that the latter is a simple orthorhombic crystal. The translational symmetry of the problem is now given by the vectors:

$$\vec{R} = (2\vec{a})n + \vec{b}m$$

and the reciprocal vectors:

$$\vec{g}_{Hk} = H\left(\frac{\vec{a}}{2}\right)^* + \vec{kb}^* = \frac{H}{2}\vec{a}^* + \vec{kb}^*$$

If we put the origin in the initial clean surface, each crystal plane is split into two subplanes with $\vec{d}^{(1)} = 0$, $\vec{d}^{(2)} = \vec{a}$. We calculate the diffraction from the clean crystal according to (18'):

$$F_{Hk}(\text{clean}) = \{1 + \exp[i\vec{a}(\vec{k}_i - \vec{k}_f)]\} \langle \vec{k}_f | t | \vec{k}_i \rangle$$

but

$$\vec{a}(\vec{k}_i - \vec{k}_f) = \vec{a}\left[H\frac{\vec{a}^*}{2} + k\vec{b}^*\right] = \pi H$$

where H is an integer. We see that in the new notation the equality $F_{Hk} = 0$ holds for H odd, i.e. the new reciprocal net is denser in the direction $\vec{a}^*$. If we add to the structure factor the contribution from the adsorbed layer, characterized by $d^{(0)}$ and $\langle \vec{k}_f | t^{(0)} | \vec{k}_i \rangle$, then (18) assumes the form:

$$F'_{Hk} = \exp[i\vec{d}^{(0)}(\vec{k}_i - \vec{k}_f)] \langle \vec{k}_f | t^{(0)} | \vec{k}_i \rangle + \sum_{q=1}^{\infty} \exp[i\vec{c}q(\vec{k}_i - \vec{k}_f)] F_{Hk}$$

For the condition for H odd, we have

$$F'_{(2H'+1),k} = \exp[i\vec{d}^{(0)}(\vec{k}_i - \vec{k}_f)] \langle \vec{k}_i | t^{(0)} | \vec{k}_f \rangle$$

$$= \exp[i\vec{d}_p^{(0)}\vec{g}_{(2H'+1),k}] \cdot \exp[id_z^{(0)}(k_{i,z} - k_{f,z})] \langle \vec{k}_f | t^{(0)} | \vec{k}_i \rangle$$

From this we see that new diffraction beams ('extra' beams) stem from the adsorbed layer and their intensity, proportional to $\sim |\langle \vec{k}_f | t^0 | \vec{k}_i \rangle|^2$, is a slowly varying function of $\vec{k}_i$. The other beams, on the other hand, can show considerable changes with respect to $\vec{k}_i$. Although they are influenced by the presence of the surface layer, their positions are nevertheless in agreement with the positions of the initial clean crystal.

In conclusion, we can say that from the (pseudo)kinematical theory we obtain:

(a) The reflections (hk) on the right places;
(b) The evidence about the formation of the surface layer (selvedge) with a periodicity other than that of the substrate.

8. DYNAMICAL THEORY

As already mentioned, the complete solution of the problem consists in solving the system of Eqs (11). Let us start with the first of them, (11a), which gives

$$\langle \vec{k}' | t_\alpha | \vec{k} \rangle = \langle \vec{k}' | v_\alpha | \vec{k} \rangle + \langle \vec{k}' | v_\alpha G^+ t_\alpha | \vec{k} \rangle$$

This equation describes the scattering on an isolated 'centre'. The corresponding Lippmann-Schwinger equation:

$$|\psi_k^+\rangle = |\vec{k}\rangle + G^+ v_\alpha |\psi_k^+\rangle = |\vec{k}\rangle + G^+ t_\alpha |\vec{k}\rangle$$

gives

$$\langle \vec{k}'|t_\alpha|\vec{k}\rangle = \langle \vec{k}'|v_\alpha|\psi_k^+\rangle = \int e^{-i\vec{k}'\cdot\vec{r}} v_\alpha(\vec{r}) \psi_k^+(\vec{r}) d^3r$$

Let us suppose that the potential $v_\alpha(\vec{r})$ is equal to zero outside a certain volume Ω_α and that the individual Ω_α do not overlap. Then $\psi_k^+(\vec{r})$ in the above integral is also the solution of the equation $(H_0 + u_\alpha)|\varphi_k\rangle = E_k|\varphi_k\rangle$, where $u_\alpha = v_\alpha$ inside Ω_α. It seems that if we are able to solve this equation, the quantity $\langle \vec{k}'|t_\alpha|\vec{k}\rangle$ has been found. However, we have to find such solutions which fulfil the new boundary condition on the surface of Ω_α. For this purpose, we further re-arrange the expression:

$$\langle \vec{k}'|t_\alpha|\vec{k}\rangle = \int_{\Omega_\alpha} e^{-i\vec{k}'\cdot\vec{r}} u_\alpha(\vec{r}) \varphi_k(\vec{r}) d^3r$$

$$= \int_{\Omega_\alpha} e^{-i\vec{k}'\cdot\vec{r}} (E - H_0) \varphi_k(\vec{r}) d^3r$$

$$= \int_{\Omega_\alpha} \left(\varphi_k(\vec{r}) E_k e^{-i\vec{k}'\cdot\vec{r}} - e^{-i\vec{k}'\cdot\vec{r}} H_0 \varphi_k(r) \right) d^3r$$

This can be further modified by means of the relation:

$$E_{k'} e^{-i\vec{k}'\cdot\vec{r}} = H_0 e^{-i\vec{k}'\cdot\vec{r}}$$

to a form:

$$\langle \vec{k}'|t_\alpha|\vec{k}\rangle = \int_{\Omega_\alpha} \left(\varphi_k(\vec{r}) H_0 e^{-i\vec{k}'\cdot\vec{r}} - e^{-i\vec{k}'\cdot\vec{r}} H_0 \varphi_k(\vec{r}) \right) d^3r$$

$$+ (E_k - E_{k'}) \int_{\Omega_\alpha} \varphi_k(\vec{r}) e^{-i\vec{k}'\cdot\vec{r}} d^3r$$

Since we are interested in the scattering on the energy shell, the second term is equal to zero. The first integral on the right-hand side will be re-arranged by means of the Green theorem to give

$$\langle \vec{k}'|t_\alpha|\vec{k}\rangle = \int_{\Sigma_\alpha} \left(\varphi_k(\vec{r}) \frac{\partial}{\partial n} e^{-i\vec{k}'\cdot\vec{r}} - e^{-i\vec{k}'\cdot\vec{r}} \frac{\partial}{\partial n} \varphi_k(\vec{r}) \right) d\sigma$$

where $\partial/\partial n$ denotes the derivative in the direction of the normal to the surface Σ_α. Accordingly the result depends on the knowledge of φ_k and $(\partial/\partial n)\varphi_k$ on the surface Σ_α. Since the solution of the equation

$$(H_0 + v_\alpha)|\psi\rangle = E|\psi\rangle$$

coincides outside the volume Ω_α with the solution of the equation $H_0 \chi = E\chi$ and inside Ω_α with that of $(H_0 + u_\alpha)\varphi = E\varphi$, on the boundary surface Σ_α, it must satisfy the relations $\chi = \varphi$ and

$$\frac{\partial}{\partial n} \chi = \frac{\partial}{\partial n} \varphi$$

From these matching conditions we can determine $\varphi_k(\vec{r})$ as well as $(\partial/\partial n)\varphi_k(\vec{r})$ and thus calculate the amplitude $\langle\vec{k}'|t_\alpha|\vec{k}\rangle$. It is therefore evident that we obtain different problems according to how we divide the potential V and this is where the various dynamical theories actually differ.

(a) If V is undivided ($v_\alpha = V$), then only two of the equations (11) are left: $T = V + VG^+T$ and $I = |\langle\vec{k}_f|T|\vec{k}_i\rangle|^2$. $u_\alpha = U$ is the potential of an infinite crystal and we obtain the solution from the band theory, the matching taking place on the plane of the surface. This approach is called the band-structure approach [21-23], because the main problem consists in solving the band structure of the crystal. In its original form this approach resumes Bethe's work. The others are generally called multiple-scattering approaches.

(b) (i) If we divide the potential into translationally equivalent columns of reference perpendicular to the surface we obtain multiple scattering as treated, for example, by Kambe [24] and Kerre and Phariseau [25]. The main problem here is to find the Green function for the column that takes into account crystal periodicity.

(ii) If we divide the potential into layers parallel to the surface, we obtain the method of McRae [26], which resembles very much Darwin's approach [27]. The problem here is to find the reflection and transmission coefficients for each layer.

(iii) Finally, we can divide the potential into individual atomic potentials ('muffin-tins'), whereby we arrive at the method of Beeby [20] or Duke [28]. This method is close to Ewald's approach [29]. The problem lies in the determination of phase shifts.

Each of these approaches naturally has its advantages and drawbacks, and in particular cases one or the other is better suited. The band approach and the 'atomic' multiple scattering evidently represent the extreme cases. If a 'cut-off' potential is expected to be a good approximation (e.g. with pure metals), the band approach is suitable because of wide experience with band structure calculation. On the other hand, for determining unknown structures, the multiple-scattering approach seems more convenient as it directly involves positions of the individual atoms and enables a more feasible parametrization. Capart [22] has demonstrated that for the same model all methods lead to the same results.

9. INCLUSION OF INELASTIC COLLISIONS

This section contains only some remarks on this important question. From X-ray diffraction it is known that for the dynamical diffraction in the Bragg case the penetration depth of the radiation is usually by an order of magnitude smaller than follows from the absorption coefficient; since normal absorption is the consequence of inelastic collisions, the latter influence diffraction profiles relatively little in the dynamical case. From considerable changes of the diffraction profiles in LEED it can be inferred that, on the contrary, the penetration depth is limited to a larger extent by inelastic collisions than by dynamical effects and that, consequently, inelastic collisions play an essential role. This is an argument permanently emphasized by Duke [28, 5]. It is therefore obvious that quantitative agreement between theory and experiment can be achieved only if inelastic collisions are included in the model. One can imagine a number of mechanisms for achieving this: by exciting individual atoms in the crystal, upon which the principle of Auger spectroscopy is based (see e.g. Ref.[30]); and further, by collective excitations, i.e. plasma oscillations, exciton formation, etc. These processes manifest themselves partly in the reduction of the elastic intensity, partly in the broadening of the diffraction spots, in blurring of the fine structure of the measured profiles, etc.

In Section 4 it was mentioned that the influence of inelastic collisions is included in the optical potential V, but only in a formal way, because the latter is not known. It is therefore necessary to analyse the respective contributions further. One possibility is to divide the inelastic collisions into individual and collective excitations. The first can be considered localized more in the region of individual atoms, the second in the electron gas filling the space between the 'atoms'. In the potential V, the part representing the electron gas is then practically given by its mean value. If we choose this constant to be a complex number, we prefer the inelasticity in the electron gas. In the band approach this means that we simply put $E = E_r + iE_i$, where $E_i = \overline{V_i}$, without introducing any other change. In the multiple-scattering approach the necessary changes were introduced for the first time by Duke [28]; he redefined G^+ (see Eq.(7)) and instead of $(k^2 - H + i\epsilon)^{-1}$ he put $(k^2 - H + iE_i)^{-1}$. This leads to a faster convergence of the Born series (physically to the reduction of the probability of multiple scattering). The inclusion of the absorption in the region of atoms is respected in the band approach by considering V to be a complex quantity (not only its mean value). In the multiple-scattering approach it is respected by introducing complex phase shifts.

10. SUPPLEMENT: DISCUSSION OF A SIMPLE EXAMPLE BY MEANS OF THE BAND-STRUCTURE APPROACH

To illustrate at least one of the dynamical theories, we now use the band approach to investigate the scattering of electrons on a simple crystal model with an ideal surface [21]. The selected model allows the necessary calculations to be carried out explicitly and typical LEED situations to be clarified in the discussion. It was shown in Section 8 that the Schrödinger equation of a crystal with a geometrically ideal surface differs from that of an infinite crystal by a new boundary condition only. The wave function in a finite

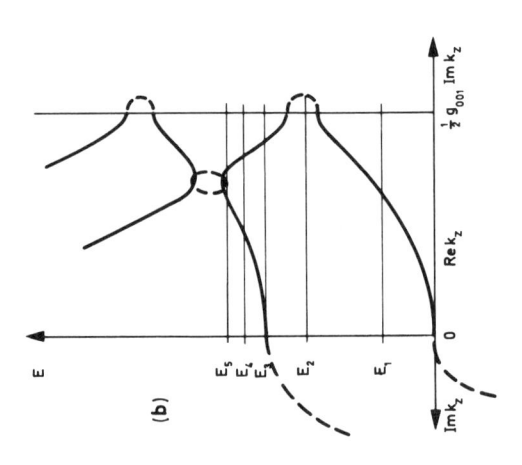

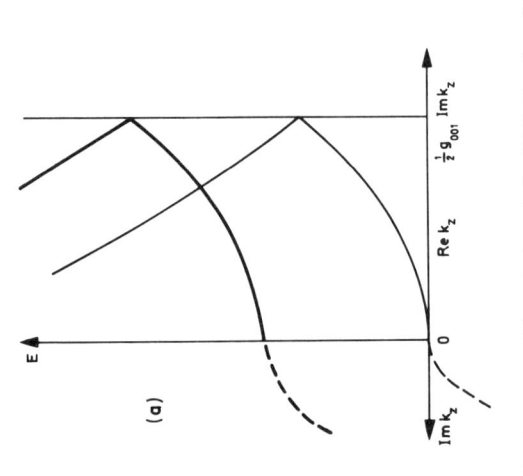

FIG. 4. (a) Band structure of free electrons in the (001) direction of the Brillouin zone. (b) Band structure in the nearly free electron approximation. Energies $E_1 \ldots E_5$ are explained in the text.

crystal can then be written as a superposition of possible solutions of the Schrödinger equation of an infinite crystal including those solutions which the conventional theory usually eliminates by means of the Born-von Kármán boundary conditions. This is naturally possible because these solutions (Bloch functions in a generalized sense) form something like a complete set of functions consistent with the new boundary condition.

The situation can also be described as follows. The new boundary condition deprives the component of the wave vector $\vec{k}$, which is perpendicular to the surface, of the property of a 'good' quantum number and leaves this property only to the remaining two components. For a given energy and given values of the two 'good' quantum numbers, the wave function of the finite crystal is a linear combination of all Bloch functions which the band theory can provide, i.e. with real as well as complex values of the wave vector component perpendicular to the surface.

In our further exposition the notation and some formulas from the paper by Capart [22] are used. These are denoted by C. (Especially important are pages 362 and 366 of Capart's work.) The energy E will be positive as everywhere in the present paper because we deal with the pure scattering.

Let us illustrate now the band approach on a simple example: we take an electron beam incident perpendicularly ($\vec{K}_t = 0$) on the (001) surface of an orthorhombic crystal. The elementary translation $\vec{b}$ of the latter along the y axis will be supposed smaller than $\vec{a}$ such that the effect connected with the former will not intervene in the energy region considered. In the (001) direction of the Brillouin zone, the crystal will be represented by a nearly-free electron (NFE) band structure, shown in Fig. 4b. Its wave function will therefore consist of two Bloch waves:

$$\psi_s = \sum_{n=1,2} B_n \Phi_n \qquad (19), (C8)$$

either of which will include two Fourier components with reciprocal vectors $g_{00} = 0$, $g_{10} \neq 0$, parallel to the surface:

$$\Phi_n = \sum_{h=0,1} C_{h0}(K_n, z) e^{ig_{h0}x} \qquad (20a), (C7)$$

Strictly speaking, we should also include in the sum in Eq. (20a) the term with h = -1. Due to the symmetry of the crystal surface, this would result in the appearance of symmetric or antisymmetric linear combinations of plane waves $e^{\pm ig_{10}x}$ in our formulas instead of plane waves alone. However, this would not lead to anything new in our model and therefore we shall not consider it.

In the case of free electrons (see Fig. 4a), the wave function (19) would be a linear combination of two plane waves Φ_n^{free} (n = 1, 2):

$$\Phi_1^{free} = e^{iK_1 z}, \quad \Phi_2^{free} = e^{iK_2 z} e^{ig_{10}x} \qquad (20b)$$

where

$$K_1 = \sqrt{E - V_{000}}, \quad K_2 = \sqrt{E - V_{000} - g_{10}^2}$$

and $V_{000} < 0$ denotes the potential step from the vacuum (zero of energy) to the bottom of the energy band. The dependence of the energy on the real or imaginary part of the component k_z of the reduced wave vector $\vec{k}$ would, respectively, for Φ_1^{free} correspond to the thin line in Fig. 4a and for Φ_2^{free} to the thick line. In this case, however, we should obtain only the specularly reflected beam from the potential step, the intensity of the former decreasing with increasing energy.

The electronic wave function in the vacuum ($z < 0$) will have the form:

$$\psi_\nu = e^{iK_\nu z} + \sum_{h=0,1} A_{h0} \exp[i(g_{h0} x - K_{\nu,h0} z)] \qquad \text{(21a), (C2)}$$

where

$$K_{\nu,h0} = \sqrt{E - g_{h0}^2} \qquad \text{(21b)}$$

and $K_\nu = K_{\nu,00}$; $C_{h0}(K_n, z)$ have a different form according to the range of the energy spectrum investigated. Let us consider five different energies, E_1 to E_5, from Fig. 4b.

In the case of the energies E_1 and E_2 it is reasonable to consider only one Bloch function and one Fourier component with $g_{00} = 0$. Only $C_{00}(K_1, z)$ will therefore be different from zero in both cases. For E_1 it will have the form:

$$C_{00}(K_1, z) = e^{iK_1 z} + \eta \exp[i(K_1 - g_{001})z]$$

where η is a very small quantity, for which the NFE perturbation calculation gives

$$\eta = \frac{V_{001}}{E - V_{000} - (K_1 - g_{001})^2}$$

and $\eta = A_{00}$ holds, as is immediately seen from Eqs (C9) and (C10) of Ref. [22] for continuous matching of the wave function. If E_1 approaches the forbidden band, the specular reflection from the crystal surface increases together with η. For energies E_1 sufficiently far apart from the forbidden band, there is practically no reflection of the incident beam from the crystal (except for the reflection from the 'inner' potential V_{000}).

For E_2 in the forbidden band, $C_{00}(K_1, z)$ has the form [31]:

$$C_{00}(K_1, z) = e^{-\kappa z} \cos(\tfrac{1}{2} g_{001} x + \delta)$$

$$\kappa^2 = -E - (\tfrac{1}{2}g_{001})^2 + \sqrt{E(g_{001})^2 + V_{001}^2}$$

$$\sin 2\delta = -\frac{g_{001}\,q}{V_{001}}$$

as can be shown by a simple NFE calculation. By using Eqs (C9) and (C10) we obtain for the reflection coefficient A_{00} the relation

$$A_{00} = \frac{iK_\nu + \kappa - \tfrac{1}{2}g_{001}\operatorname{tg}\delta}{iK_\nu - (\kappa - \tfrac{1}{2}g_{001}\operatorname{tg}\delta)}$$

which shows that the electron beam with energy in the forbidden band undergoes total reflection because $|A_{00}|^2 = 1$ holds.

For energies E_3, E_4 and E_5, both Bloch functions Φ_1, Φ_2 and both Fourier components must be considered. All four quantities $C_{h0}(K_n, z)$ are therefore different from zero and in our NFE model have the form:

$$C_{00}(K_1, z) = e^{iK_1 z}\alpha \;,\; C_{10}(K_1, z) = e^{iK_1 z}\beta$$
$$C_{00}(K_2, z) = e^{iK_2 z}\gamma \;,\; C_{10}(K_2, z) = e^{iK_2 z}\delta \tag{22}$$

where the magnitude of the coefficients $\alpha, \beta, \gamma, \delta$ follows from the calculation of the band structure of volume states of the infinite crystal in the NFE approximation. Eqs (C9) and (C10) give the following inhomogeneous system of equations for the unknown coefficients A_{00}, A_{10}, B_1 and B_2:

$$\left. \begin{array}{r} 1 + A_{00} = \alpha B_1 + \gamma B_2 \\[4pt] A_{10} = \beta B_1 + \delta B_2 \\[4pt] K_{\nu,00}(1 - A_{00}) = \alpha K_1 B_1 + \gamma K_2 B_2 \\[4pt] -K_{\nu,10} A_{10} = \beta K_1 B_1 + \delta K_2 B_2 \end{array} \right\} \tag{23}$$

and its solution reads:

$$A_{00} = [-\alpha\delta(K_1 - K_\nu)(K_2 + K_{\nu,10}) + \beta\gamma(K_2 - K_\nu)(K_1 + K_{\nu,10})]X^{-1} \tag{24a}$$

$$A_{10} = -2\beta\delta K_\nu(K_1 - K_2)X^{-1} \tag{24b}$$

$$B_1 = 2\delta K_\nu(K_2 + K_{\nu,10})X^{-1} \tag{24c}$$

$$B_2 = -2\beta K_\nu (K_1 + K_{\nu,10}) X^{-1} \tag{24d}$$

where

$$X = \alpha\delta (K_1 + K_\nu)(K_2 + K_{\nu,10}) - \beta\gamma (K_2 + K_\nu)(K_1 + K_{\nu,10}) \tag{24e}$$

Let us now consider special cases. If the conditions now to be discussed are fulfilled, surface resonance can appear in the neighbourhood of the energy E_3. In the scattering theory, the effect of resonance is often related to the existence of two simultaneously occurring phenomena: the fluctuation of the intensity of one scattering 'channel' when another 'channel' is opened. In the neighbourhood of E_3 the fluctuation appears in A_{00}, i.e. in the specularly reflected beam, and a new 'channel' opens in A_{10}, the latter corresponding to the beam propagating parallel to the crystal surface. Here for the first time A_{10} assumes non-zero (and straight large) values. Strictly speaking, the resonance should appear at an energy given by the equation

$$K_{\nu,10} = 0 \tag{25}$$

when $K_{\nu,10}$ passes from pure imaginary values to real ones, while the corresponding wave (Eq.(19)) changes from a wave damped into the vacuum to a propagating wave. Since $K_{\nu,10}$ is the z component of the wave vector, Eq.(25) says that the propagation takes place parallel to the crystal surface.

However, the energy E_3 does not correspond exactly to the condition (25), but, according to (20b) and (21b), is shifted by the 'inner' potential V_{000}. This is used for the explanation of both the width of the resonance line and its sharpness, and eventually its disappearance in some cases. All these properties of the surface resonance are obvious from Eqs (24a) and (24b). Let us re-arrange the first of them into the form:

$$\frac{1 + A_{00}}{1 - A_{00}} = \frac{\alpha\delta(q + q') - \beta\gamma(n + q)}{\alpha\delta n(q + q') - \beta\gamma q'(n + q)} \tag{26}$$

where

$$q = \frac{K_{\nu,10}}{K_\nu}, \quad q' = \frac{K_2}{K_\nu}, \quad n = \frac{K_1}{K_\nu} \tag{27}$$

As already said, q and q' are not equal to zero at the same energy, i.e. simultaneously. However, in the energy range

$$V_{000} + g_{10}^2 \le E \le g_{10}^2 \tag{28}$$

which roughly corresponds to the width of the resonance line, both q and q' may be small quantities if V_{000} is not too large. In this range q' is real and q imaginary, and hence

$$|q + q'|^2 = |q|^2 + |q'|^2 = |V_{000}| K^{-2} \tag{29}$$

If V_{000} is so large that the inequality

$$|q + q'| < \left|\frac{\beta\gamma}{\alpha\delta}(n + q)\right| \tag{30}$$

is not fulfilled, it follows from (26) that the specular reflection A_{00} is small and the resonance line is not pronounced. If we suppose that in the region of resonance

$$K_\nu \approx K_1 = g_{10}$$

holds, (30) can be simplified, because (26) can be written in the form

$$\frac{1 + A_{00}}{1 - A_{00}} = \frac{(q + q') - \frac{\beta\gamma}{\alpha\delta}(n + q)}{q + q'} \tag{31}$$

and (30) consequently reads

$$|V_{000}| < \left(\frac{\beta\gamma}{\alpha\delta} g_{10}\right)^2 \tag{32}$$

If (32) is not satisfied, we cannot expect in our model a pronounced development of the resonance line in the specularly reflected beam.

Moreover, let us consider the 'channel' A_{10} and show that immediately it opens it assumes large values. By using (27) we can write (24b) in the form:

$$A_{10} = \frac{2\beta\delta(q - n)}{\alpha\delta(n + 1)(q + q') - \beta\gamma(q' + 1)(n + q)} \tag{33}$$

and indeed if the resonance is pronounced

$$A_{10} \approx -\frac{\beta}{\alpha(q + q')}$$

is valid if the first term in the denominator prevails, or

$$A_{10} \approx -\frac{2\delta}{\gamma}$$

is valid if the second term prevails. In both cases A_{10} can take large values. However, the intensity in this 'channel' rapidly disappears with increasing distance of the energy from the resonance region (28).

At the energy E_4, the quantities q and q' are close to unity and A_{10} has practically zero values. In this energy region, the Bloch functions are almost identical with the plane waves:

$$\Phi_1 \approx \Phi_1^{free}, \quad \Phi_2 \approx \Phi_2^{free}$$

and β and γ can be considered negligible in comparison with other quantities appearing in (24), (31) and (33). From these equations we obtain (by putting $\alpha = \delta = 1$)

$$A_{00} = \frac{1-n}{1+n}, \quad A_{10} = 0$$
$$B_1 = -\frac{2}{1+n}, \quad B_2 = 0$$
(34)

with an accuracy up to the quantities of the order of β, γ. We see that in the situations where the energy bands behave as free electron bands, the incident beam passes through the crystal with practically unchanged intensity, only an insignificant specular reflection occurring on the potential V_{000}. Together with $V_{000} \to 0$ even this reflection disappears. Equation (34) strongly reminds one of optics (Fresnel formulas), if we realize that n has the meaning of the index of refraction.

If the energy (region E_5) approaches the next forbidden band, this time within the Brillouin zone, the increasing interaction between both Fourier components leads to an increase of β and γ and, in agreement with (24), both A_{10} and A_{00} increase. The growth attains its maximum in the forbidden band. A_{10} corresponds to the non-specularly reflected beam; A_{00} corresponds to the line of the respective secondary Bragg reflection in the specularly reflected beam. We do not give explicit expressions for the case with the energy in the forbidden band because they are derived elsewhere by the use of Green functions [32].

It is obvious that, within the framework of the NFE model discussed, it is possible to determine directly from the known band structure the intensities of the reflected beams or to draw conclusions from the latter on the extent of deformations of the NFE band structure as compared with the free-electron case (see Fig.4). However, it must be pointed out that the NFE approximation holds only in cases where the atomic scattering is weak. In energy regions with strong atomic scattering [33], it cannot be applied. These are those regions the energies of which are, roughly speaking, specular images of atomic bound-state energies with respect to the zero of energy.

REFERENCES

[1] BETHE, H.A., Ann. Physik 87 (1928) 55.
[2] DAVISSON, C.J., GERMER, L.H., Phys. Rev. 30 (1927) 705.
[3] STERN, R.M., PERRY, J.J., BOUDREAUX, D.S., Rev. Mod. Phys. 41 (1969) 275.
[4] STERN, R.M., TAUB, H., An Introduction to the Dynamical Scattering of Electrons by Crystals, Polytech. Inst. Brooklyn (1971).
[5] DUKE, C.B., Theory of low energy electron diffraction from single crystal solid surfaces, preprint (1970), unpublished. See also LEED — Surface Structures of Solids (Proc. Int. Summer School on the Investigation of the Surface Structures of Solids by LEED and Supplementary Methods, Prague, 1972 (LÁZNIČKA, M., Ed.), Union of Czechoslovak Mathematicians and Physicists, Prague (1972).
[6] MIYAKE, S., HAYAKAWA, K., Acta Crystallogr. Sect. A. 26 (1970) 60.
[7] FINGERLAND, A., TOMÁŠEK, M., Čs. Čas. Fys. A22 (1972) 602 (in Czech).
[8] MOLIERE, K., Čs. Čas. Fys. A19 (1969) 181 (in Czech).
[9] LÁZNIČKA, M., Pokroky MFA 15 (1970) 16 (in Czech).

[10] LANDER, J.J., MORRISON, J., J. Appl. Phys. 34 (1963) 3517.
[11] KUBĚNA, J., Czech. J. Phys. B 18 (1968) 777, 1233.
[12] STERN, R.M., BALIBAR, F., Phys. Rev. Lett. 25 (1970) 1338.
[13] CELÝ, J., Phys. Stat. Sol. 4 (1964) 521.
[14] PARK, R.L., HOUSTON, I.E., SCHREINER, D.G., Rev. Sci. Instrum. 42 (1971) 60.
[15] YOSHIOKA, H., J. Phys. Soc. Jap. 12 (1957) 618.
[16] SLATER, J.C., Phys.Rev. 51 (1937) 840.
[17] ROMAN, P., Advanced Quantum Theory, Addison-Wesley, Reading, Mass. (1965).
[18] WOOD, E.A., J. Appl. Phys. 35 (1964) 1306.
[19] LANDER, J.J., Surf. Sci. 1 (1964) 125.
[20] BEEBY, J.L., J. Phys. C 1 (1968) 82.
[21] BOUDREAUX, D.S., HEINE, V., Surf. Sci. 8 (1967) 426.
[22] CAPART, G., Surf. Sci. 13 (1969) 361.
[23] MARCUS, P.M., JEPSEN, D.W., Phys. Rev. Lett. 20 (1968) 925.
[24] KAMBE, K., Z. Naturforsch. 22a (1967) 322, 422; 23a (1968) 1280.
[25] KERRE, E., PHARISEAU, P., Physica 51 (1971) 509.
[26] McRAE, E.G., Surf. Sci. 11 (1968) 479, 492.
[27] DARWIN, C.G., Philos. Mag. 27 (1914) 675.
[28] DUKE, C.B., TUCKER, C.W., Jr., Surf. Sci. 15 (1969) 231.
[29] EWALD, P.P., Ann. Phys. 49 (1916) 1,117; 54 (1917) 519.
[30] TAYLOR, N., Čs. Čas. Fys. A21 (1971) 511 (in Czech).
[31] FORSTMANN, F., Z. Phys. 235 (1970) 69.
[32] TOMÁŠEK, M., to be published in Czech. J. Phys.
[33] PENDRY, J.B., FORSTMANN, F., J. Phys. C. 3 (1970) 59.

IAEA-SMR-15/33

LOW-ENERGY ELECTRON DIFFRACTION AND SURFACE TOPOGRAPHY

G.A. SOMORJAI
Department of Chemistry,
University of California,
Berkeley, Calif.,
United States of America

Abstract

LOW-ENERGY ELECTRON DIFFRACTION AND SURFACE TOPOGRAPHY.
Part I: 1. Surface topography. 2. Surface crystallography. 3. Structure of solid surfaces. 4. Structure of adsorbed gases on solid surfaces. Part II: 5. Auger electron spectroscopy. Part III: 6. Interaction of molecular beams with solid surfaces. 7. Results of elastic scattering studies. 8. Results of inelastic scattering studies. 9. Reactive scattering of molecular beams. 10. Future trends. Part IV: 11. Catalysis by single-crystal surfaces at low and high pressures.

PART I

1. SURFACE TOPOGRAPHY

Surfaces have two major functions that make them so important in our everyday life: (a) The surface is the first line of defence of any condensed phase against external attack by ambient gases, for example, that cause corrosion and other solid-state reactions. Therefore, passivation of the surface to prevent the occurrence of chemical reactions is one of the major preoccupations of surface scientists. (b) Chemical reactions can occur at surfaces with optimum efficiency and specificity. The activation energy for surface reactions is lowered greatly with respect to the activation energy for the same net reaction in the gas phase. In addition, out of many competing thermodynamically feasible reactions, the proper surface selects one or two while rejecting the others.

In the past it has been difficult to study the topmost layer of solids on an atomic scale. The surface layer contains roughly 10^{15} atoms/cm^2, while bulk densities are of the order of 10^{22} atoms/cm^3. Any surface-sensitive experimental probe should be able to detect the properties of the few surface atoms in the background of the very large concentration of atoms in the underlying layers. In recent years a great variety of techniques have become available, mainly involving scattered electrons and atoms, which permit us to study various surface properties on an atomic scale. Elastically scattered low-energy electrons and atoms provide information about the atomic surface structure. Inelastically scattered beams of electrons (electron spectroscopy) provide spectroscopic information on the electronic structure and bonding of surface atoms, while inelastic scattering of atoms or molecules reveals the nature of energy transfer between the incident particle and the surface. Several of these techniques will be discussed in later sections.

What sort of information would we like to have about surfaces? First, we ought to know the atomic surface structure, and this is obtained by low-energy electron diffraction (LEED). We can now determine the structure of surfaces and of adsorbed gases and vapours, i.e. we can carry out surface crystallography. We must also know the chemical composition at the surface to make sure that impurities segregated at the surface do not influence our experimental results. Auger electron spectroscopy or photo-electron spectroscopy provides this information. With knowledge of the atomic structure and surface composition we can study atomic and electron transport at surfaces (atomic surface diffusion or ion diffusion and electronic or ionic surface conductance). Finally, using a combination of experimental techniques, we can then proceed to study complex chemical reactions at surfaces and study the reactivity as related to the atomic structure and chemical composition.

Let us now review the topography of solid surfaces. Figure 1 shows an optical microscope picture of a freshly etched (100) surface of a NaCl crystal. The pyramidal pits that appear on the surface are called etch pits. These form when line defects, dislocations, intercept the surface. Dislocation densities of $10^5/cm^2$ are common for ionic solids (like alkali halides), while dislocation densities as high as $10^8/cm^2$ can be present in metals. An electron microscope picture of a (111) face of a platinum single crystal is shown in Fig.2. This surface shows a great deal of roughness; hills and valleys and atomic terraces are present. These features are more pronounced in Fig.3, which shows a scanning electron microscope picture of a freshly vaporized zinc surface. Atomic terraces 1-2 μm in width separated by ledges are clearly visible. Thus, these and other studies have

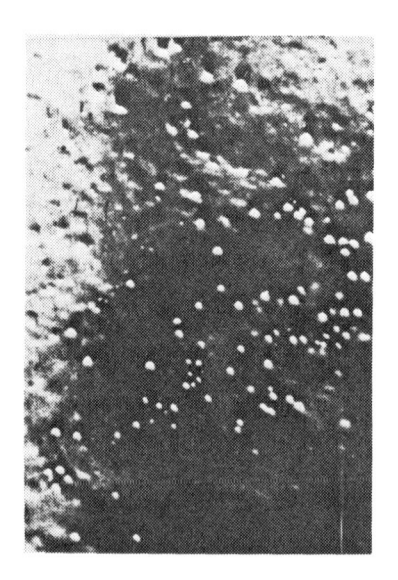

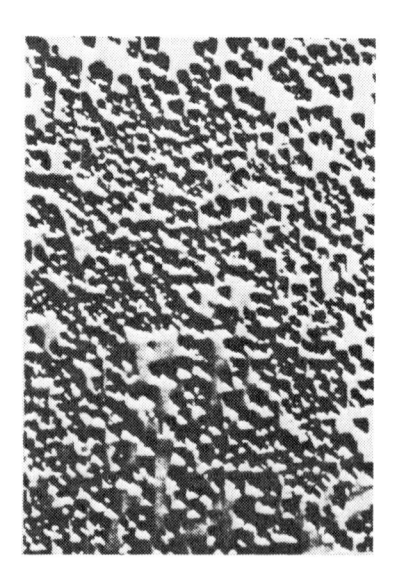

FIG. 1. Etch pits on the (100) crystal face of NaCl.

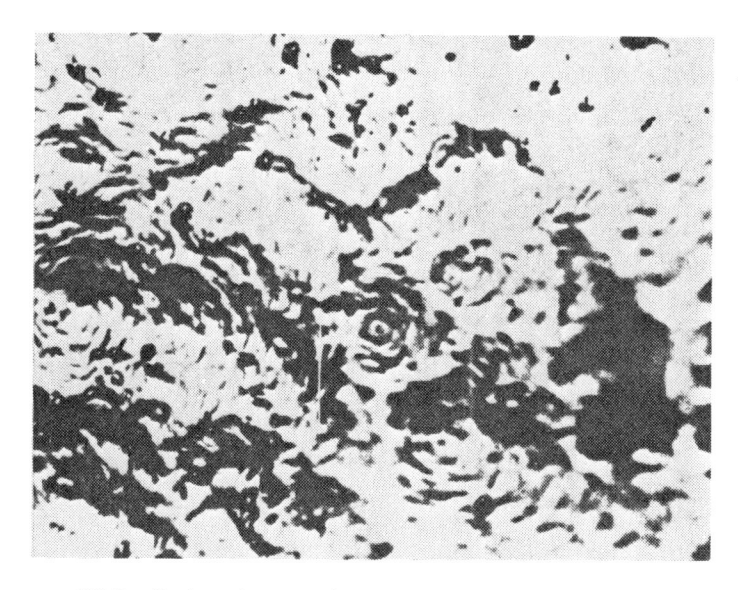

FIG.2. Electron microscope picture of (111) crystal face of platinum.

clearly indicated that the surface is heterogeneous. The model that was developed depicting surface atoms in various positions on a heterogeneous surface is shown in Fig.4. This model, which assumes only nearest-neighbour interactions, was developed over thirty years ago and has been used to advantage in explaining crystal growth characteristics and the kinetics of vaporization.

What is the experimental evidence for the presence of various surface sites, ordered terraces, steps and kinks on an atomic scale? Much of the experimental evidence comes from LEED studies. The principle of the LEED experiment is shown in Fig.5. Electrons of energy 30-150 eV are back-scattered from one face of a single crystal and the elastically scattered fraction that contains the diffraction information is accelerated onto a fluorescent screen where the diffraction pattern is displayed. It should be noted, however, that most of the incident electrons lose energy in the scattering process, as shown by Fig.6. At 100 eV no more than 5% of the scattered beam is elastic. A suitable grid system with a retarding potential is employed to retard the large inelastic fraction and thereby separate them from the diffracted electrons. Figure 7 shows a typical diffraction pattern obtained from the (111) crystal face of platinum. The surface has a sixfold rotational symmetry, which is reflected in its diffraction pattern. Thus the surface structure is that expected from the projection of the bulk X-ray unit cell to the (111) crystal face. The sharpness of the diffraction spots indicates that the surface is well ordered; the size of the ordered domains is larger than the coherence length (2-500 Å) of the low-energy electrons.

In addition to low Miller index surfaces, the surfaces characterized by high Miller index have also been studied. One crystallographic zone of platinum is shown in Fig.8. If we cut the surface at 6.2° from the (111) face in the direction of the (100) face, we obtain a surface which is

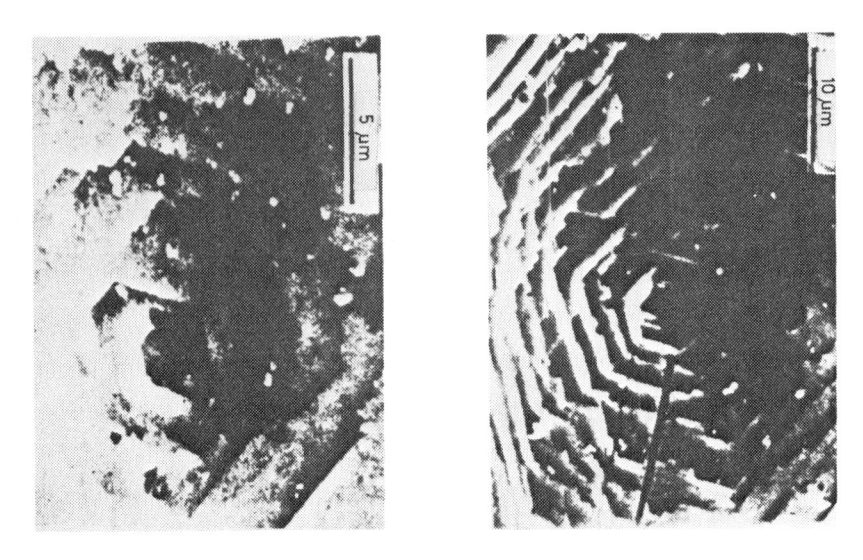

FIG. 3. Scanning electron microscope picture of freshly vaporized zinc crystal surface.

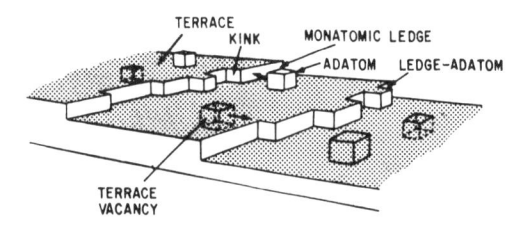

FIG. 4. Schematic representation of a heterogeneous surface.

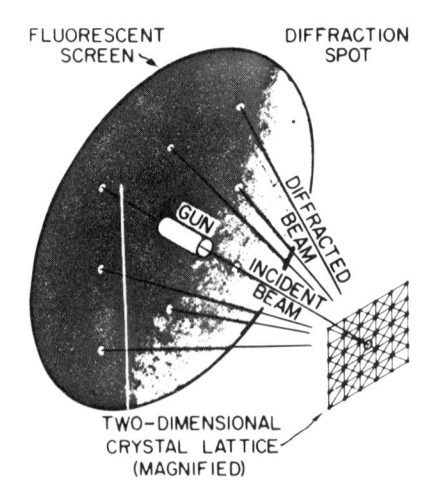

FIG. 5. Principle of LEED experiment.

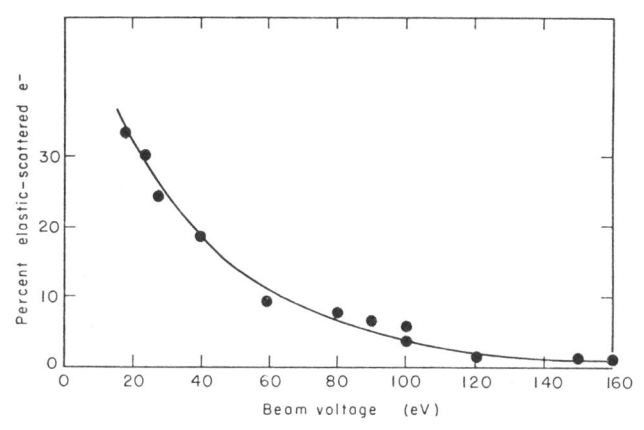

FIG. 6. Fraction of elastically scattered electrons as a function of electron energy for a platinum crystal surface.

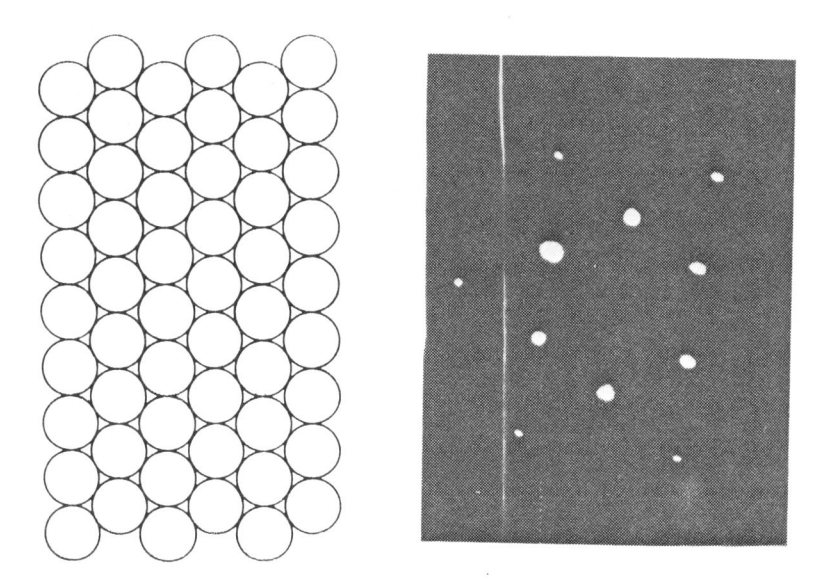

FIG. 7. Diffraction pattern and schematic representation of the (111) crystal face of platinum.

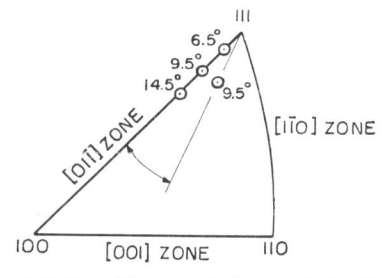

FIG. 8. One crystallographic zone of a face-centred cubic crystal.

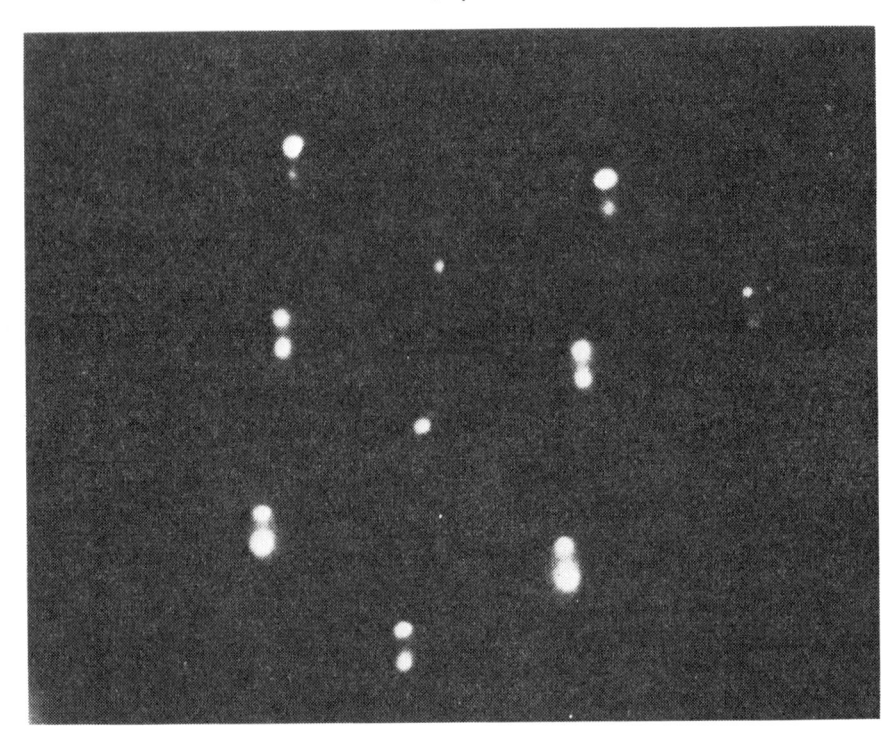

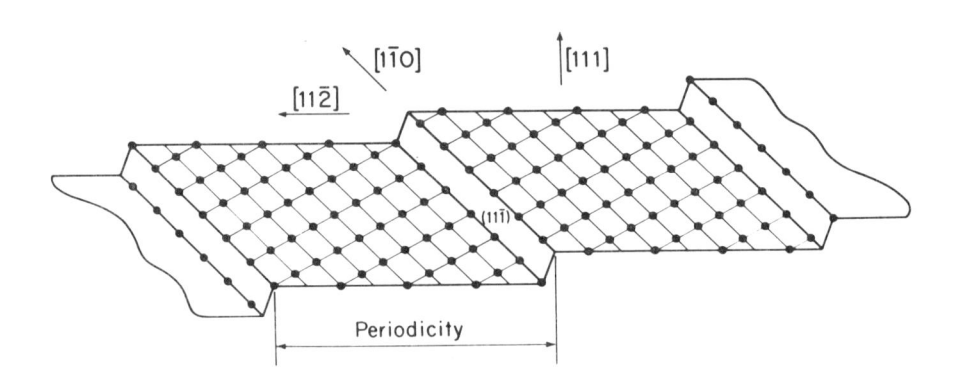

FIG. 9. Diffraction pattern and schematic representation of (a) a stepped platinum surface and

LEED AND SURFACE TOPOGRAPHY 179

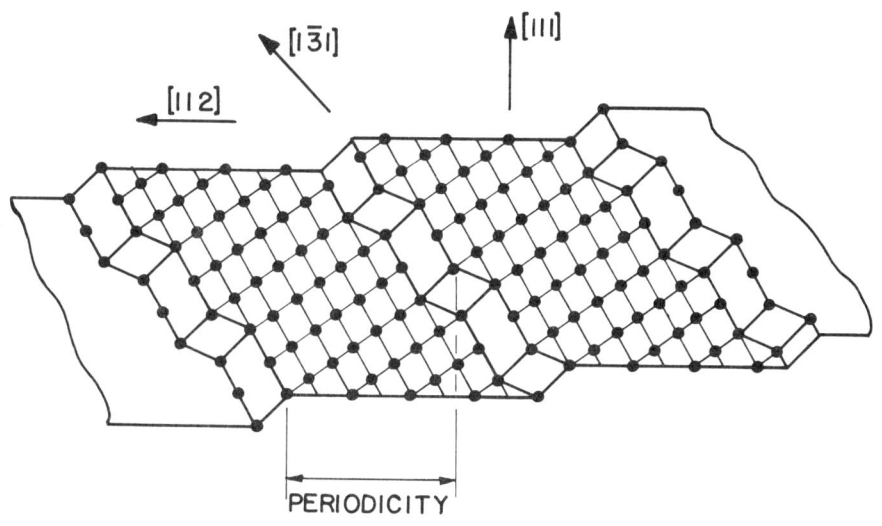

(b)
Pt(S)-[7(111)X(310)]

(b) a stepped platinum surface with high kink density.

characterized by a high Miller index of (544). The LEED pattern and the schematic representation of this surface is shown in Fig.9(a). The diffraction pattern shows doublets at well-defined electron energies in the direction of the cut, which indicates the appearance of a new periodic structure on the surface. Detailed analysis indicates that this surface has a stepped structure; atomic terraces of (111) orientation are separated by periodically arranged steps of one atom in height. The ordered arrangement of the atomic steps gives rise to the new diffraction features. Thus we have experimental evidence not only for the presence of well-ordered large domains of surface atoms but also for atomic steps. Experimental evidence for the presence of kinks can also be obtained if, for example, we cut a crystal at 9.5° of the (111) face and 20° away from the (100) direction, i.e. in the middle of the crystal zone. The resulting surface has a fairly high Miller index step orientation (310). The LEED pattern and schematic representation of this surface, shown in Fig.9(b), is also a stepped surface with (111) orientation terraces. The steps, however, must have a high density of kinks.

A heterogeneous surface not only has ordered areas where most atoms are located at their equilibrium positions, but it has disordered areas as well. Disorder may be introduced by mechanical treatment of various kinds that places atoms in non-equilibrium positions. Due to the appreciable activation energy for surface diffusion at low temperatures, the disorder is retained and it is only annealed at elevated temperatures where surface diffusion rates are appreciably large. Figure 10 shows a LEED pattern from a highly disordered ion-bombarded surface. The broad diffraction spots and their low intensity indicate the presence of surface disorder that have decreased the size of the ordered atomic domains at the surface. Thus, some information about the surface disorder is obtainable from the diffraction pattern. However, a more versatile tool to study surface disorder is the scattering of helium atoms from surfaces. The scheme of the experiment is shown in Fig.11. A beam of helium atoms is scattered from a crystal surface and the intensity of the specularly scattered helium beam is detected. For this experiment the angle of incidence and the angle of scattering with respect to the surface normal are identical. The experimental apparatus is shown in a schematic diagram in Fig.12. When the helium beam is scattered specularly, the rectilinear plot, shown in Fig.13, shows a large peak at the scattering angle and the intensity drops off rapidly away from the angle of specularity. This happens if the incident atom spends very little time on the surface, less than atom vibration times of the order of 10^{-12} seconds, so that energy exchange between the incident helium atom and the surface does not take place. If the incident atom stays at the surface longer (many vibrational periods), energy exchange with the surface can take place and then the helium atom is re-emitted with a Maxwellian distribution characteristic of the surface temperature. Under these conditions, cosine angular distribution shown by line (b) results, where the intensity drops off as the cosine of the angle of scattering away from the surface normal.

Figure 14 shows the angular distribution of scattered helium from a well-ordered platinum (100) surface. The scattered beam is highly specular; the intensity at the specular angle drops off sharply away from the specular angle. Figure 15 shows the angular distribution of helium scattered from a highly disordered etched platinum surface. Under these conditions, the

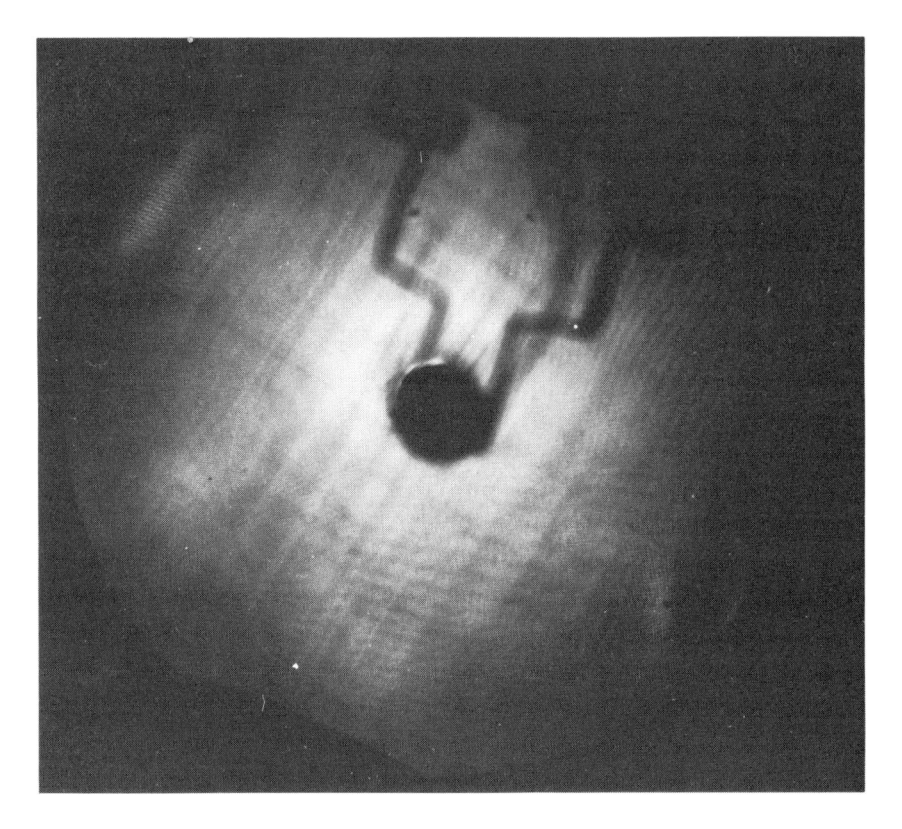

FIG. 10. LEED pattern from an ion-bombarded, disordered surface.

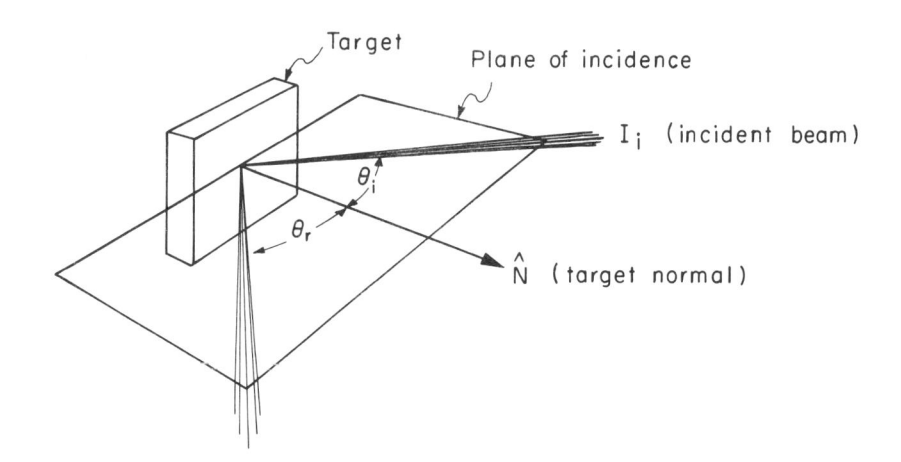

FIG. 11. Scheme of the atomic beam-surface scattering experiment.

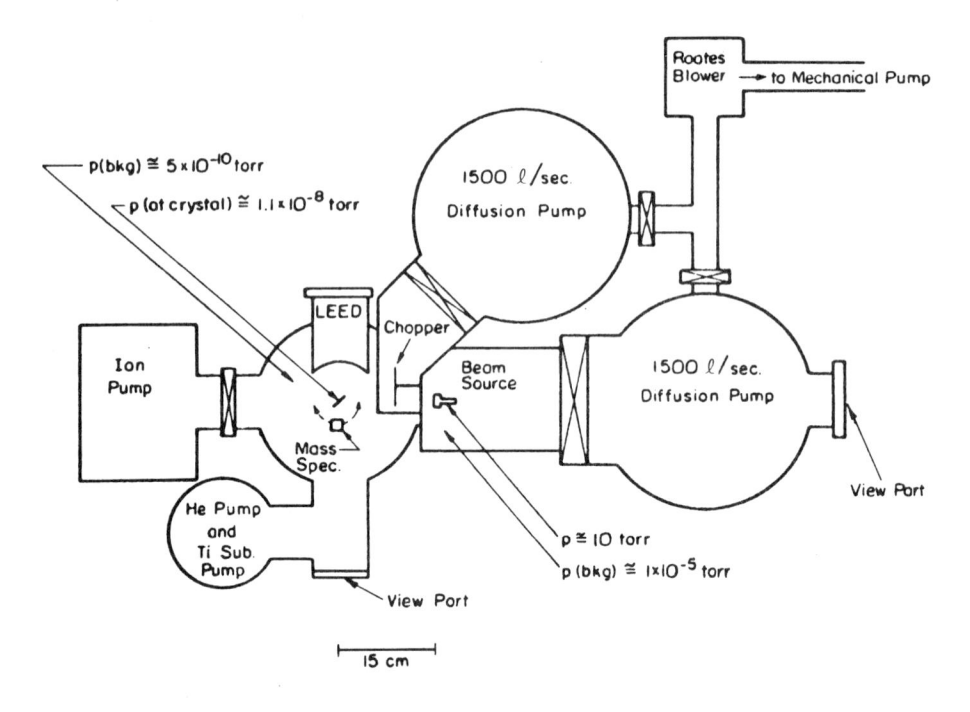

FIG. 12. Experimental apparatus for molecular beam-surface scattering.

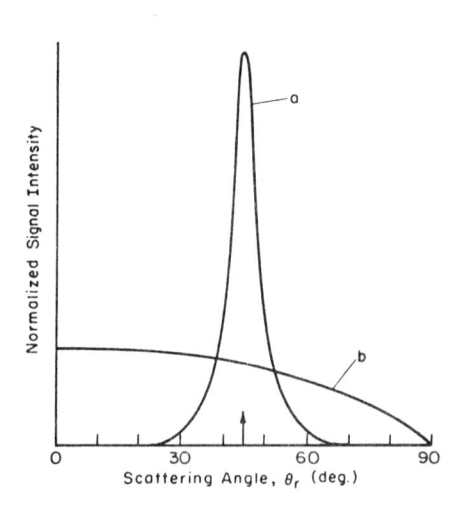

FIG. 13. Rectilinear plot depicting the angular distribution of (a) specularly scattered and (b) surface equilibrated scattered beams.

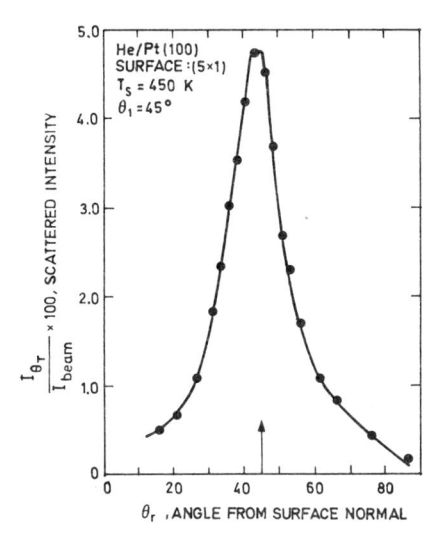

FIG. 14. Angular distribution of helium scattered from a well ordered Pt(100) crystal face.

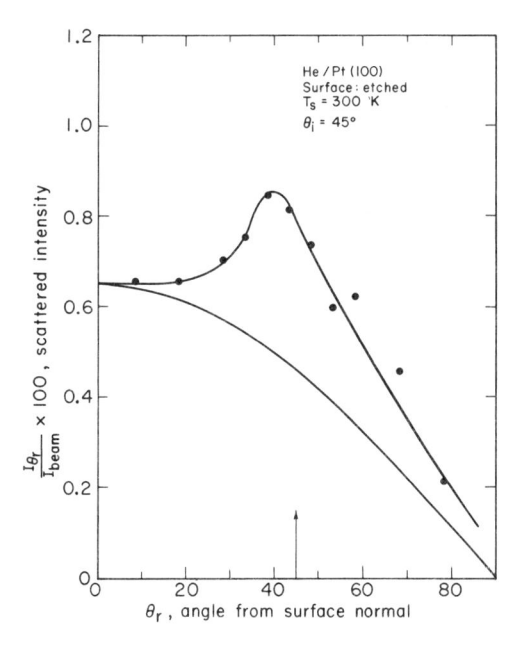

FIG. 15. Angular distribution of helium scattered from a disordered platinum surface.

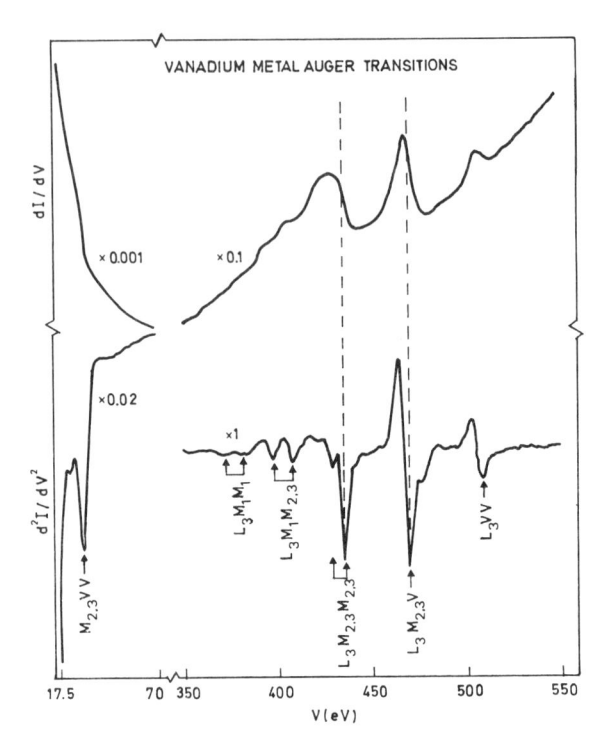

FIG. 16. First and second derivative Auger spectra from a V(100) surface.

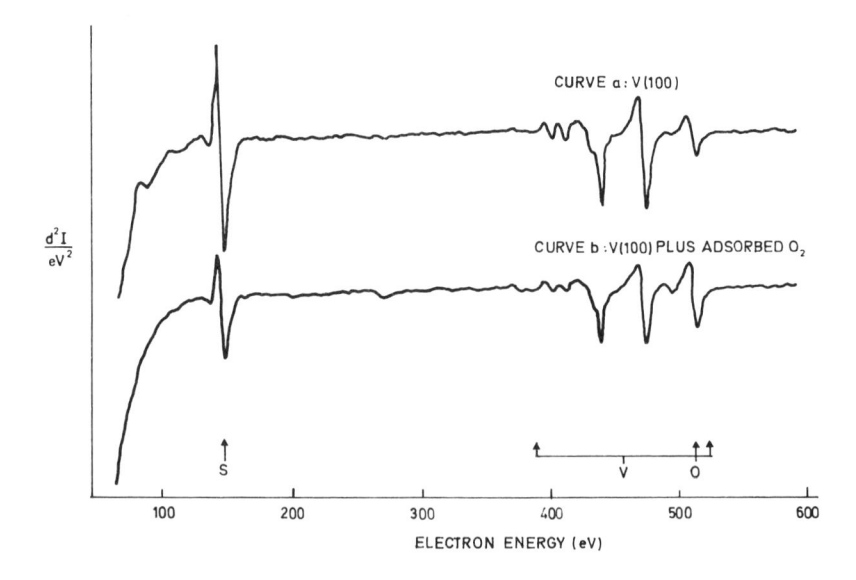

FIG. 17. Second derivative Auger spectrum from a vanadium surface.

scattering distribution is almost cosine-like, with a small peak at the specular angle. By monitoring the angular distribution of scattered helium beams, the degree of roughness of the surface, on an atomic scale, can be detected. At present, helium atom scattering is our most sensitive tool to study atomic disorder at surfaces.

One of the most important techniques for the study of surfaces is Auger electron spectroscopy. This tool provides us with a surface chemical analysis, both qualitative and quantitative. By suitable electronic circuitry, the Auger electrons emitted from surface atoms can be separated from other inelastically scattered electrons, and first and second derivatives of the inelastic current as a function of electron energy are displayed in Fig.16 for a vanadium (100) surface. The peaks are due to Auger transitions within the surface atoms, and these transitions 'fingerprint' each element present on the surface. The sensitivity of Auger analysis is about 1% of a monolayer or 10^{13} atoms/cm^2 and thus the presence of impurities can readily be discovered by this technique. Figure 17 shows the same Auger spectrum as displayed in Fig.16, but on a more extended scale. The presence of sulphur and oxygen on the vanadium surface is clearly discernable by their Auger transitions. The height of the Auger peaks can be related to the amount of sulphur and oxygen present on these surfaces. Auger electron spectroscopy is always an integral part of a clean surface experiment since we need a control of surface cleanliness, and this is provided by electron spectroscopy.

2. SURFACE CRYSTALLOGRAPHY

The development of modern chemistry owes a great deal to techniques that permit the study of chemical phenomena on an atomic or molecular scale. Great advances in our understanding of the structure of molecules in the gas phase have come from the development of spectroscopy, from electron spectroscopy (ESCA) to microwave spectroscopy. These studies served as the basis for deciphering the molecular dynamics of gas phase reactions. In the solid state, X-ray, electron, and neutron diffraction studies of the atomic structure laid the foundation for the understanding of many solid-state transport phenomena. In general, solving the structure, atomic or electronic, is prerequisite to unravelling the chemical reaction dynamics.

In surface science the technique of low-energy electron diffraction (LEED) yields the structure of surfaces on an atomic scale. During the past ten years this technique has been providing detailed information on the structure of solid surfaces and on the structure of adsorbates. In the past five years an understanding of the nature of the diffraction of low-energy electrons has been achieved. That is, a theory has been developed that enables computation of the diffraction beam properties (intensity versus energy) if the positions of the surface atoms are specified. Thus, surface crystallography has emerged as a new field of surface science that permits the determination of the unique positions of atoms on the surface from the intensities of the diffraction beams.

From these studies a physical picture of the surface atomic structure is emerging. We know now that, for clean surfaces, atoms may occupy sites that are different from those expected from the projection of the bulk

(X-ray) unit cell. High Miller index surfaces have a unique stepped surface structure and atoms in steps exhibit exceptional reactivity. The chemical composition of diatomic and polyatomic solids may be very different from the bulk stoichiometry. Moreover, adsorbed gases and vapours form ordered surface structures and may undergo order-disorder transformations under the proper conditions of temperature and pressure.

In this section we review the present status of surface crystallography and much of our knowledge of the atomic structure of surfaces and of adsorbed molecules. The method of surface structure analysis will be presented, and then we shall discuss the structure of clean solid surfaces of low and high Miller index. Then the structure of adsorbed molecules of small and large size relative to the interatomic distance in the substrate plane will be discussed.

Determination of the unique atomic position of adsorbates, their distance from the underlying plane of atoms and their bond angle, yields the fundamental experimental data needed to unravel the surface chemical bond. The structures of adsorbed molecules reveal the nature of adhesion or lubrication on a molecular scale. Finally, the ordered adsorbate structures often play a rate-controlling role and are intermediates in catalysed surface chemical reactions.

2.1. Surface geometry and nomenclature

In this section some of the general features of the diffraction of low-energy electrons from crystal surfaces are discussed and a notation appropriate to the description of surface structures reviewed. In subsequent sections a detailed discussion of LEED theory and application is presented.

A typical LEED experiment consists of a mono-energetic beam of electrons ($10 \text{ eV} \leqslant E \leqslant 500 \text{ eV}$) incident on one face of a single crystal. Roughly one-half of the electrons are backscattered, and the elastically scattered fraction is allowed to impinge on a fluorescent screen. If the crystal surface is well ordered, a diffraction pattern (Fig.18) consisting of bright, well-defined spots will be displayed on the screen. The sharpness and overall intensity of the spots is related to the degree of order on the surface. Although the surface may be irregular on a microscopic and sub-microscopic scale (e.g. consisting of atomic terraces and ledges), the presence of sharp diffraction features indicates that the surface is ordered on an atomic scale, the atoms lying in planes parallel to the surface characterized by a two-dimensional lattice structure. The size of these ordered domains determines the quality of the diffraction pattern. Because of experimental limitations on the coherence width of the electron beam, ordered domains larger than approximately 500 Å in diameter are not detectable. However, if the ordered domains become significantly smaller than 500 Å, the diffraction spots broaden and become less intense.

The presence of sharp diffraction features in the LEED patterns establishes that crystal surfaces are ordered on an atomic scale. In addition, the positions and symmetry of the diffraction spots can be used to determine the two-dimensional periodicity of the ordered arrangement of surface atoms. We can imagine for the moment that the surface structure will be rather like the termination of the bulk structure along a crystal plane (e.g. the (100) plane of a cubic crystal) although there may be a

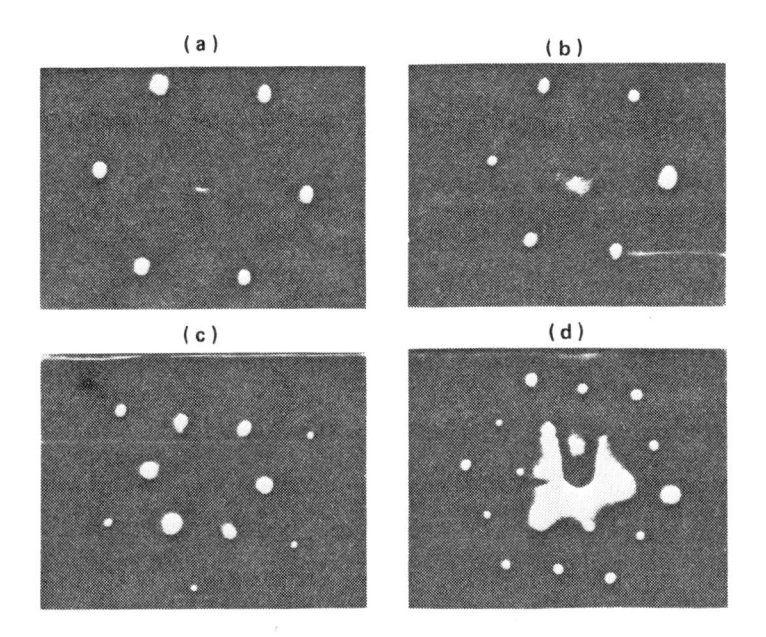

FIG. 18. Diffraction pattern of the (111) face of a platinum single crystal at four different incident electron beam energies: (a) 51 eV; (b) 63.5 eV; (c) 160 eV; (d) 181 eV.

re-arrangement or reconstruction of the surface atoms from the bulk structure. The presence of the surface destroys the bulk translational periodicity in the direction normal to the presumed planar surface. The translational periodicity of the solid parallel to the surface is retained and will be one of the five two-dimensional Bravais lattices [1]. The atoms lie in planes parallel to the surface and a translation $\vec{T}$ in the plane of the form

$$\vec{T} = n_1 \vec{a} + n_2 \vec{b} \tag{1}$$

takes each atom to an equivalent site. Here n_1 and n_2 are integers, and $\vec{a}$ and $\vec{b}$ are the primitive translation vectors that define the surface unit cell. The periodicity of the solid in the direction parallel to the surface is responsible for the basic momentum-conservation law of diffraction theory that

$$\vec{k}_\parallel' = \vec{k}_\parallel + \vec{g} \tag{2}$$

where $\vec{k}_\parallel$ and $\vec{k}_\parallel'$ are, respectively, the components of the incident and outgoing wave vector of the scattered electron in the direction parallel to the surface. The discrete set of vectors $\vec{g}$ comprise the surface reciprocal mesh so that, together with energy conservation ($E(\vec{k}') = E(\vec{k})$), Eq.(2) defines the directions of the allowed diffraction beams appearing in the diffraction pattern. In analogy with the X-ray crystallography of the bulk (see e.g. Ref.[2]), the primitive vectors $\vec{a}^*$ and $\vec{b}^*$ of the surface reciprocal mesh are related to the direct space vectors $\vec{a}$ and $\vec{b}$ by the equations:

$$\vec{a}^* = \frac{\vec{b} \times \vec{z}}{(\vec{a} \cdot \vec{b} \times \vec{z})} \tag{3a}$$

$$\vec{b}^* = \frac{\vec{z} \times \vec{a}}{(\vec{a} \cdot \vec{b} \times \vec{z})} \tag{3b}$$

where $\vec{z}$ is the surface normal. The allowed diffraction beams $\vec{g}$ are formally labelled by beam indices (hk) according to the equation:

$$\vec{g} = 2\pi(h\vec{a}^* + k\vec{b}^*) \tag{4}$$

It is evident from Eq.(2) that the diffraction pattern gives a representation of the surface reciprocal lattice, and indeed the vectors $\vec{a}^*$ and $\vec{b}^*$ may be determined from a measurement of the beam angles. Then by application of relations inverse to Eq.(3) one may solve for the vectors $\vec{a}$ and $\vec{b}$ that define the surface unit cell. Figure 19, for example, illustrates the relation of the direct and reciprocal space vectors for the case of the hexagonal Bravais lattice.

CRYSTAL LATTICE   RECIPROCAL LATTICE

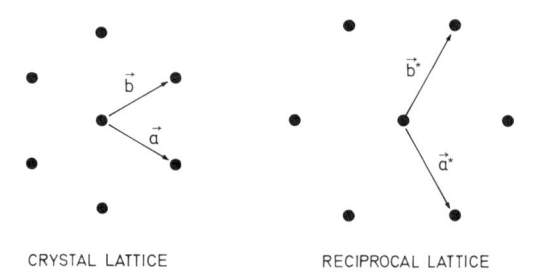

FIG.19. Translation vectors $\vec{a}$, $\vec{b}$ for the hexagonal Bravais lattice and corresponding reciprocal lattice vectors $\vec{a}^*$, $\vec{b}^*$. Note that the reciprocal lattice is a hexagonal lattice rotated by 30° with respect to the crystal lattice.

The basic complication of surface structure analysis via LEED comes from the fact that observation of the diffraction pattern geometry serves only to determine the size and shape of the two-dimensional unit cell which characterizes the translational periodicity parallel to the surface. Critical information about structural variations in the direction normal to the surface must be extracted from an analysis of the intensity of the diffracted beams. Such an intensity analysis is in principle required, for example, to determine the packing sequence and interlayer spacings of the top few atomic layers of a single-crystal surface. These types of analysis constitute the fundamental motivation and application of LEED theory and are reviewed in Section 2.2.

It is appropriate at this point to discuss notational conventions [1,3] for classifying surface structures. In the simplest case, the surface structure is given by the termination of the bulk structure along a given crystal plane. The surface unit mesh in such cases is briefly referred to as (1 × 1), indicating that the lattice vectors $\vec{a}$ and $\vec{b}$ in the surface region are identical to those of the underlying bulk substrate (e.g. Pt(111) - (1 × 1)). In more general cases the surface structure will differ greatly from that of the bulk substrate. An interesting example is the deposition of layers of foreign atoms (adsorbate) on the substrate material. Such adsorbed overlayer structures will in general have a periodicity different from that of the substrate and may even cause reconstruction of the substrate atoms near the surface. Another case is the reconstruction of the surface region of a chemically clean material, as occurs, for example, in silicon. It is clear that the structure of the first several outer atomic layers of a particular system may be quite complex. Surface scientists often refer to this surface and near-surface region as the selvedge [1]. It is reasoned that the selvedge region extends only a relatively short distance into the surface before the space group symmetry of the bulk substrate is regained. The problem of surface structure analysis thus involves determination of the structure of the selvedge region and its orientation with respect to the undistorted bulk substrate.

It is frequently the case that the surface structures have unit cells that are integral multiples of the substrate unit cell. For example, the notation W(211) - (2 × 2) - H is conveniently used to refer to the adsorption of hydrogen on a (211) face of tungsten characterized by each axis of the adsorbate unit cell being twice that of the substrate. Such concise notation may also be profitably employed in cases where the surface structure is rotated in a simple fashion with respect to the substrate. As shown in Fig. 20, the c(2 × 2) structure on a square lattice substrate (where the symbol c indicates a 'centred' mesh) may equivalently be designated as p($\sqrt{2}$ × $\sqrt{2}$) - R45°. The latter notation indicates that the primitive cell of the surface structure is rotated by 45° with respect to that of the substrate, and the sides of the surface mesh are in the ratio of $\sqrt{2}$ to those of the substrate.

The notation scheme outlined above is suitable for simple surface structures and is commonly used in the literature. However, a more

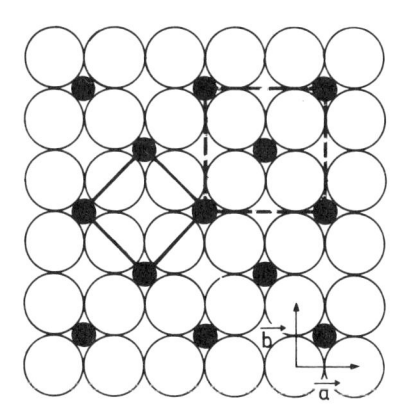

FIG.20. Schematic of the c(2 × 2) adsorbate structure on a square lattice substrate illustrating the primitive ($\sqrt{2}$ × $\sqrt{2}$) - R45° adsorbate unit cell (solid lines) and the centred (2 × 2) adsorbate unit cell (dashed lines) in terms of the substrate unit cell vectors $\vec{a}$, $\vec{b}$. The adsorbate atoms (small black circles) are shown in fourfold co-ordinated bonding sites above substrate atoms (large open circles).

general matrix notation [3, 4] is appropriate for complex structures. We consider a substrate primitive cell with translation vectors $\vec{a}$ and $\vec{b}$ and an overlayer structure with corresponding vectors $\vec{a}_s$ and $\vec{b}_s$. Defining a set of cartesian unit vectors $\hat{x}$ and $\hat{y}$, we can construct matrices $\vec{A}$ and $\vec{A}_s$ such that

$$\begin{bmatrix} \vec{a} \\ \vec{b} \end{bmatrix} = \vec{A} \begin{bmatrix} \hat{x} \\ \hat{y} \end{bmatrix} \tag{5a}$$

and

$$\begin{bmatrix} \vec{a}_s \\ \vec{b}_s \end{bmatrix} = \vec{A}_s \begin{bmatrix} \hat{x} \\ \hat{y} \end{bmatrix} \tag{5b}$$

The relationship between the unit meshes is compactly described by the transformation matrix $\vec{G}$ satisfying

$$\vec{A}_s = \vec{G}\,\vec{A} \tag{6}$$

As discussed by Estrup and McRae [3], the determinant G of the matrix $\vec{G}$ may be used to define the possible relationships between the substrate and overlayer nets. The nets are designated as simply related, rationally related, or irrationally related according to whether G is respectively an integer, a rational number, or an irrational number. Furthermore, the composite system formed by the superposition of the two nets is respectively designated above as a simple structure, a coincidence-site structure, or an incoherent structure. The incoherent superposition is not in itself describable by a net. However, the superposed system of either the simple or coincidence site structure is itself characterized by a Bravais net, and in these cases the overlayer and substrate are said to be in register. Returning then to the notational example of the c(2 × 2) structure on a square lattice substrate, we find that the transformation matrix $\vec{G}$ is

$$\begin{pmatrix} 1 & 1 \\ -1 & 1 \end{pmatrix}$$

and the nets are simply related. Thus, the notation

$$\text{Ni}(100) - \begin{pmatrix} 1 & 1 \\ -1 & 1 \end{pmatrix} - S(1/2)$$

could be used to denote the c(2 × 2) adsorption of one-half monolayer of sulphur on a (100) face of nickel.

2.2. Theory of LEED from crystalline surfaces

In Section 2.1, some of the general features of elastic LEED were set forth and the notation of surface crystallography discussed. It was shown how observation of intensity pattern geometry may be used to determine the periodicity of the surface structures in the direction parallel to the surface. However, a principal result of this discussion was the necessity of an intensity analysis for elucidation of the surface structure in the direction normal to the surface. A case of great current interest, for example, is the determination of the registry of a chemisorbed overlayer with respect to the bulk substrate by means of LEED intensity analyses. The standard analysis in elastic LEED involves the variation in the intensity of a given diffraction beam (spot) as a function of the incident beam energy. Experimentally, the intensity data are normally taken in the energy range $20 \leqslant E \leqslant 300$ eV by means of a spot photometer or Faraday collector. It is in this energy range that the scattered electrons are most surface-sensitive. Theoretical calculations employing models of the surface atomic geometry are then compared with the experimental intensity-energy (I-V) profiles. Surface crystallography is, in principle, carried out by finding the (presumably unique) surface structure which optimizes the agreement between theory and experiment over a significantly large range of diffraction conditions (i.e. energy range, number of diffracted beams, and incident beam angles). In practice, such a procedure is rather difficult because the theoretical calculations must provide an adequate description of the complex nature of the electron-solid interaction in the energy range relevant to LEED. Nevertheless, important strides have been made in recent years in constructing a suitable 'microscopic' LEED theory which is accompanied by manageable computation times and computer storage requirements. In addition, intensity-averaging procedures and intensity transform techniques (so-called data reduction methods) have been proposed as possible alternatives to the microscopic-model calculations. In the remainder of this section the microscopic LEED theory is reviewed and the essential features of the at present more tentative data reduction methods are outlined. In Section 2.3, recent applications of intensity analysis to surface crystallography are examined.

2.2.1. Microscopic-model theory

The goal of microscopic LEED theory is a quantum-mechanical formulation of the electron-solid scattering process and, in particular, calculations of the I-V profiles in which the main 'adjustable parameter' is the surface atomic geometry itself. There are two major features of the electron-solid interaction evidenced in the I-V profiles and in other scattering data that the theory must provide for: (a) in contrast to the case of X-ray scattering, the elastic scattering cross-sections [5, 6] for low-energy electrons from atoms are large — on the order of 10 $Å^2$/atom; and (b) the incident electrons interact strongly with the valence electrons in the solid, resulting in a high probability of inelastic scattering [6, 7] (e.g. plasmon, particle-hole, and ion-core excitations) within the first few atomic layers of the surface. Features (a) and (b) taken together with the wavelike behaviour of the electrons make LEED a sensitive probe of the surface atomic structure. Feature (a), however, renders the use of the simple kinematical [6] (single-scattering) theory inadequate in LEED and necessitates the use of multiple scattering or so-called dynamical theories. Feature (b), on the other hand, means that electrons are removed from the elastic electron beam due to 'inelastic-collision damping' [7] with a characteristic mean free path $\lambda_{ee} \sim 3$-10 Å. The inelastic-collision damping tends to reduce, though by no means eliminates, the effect of multiple scattering. The presence of multiple scattering introduces 'secondary' maxima in the I-V profiles in addition to the Bragg peaks (from the analogous integral-order peaks of X-ray diffraction) anticipated from kinematical theory. Moreover, the effect of inelastic scattering has been shown to limit the amount of secondary structure — generally smoothing and broadening the diffraction peaks.

There have been a number of formulations of LEED theory which take multiple scattering and inelastic scattering into account [7-16]. Here we shall briefly outline the theory originally proposed by Beeby [11]. The theory was subsequently modified to include inelastic-collision damping by Duke and Tucker [7] and the approximate effects of lattice vibrations by Duke and Laramore [16]. The theory has much in common with other treatments, of course, and its presentation serves to illustrate both the degree of sophistication and limitations of the current theories.

The incident electron is taken to be in a plane-wave state with energy E and wave vector $\vec{k}$. The elastically scattered electrons are observed with the energy E and wave vector $\vec{k}'$. Following Beeby's formulation, the scattered intensity is given by

$$I(\vec{k} \to \vec{k}') \propto \left| \int \exp(-i\vec{k}' \cdot \vec{r}') T(\vec{r}', \vec{r}) \exp(i\vec{k} \cdot \vec{r}) \, d\vec{r} \, d\vec{r}' \right|^2 \tag{7}$$

where T is the total scattering matrix satisfying

$$T(\vec{r}', \vec{r}) = V(\vec{r}') \delta(\vec{r} - \vec{r}') + \int V(\vec{r}') G_0(\vec{r}' - \vec{r}'') T(\vec{r}'', \vec{r}) \, d\vec{r}'' \tag{8}$$

and $G_0(\vec{r})$ is the usual outgoing free-particle propagator (see, e.g., Ref.[17]):

$$G_0(\vec{r}) = \frac{1}{8\pi^3} \int \frac{\exp(i\vec{k} \cdot \vec{r}) \, d\vec{k}}{E - \frac{\hbar^2 k^2}{2m} + i\epsilon} \tag{9}$$

The crystal potential $V(\vec{r})$ is taken to be a superposition of non-overlapping ion-core potentials situated at lattice sites $\vec{R}$:

$$V(\vec{r}) = \sum_{\vec{R}} v_{\vec{R}}(\vec{r} - \vec{R}) \tag{10}$$

where $v_{\vec{R}}(\vec{r})$ is the ion-core potential at position $\vec{R}$. Using Eqs (8) and (10), it can be shown that [18]

$$T(\vec{r}', \vec{r}) = \sum_{\vec{R}} t_{\vec{R}}(\vec{r}' - \vec{R}, \vec{r} - \vec{R})$$

$$+ \sum_{\vec{R} \neq \vec{R}'} \int t_{\vec{R}}(\vec{r}' - \vec{R}', \vec{r}'' - \vec{R}) G_0(\vec{r}'' - \vec{r}''') t_{\vec{R}'}(\vec{r}''' - \vec{R}, \vec{r} - \vec{R}) \, d\vec{r}'' \, d\vec{r}'''$$

$$+ \ldots \tag{11}$$

where the multiple scattering from the atom at $\vec{R}$ is given by the single-site t-matrix:

$$t_{\vec{R}}(\vec{r}', \vec{r}) = v_{\vec{R}}(\vec{r}') \delta(\vec{r} - \vec{r}') + \int v_{\vec{R}}(\vec{r}') G_0(\vec{r}' - \vec{r}'') t_{\vec{R}}(\vec{r}'', \vec{r}) \, d\vec{r}'' \tag{12}$$

Equation (11) is a useful result because it expresses the total scattering matrix in terms of successively higher-order scatterings between single atomic sites with the important provision that no two successive scatterings be off the same atom. It also allows for a precise treatment of the scattering from a single atom in terms of the t-matrix $t_{\vec{R}}(\vec{r},\vec{r}')$ defined in Eq.(12). This is a non-trivial point, in view of the general inadequacy of Born approximation treatments of low-energy electron scattering from atomic potentials.

The reduction of Eq.(11) to an algebraic form and the performance of the sums over scattering paths is accomplished by a series of manipulations prescribed by Beeby. The reader is referred to the original papers [11, 19] for details of the methods, but we mention here the essential points. The atomic potentials are assumed to be spherically symmetric so that the single-site t-matrices are appropriately expanded in an angular momentum representation. Each scattering is on the energy shell and may be defined in terms of the phase shifts [17]. Taking the atoms to lie in planes parallel to the surface, the planes are further divided into subplanes all of which have an identical structure termed the substructure. The substructure is essentially the smallest structure common to the primitive cells of all the atomic planes parallel to the surface. Furthermore, each subplane, by construction, contains only one type of atom and thereby has scattering properties described by a single t-matrix. The subplane concept greatly facilitates the performance of the sums over scattering paths and naturally leads to the conceptual separation of scattering events taking place solely within subplanes and those linking subplanes. The final result for the scattered intensity is [11]

$$I(\vec{k} \rightarrow \vec{k}') \propto \left| \sum_{LL'} Y_L(\vec{k}') Y_{L'}(\vec{k}) \left[\sum_{\nu} \exp\left(i(\vec{k} - \vec{k}') \cdot \vec{d}_{\nu} \right) \vec{T}_{\nu} \right]_{LL'} \right.$$

$$\left. \times \frac{(2\pi)^2}{A} \sum_{\vec{g}} \delta(\vec{k}_{\parallel} - \vec{k}'_{\parallel} + \vec{g}) \right|^2 \quad (13)$$

where the index L denotes the pair of indices (ℓ, m), the $Y_L(\vec{k})$ are the (real) spherical harmonics, and A is the area of the surface unit cell. The sum over delta functions expresses the conservation law of Eq.(4) arising from the two-dimensional periodicity of the subplanes. The variable ν indexes a given subplane with its (fixed) origin centred on an atom at $\vec{d}_{\nu}$. The matrix $\vec{T}_{\nu}$ is defined by the equations:

$$\vec{T}_{\nu} = \vec{\tau}_{\nu} + \vec{\tau}_{\nu} \sum_{\nu' (\neq \nu)} \vec{G}^{\nu\nu'}(\vec{k}) \vec{T}_{\nu'} \quad (14a)$$

$$\vec{\tau}_{\nu} = \vec{t}_{\nu}(\kappa) \left\{ \vec{I} - \vec{G}^{SP}(\vec{k}) \vec{t}_{\nu}(\kappa) \right\}^{-1} \quad (14b)$$

where $\kappa^2 = 2mE/\hbar^2$. The matrices $\vec{G}^{\nu\nu'}$ and $\vec{G}^{SP}$ are appropriately defined structure factors similar to those found in the Korringa-Kohn-Rostoker method of energy band theory and are relatively straightforward to calculate.

The matrix $\vec{t}_\nu(\kappa)$ is a diagonal matrix for single-site scattering with elements

$$t_{\ell,\nu}(\kappa) = -\left(\frac{\hbar^2}{2m}\right)\left[\frac{\exp(2i\delta_\ell(\kappa))-1}{2i\kappa}\right] \tag{15}$$

where $\delta_\ell(\kappa)$ are the phase shifts evaluated for the type of scatterer in subplane ν. The matrix $\vec{t}_\nu$ represents all of the scattering events taking place solely within the subplane ν. Finally, the matrix $\vec{T}_\nu$ represents all of the scattering events which end in plane ν.

Although Eqs (13) and (14) are not particularly transparent from a physical point of view, they do provide a straightforward mathematical procedure for calculating the intensities. Choosing ℓ phase shifts to describe the electron-ion-core scattering and dividing the model structure into N subplanes essentially requires the inversion of an ($N\ell^2 \times N\ell^2$) complex matrix for exact solutions to Eq.(14a). As an example of the total computation time involved for such 'exact' intensity calculations, Tong and Kesmodel [20] reported results with a computer program requiring approximately 30 seconds per energy point on a CDC 6600 machine and 6 seconds per point on a CDC 7600 machine for a model calculation utilizing 5 phase shifts and 5 atomic layers. The computer time and storage requirements rise rapidly as the number of layers or phase shifts needed for reasonable accuracy are increased. This fact has led to the proposal of various perturbation approaches to the solution of Eqs (14) or similar equations [21-24].

Duke and co-workers [7, 16] modified Beeby's results to include an approximate treatment of inelastic-collision damping and the effects of lattice vibrations, although the structure of the equations is unchanged. Briefly, the damping effects are introduced through the inclusion of an electronic self-energy term $\Sigma(\vec{k},E)$ in the propagator of Eq.(9) and a boundary condition on the incident electron wave vector inside the solid which makes it a complex number. Physically, these prescriptions cause the electron wave fields inside the solid to be exponentially decaying, thereby reducing the effects of multiple scattering. The effects of lattice vibrations are included by means of a renormalization of the single-site t-matrices. In essence, the scattering at each site is modified by a Debye-Waller factor that accounts for the loss of electrons by quasi-elastic phonon scattering. The phonon scattering is termed quasi-elastic because the energy changes involved are very small ($\Delta E = \hbar\omega \approx 0.01$ eV), and the phonon-scattered electrons are not separated from the purely elastic component in conventional LEED experiments. The phonon scattering, however, can change the momentum of the electrons significantly, thereby causing a fraction of the electrons to be scattered (thermal diffuse scattering) into the background between the diffraction spots. The Debye-Waller corrections to the I-V profiles have the general effect of reducing the heights of the higher energy peaks in relation to those at lower energies, but such corrections may also change peak shapes significantly.

For computational reasons a number of simplifying assumptions are inevitably made. The atomic potentials are in principle obtained from self-consistent calculations appropriate to the surface region, but in practice bulk band structure potentials or potentials obtained from the overlap of atomic charge densities are employed. The effects of inelastic damping are normally included by a suitably parametrized electron self-energy term or by an optical potential model. Finally, the lattice vibrational amplitudes are taken to be spherically symmetric and for most calculations independent of distance from the surface. Several systematic calculations, primarily for clean metal surfaces, have indicated that the above approximations are acceptable for achieving adequate agreement with experimental I-V profiles taken at constant temperature. Obtaining the observed temperature dependence of the intensities requires more accurate treatments of the lattice vibrations.

2.2.2. Data reduction methods

Two types of data reduction methods for the analysis of LEED intensity profiles and the extraction of surface atomic geometry have been proposed in recent years [25-28]. Since the general validity of these methods is currently under study, we shall only briefly outline the principles. The data reduction methods have their basis in the fact that kinematical (single-scattering) features of I-V profiles are generally predominant. Although the multiple-scattering features are rarely insignificant, it is argued that the effects of multiple scattering may be greatly reduced or eliminated by suitable averaging or transform procedures.

The first type of data reduction method involves the averaging of a large number of intensity profiles at constant momentum transfer $\vec{S} = \vec{k}' - \vec{k}$. It is well known that the kinematical scattering intensity is a function only of the energy E and $\vec{S}$, while the multiple scattering involves intermediate scattering variables and is not simply a function of $\vec{S}$ and E. It is therefore proposed that averaging the intensity over a suitable diffraction parameter (such as the azimuthal angle), which keeps $\vec{S}$ and E constant, will retain the kinematical peaks in the intensity profiles while averaging out the dynamical features. One would then extract the surface atomic geometry from trial calculations using a fitting procedure with the relatively simple kinematical theory. Lagally et al. [25-26] have applied such a method in the analysis of data from Ag(111) and Ni(111), and their results demonstrate the kinematical appearance of the averaged profiles. As emphasized by Duke and Smith [29] and Pendry [30], however, the central question is whether the resulting smoothed curves are sufficiently kinematical in character to be useful for accurate surface structure determinations. This question has yet to be fully explored.

The second type of data reduction method [27-28] uses the Patterson function or Fourier transform of the intensities $I_{hk}(S)$, where (hk) denotes a given diffraction beam and S is the momentum transfer in the direction normal to the surface. In LEED applications these transforms take the form of a complex function:

$$P(x, y, z) = \sum_{h,k=-\infty}^{\infty} \int_0^{\infty} I_{hk}(S) \exp[2\pi i(hx + ky + Sz)] \, dS \qquad (16)$$

Each point (x, y, z) in the space represents a position vector connecting two scattering centres translated to an arbitrary common origin, and local maxima in P essentially correspond to probable locations of scattering centres provided that data truncation errors and dynamical scattering do not introduce anomalous peaks. Clarke et al. [27] have used a real cosine transform to examine the structure of the Pt(100) surface. Buchholz and Lagally [28] have applied a transform method to LEED data from Ni(111) with the conclusion that transforms of individual LEED profiles taken at different diffraction geometry (e.g. different angles of incidence) are not the same, but transforms of averaged LEED data give expected auto-correlation functions. Their conclusion points to a central problem with the use of transform methods in LEED: that the available range in $\vec{S}$ may not be sufficiently large to eliminate data truncation errors and the effects of multiple scattering. Nevertheless, if these difficulties are surmounted, the transform method offers the advantage of being a fast, automated procedure for surface structure analysis.

2.3. Recent analyses of clean and overlayer systems

In this section recent applications of microscopic LEED theory to the structure analysis of clean and simple overlayer systems are described. The data reduction procedures of intensity averages or transforms have not yet been extensively applied and will not be discussed here. Although the analysis of clean surfaces appears to be in a rather satisfactory state of development, we shall see that the extension to adsorbed overlayer systems has met with several difficulties. Specifically, disagreements between theoretical analyses have arisen about the structures, both in a qualitative and quantitative sense, of certain overlayer systems. However, these contradictory analyses used independent sets of experimental data, and discrepancies between the data are seemingly a major source of the difficulty. We do not judge these problems to be fundamental but rather regard them as temporary setbacks in the rapidly developing methodology of surface crystallography via LEED.

The analysis of clean crystal surfaces has provided the testing ground for microscopic LEED theories. Work has generally centred on the analysis of low-index faces of clean metals: aluminium [13, 23, 31-33], copper [13, 15, 34], nickel [20, 35, 36] and silver [13, 37]. The importance of systematic studies over a wide range of incident angles and diffracted beams has been emphasized in order to examine the sensitivity and range of applicability of model calculations. Such studies have been recently carried out for nickel [20, 38, 39] and copper [34]. As discussed in Section 2.2, current theories employ reasonably accurate descriptions of both the elastic electron-ion-core scattering and inelastic electron-electron collisions as well as making approximate provisions for the effect of lattice vibrations. Figure 21 is representative of the kind of agreement achieved in recent model calculations [20] on clean metal surfaces. One notes the good agreement between theory and experiment in terms of peak positions, peak widths, and the angular evolution of the I-V profiles. The agreement in terms of absolute reflectivities and relative peak heights is less satisfactory but certainly adequate. Indeed, uncertainties in the experimental data and models of the electron-solid potential limit the general agreement in peak positions to within 2-4 eV and peak intensities to within approximately 50%. The precision in peak positions should allow the determination of atomic distances to within approximately 0.1 Å. The peak heights and shapes can also be greatly affected by small changes in atomic positions, thereby providing an additional criterion for optimizing trial surface structures.

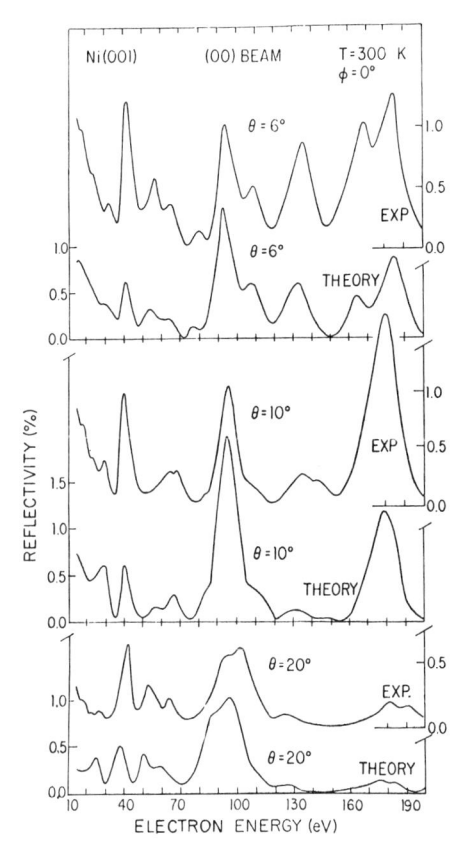

FIG. 21. Comparisons between theory and experiment of LEED intensity-energy spectra for the (00) beam, Ni(001) at room temperature and for three incident beam angles. The theoretical calculations are from Tong and Kesmodel [20] using a five-phase-shift, five-layer multiple-scattering computer program. The experimental results are from Demuth and Rhodin [38, 39].

The structures of the uppermost atomic layers of the low index faces of aluminium, copper and nickel appear to be very similar to the bulk structures. The intensity patterns exhibit the two-dimensional unit cells expected from the termination of the bulk structure. Nevertheless, several researchers have investigated the possible expansion or contraction of the outermost atomic layer in the direction normal to the surface. For the (100), (110) and (111) faces of nickel and for the (100) and (111) faces of copper, the outer layer spacing was found to be equal to the bulk interplanar spacing to within 5%. Similar conclusions hold for the (100) and (111) faces of aluminium but the outer layer of the (110) face of aluminium is apparently contracted or moved inward 10-15% relative to the bulk spacing [33]. The more challenging cases of the reconstruction of clean surfaces such as occur for Pt(100) and Si(111) will no doubt be thoroughly studied with the attainment of extensive experimental intensity data on these systems.

The most interesting and technologically relevant applications of LEED theory have been to the analysis of ordered adsorbate-substrate or 'overlayer' systems. Such ordered overlayers may be formed, for example, by the introduction of foreign gas atoms or molecules to the surface of an initially clean single crystal. As discussed in Section 2.1, these overlayer structures are very often characterized by a two-dimensional lattice periodicity different from that of the underlying substrate, thereby leading to the occurrence of additional spots in the diffraction patterns. Model analyses have now been applied to low-coverage coincidence structures. For these cases the adsorbed overlayer atoms are usually regarded as lying in a plane above an undistorted substrate lattice, and the problem reduces to that of determining the vertical and horizontal registry of the overlayer with respect to the substrate.

Several such analyses have recently been reported, most claiming accuracy to within 0.1 Å in adsorbate substrate distances [40-46]. Andersson and Pendry [40] examined the Ni(100) - c(2 × 2) - Na system with the conclusion that the Na atoms occupy fourfold co-ordinated sites at a distance of 2.87 Å above the topmost Ni layer. Forstmann et al. [41] reported an analysis of the Ag(111 - ($\sqrt{3} \times \sqrt{3}$)R30° - I structure with the conclusion that the iodine atoms occupy threefold sites a distance 2.25 Å above the topmost silver layer. Demuth et al. [42] have examined the c(2 × 2) overlayer structures of O, S, Se and Te on Ni(100), finding the adsorbate atoms to occupy fourfold co-ordinated bonding sites at displacements of 0.90 Å, 1.30 Å, 1.45 Å and 1.90 Å, respectively, from the centre of the top layer of nickel atoms. Andersson et al. [43] claim a similar structure for the oxygen on nickel system but place the oxygen atoms at 1.5 Å above the nickel layer, at variance with the results of Demuth et al. Finally, Duke et al. [44] suggest that the oxygen-nickel structure is a fourfold co-ordinated reconstructed square overlayer with both Ni and O atoms lying in the range of 1.75 - 1.90 Å above the Ni substrate. Thus, considerably different structures have been proposed for the Ni(100) - c(2 × 2) - O system. Similar problems have arisen in the analysis of the Ni(100) - c(2 × 2) - S system [42, 45]. However, in both of the above cases, different researchers analysed different sets of experimental data. The data were sufficiently different to cause the differing conclusions. These results, of course, point to the need for a large base of reproducible intensity data for such overlayer systems before accuracy can be achieved in terms of absorbate-substrate distances.

3. STRUCTURE OF SOLID SURFACES

3.1. Structure of clean unreconstructed and reconstructed solid surfaces

Surface reconstruction is defined as the state of the clean surface when its LEED pattern indicates the presence of a surface unit mesh different from the bulklike (1 × 1) unit mesh that is expected from the projection of the bulk X-ray unit cell. Conversely, an unreconstructed surface has a surface structure and a (1 × 1) diffraction pattern expected from the projection of the X-ray unit cell to that particular surface. Such a definition of surface reconstruction does not tell us anything about possible changes in the interlayer distances between the first and second layers of atoms at the surface by contraction or expansion in the z direction perpendicular to the surface that can take place without changing the (1 × 1) two-dimensional surface unit cell size or orientation. Indeed, several low Miller index surfaces of clean monatomic and diatomic solids exhibit unreconstructed surfaces, but the surface structure also exhibits contraction or expansion perpendicular to the surface plane in the first layer of atoms.

Over the past several years the intensities of the various LEED beams have been measured for clean aluminium [47], nickel [48], silver [49], copper [50], and tungsten surfaces, as well as for lithium fluoride [51]. In all of these studies, low Miller index ((100), (110) or (111)) crystal faces have been investigated. Using these experimental intensity data, calculations have been performed to determine the position of surface atoms based on theories in which the only adjustable parameters are the atomic positions at the surface. Diffraction beam intensity data are available for several other monatomic and diatomic solids, but with these the structure analysis has been lacking. The calculations indicate X-ray unit cell to within 5% of the interatomic distance for atoms in the Al(100) and (111) and Cu(100) and (111) crystal faces as well as for Ni(100), (110) and (111) crystal faces. These calculations can determine the atomic position in the surface layer within 0.1 Å. However, the (110) face of aluminium was found to be contracted by about 10-15% from the bulk interlayer spacing [33]. The best agreement between calculations and experimental intensities for Al(110) are obtained when the surface atoms are allowed to move closer to the second layer. Since the Al(110) crystal face is of somewhat lower atomic density than the (111) or (100) crystal faces, this observation may signify a trend that would indicate that surface re-arrangement without reconstruction by expansion or contraction of atoms in the z direction may take place in more open crystal faces while such an occurrence is not likely in high-density, low-surface free-energy crystal faces.

Similar changes in the interlayer spacing have been calculated by Laramore and Switendick for the (100) face of lithium fluoride [52]. According to their calculations, the top lithium and fluoride ion sublayers were separated by about 0.25 Å. The lithium ion sublayer appears to be contracted by a greater amount towards the bulk. LEED experiments indicate that various alkali metal halide (100) surfaces have the (1 × 1) surface structure expected from the projection of the X-ray unit cell. However, Gallon et al. [53] have shown that the stoichiometry of the alkali metal halide at the surface may be very different from the composition in the bulk. There may be surface excess of either the alkali metal atom or the halogen. While such changes do not seem to affect the unit cell size at the surface, this non-stoichiometry may be responsible for the magnitude of the contraction of the sublayers at the surface. The change of interatomic distance at alkali-halide surfaces has been calculated by Benson [54] and others without the need of assuming non-stoichiometry. However, the magnitude of their predicted values are different from those calculated in the LEED analysis of Laramore and Switendick. LEED studies of the surface structure of lithium hydride by Holcomb et al. [55] indicated that the composition at the surface is different from that of the bulk composition. There is evidence of precipitation of the alkali metal on the alkali hydride surface. During these chemical changes, however, the surface diffraction pattern remained characteristic of (1 × 1) unreconstructed surface structure. Thus it appears that, at least for alkali halides, the surface free energy is lowered by the introduction of excess defects, positive or negative ion vacancies, which will change the chemical composition and result in a marked non-stoichiometry in the surface layer. In this way the surface free energy is to be lowered more than by suitable reconstruction of the surface by which atoms occupy new equilibrium positions. The surface free energy, of course, may also be lowered by changes of interatomic distances perpendicular to the surface plane. We shall observe a combination of these effects on these unreconstructed ionic crystal surfaces. It is expected that compound semiconductors, those formed from elements in the II-VI and III-V groups of the periodic table, may also show similar effects [56]. There are several cases where both the surface composition and the surface unit cell change simultaneously. These changes that appear on various oxide surfaces will be discussed later.

3.1.1. Surface reconstruction

There are several low Miller index surfaces that exhibit reconstruction, i.e. the surface unit mesh is different from the usual bulk-like (1 × 1) mesh [57]. These reconstructed surfaces are the silicon (111), (100) and (110) [58, 59], the germanium (111), (100) and (110) surfaces [58], the diamond (111) [60], the platinum (100) and (110) [61], the gold (100) and (110) [62], and iridium (100) and (110) surfaces [63], the bismuth (11$\bar{2}$0) [64], the antimony (11$\bar{2}$0) [64], and the tellurium (0001) crystal faces [65]. Various diatomic solids, the gallium arsenide (111), ($\bar{1}\bar{1}\bar{1}$) crystal faces [66], the gallium antimonide (111) and ($\bar{1}\bar{1}\bar{1}$) [66], as well as the cadmium sulphide [67] and zinc oxide (0001) [68] faces and oxides under suitable conditions, i.e. vanadium oxide [69], aluminium oxide [70], and barium titanate [71], also exhibit surface reconstruction.

One of the most detailed studies of surface reconstruction was carried out on the Si(111) surface [72]. Upon cleaving at 25°C, the surface exhibits a (2 × 1) surface structure. On heating to about 300-400°C, the surface structure changes, according to Mönch, the (2 × 1) structure converting to the (7 × 7) structure. The (7 × 7) structure is then the stable structure of the (111) crystal face. Joyce [73], however, reported that in the presence of trace impurities, such as iron or nickel, the (2 × 1) surface is converted first to a (1 × 1) structure at 400°C, and the (7 × 7) structure forms only upon heating to 700°C. There is enough evidence to indicate that the temperature at which the impurity stabilized (1 × 1) surface structure transforms into the (7 × 7) structure depends markedly on the amount and the nature of the trace impurities on the surface.

There are several theories that can explain surface reconstruction in the absence of any major change in chemical composition at the surface. Taloni and Haneman [74] showed that relaxation of surface atoms out of the surface plane increases the overlap of localized electron orbitals, thereby lowering the surface free energy. Trullinger and Cunningham [75] proposed that the softening of phonon modes at the surface gives rise to the periodic relaxation of surface atoms. All these models indicate that surface reconstruction is indeed possible and results in a lowering of the surface free energy, but they do not predict the unique surface structure that is likely to be most stable. Since transformation from one surface structure to another can take place on both silicon and germanium surfaces as a function of temperature, the magnitude of the surface energies associated with the two structures are within kT of each other. Such a small energy difference should make it difficult to predict the relative stability.

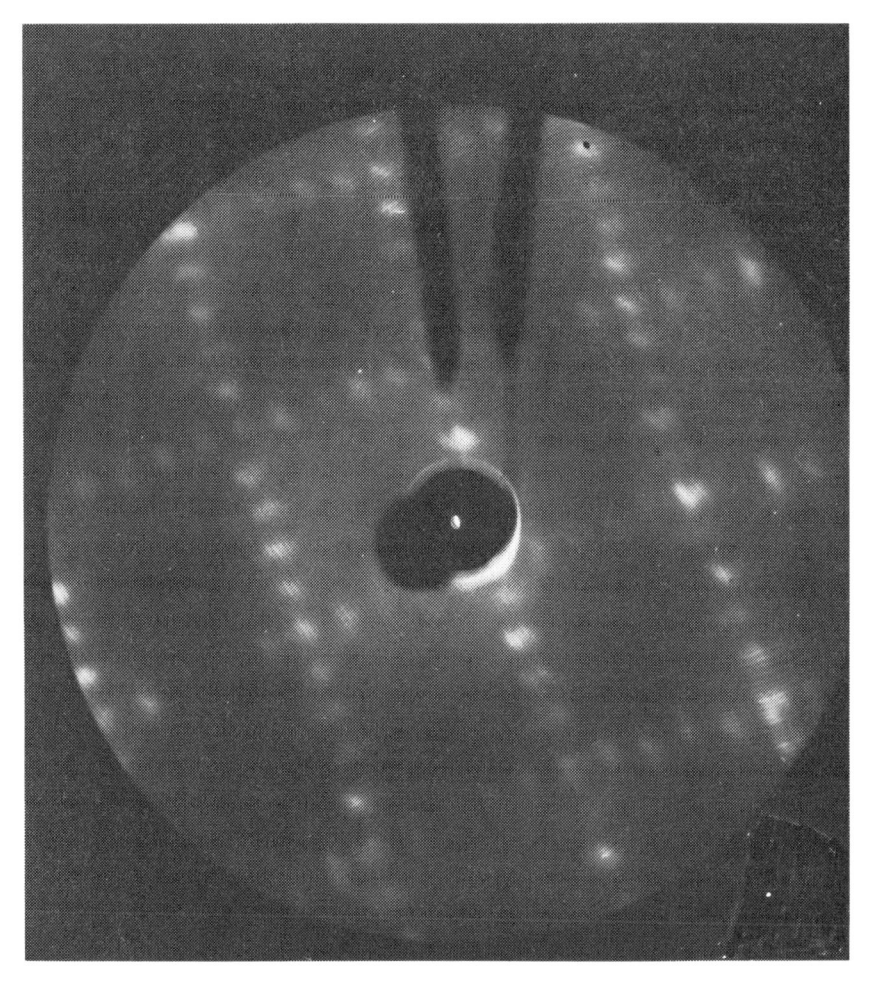

FIG. 22. Diffraction pattern of the Pt(100) - (5 × 1) structure at 124 eV.

Among metals, the most consistent changes of the surface structure were observed for the (100) crystal faces of three 5d transition metals which are neighbours in the periodic table. These metals are gold, platinum and iridium. All three metals exhibit the so-called (5 × 1) surface structure that is shown in Fig. 22. There are two perpendicular domains of this structure, and there are 1/5, 2/5, 3/5 and 4/5 order spots between the (00) and (10) diffraction beams. The surface structure is not quite as simple as the short-hand notation indicates, as is shown by the splitting of the fractional order beams. The surface structure appears to be stable at all temperatures from 25°C to the melting point although at elevated temperatures impurities from the bulk can come to the surface and cause a transformation of this structure to the impurity stabilized (1 × 1) surface structure [61]. Carbon at the surface that may diffuse out of the bulk in minute quantities or adsorbed gases of various types (CO, C_2H_2, etc.) can cause the surface atoms to relax back to their bulk-like (1 × 1) atomic positions [61]. The diffraction beam intensities of the (5 × 1) structure are under close investigation in many laboratories.

Preliminary calculations by Clarke et al. [27] and in this laboratory (Berkeley) indicate that a model for Pt(100) in which the surface atoms assume a distorted hexagonal configuration by out-of-plane buckling is favoured. The apparent (5 × 1) unit cell is then the result of coincidence of the atomic position of atoms in the surface, i.e. in the distorted hexagonal layer, with atoms of the undistorted second layer below. Surface atoms in any crystal face are in an anisotropic environment which is very different from that around bulk atoms. The crystal symmetry that is experienced by each bulk atom is markedly higher than for atoms placed on the surface. The change of symmetry and the lack of neighbours in the direction perpendicular to the surface permit displacements of the surface atoms in ways that are not allowed in the bulk. Surface relaxation can give rise to a multitude of surface structures depending on the electronic structure of a given substance. It is indeed surprising that there are so many solid surfaces which do not exhibit surface reconstruction. The adsorption of gases, such as oxygen or hydrogen, or the presence of impurities that segregate on the surface from the bulk, may cause or inhibit surface reconstruction, as indicated by many recent experiments.

Changes of chemical composition at the surface can produce marked changes in surface structure and cause the formation of a new surface unit cell. The (0001) crystal face of Al_2O_3 has been studied by several researchers. Upon heat treatment to elevated temperature there is an apparent reconstruction of the surface oxides evidenced by the formation of a new diffraction pattern accompanied by oxygen evolution [70]. In ultrahigh vacuum such heat treatment has resulted in the transformation of the (1 × 1) surface structure to one characterized by a $(\sqrt{31} \times \sqrt{31})R \pm 9°$ unit mesh. Structural re-arrangement was accompanied by the loss of oxygen; therefore, it has been interpreted as an oxygen-poor or a reduced oxide surface structure. The structural transformation is reversible, however, depending either on the partial pressure of oxygen or on the presence of excess aluminium on the surface. The complex surface structure whose formation is observed can be explained assuming the formation of AlO or Al_2O at the surface. Fiermans and Vennik reported on some interesting observations on vanadium pentoxide [69]. Under the influence of the low-energy electron beam incident on the surface, the transformation of $V_2O_5(010)$ to $V_{12}O_{16}(010)$ was observed in the surface layer accompanied by the loss of oxygen. They have demonstrated that this proceeds by domain formation on the surface and the two different structures $V_{12}O_{16}(010)$ - (4 × 1) and $V_{12}O_{16}(010)$ - (1 × 2) are involved depending on the degree of non-stoichiometry of the sample. Studies of Szalkowski and Somorjai have confirmed that the surface of V_2O_5 is unstable and it is reduced in vacuum [76]. There are lower oxides, $VO_{0.9}$, V_2O_3 and VO_2, that retain their surface composition, which is the same as that of bulk composition within the experimental accuracy (5%).

Aberdam and co-workers [71] have observed the diffraction pattern from the (001) face of barium titanate, $BaTiO_3$, prepared by different heat treatments. Near 1120 K the (1 × 1) mesh is noted which changes after a long period to the $(\sqrt{3} \times \sqrt{3})$ structure. The surface arrangement is considered to be due to the ordering of vacancies at the surface. A hysteresis in the temperature curve was found between 370 and 700 K and this could be associated with a cubic-tetragonal surface phase transition.

So far, three examples have been noted in which the reduction of the surface oxide causes a change of surface structure and surface unit cell. The effect of the reverse process, oxidation, has caused similar re-arrangement. Oxidation of nickel and other metal surfaces may cause reconstruction of the surface layer and the surface layer is then characterized by a mixed layer containing both oxygen and metal atoms, although the evidence is still circumstantial as to the chemical character of the structure of the reconstructed layer. Most experiments indicate that such a re-arrangement is likely to take place during highly exothermic surface reactions such as oxidation, nitridation, or during the formation of carbides.

3.1.2. Stepped, high Miller index surfaces

LEED studies have been applied, in general, to study the surface structure of close-packed faces of solids of low Miller index. These surfaces are chosen for structural investigation since they have the lowest surface free energy, and they are therefore stable with respect to re-arrangement of crystal faces or to disordering up to or near the melting point. Studies of surfaces of high Miller index and higher surface free energy are important in their own right. It is important to elucidate their atomic structure and stability under a variety of experimental conditions in the presence of reactive and inert gases and in vacuum.

The earliest diffraction observation from a high index surface is probably that of niobium [77]. The first detailed study of a stepped surface of this type is that of Ellis and Schwoebel [78]. They had examined a uranium dioxide, UO_2, crystal cut at 11.4° from the (111) plane in the (112) zone. Heating this sample at 1100 K in ultra-high vacuum for one hour produces a diffraction pattern resembling that from a UO_2(111) crystal face except that the spots were elongated and appeared to be split into multiplets.

TABLE I. ANGLES OF CUT, MILLER INDEX, AND DESIGNATION OF STEPPED PLATINUM SURFACES

Angle of cut	Miller index	Designation
6.2° from (111)	(533)	Pt(S)-[9(111) × (100)]
9.5° from (111)	(755)	Pt(S)-[6(111) × (100)]
14.5° from (111)	(544)	Pt(S)-[4(111) × (100)]
9.5° from (111) rotated 20°	(976)	Pt(S)-[7(111) × (310)]

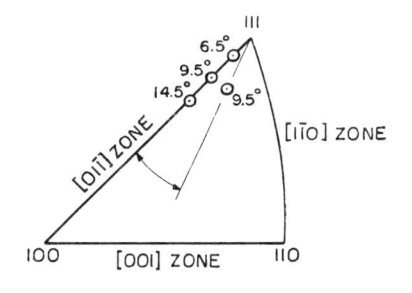

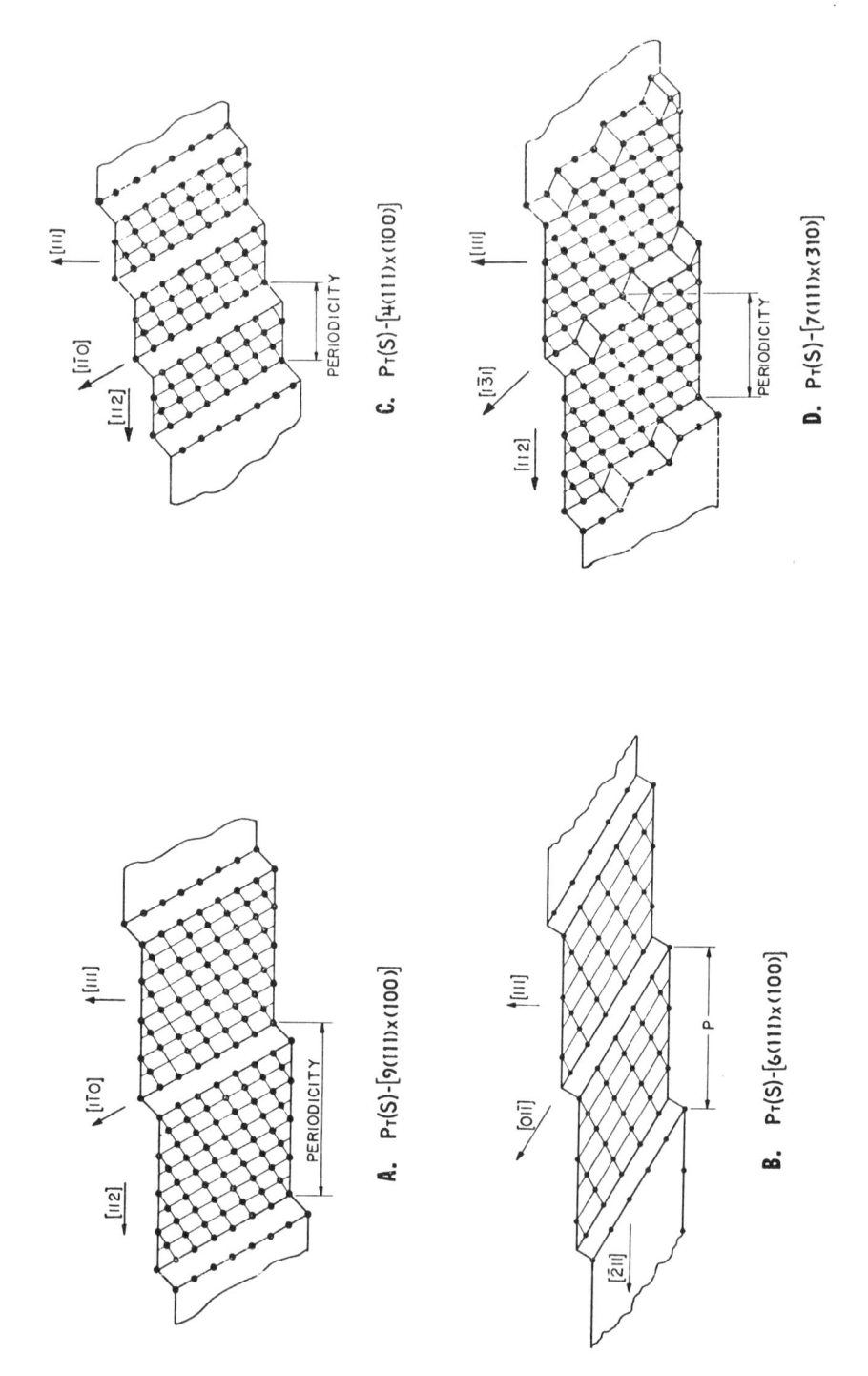

FIG. 23. Schematic representations of stepped platinum surfaces.

Heating at 1200 K in 10^{-7} Torr oxygen generated a pattern with each (111) spot resolved into a well defined doublet at certain electron energies. This behaviour with doublets appearing in place of single spots characteristic of a terrace geometry has been reported for all the stepped surfaces examined. Recent low-energy electron diffraction investigations of copper [79], germanium [80], gallium arsenide [80], and platinum [81] surfaces indicate that the surfaces of crystals characterized by high Miller index consist of terraces of low index planes separated by steps often one atom in height. The ordered stepped surfaces displayed varying degrees of thermal stability. Figure 8 shows one crystallographic zone of a face-centred cubic crystal. The circles indicate the direction and angle of cut of the various high Miller index surfaces and Table I indicates the Miller indexes associated with these crystal surfaces. (The diagram of Fig. 8 is repeated as part of Table I for easier reference.) The surface structure that can be derived from the diffraction patterns[1] obtained from the various high Miller index surfaces are indicated in Fig. 23. The terrace width is calculated from the doublet separation while the step height is obtained from the variation of the intensity maximum of the doublet diffraction beam features with electron energy. Let us consider the analysis of the diffraction patterns.

Several approaches are available in the literature, all kinematic and all yielding the same results. Henzler [80], extending the derivation by Ellis and Schwoebel [78], has shown that the scattered intensity I at an angle ϕ, with electron beam incidence normal to the terraces, is given by

$$I = \text{const} \frac{\sin^2[\tfrac{1}{2}k\, a(N+1)\sin\phi]}{\sin^2[\tfrac{1}{2}k\, a\sin\phi]}$$

$$\times \sum_{i=-\infty}^{+\infty} \delta[\tfrac{1}{2}k(Na+g)\sin\phi + \tfrac{1}{2}kd(1+\cos\phi) - i\pi] \qquad (17)$$

where the terrace has (N+1) rows, $k = 2\pi/\lambda$, a is the separation of the atomic rows, d is the step height, and g is the horizontal shift of one terrace compared to that below it. The first term is the intensity distribution for a grating of (N+1) slits, and the maxima are given by the Bragg equation:

$$\tfrac{1}{2}k\, a \sin\phi = n\pi \qquad (18)$$

The second term is the sum of δ functions with a separation $\Delta\phi$ given (near $\phi = 0$) by $\Delta\phi = \lambda/(Na+g)$, in other words, dependent only on the width and the displacement of the terraces. When two δ functions fall on a maximum of the intensity curve, a doublet arises and when only one δ function falls on the maximum of the intensity function, a singlet is observed. The δ functions converge towards the specular reflection of the high-index plane. The spot pattern itself, however, converges towards a (00) spot of the terrace plane. It has been shown that the separation of the doublet is

---

[1] It is regretted that the diffraction patterns themselves are not reproduced here.

inversely proportional to (Na + g), the terrace width, which is therefore easily determined. Also, the step height can be found from

$$V_{00 \text{ (singlet max)}} = \frac{150}{4d^2} s^2 \qquad (19)$$

where V_{00} are the voltages where a singlet of maximum intensity is observed, d is the step height and s is an integer. This method has been applied to the determination of step height by Henzler [82] and by Lang, Joyner and Somorjai [81]. The diffraction patterns to be expected from stepped surfaces have also been examined using laser simulation by Campbell and Ellis [83], who have shown that the single scattering diffraction pattern is potentially very informative. The terrace width does not have to be very precise to obtain satisfactory diffraction patterns. Houston and Park [84], in a theoretical study, have shown that there may be a great deal of uncertainty in the step width. All that is needed is that on an average the step width is well defined to obtain a diffraction pattern of satisfactory quality. That is, if the diffraction pattern indicates that the terrace width is 6 atoms wide, this does not rule out the presence of a large number of terraces 4, 5, 7 or 8 atoms wide. Since the re-arrangement of high Miller index surfaces to ordered low-index terraces separated by step takes place regardless of the chemical bonding in the crystal, it may be regarded as a general structural property of high-index surfaces. It is therefore of value to have a standardized nomenclature to identify stepped surface structures. Stepped surfaces are indicated by the postscript S so that Pt(S) indicates a stepped platinum crystal surface. The ordered step array can then be completely designated by the width and the orientation of the terraces and the height and the orientation of the steps. Thus, a stepped surface may be designated as Pt(S)[m(111) × n(100)], where m(111) designates a terrace of (111) orientation and m atomic rows in width and n(100) indicates a step of (100) orientation and n atomic layers high. Pt(S)-[m(111) × (100)] indicates the structure of various high Miller index platinum stepped surfaces having step heights of one atomic layer. (The 1 is not shown in front of the step orientation.) A more detailed description of nomenclature of more complex stepped structures is given elsewhere [81].

The thermal stability of the steps is of great interest; however, only a few studies have been directed to probe the high-temperature structural properties of high Miller index surfaces. For semiconductors, where the surface is generated by cleavage, the steps may be removed at elevated temperature, and faceting occurs. But, in metals, the stepped high index surfaces are found to be stable close to the melting temperature.

Perhaps the most significant property of stepped surfaces is their great reactivity compared to low-index crystal faces. The chemisorption of hydrogen, oxygen and carbon monoxide was studied by LEED on ordered stepped surfaces of platinum [85]. The stepped surfaces behave very differently during chemisorption from those of low-index platinum surfaces, and the various stepped surfaces also behave differently from each other. Hydrogen and oxygen, which do not chemisorb easily on the (111) and (100) crystal faces of platinum, chemisorb readily at relatively low temperature on the stepped platinum surface. In contrast to the ordered adsorption of CO on low-index platinum surfaces, where several ordered-surface structures have been detected, the adsorption is disordered on stepped surfaces, and there is evidence of dissociation of the molecule. Perhaps the best evidence of the enhanced reactivity of stepped surfaces comes from molecular beam studies on platinum surfaces. The hydrogen-deuterium exchange to form hydrogen deuteride, HD, was studied on (111) and stepped platinum surfaces [86]. While the scattering of both H_2 and D_2 was highly specular from the (111) crystal face and no HD signal could be detected at any surface temperature between 300 K and 1000 K for any angle from the surface normal, HD is readily detected from stepped surfaces over this temperature range. Between 5% and 10% of the incident deuterium is converted to HD on a stepped surface with 9-atom-wide (111) orientation terraces. It appears that, at least on platinum surfaces, the dissociation of large-binding-energy diatomic molecules takes place at steps, or at least that steps play a rate-determining role in the chemical process. The reactivity of stepped surfaces was also investigated during the chemisorption of various hydrocarbons on platinum surfaces. Aromatic and aliphatic hydrocarbons adsorbed on low-index (111) and (100) platinum crystal faces without any apparent decomposition or dehydrogenation in the temperature range of 300-500°C [87]. LEED and work-function change measurements both indicate that these molecules remain intact on the low-index platinum surfaces. Therefore, their surface crystallography may be studied conveniently in this temperature range. The chemisorption of over 25 hydrocarbons has been studied by LEED on four different stepped crystal faces of platinum: the Pt(S)-[9(111) × (100)], Pt(S)-[6(111) × (100)], Pt(S)-[7(111) × (310)] and Pt(S)-[4(111) × (100)] structures [88]. These surface structures are shown in Fig.23. The chemisorption of hydrocarbons produces carbonaceous deposits with characteristics which depend on the substrate structure, the type of hydrocarbon chemisorbed, the rate of adsorption and the surface temperature. Thus, in contrast to the chemisorption behaviour on low Miller index surfaces, breaking of carbon-hydrogen and carbon-carbon bonds can readily take place at stepped surfaces of platinum. Hydrocarbons on the [9(100) × (100)] and [6(111) × (100)] crystal faces mostly form ordered, partially dehydrogenated carbonaceous deposits while disordered carbonaceous layers are formed on the [7(111) × (310)] surface, which has a high concentration of kinks in the steps. The distinctly different chemisorption characteristics of these stepped platinum surfaces can be explained by considering the interplay of four competing processes:

(a) The nucleation and growth of ordered carbonaceous surface structures;
(b) Dehydrogenation, i.e. breaking of carbon-hydrogen bonds in the adsorbed organic molecules;
(c) Decomposition of the organic molecules, i.e. breaking of both carbon-hydrogen and carbon-carbon bonds at steps; and finally
(d) Re-arrangement of the substrate by faceting.

On the [9(111) × (100)] and [6(111) × (100)] crystal faces, processes (a) and (b) predominate. On the [7(111) × (310)] face, process (c) predominates, while process (d) is the most important on the [4(111) × (100)] face. The importance of atomic steps in surface chemical reaction on platinum cannot be emphasized strongly enough. In many reactions the dissociation of large binding energy diatomic molecules is a rate-limiting step. Atomic steps appear to catalyse this process. The lack of reactivity of low Miller index surfaces in hydrocarbon reactions indicates the importance of steps in breaking carbon-hydrogen and carbon-carbon bonds so important in various surface reactions of hydrocarbons. The nucleation and growth of ordered carbon structures that appear only on stepped platinum surfaces are important in catalysing complex structure-sensitive organic reactions such as isomerization and dehydrocyclization [89]. It appears that by cutting the right stepped surface one can produce prototypes of surfaces present on most catalysts under industrial conditions, and, using these well-defined, well-characterized stepped surfaces, one can establish correlations between chemical reactivity and surface structure under controlled conditions. It appears that stepped surfaces are more characteristic of the structure of real surfaces that participate in crystal growth or vaporization or surface chemical reactions. Hence, the electronic and atomic properties of stepped surfaces, their chemisorption and reactivity, will be a topic for intense investigations in the future.

4. STRUCTURE OF ADSORBED GASES ON SOLID SURFACES

Much of the thrust of surface crystallography is aimed at understanding the structure of adsorbed gases on surfaces. Experimental information on the structure of adsorbed gases from LEED studies has been accumulating rapidly since the late 1950s. Most of the experiments initially concentrated on the adsorbed structures formed by monatomic and diatomic gases on low Miller index surfaces of monatomic solids [90]. The gas species in most cases has molecular dimensions smaller than the interatomic distances in the substrate. Only in a few recent instances, such as in the cases of oxygen, selenium and sulphur adsorption on Ni(100) surfaces, have LEED intensity analyses been used in efforts to identify the unique atomic positions of these adsorbed atoms or molecules on the surface as well as the distances of separation from the metal atoms. In the over 200 surface structures that have been reported, the surface structures were identified only by viewing the diffraction pattern without making use of the intensities of the various diffraction beams. The diffraction pattern, of course, reveals the rotational symmetry and the size of the unit cell of the surface structure with respect to that of the substrate. This information, however, does not define a unique atomic site for each adsorbed species as the diffraction pattern may be assigned to several surface structures, all of them characterized by the same size and symmetry of unit cell. This section will be concerned only with a discussion of the surface structure of adsorbed gases as determined from the symmetry and separation of the diffraction beams in the diffraction pattern. The discussion will be restricted to gases that do not undergo chemical reactions on and with the substrate. Thus, the adsorbed structure observed on stepped surfaces will not be discussed but the papers that discuss the structures of these partially dissociated organic molecules adsorbed on stepped surfaces are referred to in Section 3. Such chemical reactions are quite common and may follow the chemisorption of oxygen, hydrogen or other reactive molecules. The surface structure of adsorbed gases on low Miller index surfaces will be discussed in two parts. First, the surface structures of small molecules will be discussed, and then in a separate section the surface structures of adsorbed hydrocarbons, large aliphatic and aromatic molecules will be reviewed.

4.1. Principles of ordered adsorption

Practical studies of adsorption require surface coverage of the adsorbed gas greater than about 5% of the number of available surface sites, which is approximately 10^{15} atoms/cm^2 under the conditions of most experiments [57]. The coverage σ is determined primarily by the residence time τ of the incident atoms or molecules and by the incident flux F:

$$\sigma = \tau F \qquad (20)$$

The flux is given by

$$F\left(\frac{\text{molecules}}{\text{cm}^2\text{sec}}\right) = 3.52 \times 10^{22} \frac{P_{\text{Torr}}}{(MT)^{1/2}} \qquad (21)$$

using the kinetic theory of gases, while the residence time can be expressed as

$$\tau = \tau_0 \exp\left[\frac{\Delta H_{\text{ads}}}{RT}\right] \qquad (22)$$

where P is the vapour pressure and ΔH_{ads} is the heat of adsorption. τ_0 is related to the period of a single-surface atom vibration, and the other symbols have their usual meanings. Frequently, $\Theta = \sigma/\sigma_0$, the degree of covering, is defined in discussing the properties of the adsorbed layers where σ_0 is the number of surface sites available for adsorption. The experimental conditions are adjusted by manipulation of the flux or the temperature to obtain measurable adsorption rates or coverages. The heat of adsorption depends on the coverage and therefore it is customary to define

$$\Delta H_{\text{ads}}^{\text{diff}} = \left(\frac{d\Delta H_{\text{ads}}}{dN}\right)_T \qquad (23)$$

where $\Delta H_{\text{ads}}^{\text{diff}}$ is the increase in the heat liberated by the adsorption of an additional amount of gas dN. Recently, work-function measurements and ellipsometry have been used to obtain the differential heats of adsorption as a function of coverage using single-crystal surfaces. Such measurements reveal the nature of the molecular interaction in the adsorbed layer, whether it is attractive or repulsive. The ordering of rare gases, xenon and argon, which have low heats of adsorption (2-8 kcal/mole) have been studied successfully in the temperature range of 10-78 K. On the other hand, most molecules that chemisorb, i.e. have high heats of adsorption ($\geq$ 15 kcal/mole), can be studied readily at 300 K and even at low gas pressures that are commonly used in LEED studies ($< 10^{-4}$ Torr).

Ordering of the molecules on the surface requires that the adsorbed species have sufficient mobility [91]; thus, the adsorbed molecules have to be able to overcome the activation energy ΔE_D associated with surface diffusion. Fortunately, the heat of adsorption, ΔH_{ads}, is in general greater than ΔE_D by at least a factor of two in most cases, so that the coverage remains relatively unchanged while surface ordering proceeds. Indeed, two-dimensional gas-liquid-solid transitions are possible because the adsorbed layer is now protected from desorption or diffusion into the bulk by large activation energy barriers for many of the systems of interest. It has also been found that suitable impurities adsorbed on surfaces can reduce the activation energy of surface diffusion so that ordering may commence in the presence of certain catalysts at lower temperatures. LEED studies of ordering as a function of coverage have detected order-disorder transitions that were likened to liquid-solid transitions in two dimensions as a function of coverage and temperature. Disordered adsorption at low coverages can be followed by ordering with increasing coverage as the motion of the molecules in the adsorbed phase becomes restricted. To overcome the decrease in entropy associated with the formation of the ordered layer on the surface, there is likely to be a large heat of ordering (exothermic) similar to the heat liberated in freezing a liquid. If the attractive interaction between the adsorbed molecules is large, ordered islands of the adsorbate may appear at low coverages. If the attractive interaction between the adsorbed molecules is weak compared to RT, the thermal energy at the temperature of the experiment, the adsorbed layer remains disordered. Ordering of these disordered layers may be controlled by changing the temperature of the substrate and/or changing the coverage. In general, the important system parameters that control ordering are the heat of adsorption as a function of coverage and the activation energy of surface diffusion, while the important experimental parameters are the coverage and the temperature.

4.2. Structure of small adsorbed molecules on low Miller index surfaces

The substrates which have been most frequently used in adsorption studies by LEED are the highest atomic density crystal faces of monatomic solids with face-centred cubic or body-centred cubic crystal structure. These crystal faces also have the lowest surface free energies and are therefore the most stable. The surface structures of small adsorbed gas molecules have been reported and have been tabulated [91]. A previous compilation [92] was revised to include the new surface structures reported in the past two years. Surface structures of gases adsorbed on substrates with 2-fold, 4-fold and 6-fold rotational symmetry are tabulated separately, since this classification permits useful correlation of the various structures. Inspection of these tables reveals that most of the surfaces are characterized by (a) the smallest unit cell permitted by the molecular dimensions and adsorbate-adsorbate and adsorbate-substrate interactions, and (b) the molecules adsorbed on the surface are likely to form ordered structures that have the same rotational symmetry as the substrate. These correlations were expressed as the rule of 'close-packing' and the rule of rotational symmetry, and their judicious application permits the prediction of surface structures or surface unit cells with a reasonable degree of accuracy. There are, of course, exceptions to these rules of ordering. These arise if there is a chemical reaction between the substrate and the adsorbed molecule. The presence of multiple binding states detectable, for example, during the chemisorption of CO on several metal surfaces, also makes the application of these simple rules difficult. It appears that for small molecules, whose dimensions are smaller than or similar in size to the interatomic distances in the substrate, violation of the rules of ordering is indicative of chemical interaction with the substrate that results in the formation of coincidence lattices with large apparent unit cells. It should be noted that such rules are either not applicable or rarely applicable to molecules whose size is greater than the interatomic distance in the substrate plane. These adsorbates may overlap several substrate atoms and their interaction with the substrate may be described only by a complex potential energy surface that contains contributions from many surface atoms. Also, in this case, the nature of the attractive interaction between the adsorbed molecules should play a more important role in determining the adsorbate structure. Thus, these surface structures will be discussed separately in the next section.

4.3. Structure of large adsorbed molecules on solid surfaces

The surface structure of adsorbed xenon has been studied on various substrate surfaces. Early evidence from studies of xenon adsorption on a graphite substrate [93] provided support for site adsorption by demonstrating the existence of $(\sqrt{3} \times \sqrt{3})$ - R30° xenon structure at 90 K. At lower equilibrium surface coverages, fuzzy ring-like diffraction features were observed and were considered as indicative of close-packed arrangement. Palmberg [94] has examined the adsorption of xenon on Pd(100) at 77 K in combination with work function measurements and Auger electron spectroscopy. Again, extra diffraction features appear only as the monolayer coverage is reached and the xenon structure has the symmetry of the substrate with unit cell vectors parallel to those of the underlying metal. The unit cell size is, however, not related to the palladium unit cell and yields a xenon-xenon spacing of 4.4 Å, close to the solid xenon value of 4.37 Å. The packing of the physically adsorbed layer is therefore dominated by the xenon-xenon attractive interaction. Xenon adsorption on the (100) crystal face of copper at 77 K was studied by Chesters and Pritchard [95], who again observed disorder at low coverages giving way to domains of ordered xenon at close to the monolayer. Here the physically adsorbed layer has 6-fold symmetry rather than the 4-fold symmetry of the (100) copper substrate and the xenon-xenon distance of 4.5 ± 0.1 Å is found. Ignatiev et al. [96] have demonstrated the growth of ordered (111) orientation xenon films on Ir(100) substrate. Thus, it appears that regardless of the substrate structure and rotational symmetry, xenon forms a (111) orientation overlayer on the various substrate surfaces. Similar results were obtained in a systematic study by Dickey et al. [97] at 8 K where ordered structures were reported for the physical adsorption of argon and neon on the (100) plane of niobium.

LEED and work function change studies of the adsorption of a large number of substituted aromatic molecules were carried out by Gland and Somorjai [87, 98] on the (111) and (100) crystal faces of platinum. These studies were carried out at low pressures (10^{-9} to 10^{-7} Torr) and at temperatures of 20-300°C. After adsorption, reorientation of the molecules in the adsorbed layer is necessary to form the ordered structures. Molecules that have either higher rotational symmetry, e.g. mesitylene, or have only small size substituents on the benzene rings, exhibit better ordering if the adsorption is carried out at low incident flux. The adsorbed layers are more ordered on the (111) crystal face than on the (100) crystal face of platinum. The work function changes upon adsorption range from

TABLE II. WORK FUNCTION CHANGES AND STRUCTURAL INFORMATION FOR ADSORPTION OF ORGANIC COMPOUNDS ON THE Pt(111) AND Pt(100)-(5 × 1) SURFACES

Adsorbate	Temp. (°C)	Pt(111) Pressure (Torr)	Pt(111) WFC (V)	Adsorbate diffraction features or surface structure	Pt(100)-(5×1) Pressure (Torr)	Pt(100)-(5×1) WFC (V)	Substrate structure after adsorption	Adsorbate diffraction features or surface structures
Acetylene	20	1×10^{-8}	-1.5	(2 × 2)	4×10^{-7}	-1.65	(1 × 1)	$(\sqrt{2} \times \sqrt{2})R45°$
	20	1×10^{-8} (10 min)	-1.65	disordered				
	150	4×10^{-7}	-1.8	disordered	4×10^{-7}	-1.7	(1 × 1)	$(\sqrt{2} \times \sqrt{2})R45°$
Aniline	20	1×10^{-8}	-1.8	streaks at 1/3 order diffuse (1/2 0) features	1×10^{-8}	-1.75	(1 × 1)	disordered
Benzene	20	4×10^{-7}	-1.8	poorly ordered	3×10^{-7}	-1.6	(1 × 1)	diffuse ringlike 1/2 order streak
	20	4×10^{-7} (5 min)	-1.4	$\begin{vmatrix} 4 & -2 \\ 0 & 4 \end{vmatrix}$				
	20	4×10^{-7} (40 min)	-0.7	$\begin{vmatrix} 4 & -2 \\ 0 & 5 \end{vmatrix}$	3×10^{-7} (2 hrs)	-1.3	(1 × 1)	diffuse 1/2 order streak
Biphenyl	20	2×10^{-9}	-1.85	very poorly ordered	2×10^{-9}	-1.8	(1 × 1)	disordered
n-Butylbenzene	20	8×10^{-9}	-1.5	disordered	8×10^{-9}	-1.5	(1 × 1)	disordered
t-Butylbenzene	20	5×10^{-8}	-1.7	disordered	5×10^{-8}	-1.75	(1 × 1)	disordered
Cyanobenzene	20	1×10^{-8}	-1.6	diffuse (1/3 0) features	1×10^{-8}	-1.5	faint (5 × 1)	disordered
1,3-Cyclohexadiene	20	2×10^{-8}	-1.75	poorly ordered	2×10^{-8}	-1.7	(1 × 1)	diffuse 1/2 order streak
	20	2×10^{-8} (1 hr)	-1.3	$\begin{vmatrix} 4 & -2 \\ 0 & 4 \end{vmatrix}$	2×10^{-8} (1 hr)	-1.6	(1 × 1)	diffuse 1/2 order streak
	20	3×10^{-7} (5 hrs)	-0.8	$\begin{vmatrix} 4 & -2 \\ 0 & 5 \end{vmatrix}$	2×10^{-8} (5 hrs)	-1.4	(1 × 1)	diffuse 1/2 order streak
Cyclohexane	20	6×10^{-9}	-1.2	(1 × 1) low background	6×10^{-9}	-0.75	(5 × 1)	low background
	20	4×10^{-7}	-0.7	very poorly ordered	4×10^{-7}	-0.4	(1 × 1)	diffuse streaked (2 × 1) pattern
	150	4×10^{-7}	-1.1	apparent (2 × 2)	4×10^{-7}	-1.2	(1 × 1)	streaked (2 × 1) pattern
	300	4×10^{-7}	-1.4	disordered	4×10^{-7}	-1.5	(1 × 1)	disordered
Cyclohexene	20	6×10^{-7}	-1.7	$\begin{vmatrix} 2 & 2 \\ 4 & -2 \end{vmatrix}$	6×10^{-7}	-1.6	(1 × 1)	diffuse (1/2 0) features
	150	6×10^{-7}	-1.6	apparent (2 × 2)	6×10^{-7}	-1.5	(1 × 1)	streaked (2 × 1) pattern
Cyclopentane	20	7×10^{-9}	-0.95	(1 × 1) low background	7×10^{-9}	-0.4	(5 × 1)	low background
	20	4×10^{-7}	-0.7	disordered	4×10^{-7}	-0.3	(1 × 1)	diffuse features at 1/2 order
Cyclopentene	20	—	—	---	2×10^{-7}	-1.4	(1 × 1)	diffuse streaked (1/2 0) features
2,6-Dimethylpyridine	20	4×10^{-8}	-1.6	diffuse 1/3.2 2/3.2 order streaks	4×10^{-8}	-1.5	faint (5 × 1)	disordered
3,5-Dimethylpyridine	20	6×10^{-8}	-2.3	diffuse 1/2 order streak	6×10^{-8}	-2.2	(1 × 1)	disordered

TABLE II (continued)

Adsorbate	Temp. (°C)	Pt(111) Pressure (Torr)	Pt(111) WFC (V)	Pt(111) Adsorbate diffraction features or surface structure	Pt(100)-(5×1) Pressure (Torr)	Pt(100)-(5×1) WFC (V)	Pt(100)-(5×1) Substrate structure after adsorption	Pt(100)-(5×1) Adsorbate diffraction features or surface structures
Ethylene	20	1×10^{-8}	-1.5	diffuse (1/2 0) features	1×10^{-8}	-1.2	(1 × 1)	($\sqrt{2} \times \sqrt{2}$)R45°
Ethylene	250	1×10^{-8}	-1.7	disordered	1×10^{-8}	-1.5	(1 × 1)	disordered
Graphitic overlayer	950		-1.1	ringlike diffraction features		-1.0	(1 × 1)	ringlike diffraction features
n-Hexane	20	5×10^{-8}	-1.1	disordered	5×10^{-8}	-0.8	(1 × 1)	disordered
n-Hexane	20	5×10^{-8} (5 hrs)	-0.9	disordered	5×10^{-8} (5 hrs)	-0.6	(1 × 1)	disordered
n-Hexane	250	5×10^{-8}	-1.5	disordered	5×10^{-8}	-1.2	(1 × 1)	disordered
Isoquinoline	20	6×10^{-8}	-1.9	diffuse (1/3 0) and (2/3 0) features	6×10^{-8}	-2.1	(1 × 1)	disordered
Mesitylene	20	4×10^{-8}	-1.7	streaks at 1/3.4 order diffuse (2/3.4 0) features	4×10^{-8}	-1.7	(5 × 1)	1/3 order streaks
Mesitylene	20	4×10^{-7}	-1.35	disordered	4×10^{-7}	-1.2	(1 × 1)	disordered
2-Methyl-naphthalene	20	6×10^{-8}	-2.0	very poorly ordered	4×10^{-9}	-1.6	faint (5 × 1)	disordered
Naphthalene	20	9×10^{-9}	-1.95	apparent (3 × 1)	9×10^{-9}	-1.7	(1 × 1)	disordered
Naphthalene	150	9×10^{-9}	-2.0	(6 × 6)	9×10^{-9}	-1.65	(1 × 1)	disordered
Nitrobenzene	20	9×10^{-9}	-1.5	diffuse (1/3 0) features (pattern electron beam sensitive)	9×10^{-9}	-1.4	(1 × 1)	disordered
Piperidine	20	8×10^{-8}	-2.1	disordered	8×10^{-8}	-2.05	faint (5 × 1)	disordered
Propylene	20	2×10^{-8}	-1.3	(2 × 2) (pattern electron beam sensitive)	2×10^{-8}	-1.2	(1 × 1)	1/2 order streaks (pattern electron beam sensitive)
Pyridine	20	1×10^{-8}	-2.7	diffuse (1/2 0) features	1×10^{-8}	-2.4	(1 × 1)	disordered
Pyridine	250	1×10^{-8}	-1.7	well defined streaks at 1/3, 2/3, 3/3 order	1×10^{-8}	-	(1 × 1	($\sqrt{2} \times \sqrt{2}$)R45°
Pyrrole	20	6×10^{-8}	-1.45	diffuse (1/2 0) features (pattern electron beam sensitive)	6×10^{-8}	-1.6	(1 × 1)	diffuse (1/2 0) features
Quinoline	20	3×10^{-8}	-1.45	diffuse 1/3 order streaks	3×10^{-8} (6 min)	-	(5 × 1)	diffuse 1/3 order streaks
Quinoline	20				3×10^{-8} (14 min)	-1.7	(1 × 1)	disordered
Styrene	20	6×10^{-8}	-1.7	streaks at 1/3 order	6×10^{-8}	-1.65	(1 × 1)	very poorly ordered
Toluene	20	1×10^{-9}	-1.7	streaks at 1/3 order	1×10^{-9}	-1.55	(5 × 1)	streaks at 1/3 order
Toluene	150	1×10^{-9}	-1.65	(4 × 2)	1×10^{-9}	-1.5	(1 × 1)	disordered
m-Xylene	20	1×10^{-8}	-1.8	streaks at 1/2.6 order	1×10^{-8}	-1.65	(5 × 1)	streaks at 1/3 order

-0.3 eV for cyclopentane to -2.7 eV for pyridine. The surface structure and the corresponding work function changes observed during adsorption are shown in Table II. Both the diffraction and work function change data indicate that, under the conditions of these experiments, all of the molecules chemisorb with their benzene ring parallel to the surface and interact with the metal surface primarily via the π electrons in the benzene ring. The substituent groups play an important role in determining the ordering characteristics of the overlayer but do not markedly affect the strength of the chemical bond between the substrate and the adsorbate. An interesting case history of change of chemical bonding with increasing coverage is that of benzene adsorbed on Pt(111). Benzene first forms a disordered layer on the Pt(111) surface but with further exposure the Pt(111)-$\begin{pmatrix} 4 & -2 \\ 0 & 4 \end{pmatrix}$-benzene structure is formed. (We use matrix notation in terms of the vectors of the substrate mesh.) Continued exposure results in the transformation of the surface structure to another ordered surface structure Pt(111)-$\begin{pmatrix} 4 & -2 \\ 0 & 4 \end{pmatrix}$. The first structure forms shortly after the minimum in the work function change has been reached, $\Delta\phi$ = -1.4 V. After the minimum has been passed, the work function change increases towards a steady-state value of -0.7 V. The second ordered structure ($\begin{pmatrix} 4 & -2 \\ 0 & 5 \end{pmatrix}$) forms when the work function change is about -1.1 V. This correlation between the transformation of the benzene surface structure and the change in the work function suggests that the orientation of the adsorbed benzene molecules is changing markedly as a function of increased exposure. A decrease in the density of the adsorbed layer during the order-order phase transformation is not possible because of the high flux that is continuously incident on the crystal throughout the experiment. In fact, the density of the adsorbed layer is increasing during continued exposure, as indicated by the observation that higher incident benzene flux causes the transformation of one benzene surface structure to another to occur more rapidly. The work function change indicates that the magnitude of the charge transfer is decreasing as the density of the adsorbed layer increases. If the adsorbed species has the same bonding characteristics during the transformation and the coverage increases, the work function would be further decreased. Thus, the increasing density accompanied by a decreasing magnitude of work function change can only be explained by assuming that the area of the adsorbed molecule must be decreasing. A likely model consistent with these criteria is that initially benzene is adsorbed with its ring parallel or at a small angle to the surface. The final adsorbed state may involve reoriented benzene molecules adsorbed with their rings at a large angle or perpendicular to the surface. The initial adsorbed species would be held on the surface by π bonds of the aromatic ring similar to the bonds in the so-called sandwich compounds. Since the metal surface is highly electron deficient, a large induced dipole would be expected in the adsorbed layer. The second structure that forms at large exposures may involve benzene molecules adsorbed with their rings perpendicular to the surface. For this type of adsorption to occur, the benzene must lose either a hydrogen to form a σ bond or its aromaticity. Recent exchange studies between deuterobenzene and benzene on platinum films have shown rapid exchange of hydrogen and deuterium between these species [99]. These workers postulate loss of hydrogen in benzene without loss of aromaticity to form a singly bonded adsorbed species. Thus, the adsorbed species that gives the second ($\begin{pmatrix} 4 & -2 \\ 0 & 5 \end{pmatrix}$) structure is most likely a singly dehydrogenated benzene molecule covalently bonded to the surface.

This type of reorientation satisfies both criteria for the surface transformation, i.e. the surface area occupied by the adsorbed species decreases and the amount of charge transfer decreases as well. LEED and work function changes indicate that pyridine bonds primarily through its nitrogen to the platinum surface. The corresponding work function change is very large, of the order of -2.5 eV. Naphthalene forms a very well ordered Pt(111)-(6 × 6) naphthalene structure when adsorbed on the (111) surface at 150°C, and the work function change upon adsorption is about -2.0 eV. Both the surface structure and the work function change on adsorption of naphthalene indicate that the naphthalene molecules lie parallel to the platinum surface and π bond to the metal substrate. Another interesting study was the LEED and work function investigation of the adsorption of cyclohexane, cyclohexene and cyclohexadiene on the Pt(111) and (100) surfaces [100]. Both the surface structure and the work function change data can be correlated to the various chemical bonds that these saturated and partly dehydrogenated molecules form with the metal surface. Both cyclohexane and cyclohexene adsorb on the metal surface without strong chemical interaction that would lead to dehydrogenation. These molecules stay intact and their structural characteristics can be rationalized from their molecular geometry and bonding abilities. Cyclohexane forms a single σ bond while cyclohexene π bonds through its olefinic double bond to the platinum surface. Cyclohexadiene appears to be unstable on platinum surfaces. It dehydrogenates rapidly at 25°C and the surface structures that form are those characteristic of benzene.

Studies of the surface structure and chemical bonding of organic molecules to various solid surfaces is an important field that underlies the phenomena of adhesion and lubrication. Future studies in this field will be extended to larger organic molecules of greater complexity. It appears that detailed studies of molecular crystals of various types to determine the surface structure and surface crystallography can also be carried out. As surface-structure analysis using the diffraction beam intensities allows routine determination of the surface structures of small molecules, surface crystallography will turn to more complex structures. Such development is certainly expected during the next decade.

REFERENCES (PART I)

[1] WOOD, E.A., J. Appl. Phys. 35 (1964) 1306.
[2] KITTEL, C., Introduction to Solid State Physics, 4th ed., Wiley, New York (1971).
[3] ESTRUP, P.J., McRAE, E.G., Surf. Sci. 25 (1971) 1.
[4] PARK, R.L., MADDEN, H.H., Jr., Surf. Sci. 11 (1968) 188.
[5] MOTT, N.F., MASSEY, H.S.W., The Theory of Atomic Collisions, Oxford University Press, London (1965).
[6] LANDER, J.J., in Progress in Solid State Chemistry 2, Pergamon, New York (1965).
[7] DUKE, C.B., TUCKER, C.W., Jr., Surf. Sci. 15 (1969) 231.
[8] McRAE, E.G., J. Chem. Phys. 45 (1968) 3258; Surf. Sci. 8 (1967) 14.
[9] BOUDREAUX, D.S., HEINE, V., Surf. Sci. 8 (1967) 426.
[10] KAMBE, K., Z. Naturforsch. A 22 (1967) 422.
[11] BEEBY, J.L., J. Phys. C 1 (1968) 82.
[12] PENDRY, J.B., J. Phys. C 2 (1969) 2273, 2283.
[13] JEPSEN, D.W., MARCUS, P.M., JONA, F., Phys. Rev. Lett. 26 (1971) 1365; Phys. Rev. B 5 (1972) 3933.
[14] STROZIER, J.A., JONES, R.O., Phys. Rev. B 3 (1971) 3228.

[15] CAPART, G., Surf. Sci. 26 (1971) 429.
[16] DUKE, C.B., LARAMORE, G.E., Phys. Rev. B 2 (1970) 4765.
[17] MERZBACHER, E., Quantum Mechanics, Wiley, New York (1967).
[18] BEEBY, J.L., EDWARDS, S.F., Proc. R. Soc. (London) A 274 (1963) 395.
[19] BEEBY, J.L., Proc. R. Soc. (London) A 279 (1964) 82; A 302 (1967) 113.
[20] TONG, S.Y., KESMODEL, L.L., Phys. Rev. B 8 (1973) 3753.
[21] TONG, S.Y., RHODIN, T.N., Phys. Rev. Lett. 26 (1971) 711.
[22] TAIT, R.H., TONG, S.Y., RHODIN, T.N., Phys. Rev. Lett. 28 (1972) 553.
[23] TONG, S.Y., RHODIN, T.N., TAIT, R.H., Phys. Rev. B 8 (1973) 421, 430.
[24] PENDRY, J.B., Phys. Rev. Lett. 27 (1971) 856.
[25] LAGALLY, M.G., NGOC, T.C., WEBB, M.B., Phys. Rev. Lett. 26 (1971) 1557; J. Vac. Sci. Technol. 9 (1972) 645.
[26] NGOC, T.C., LAGALLY, M.G., WEBB, M.B., Surf. Sci. 35 (1973) 117.
[27] CLARKE, T.A., MASON, R., TESCARI, M., Surf. Sci. 30 (1972) 553; Proc. R. Soc. (London) A 331 (1972) 321; Surf. Sci. 40 (1973) 1.
[28] BUCHHOLZ, J.C., LAGALLY, M.G., Surf. Sci. 41 (1974) 248.
[29] DUKE, C.B., SMITH, D.L., Phys. Rev. B 5 (1972) 4730.
[30] PENDRY, J.B., J. Phys. C 5 (1972) 2567.
[31] LARAMORE, G.E., DUKE, C.B., Phys. Rev. B 5 (1972) 267.
[32] JEPSEN, D.W., MARCUS, P.M., JONA, F., Phys. Rev. B 6 (1972) 3684.
[33] MARTIN, M.R., SOMORJAI, G.A., Phys. Rev. B 7 (1973) 3607.
[34] LARAMORE, G.E., Phys. Rev. B (to be published).
[35] DEMUTH, J.E., MARCUS, P.M., JEPSEN, D.W. Phys. Rev. B 11 (1975) 1460.
[36] LARAMORE, G.E., Phys. Rev. B 8 (1973) 515.
[37] JEPSEN, D.W., MARCUS, P.M., JONA, F., Phys. Rev. B (to be published).
[38] DEMUTH, J.E., Ph.D. dissertation, Cornell Univ. (1973).
[39] DEMUTH, J.E., RHODIN, T.N. Surf. Sci. 42 (1974) 261.
[40] ANDERSSON, S., PENDRY, J.B., J. Phys. C 5 (1972) L41.
[41] FORSTMANN, F., BERNDT, W., BÜTTNER, P., Phys. Rev. Lett. 30 (1973) 17.
[42] DEMUTH, J.E., JEPSEN, D.W., MARCUS, P.M., Phys. Rev. Lett. 31 (1973) 540; J. Phys. C 6 (1973) L307.
[43] ANDERSSON, S., KASEMO, B., PENDRY, J.B., VAN HOVE, M.A., Phys. Rev. Lett. 31 (1973) 595.
[44] DUKE, C.B., LIPARI, N.O., LARAMORE, G.E., Nuovo Cim. (to be published).
[45] DUKE, C.B., LIPARI, N.O., LARAMORE, G.E., THEETEN, J.B., Solid State Commun. 13 (1973) 579.
[46] IGNATIEV, A., JONA, F., JEPSEN, D.W., MARCUS, P.M., Surf. Sci. 40 (1973) 439.
[47] JONA, F., IBM J. Res. Dev. 14 (1970) 4.
[48] (a) LAGALLY, M.G., NGOC, T.C., WEBB, M.B., J. Vac. Sci. Technol. 9 (1972) 645;
(b) DEMUTH, J.E., TONG, S.Y., RHODIN, T.N., J. Vac. Sci. Technol. 9 (1972) 639.
[49] LAGALLY, M.G., Z. Naturforsch. 25 (1970) 1567.
[50] (a) ANDERSSON, S., Surf. Sci. 18 (1969) 325;
(b) REID, R.J., Surf. Sci. 29 (1972) 603.
[51] McRAE, E.G., CALDWELL, G.W., Surf. Sci. 2 (1967) 509.
[52] LARAMORE, G.E., SWITENDICK, A.C., Phys. Rev. B 7 (1973) 3615.
[53] GALLON, T.E., HIGGINBOTHOM, I.G., PRUTTON, M., TOKUTAKA, H., Surf. Sci. 21 (1970) 224.
[54] BENSON, G.C., J. Chem. Phys. 35 (1961) 2113.
[55] HOLCOMBE, C.E., Jr., POWELL, G.L., CLAUSING, R.E., Surf. Sci. 30 (1972) 561.
[56] BOTTOMS, W.R. (private communication).
[57] SOMORJAI, G.A., Principles of Surface Chemistry, Prentice Hall, New Jersey (1972).
[58] HERON, D.L., HANEMAN, D., Surf. Sci. 21 (1970) 12.
[59] LANDER, J.J., MORRISON, J., J. Chem. Phys. 33 (1962) 729.
[60] MARSH, J.B., FARNSWORTH, H.E., Surf. Sci. 1 (1964) 3.
[61] MORGAN, A.E., SOMORJAI, G.A., Surf. Sci. 12 (1968) 405.
[62] PALMBERG, P.W., RHODIN, T.N., Phys. Rev. 161 (1967) 586.
[63] GRANT, J.T., Surf. Sci. 18 (1969) 228.
[64] JONA, F., Surf. Sci. 8 (1967) 57.
[65] ANDERSSON, S., MARKLUND, I., ANDERSSON, D., in the Structure and Chemistry of Solid Surfaces (SOMORJAI, G.A., Ed.), Wiley, New York (1969).
[66] MacRAE, A.U., GOBELI, G.W., J. Appl. Phys. 35 (1964) 1629.

[67] CAMPBELL, B.D., HAQUE, G.A., FARNSWORTH, H.E., in The Structure and Chemistry of Solid Surfaces (SOMORJAI, G.A., Ed.),Wiley, New York (1969).
[68] CHUNG, M.F., FARNSWORTH, H.E., Surf. Sci. 22 (1970) 93.
[69] FIERMANS, L., VENNIK, J., Surf. Sci. 18 (1969) 317.
[70] FRENCH, T.M., SOMORJAI, G.A., J. Phys. Chem. 74 (1970) 2459.
[71] ABERDAM, D., GAUBERT, C., Surf. Sci. 27 (1971) 571.
[72] MÖNCH, W., Adv. Solid State Phys. 13 (1973) 241.
[73] JOYCE, B.A., Surf. Sci. 35 (1973) 1.
[74] TALONI, A., HANEMAN, D., Surf. Sci. 10 (1968) 215.
[75] TRULLINGER, S.E., CUNNINGHAM, S.L., Phys. Rev. Lett. 30 (1973) 913.
[76] SZALKOWSKI, F.J., SOMORJAI, G.A., J. Chem. Phys. 56 (1972) 6097.
[77] HAAS, T.W., Surf. Sci. 5 (1966) 345.
[78] ELLIS, W.P., SCHWOEBEL, R.L., Surf. Sci. 11 (1968) 82.
[79] PERDEREAU, J., RHEAD, G.E., Surf. Sci. 24 (1971) 555.
[80] HENZLER, M., Surf. Sci. 19 (1970) 159.
[81] LANG, B., JOYNER, R.W., SOMORJAI, G.A., Surf. Sci. 30 (1972) 440.
[82] HENZLER, M., Surf. Sci. 22 (1970) 12.
[83] CAMPBELL, B.D., ELLIS, W.P., Surf. Sci. 10 (1968) 118.
[84] HOUSTON, J.E., PARK, R.L., Surf. Sci. 26 (1971) 269.
[85] LANG, B., JOYNER, R.W., SOMORJAI, G.A., Surf. Sci. 30 (1972) 454.
[86] BERNASEK, S.L., SIEKHAUS, W.J., SOMORJAI, G.A., Phys. Rev. Lett. 30 (1973) 1202.
[87] GLAND, J.L., SOMORJAI, G.A., Surf. Sci. 38 (1973) 157.
[88] BARON, K., BLAKELY, D.W., SOMORJAI, G.A., Surf. Sci. 41 (1974) 45.
[89] LANG, B., JOYNER, R.W., SOMORJAI, G.A., J. Catal. 27 (1972) 405.
[90] SOMORJAI, G.A., FARRELL, H.H., Adv. Chem. Phys. 20 (1971) 215.
[91] SOMORJAI, G.A., Surf. Sci. 34 (1973) 156.
[92] SOMORJAI, G.A., SZALKOWSKI, F.J., J. Chem. Phys. 54 (1971) 389.
[93] LANDER, J.J., MORRISON, J., Surf. Sci. 6 (1967) 1.
[94] PALMBERG, P.W., Surf. Sci. 25 (1971) 598.
[95] CHESTERS, M.A., PRITCHARD, J., Surf. Sci. 28 (1971) 460.
[96] IGNATIEV, A., JONES, A.V., RHODIN, T.N., Surf. Sci. 30 (1972) 573.
[97] DICKEY, J.M., FARRELL, H.H., STRONGIN, M., Surf. Sci. 23 (1970) 448.
[98] GLAND, J.L., SOMORJAI, G.A., Surf. Sci. 41 (1974) 387.
[99] BARON, K., MOYES, R.B., SQUIRE, R.C., in Proc. 5th Int. Congr. on Catalysis, Palm Beach, Florida, Aug. 1972.
[100] GLAND, J.L., BARON, K., SOMORJAI, G.A., J. Catal. 36 (1975) 305.

PART II

5. AUGER ELECTRON SPECTROSCOPY

Auger electron spectroscopy (AES) provides a means to measure surface composition directly by experiment. This non-destructive technique detects the Auger electrons emitted from atoms of the condensed phase, liquid or solid. As long as low-energy Auger peaks are studied, the Auger electrons detected are emitted primarily by the atoms of the topmost atomic layer [1]. It has been shown that the Auger peak heights are proportional to concentrations of the corresponding atomic species. Thus, AES allows qualitative and quantitative chemical analysis of the surface with a sensitivity of approximately 1% of the monolayer, or about 13 atoms/cm^2.

The principle of AES is simple enough. A beam of electrons of high enough energy to eject bound electrons from inner-shell electronic states is incident on the surface of a solid or a liquid. As the inner-shell bound electrons are ejected, an excited-state atom which is ionized in one of its inner levels is created. The electron vacancy thus formed is filled by a de-excitation process, in which an electron from a higher energy state falls into the vacancy, the process continuing until an electron from the conduction band is the one involved in the de-excitation. The energy released in each of these electronic transitions can be dissipated in one of two ways, depending on the magnitude of the de-excitation energy. One way is through the creation of a photon of the appropriate wavelength, the process then being known as X-ray fluorescence. The alternative method is for the de-excitation electron to transfer the energy to another electron through Coulombic interaction. If this second electron possesses a binding energy less than the de-excitation energy transferred to it, it will be ejected from the atom, leaving behind a now doubly ionized species. The electron emitted as a result of this process is called an Auger electron, in honour of Pierre Auger, who first saw their tracks in a Wilson cloud chamber in 1925 and correctly explained their origin [2].

It is obvious from the preceding discussion that the energy of an Auger electron is primarily a function of the bound-state energy levels existing in the atom and therefore contains qualitative analysis information about the atom from which it originated.

When atoms are brought together to form a solid, the atomic energy levels broaden into effectively continuous bands. In Fig. 24 the Auger mechanism of de-excitation is illustrated upon a schematic diagram of the electronic band structure as a typical metallic solid of atomic number Z. The shaded areas represent the filled portions of the bands, three of which have been designated by the generalized notation W_0, X_p and Y_q, with the respective mean energies: $-\bar{E}_{W_0}(Z)$, $-\bar{E}_{X_p}(Z)$, and $-\bar{E}_{Y_q}(Z)$ relative to the chosen zero of energy — the Fermi level. ϕ_c is the work function of the crystal. In drawing the schematic diagram it is assumed that a single electron vacancy has already been produced in the W_0 band. If an electron from the X_p band fills that vacancy, energy of the magnitude:

$$\Delta \bar{E}_{X_p, W_0}(Z) = \{[-\bar{E}_{X_p}(Z)] - [-\bar{E}_{W_0}(Z)]\} = [\bar{E}_{W_0}(Z) - \bar{E}_{X_p}(Z)]$$

is released. If this energy is transferred to an electron in the Y_q band, this electron must lose the energy:

$$\Delta \bar{E}_{\phi_c, Y_q}(Z') = \{\phi_c - [-\bar{E}_{Y_q}(Z')]\} = [\bar{E}_{Y_q}(Z') + \phi_c]$$

in order to escape from the crystal, i.e. the Y_q band electron will be ejected from the atom provided that

$$[\bar{E}_{W_0}(Z) - \bar{E}_{X_p}(Z)] > [\bar{E}_{Y_q}(Z') + \phi_c]$$

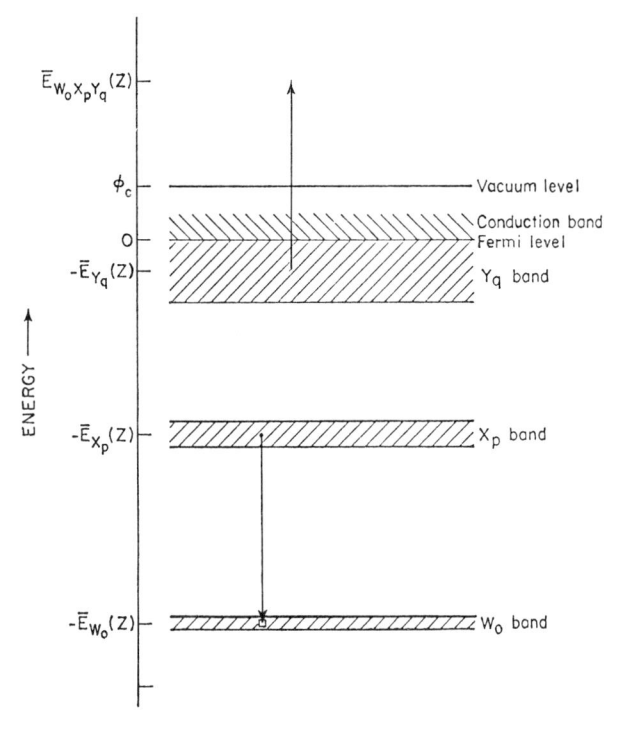

FIG. 24. A schematic representation of the Auger electron emission process from a metallic solid containing an initial electron vacancy in the W_0 energy band, an X_p band electron undergoing de-excitation to fill the vacancy, and a Y_q band electron being ejected from the sample.

Thus, the binding energy (relative to the vacuum level) of the electron in the Y_q energy band must be smaller than the energy transferred to it during the de-excitation process if Auger electron emission is to occur. The emitted electron appears outside the crystal with the energy:

$$\bar{E}_{W_0 X_p Y_q}(Z) = \bar{E}_{W_0}(Z) - \bar{E}_{X_p}(Z) - \bar{E}_{Y_q}(Z') - \phi_c \qquad (24)$$

relative to the crystal Fermi energy. The term $\bar{E}_{Y_q}(Z')$ has been substituted for $\bar{E}_{Y_q}(Z)$ because the latter refers to the energy level of the singly ionized atom, and as the Auger electron is ejected we are simultaneously creating a doubly ionized atom. It is therefore obvious that as the Auger process occurs the diagram of Fig. 24 becomes, strictly speaking, invalid since the energy band levels will re-arrange to their most stable state under the new electrostatic conditions. It has been postulated that $\bar{E}_{Y_q}(Z')$ should be the ionization energy of an electron from the Y_q band of the Z^+ ion and

$$\bar{E}_{Y_q}(Z') = \bar{E}_{Y_q}(Z+1)$$

If we rewrite Eq. (24) as

$$\overline{E}_{W_0 X_p Y_q}(Z) = \overline{E}_{W_0}(Z) - \overline{E}_{X_p}(Z) - \overline{E}_{Y_q}(Z+\delta) - \phi_c \qquad (25)$$

where δ is some incremental charge, in most cases the observed Auger energies have been intermediate between those calculated using Eq. (25) with δ = 0.50 and δ = 0.75 [3]. It perhaps seems more reasonable to use Slater's rules to determine the effective screening constant on the Y_q electron by the X_p electron and, in this case [4], Eq. (24) becomes

$$\overline{E}_{W_0 X_p Y_q}(Z) = \overline{E}_{W_0}(Z) - \overline{E}_{X_p}(Z) - \{\overline{E}_{Y_q}(Z) + S[E_{Y_q}(Z+1) - \overline{E}_{Y_q}(Z)]\} - \phi_c \qquad (26)$$

where S = Slater's formulae constant for the energy levels involved.

At any rate, the emitted Auger electron has the kinetic energy $\overline{E}_{W_0 X_p Y_q}(Z)$ and is associated with what is commonly labelled as a $W_0 X_p Y_q$ Auger process. Because they are energetically feasible only in parts of the periodic table, Auger transitions which result in a final vacancy in the same major shell as the initial vacancy (e.g. a $L_1 L_3 M_2$ transition) are referred to as Coster-Kronig transitions.

It is of interest to note here, and it will become quite important later on, that the Auger energy given by Eq. (26) is not the electron energy measured by the analyser. As an electron traverses the region between the crystal and the analyser it experiences a slight acceleration or deceleration, as shown in Fig. 25 [5], due to the difference in work functions of the analyser and crystal. This difference is known as the contact potential ϕ_{CP} and its magnitude is $\phi_{CP} = \phi_A - \phi_C$, where ϕ_A is the analyser work function. Equation (24) therefore becomes

$$\overline{E}_{W_0 X_p Y_q}(Z) = \overline{E}_{W_0}(Z) - \overline{E}_{X_p}(Z) - \overline{E}_{Y_q}(Z') - \phi_C - \phi_{CP}$$

$$= \overline{E}_{W_0}(Z) - \overline{E}_{X_p}(Z) - \overline{E}_{Y_q}(Z') - \phi_A \qquad (27)$$

In the retarding field device employed in LEED, ϕ_A is the work function of the grid material.

5.1. Peak energy analysis [6]

If the simplest case is considered, i.e. the KLL transition series, the $W_0 X_p Y_q$ nomenclature suggests that there should be six transitions: $KL_1 L_1$, $KL_1 L_2$, $KL_1 L_3$, $KL_2 L_2$, $KL_2 L_3$, and $KL_3 L_3$ (since transitions involving the same two L subshells in a different order, e.g. $KL_1 L_2$ and $KL_2 L_1$, must be considered identical as required by energy conservation). In reality, the actual number of radiationless transitions observed from an initial K state to a final two-vacancy L state is not six but nine. The fact that we assign definite values of 0, p and q means that a good quantum

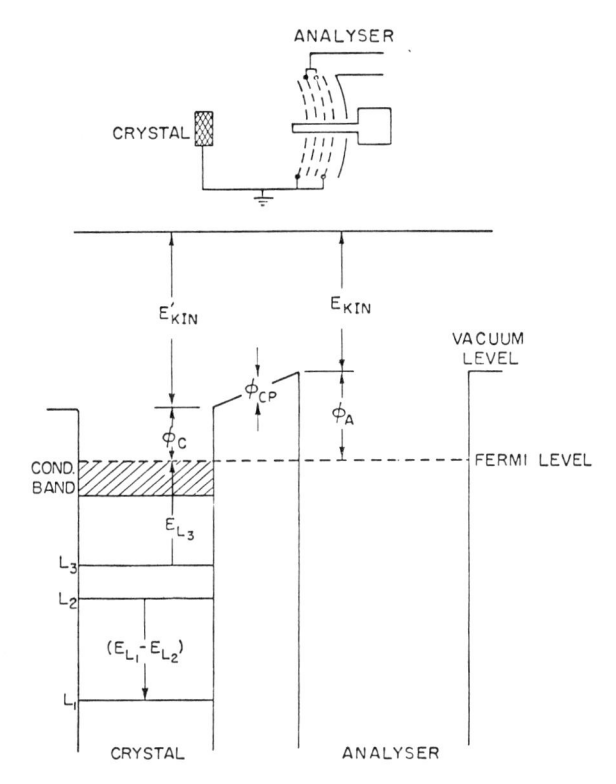

FIG. 25. Schematic diagram illustrating modification of the kinetic energy of an emitted Auger electron by the sample crystal-analyser contact potential ϕ_{CP}.

number is ascribed to the total angular momentum of the individual electrons in the various subshells, which is appropriate only in the j-j coupling scheme. A complete description of the Auger process can only be given in the total number of final states which result from having two electron vacancies. For the KLL case, an atom in this final state may have one of the following electron configurations in the L shell: $2s^2 2p^4, 2s^1 2p^5$, or $2s^0 2p^6$. The qualitative splitting effects leading to the evolution of the final Auger states are schematically drawn in Fig. 26, as the following (progressively restrictive) assumptions are made:

(a) If the electrons move in a $1/r$ potential, the screening and exchange interactions are negligible, and there is no spin-orbit interaction, the three electron configurations are completely degenerate in energy;
(b) If screening and Coulombic interaction among the electrons are allowed, the degeneracy is annihilated and the resulting levels possess different angular momenta;
(c) Exchange interaction causes a splitting into triplet and singlet states, with some of them being forbidden by the Pauli principle (indicated by broken lines in Fig. 26);
(d) The introduction of spin-orbit coupling causes splitting of the triplet terms as states of definite total angular momenta are produced.

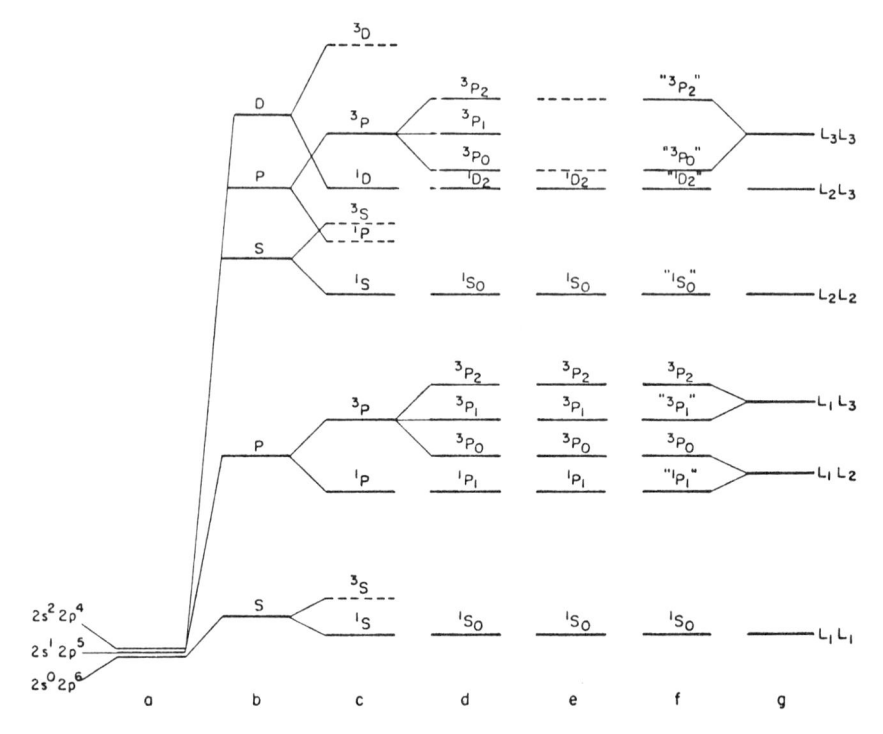

FIG. 26. Final state of Auger transitions in the KLL group under various assumptions for electron interaction and type of coupling (level positions not to scale).
(a) No electron interaction; central potential of 1/r shape; completely degenerated energies.
(b) Screening and Coulomb interaction between electrons generates levels of different orbital angular momentum.
(c) Exchange interaction causes a splitting in singlet and triplet terms. Those excluded by the Pauli principle are indicated by broken lines.
(d) Spin-orbit interaction decomposes terms into individual levels of definite angular momentum.
(e) Allowed final states (conservation of angular momentum and parity); pure L-S coupling.
(f) Allowed final states; intermediate coupling; no configuration interaction. Quotation marks indicate that the states are mixed.
(g) Allowed final states; pure j-j coupling.

To determine which energy levels are allowed as final states for an Auger transition, it is necessary to include consideration of the laws of conservation of angular momentum and parity. For example, a transition from the initial $^2S_{1/2}$ state of the system (i.e. from the $1s^1$ configuration) to any of the ^3P states of the $2s^2 2p^4$ configuration violates the law of conservation of parity and is therefore disallowed. Assuming pure L-S coupling, this reduces the number of final states to seven. For very low atomic number elements the spin-orbit interaction is negligible and the number of allowed states is five. If, however, an intermediate coupling situation is assumed, this allows the final state eigenfunction ψ of the system to be written down as linear combinations of the Russell-Saunders eigenfunctions ψ_0 which possess the same total angular momentum. The $^3P_0$ and $^3P_2$ states of the $2s^2 2p^4$ configuration will then be allowed since their eigenfunctions will have contributions from the allowed $^1S_0$ and $^1D_2$ states, respectively,

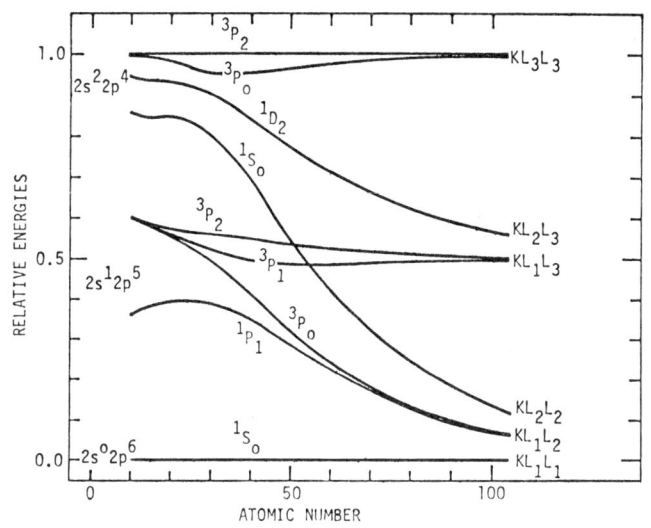

FIG. 27. Relative line positions in the KLL Auger transition group as a function of atomic number Z. The energy difference between the lines of highest and lowest energy ranges from 55 eV at Z = 10 to 17 keV at Z = 104.

and parity can therefore be conserved in transitions to these mixed states. (If configuration interaction is allowed, the $^3P_1$ term of the $2s^2 2p^4$ configuration will receive contributions from the $^3P_1$ and $^1P_1$ eigenfunctions of the $2s^1 2p^5$ configuration and a ten-line KLL spectrum should be observed. Since only nine lines have been thus far experimentally measured, this effect is ignored here.) The spin-orbit interaction dominates for the heavy elements, these two electronic motions coupling together to form a total angular momentum $\vec{j}$ for each electron which will now exist in a definite $\vec{l}, \vec{j}$ state.

A two-electron state is described by the $\vec{l}, \vec{j}$ values of both electrons, the j-j coupling scheme therefore predicting six final states for the KLL Auger transition series: L_1L_1, L_1L_2, L_1L_3, L_2L_2, L_2L_3, and L_3L_3. These states are shown in Fig. 27, along with the manner in which they approach the L-S coupling states.

A more graphic presentation of this relationship is shown in Fig. 27, where the relative energies of the experimentally observed KLL transitions (normalized to the KL_1L_1-KL_1L_3 energy interval) are plotted as a function of atomic number. It is obvious that the j-j coupling notation is becoming inadequate around an atomic number of 80, although its use in the intermediate coupling region may be justified in the cases where the finer details of the Auger spectrum cannot be observed due to insufficient resolution of the analyser or low intensity of some of the transitions. In the final analysis, however, the transitions can only be uniquely specified by using a notation in which the j-j and L-S coupling limits to which each transition tends are specified. The KLL transitions would then be labelled as follows:

$KL_1L_1(^1S_0)$, $KL_1L_2(^1P_1)$, $KL_1L_2(^3P_0)$, $KL_1L_3(^3P_1)$, $KL_1L_3(^3P_2)$,

$KL_2L_2(^1S_0)$, $KL_2L_3(^1D_2)$, $KL_3L_3(^3P_0)$ and $KL_3L_3(^3P_2)$

It must be emphasized, however, that at the present state of development of Auger spectroscopy the assignment of a $W_0 X_p Y_q$ process to an observed transition is tentative since it is primarily based on the agreement obtained when compared with calculated Auger transition energies. The preferred and unambiguous treatment, as already alluded to, of calculating the energy differences between the possible doubly and singly ionized states by quantum-mechanical means and comparing these values with the experimentally observed energy is difficult, if not generally impossible, at this time. Consequently, researchers in the field have relied on the approximations presented above. The assignment of an experimentally obtained peak to a particular Auger transition can nevertheless frequently be substantiated by applying additional experimental tests.

5.2. Intensity analysis

The accurate calculation of the peak intensity to be expected from an Auger transition turns out to be at least as difficult as determining what its energy should be. Concisely stated, the problem reduces to the question of determining (a) the probability of producing a particular subshell vacancy through the influence of the incident beam, (b) what the probability is of filling that vacancy by the Auger process of interest, and (c) what the probability is of that Auger electron escaping from the solid and being detected. These problems are discussed separately in the following sections and then merged to provide an estimate of the magnitude of the Auger yield to be expected.

5.2.1. Ionization probability

As may be easily imagined, the distribution of primary vacancies is a function of the nature of the atomic excitation process, the ideal method being one which produces vacancies in a single atomic subshell. The techniques that have been employed to produce inner-shell vacancies include orbital electron capture by the nucleus and the internal conversion of electrons in radioactive atoms, photo-ionization by an X-ray or low-energy γ-ray beam, and excitation by charged-particle (electron or proton) impact.

Since electron impact excitation was used in the investigation we are discussing, we shall limit ourselves to that method of ionization. In the range of intermediate beam energies (1-10 keV) such as was used, little work has been done on the determination of the energy dependence of the ionization cross-section for a given atomic energy level. Most of the theoretical work has used variations of the Bethe expression:

$$Q_{W_0} = \frac{2\pi e^4}{E_p E_{W_0}} b \ln\left(\frac{4 E_p}{B}\right)$$

where

Q_{W_0} = the cross-section for ionization of the W_0 subshell
e = the electronic charge
E_p = the primary electron beam energy
E_{W_0} = the binding energy of the W_0 subshell
b, B = empirical constants

which is valid for higher energy collisions. However, the most accurate values for these cross-sections seem to be based on the classical calculations done by Gryzinski [7] for inelastic atomic collisions. These computations were based on the relations for binary collisions (i.e. the independent pair interactions of the individual elements of the colliding systems) as well as for the Coulomb collisions derived in the laboratory system of coordinates. For the process of single ionization by electron impact, Gryzinski's equation for the ionization cross-section assumes the simple form:

$$Q_{W_0} = \left[\frac{\sigma_0}{(E_{W_0})^2}\right] g_{W_0}(U) \tag{28}$$

$$g_{W_0}(U) = \frac{1}{U} \left(\frac{U-1}{U+1}\right)^{3/2} \left\{1 + \frac{2}{3}\left(1 - \frac{1}{2U}\right) \ln[2.7 + (U-1)^{1/2}]\right\}$$

where

$\sigma_0 = \pi e^4 Z_p^2 = 6.5 \times 10^{-14} \, Z_p^2 \, (eV^2 \, cm^2)$
Z_p = charge of bombarding particle in units of the elementary charge
$U = E_p/E_{W_0}$

5.2.2. Transition probability

The rate of a particular Auger process is determined by the energy transfer probability between electrons in the various energy levels. As mentioned, the interaction between the electrons that participate in the Auger process is essentially electrostatic in nature, and theoretical Auger transition probability calculations therefore require the evaluation of transition matrix elements of the form:

$$D = \int\int \psi_f^*(r_1) \psi_f(r_2) \frac{e^2}{|r_1 - r_2|} \psi_i(r_1) \psi_i(r_2) \, dr_1 \, dr_2 \tag{29}$$

where $\psi_i(r_1)$ and $\psi_f(r_1)$ are the initial and final state wave functions, respectively, of one electron; $\psi_i(r_2)$ and $\psi_f(r_2)$ are the equivalent wave functions of the second electron; and r_1 and r_2 are the respective spacial co-ordinates of the two electrons. $\psi_f^*(r_2)$ is, of course, a continuum wave function whereas the others are bound-state wave functions. The exchange matrix elements are therefore

$$E = \int\int \psi_f^*(r_2) \psi_f(r_1) \frac{e^2}{|r_1 - r_2|} \psi_i(r_1) \psi_i(r_2) \, dr_1 dr_2$$

and the transition probability per unit time may be shown to be

$$w_{fi} = (\hbar)^{-2} |D - E|^2$$

The total transition rate is the properly weighted sum of the probabilities for the individual transitions.

The various models that have been used to carry out these calculations have run the gamut from non-relativistic unscreened hydrogenic bound-state/ unscreened Coulomb continuum wave functions to relativistic Hartree-Fock-Slater numerical methods incorporating intermediate coupling and configuration interaction. A discussion of the work carried out in this field is beyond the scope of this paper, and the interested reader is invited to peruse the excellent review by Bambynek et al. [8] for information concerning the existing state of the art in theoretical techniques.

5.2.3. Surface composition of alloys

For an ideal binary solution at a constant temperature, using the Gibbs equation, we can write for the surface composition [9]

$$\frac{x_1^s}{x_2^s} = \frac{x_1^b}{x_2^b} \exp\left[\frac{(\gamma_2 - \gamma_1)a}{RT}\right] \qquad (30)$$

The x_1^s and x_1^b are the atom fractions of the component 1 in the surface and in the bulk, respectively. a is the average surface area occupied by the molecules in the two different components and is assumed to be the same for both components. R and T have their usual meanings and γ_1 and γ_2 are the surface tensions of the pure components. According to this equation, the component with the smaller surface tension will accumulate on the surface. The model that gave rise to Eq. (30) is the so-called monolayer model. It is assumed that the surface consists of one atomic layer; the rest belongs to the bulk. The exponential dependence of the surface atom fraction ratio on the surface tension difference ensures that if this model has any validity the surface composition will always be quite different from the bulk composition.

The surface composition must be different for metal alloys, since the surface tensions are in the range of 0.5 - 1.5 newton/metre. Assuming a surface tension difference as small as 0.05 newton/metre, one obtains at 1000 K and a mean surface area of 10^{-19} m^2/atom

$$\frac{x_1^s}{x_2^s} = \frac{x_1^b}{x_2^b} \times 1.44$$

a considerable surface excess of component 1, which was assumed to have the lower surface tension.

There are many surface-active organic liquids that demonstrate the accumulation of the component with lower surface tension at the surface. Many alloys that form solid solutions that are nearly ideal also obey Eq. (30).

AES provides a means to measure the surface composition directly by experiment. The surface composition of several alloys has already been studied by AES. These include the nickel-copper [10] and silver-paladium [11] systems, as well as the lead-indium system [12]. Perhaps the lead-indium alloy system has been studied in the greatest detail. For a given bulk composition, the lead-indium ratio on the surface is found to be about five times that in the bulk. For the various bulk compositions studied, it is apparent that the surface is always considerably richer in lead than the bulk, as predicted by the monolayer model.

Perhaps a more sensitive probe of the concentration changes in the surface phase is a measurement of the temperature dependence of the Auger peak intensities. In the bulk phase the composition remains unchanged as a function of temperature for a given sample. In the surface phase, however, the composition should change as a function of temperature, as indicated by Eq. (30). Such temperature dependence of the surface composition was borne out by AES experiments on alloy surfaces. The presence of adsorbed gases that form chemical bonds of various strengths with the constituents of the alloy may markedly change the surface composition. It has been shown [13] that the surface of a gold-platinum alloy becomes enriched in platinum and the surface of a silver-paladium alloy becomes enriched in paladium in the presence of carbon monoxide. In the absence of carbon monoxide, in vacuum, the gold-platinum surface has excess gold and the silver-paladium surface has excess silver. The formation of the strong carbonyl bonds with platinum and paladium provides a driving force for the migration of these metal atoms to the surface and changes of the surface composition. Thus, in the presence of adsorbed gases that form chemical bonds with the surface atoms, when in the presence of impurities segregated at the surface, the binary surface phase is converted to a ternary system and should be studied accordingly.

It has been noted that the segregation of carbon or sulphur at the surface may have a marked effect on the surface composition of the alloy constituents.

5.2.4. Chemical shifts

A simple model attributable to Siegbahn et al. [5] may be invoked to demonstrate that the change in valence shell charge distribution, which occurs when an atom changes its valence state, is relayed to all the inner shell electrons and modifies the binding energy of each. First of all, it may be argued that the atomic valence electron orbitals essentially define a spherical valence shell of electric charge and the inner-shell electrons reside within this charged shell. If electrons are added to or removed from this valence shell, using purely classical considerations, the electric potential inside the shell is changed. Adopting the convention of defining zero potential energy of a system when the particles that comprise the system are separated by an infinite distance, the Coulombic potential energy existing in a system of two charges of magnitude q_1 and q_2, which are separated by a distance r_{12}, is given by:

$$E = \frac{1}{4\pi\epsilon_0} \frac{q_1 q_2}{r_{12}} \tag{31}$$

where ϵ_0 is a permittivity constant. Considerations of this kind indicate that the binding energy of the remaining inner-shell electrons will change on removal of an electron from the valence shell. Thus, if an atom is oxydized, there is a change in the inner-shell binding energies that should be detectable by AES.

TABLE III. CHEMICAL SHIFTS OF VANADIUM $L_3M_{2,3}M_{2,3}$ AUGER TRANSITION FOR VARIOUS VANADIUM COMPOUNDS

Compound	$\Delta(d^2I/dV^2)$ (eV)
1. V_2O_4	-2.30
2. V_2O_3	-1.45
3. $VO_{0.92}$	-0.65
4. $VO_{0.83}$	-0.95
5. V_2O_3	-1.40
6. $VO_{0.83}$	-1.05
7. VN	-1.00
8. VC	-0.85
9. V_2S_3	-0.55
10. VSi_2	-0.55

The chemical shifts of the inner shell $L_3M_{2,3}M_{2,3}$ Auger peak for various vanadium compounds have been measured with respect to the binding energy of this Auger transition in vanadium metal [14]. Table III shows the chemical shifts of this Auger transition for the various vanadium compounds. Since the experimental energy resolution is about 0.12-0.2 eV, the observed shifts are outside the experimental uncertainty. It has been found that these shifts are logically related to the chemical bond ionicity [15], as determined by the Phillips-Van Vechten ionicity scale. In addition, the ratio of the anion to cation Auger peak intensities observed for the compounds can also be successfully utilized in determining the surface composition.

REFERENCES (PART II)

[1] SOMORJAI, G.A., SZALKOWSKI, F.J., in Advances in High Temperature Chemistry 4 (EYRING, L., Ed.), Academic Press, New York (1971) 137.
[2] AUGER, P., Phys. Radium 6 (1925) 205.
[3] BERGSTRÖM, I., HILL, R.D., Ark. Fys. 8 (1954) 21.
[4] CLARKE, T.A., et al., J. Chem. Soc. A (1971) 1156.
[5] SIEGBAHN, K., et al., in ESCA: Atomic, Molecular, and Solid State Structure Studied by Means of Electron Spectroscopy, Almquist and Wiksalls, Uppsala (1967).
[6] BERGSTRÖM, I., NORDLING, C., in Alpha-, Beta-, and Gamma-Ray Spectroscopy 2 (SIEGBAHN, K., Ed.), North-Holland, Amsterdam (1965) 1523.
[7] GRYZINSKI, M., Phys. Rev. 138 (1965) A336.
[8] BAMBYNEK, W., et al., Rev. Mod. Phys. 44 (1972) 716.
[9] SOMORJAI, G.A., in Principles of Surface Chemistry, Prentice-Hall, New Jersey (1972).
[10] TARNG, M.L., WEHNER, G.K., J. Appl. Phys. 42 (1971) 2449.
[11] CHRISTMANN, K., ERTL, E., Surf. Sci. 36 (1973) 254.
[12] BERGLUND, S., SOMORJAI, G.A., J. Chem. Phys. 59 (1973) 5537.
[13] BOUWMAN, R., SACHTLER, W.M.H., J. Catal. 19 (1970) 127.
[14] SZALKOWSKI, F.J., SOMORJAI, G.A., J. Chem. Phys. 56 (1972) 6097.
[15] SZALKOWSKI, F.J., SOMORJAI, G.A., J. Chem. Phys. 61 (1974) 2064.

PART III

6. INTERACTION OF MOLECULAR BEAMS WITH SOLID SURFACES

6.1. Introduction

In recent years, a great variety of experimental techniques has become available that permits study of the clean surface and the solid-gas interface on an atomic scale. The structure of surfaces and of adsorbed gases and the chemical composition of the topmost layer at the surface are being studied by LEED and electron spectroscopy (photoelectron spectroscopy, AES, appearance potential spectroscopy, ion neutralization spectroscopy, etc.). Low-energy electron beams (1-1000 eV) are particularly applicable to investigating the atomic properties of the surface because of their low penetration and large cross-sections for excitation of both the electrons and phonons of the solid surface. Atomic and molecular beams are perhaps even more surface sensitive than low-energy electrons. Emanating from a room-temperature source, their kinetic energy is about 0.02 eV, 2-4 orders of magnitude lower than that of low-energy electrons. Since chemical bond energies are in the range of 0.5 - 10 eV, collision of incident atoms with atoms in the surface will not result in breaking chemical bonds. Like electrons, atoms are scattered by the atomic potential, and their penetration below the surface is negligible. If atomic beams are surface sensitive, then why not use them to the same extent as low-energy electrons for surface studies? In present-day technology it is much easier to generate and control (collimate, scatter and detect) charged particles than a beam of neutral species. Atomic beam scattering studies at present require special techniques newly developed to measure the energy and spatial distribution of particles. Nevertheless, generation and detection of atomic beams has been developed in the past few years to the point where molecular beam-surface scattering experiments can be carried out in most laboratories with relative ease using commercially available apparatus.

Just as LEED and electron spectroscopy are very well suited to determine the structure and composition of solid surfaces, molecular beam scattering provides detailed information on the energy transfer during surface reactions. By measuring the velocity and angular distribution of the incident beam and the beam scattered from the surface, one can determine the partitioning of the energy evolved in the surface chemical reaction between the reactants and the surface and among the reaction products. Thus, like crossed molecular beam gas phase reaction studies, surface scattering studies reveal the elementary energy-transfer steps in surface reactions. The dynamics of surface reactions on an atomic scale are at our disposal from molecular beam-scattering studies of gas-surface interactions.

In this section, we shall describe the molecular beam-surface scattering experiments, the nature of scattering (elastic, inelastic) and the experimental information that can be obtained from detection of the angular distribution and the kinetic energy of the scattered particles. We shall discuss the types of energy exchange that take place between a gas atom or molecule and the surface atoms and the theories that have been developed to explain some of these energy-transfer processes. We shall then review the results of some molecular beam-surface scattering experiments and point out directions for future investigation. We shall discuss most of the pertinent topics only briefly but provide references for the reader who is interested in exploring this important and rapidly developing field of surface science in greater detail. Attention should be given to other recent reviews available in the literature [1-4].

6.2. Molecular beam-surface scattering experiment

6.2.1. Vacuum system

A typical molecular beam vacuum system has three basic components: a beam source, a scattering surface and a detector for scattered particles. These components may be in separate, differentially pumped chambers. The design of the vacuum system is dominated by two competing considerations. The first is the desire to keep the pressure in the scattering chamber low to maintain a scattering surface as clean as possible and to reduce the number of gas phase collisions. The second is that, because of severe signal attenuation in the scattering process, it is desirable to have an incident beam as intense as possible so that acceptable signal levels can be obtained. The separate pumping of the detector chamber can also greatly enhance the signal-to-noise ratio by removing background gases. A typical vacuum system is shown in block diagram form in Fig. 28.

6.2.2. Sample preparation and treatment

It is necessary to have a scattering surface where the chemical composition, presence of adsorbed layers, atomic structure and roughness are well defined to make a meaningful interpretation of scattering results. Several methods have been used to produce and maintain clean surfaces for scattering. One can begin with a well defined single-crystal surface and keep the scattering-chamber pressure below about 10^{-9} Torr [5-7]. This requires ultrahigh-vacuum hardware and techniques. As a means of monitoring the composition of the surface, AES can be used [5], and LEED can monitor the surface structure [5-8]. Scattering studies may be carried out from surfaces at sample temperatures high enough for impurity gases not to adsorb because of their low sticking probability and short surface resident time. This technique was investigated by Yamamoto and Stickney [9] with scattering from tungsten surfaces. Still another method is that of in situ deposition of a metal film on a substrate at a rate faster than that at which it becomes contaminated. This procedure was developed by Smith and Saltsburg [10] and has since been used in a wide variety of experiments [11-13]. While the film surface can be oriented in a particular crystallographic direction, recent experiments by Sau and Merrill [8] indicate that there is significantly more disorder on an epitaxially grown film than on a well annealed, low Miller index, single-crystal surface. This was determined by monitoring the angular distribution of scattered helium atomic beams, which proved to be very sensitive to surface disorder [14, 15].

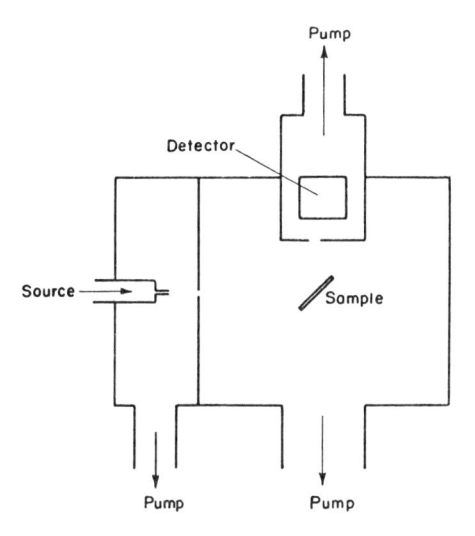

FIG.28. Block diagram of a molecular beam-surface vacuum system.

6.2.3. Beam sources

Effusion sources

The effusion oven is the classical means of obtaining a molecular beam. A gas at relatively low pressure (less than 1 Torr) is allowed to effuse through a small orifice (such that the mean free path of the gas is large compared to the dimensions of the orifice). A collimated beam is formed by pumping the gas through one or more subsequent orifices. The effusing particles have a Maxwellian velocity distribution characteristic of the temperature of the oven and have a cosine spatial distribution. The problem with such sources is their low intensity (typically $10^{13} - 10^{15}$ particles $sr^{-1} \cdot s^{-1}$) because of the limitation on the pressure in the effusion cell to maintain effusion conditions. Owing to the cosine distribution, a large fraction of the effusing molecules are pumped away in the collimating process.

Nozzle beams

When a gas at a pressure of about 100 Torr or greater is expanded adiabatically into vacuum, the enthalpy of the high-pressure gas is converted into net translational motion. If such a gas flow is properly collimated, it is possible to generate a very intense molecular beam, as discussed by Anderson et al. [16]. With such beam sources, intensities of $10^{18} - 10^{19}$ particles $sr^{-1} \cdot s^{-1}$ have been reported [17]. With the use of seeded beams [18], very high kinetic energies, in excess of 10 eV, are available. In seeded beams the velocity distribution in the beam is very narrow, corresponding to temperatures as low as 4 K [19]. The rotational temperature for a diatomic molecule in the nozzle beam is also reduced although not as much as the translational temperature [19]. Many investigators have used nozzle sources to generate beams for surface studies, and it is a particularly useful technique when scattering is being investigated as a function of incident beam energy [20]. The nozzle beam sources require very large (generally a few thousand litre · s^{-1}) pumping speeds to handle the associated large gas flows.

Multi-channel arrays

A compromise between the low-intensity effusion sources and the high-intensity nozzle beam, with its large pumping speed requirement, is the multi-channel array source [21-24]. This source consists of a small bundle of about 4000 capillary tubes, each a few μm in diameter and about 1 mm long. The array diameter is also on the order of 1 mm. The normal gas pressure behind such a source is on the order of a few Torr. This source shows peaking of intensity along the centre line of the beam considerably in excess of that for an effusion source [21]. This can increase the beam intensity by an order of magnitude over the effusion source. The velocity distribution in a multi-channel array beam is nearly, though not quite, Maxwellian [23].

6.2.4. Signal detection

Ionization detectors

The most common detector for scattered molecular beams uses electron-impact ionization. This can be a nude ionization gauge which measures total gas density or a small mass spectrometer, generally a quadrupole device which measures the partial pressure of a desired component of the scattered gas. The ionization detectors usually measure the density of the scattered gas, not the flux. Thus, while the velocity of the incident particles is considered implicitly in measuring the flux, a density-sensitive detector is sensitive to changes in the velocity of the incident molecules. This must be kept in mind when velocity analysis of scattered particles is attempted.

Generally, it is best to place the detector as close to the scattering surface as practical. If, for example, the molecules are scattered with a cosine distribution, the signal intensity will decrease with the square of the surface-detector distance. Similarly, it is also desirable to keep the incident beam-surface distance short although this is less important if the beam is well collimated. If the angular distribution of scattered particles is studied, it is customary to have the detector mounted on a rotatable feedthrough, and the scattering surface is then also made rotatable.

It should be mentioned that with present-day electron-impact ionization no more than one out of 10^3 or 10^4 atoms incident on the detector is ionized. This low detector efficiency limits the sensitivity of molecular beam-surface scattering experiments.

Electronic signal processing

The output from the ion gauge or mass spectrometer detector is generally a dc current which can be measured with a simple electrometer. However, in most realistic molecular beam experiments the dc signal level of interest will be very small and the signal-to-noise ratio rather poor (noise being due to ionization of background gas molecules), requiring a more complex approach to signal extraction. The most common technique used is that of ac phase-sensitive detection. Here the beam is modulated mechanically by a rotating slotted disc or a vibrating tuning fork between the source and the detector. A reference signal is also produced by simultaneously chopping a light beam. This ac technique allows a weak modulated signal to be detected in a relatively large background gas pressure.

6.2.5. Measurement of average velocities and velocity distributions

In addition to improving the signal-to-noise ratio in beam experiments, the phase-sensitive ac method can also give information about the time of flight of the molecules between their chopping point and the detector. This information is contained in the shift of phase angle at maximum signal intensity compared to the phase of the chopper. For instance, if one were to chop the scattered beam, assuming it is Maxwellian, then the average translational energy (temperature) of the beam can be obtained from the length of flight path, chopping frequency, molecular weight and phase shift [25-29]. Furthermore, if one can measure both the phase shift and signal amplitude as a function of modulation frequency, then a complete velocity distribution can be obtained [23]. Similarly, if the beam is chopped before striking the crystal, the time dependence of the gas-solid collision can be investigated — in particular, surface residence times can be measured [28, 29]. If a reactive scattering event is being investigated, the time dependence of the process can be measured via phase shifts to yield information about the kinetics of the reaction, and these results can be compared with model systems [28-30]. Phase-shift measurements, to be meaningful, require careful measuring of the phase and amplitude of the detector signals with respect to stable reference signals, measurements which are not at all simple in most practical beam experiments.

If one is interested in measurements of the velocity of the scattered molecular beam, an alternative to the phase-shift measurement is the 'time-of-flight' technique [31-33]. Briefly, in this method a narrow pulse of molecules is allowed to traverse a flight path and is detected by a multichannel signal-averaging instrument where the signal intensity as a function of time is measured and stored. After many cycles a complete intensity versus time-of-flight curve is obtained. The disadvantages of this method are the weak signal (due to short 'on-time' periods for the beam) and the need to know the effects of instrumentation alone on the resulting waveform. Data obtained in such an experiment are generally presented as average energy and the 'energy spread' characteristic of the beam temperature. It should be mentioned that for velocity analysis the classical method of rotating slotted-disc velocity selectors can also be used. This method, however, suffers from signal attenuation by relatively low throughputs for most velocity selectors since most of the molecules are lost in the selection process.

6.2.6. Measurement of angular distributions

A convenient means of obtaining and presenting data characterizing a molecular beam-surface scattering process is the scattered beam angular distribution. Two important features of the distribution are generally discussed: the angle of the intensity maximum of the scattered beam with respect to the angle of the incident beam and its peak width at half maximum. Most measurements are 'in-plane', which means that the angular distribution is measured in the plane defined by the incident beam and the surface normal. If the angle at which the scattered beam has maximum intensity, θ_r (measured with respect to the surface normal), equals the angle of incidence of the incoming beam, θ_i (also measured with respect to the surface normal), the scattering is said to be specular. If θ_r is between θ_i and the surface normal, the scattering is called subspecular. If $\theta_r > \theta_i$, the scattering is supraspecular. A typical specular distribution is shown in curve (a) of Fig.13. The angle of incidence is usually denoted by an arrow on the abscissa. It is also customary to plot linear signal amplitude as a function of scattering angle although polar plots are sometimes used. To compare data from different experiments, the intensities are normalized by dividing by the maximum intensity of the incident beam. In the case where the particles emitted from the surface have completely equilibrated with the surface, a cosine distribution is obtained as shown in curve (b) of Fig.13.

Finally, Fig.12 shows a molecular beam vacuum system used for scattering studies and incorporating many of the design features and equipment mentioned here [34].

6.3. Theories of beam-surface interactions

6.3.1. Types of interaction

When a gaseous particle in a molecular beam collides with a solid surface it can interact either elastically or inelastically. In an inelastic collision, energy exchange occurs and the interaction results in the creation and/or annihilation of phonons in the solid [35]. In the elastic collision there is no net energy exchange between the gas atom and the solid, and diffraction phenomena may be seen. Most of the theoretical work to date has been directed towards interpreting inelastic scattering of monatomic inert gases from solid surfaces using classical models. There have also been quantum-mechanical theoretical studies of inelastic scattering. Feuer [36] studied the interaction between rigid rotor diatomics and a solid surface. Essentially no theoretical work has been done to interpret reactive scattering.

As outlined by Goodman [35] for the inelastic case, there are three possible results of the collision. First, the molecule can lose enough of its energy to become trapped or adsorbed on the surface. Adsorbed molecules will eventually desorb and contribute to the scattered signal. Since these molecules have had a chance to equilibrate with the surface, they are likely to desorb with a cosine spatial distribution and with Maxwellian velocities characteristic of the surface temperature. Second, the molecule can lose some of its energy but still be scattered directly back into the gas phase. It is this second type of inelastic scattering which has received the greatest amount of theoretical attention. As an intermediate third case, the molecule may lose insufficient energy to adsorb but also not scatter immediately. It becomes a 'hopping' molecule along the surface [35].

Different regimes of scattering can be described as a function of the relative values of beam energy and mass and surface atomic mass, temperature and available phonon energies [35]. Associated with these different scattering regimes, different types of theoretical models appear more consistent with experimental observations. Some of the better known classical models and the conditions under which they are applicable are discussed briefly in the next section.

6.3.2. Inelastic scattering

'Thermal scattering' and the cube models

The thermal scattering regime is characterized by relatively low incident beam energies and relatively large gas-solid interaction distances (no surface penetration), resulting in scattering from an apparently smooth or flat surface [35]. This regime was first discussed by Oman [37]. In this case, the most important gas-solid interaction mechanism is through the thermal motion of the surface atoms and is most applicable to scattering from metals [38]. Because of the apparently flat surface, the thermal motion that is important during scattering is in the direction normal to the surface [35].

The theoretical models incorporating a flat surface and only perpendicular surface atom motion are the cube models. The first of these, the 'hard-cube' model, was developed by Logan and Stickney [39] and is illustrated in Fig. 29. The model assumptions are as follows: (1) the intermolecular gas-solid potential is such that the repulsive force is impulsive; (2) the scattering potential is uniform in the plane of the surface (smooth surface), and, since there is no motion of surface atoms parallel to the surface, there is no change in the tangential component of the incident particle velocity; (3) surface atoms are represented by independent cubes, and a gas particle interacts with a single surface atom by colliding with the cube once and then being scattered; (4) a temperature-dependent velocity distribution is assigned to the surface atoms. There is no attractive part to the potential. Referring to Fig. 29, the surface atom of mass M_s moves with a perpendicular velocity V_0. A gas molecule of mass M_g strikes the surface cube of mass M_s at an incident angle θ_0 (with respect to the surface normal) and with an incident normal velocity V_{n0} and a tangential velocity V_{t0}. The particle is then scattered at an angle θ_1 and with velocities $V_{t1} = V_{t0}$ and V_{n1}. The hard-cube problem can be solved exactly, and angular distributions of scattered atoms can be calculated if the velocity distribution of the incident beam is known. The model is somewhat unrealistic in that it neglects the attractive part of the gas-solid potential in the low incident-beam energy region where it is most important [35]. The interaction between solid atoms is neglected and tangential momentum exchange is not considered.

The first two failings are at least partly corrected in the 'soft-cube' model of Logan and Keck [40]. In this model the assumption of a flat surface

FIG. 29. The 'hard-cube' model [35].

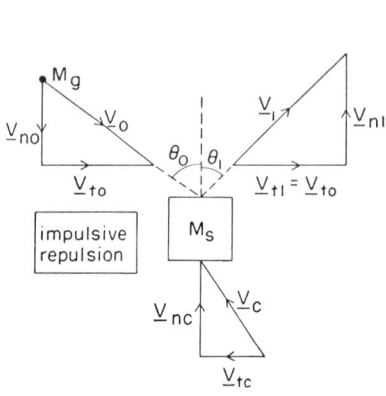

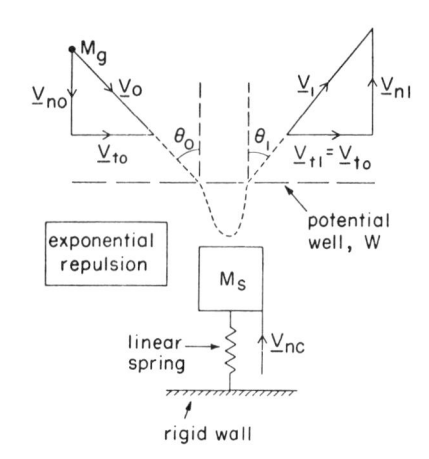

FIG. 30. The 'soft-cube' model [35].

is maintained with no exchange of tangential momentum. Now, however, there is a gas atom-solid atom potential with two parts [40]: a stationary attractive part, which increases the normal component of the gas velocity before the repulsive collision and decreases it again afterwards, and an exponential repulsive part. Also, the surface atom involved in the collision is connected by a single spring to a fixed lattice. The ensemble of oscillators making up the surface has an equilibrium distribution of vibrational energies corresponding to the temperature of the solid [40]. This model is shown in Fig. 30. The model introduces adjustable parameters for the potential well depth, range of interaction and lattice atom frequency. The solutions of the equations for angular distributions are approximate. The soft-cube model is more successful than the hard-cube model at predicting angular distributions for scattering of heavy molecules where potential attractions would be expected to be largest [40]. The model still does not include coupling between atoms of the solid.

'Thermal scattering' and general three-dimensional models

The cube models are only single-particle models and only one dimensional since they are restricted to energy transfer along the momentum component perpendicular to the surface. Classical three-dimensional lattice models have been developed by Oman [37], Lorenzen and Raff [41-44], and McClure [45]. In the lattice models an ensemble of lattice points is constructed to correspond to a particular crystal plane. Classical trajectories for scattered molecules are calculated for known incident velocities and angles by solving the equations of motion of the gas molecule and the lattice points. The gas molecule-solid atom potential is assumed to be a pairwise Morse interaction in the work of Lorenzen and Raff and a Lennard-Jones 6-12 potential in the models of Oman and McClure. Such a lattice model is illustrated in Fig. 31 in a diagram taken from Goodman [35]. Subscript 0 refers to incident molecules while 1 refers to reflected molecules. Θ is the Debye temperature of the solid. All surface atoms are

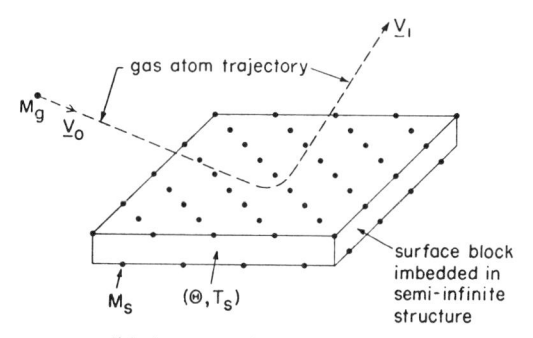

FIG.31. A three-dimensional lattice model [35].

connected to nearest neighbours by harmonic springs. To obtain reasonably reliable results, a large number of trajectories must be calculated and the incident trajectories must be chosen so that the distribution in incident angle and energy is both smooth and realistic [35]. Solving the necessary equations for the large number of trajectories needed is difficult and time-consuming and must be done numerically. The recent calculations of McClure [45-48] have been very successful in reproducing experimental results.

'Structure scattering'

In his calculations Oman [37] found that at high incident beam energies new features appeared in his scattered distributions, which he attributed to the incident molecules 'seeing' the periodic surface lattice. This is the regime of large incident energies, short interaction distances and a large ratio of incident beam energy to the thermal energy of the solid [35]. The flat-surface cube models no longer apply in this regime. One model which has been successfully applied is the hard sphere by Goodman [49]. This model will not be discussed here in detail.

An interesting phenomenon associated with the structure-scattering regime is rainbow scattering. This can be viewed as a classical mechanical result of the two-dimensional periodicity of the gas-solid interaction potential [35]. Its origins have been discussed by McClure [48], and in an extension of his earlier lattice model [45] he has carried out very high-resolution (3-5°) calculations of the rainbow scattering of Ne from LiF [47,48] in excellent agreement with experiment. These calculations appear sufficiently reliable to allow parameters for the gas-solid potential to be extracted by comparing theory with experiment [47,48]. The calculations are not applicable to gas atom-metal systems.

Since energy can only be exchanged between a gas phase particle and a solid surface through the phonons of the solid, a logical theoretical course is to treat the system quantum mechanically. There has recently been a revival of interest in the quantum theory of gas-solid interactions [50,51].

General inelastic scattering theories based on single-phonon interactions have been presented by Manson and Celli [52], Goodman [53], and Beeby [54, 55]. Ultimately, it is desirable to have a full three-dimensional multiphonon quantum treatment for a wide range of scattering phenomena. This state has not yet been reached, however.

6.3.3. Elastic scattering

The theoretical description of the elastic phenomenon of particle diffraction is a quantum-mechanical problem. Quantum treatments of elastic scattering have been formulated by Tsuchida [56] and by Cabrera et al. [57, 58].

An interesting outgrowth of the theoretical calculations for elastic scattering has been the investigation of the absence of diffraction of atomic beams by metals. This has been discussed by Beeby [59] and Weinberg [60] in terms of the Debye-Waller factor. Weinberg [61] has suggested that the intensity of scattered molecules as a function of temperature might be used to calculate gas-solid potential well depths if the surface Debye temperature is known.

7. RESULTS OF ELASTIC SCATTERING STUDIES

7.1. Diffraction

Atomic beam diffraction is in principle a very attractive technique for obtaining accurate surface structural data. Atomic de Broglie wavelengths are in the region of 0.5 - 1.5 Å, and incident energies can be made quite low (< 0.1 eV) so that there is no penetration into the surface, which is a problem associated with LEED. Similarly, atomic beams might also be very useful in investigations of the structure of adsorbed layers.

Unfortunately, as pointed out by Mason and Williams [62], atomic beam scattering as a structural probe suffers from the fact that only a small fraction of the total scattering arises from coherent events. This is particularly true for metals where diffraction has been seen only for the relatively 'rough' tungsten (112) surface [63]. Efforts by Beeby [59] and Weinberg [60,61] to correlate the lack of diffraction with the magnitude of the Debye-Waller factors have not been successful. It is likely that the absence of diffraction features from metal surfaces is simply due to a much too weak periodicity in the surface potential [62] as experienced by the incoming gas atoms.

Until very recently, only alkali halide crystals had yielded intense, unambiguous diffraction patterns. Atomic diffraction was first seen by Estermann and Stern in 1930 [64] from a LiF surface. Subsequently, their results have been reproduced and refined. O'Keefe et al. [65] have reported in their scattering work from LiF(001) that H_2O is a very likely contaminant and, in previous work, that it is very possibly a contaminant. Williams [66, 67] in a recent and comprehensive study seems to have successfully cleaned his LiF(001) surface. In one experiment [67] he used a nearly monoenergetic nozzle beam of helium. His diffraction peaks are listed in Table IV. The first-order peak is about 10% of the intensity

TABLE IV. APPROXIMATE RELATIVE INTENSITIES OF OBSERVED
DIFFRACTION PEAKS FOR He/LiF [67]

(00)	178	(-1-1)	15	(1-1)	3.1	(2-2)	2.5
(0-1)	16	(-1-2)	1.6	(1-2)	0.6	(2-4)	
(0-2)	1.4	(-1-3)	0.3	(1-3)	0.07		
(0-3)	0.05	(-2-2)	1.2				

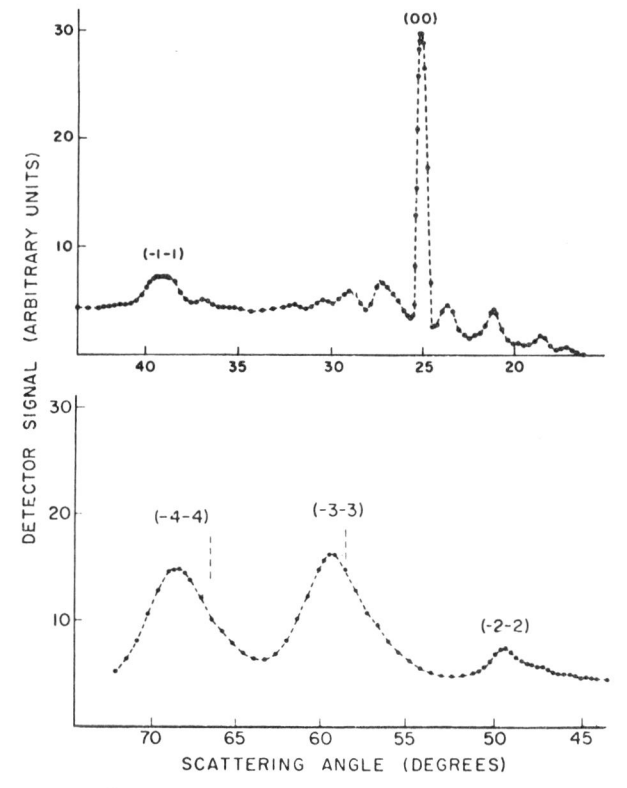

FIG. 32. In-plane scattering of neon atoms from LiF(100) showing peaks due to diffraction and phonon absorption and emission [66].

of the specular (00) peak, and the second-order peak is about 1%, in agreement with theory [56]. The relative intensities seem to be functions of incident beam angle. In a second experiment [66] Williams observed diffraction with a nozzle beam of neon from LiF(100). Figure 32 shows the resulting in-plane diffraction peaks. The total scattered intensity is much weaker than for helium, indicating a greater fraction of inelastic collisions. Also, the higher order peaks are more intense than the lower order peaks, in contrast to the helium results. The weak peaks in the vicinity of the (00) peak are attributed to inelastic scattering from phonons and will be discussed later. The complete diffraction pattern with relative intensities is summarized in Table V. Mason and Williams [62] have also

TABLE V. APPROXIMATE RELATIVE INTENSITIES OF OBSERVED DIFFRACTION PEAKS FOR Ne/LiF [66]

(00)	30	(-1-1)	2.5	(-3-3)	11
		(0-1)	14		
(1-1)	21	(0-2)	50	(-4-4)	9
		(0-4)	10		
(2-2)	20	(2-4)	8		
		(-1-3)	11		
(3-3)	4	(-2-2)	2.5		

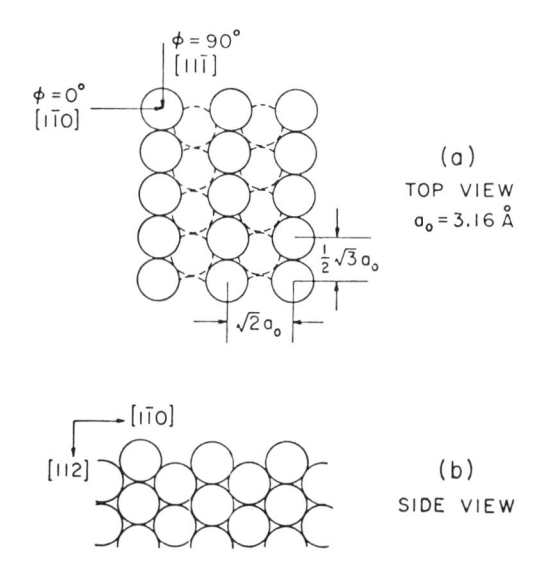

FIG. 33. Atomic structure of the (112) face of tungsten [63].

examined the diffraction by adsorbed molecules on LiF(001) and found they could obtain diffraction from an ordered overlayer of ethanol and even assign a tentative structure to this layer.

In addition to helium and neon, diffraction has been observed from alkali halide surfaces with H_2 beams and H atoms and marginally with D_2 beams [65, 68, 69].

The first observation of atomic beam diffraction from any non-alkali halide was the work on tungsten carbide by Weinberg and Merrill [70]. Diffraction peaks are well defined with helium and poorly resolved with D_2. The diffraction data are used to deduce a rather carbon-rich surface structure, possibly WC.

The first observation of diffraction by a metal surface was recently reported from clean tungsten (112) by Tendulkar and Stickney [63]. Tungsten (112) is a rather special case, however, since it has an anisotropic surface unit cell consisting of closely packed rows of exposed atoms in the [11$\bar{1}$] direction separated by open troughs as shown in Fig. 33. When the incident beam is directed across the rows, the zeroth- and first-order diffraction peaks are seen, and their positions agree with predictions. However, no diffraction peaks are observed when the beam is directed parallel to the rows and channels.

7.2. Scattering of helium and other atomic beams to monitor surface disorder and morphology

The helium-metal interaction is almost completely elastic. This is shown by the very narrow specular scattering peaks obtained from clean, well-ordered single-crystal surfaces. In one recent study, Smith and Merrill [15] examined helium scattering from single-crystal Pt(111). They observed a specular scattering peak from a smooth clean surface only slightly wider (7°) than the width of the incident beam (5°). The intensity of scattering at the specular angle was shown to be very sensitive to the amount of ethylene adsorbed on the surface. The specular intensity could be quantitatively related to ethylene exposures as low as 0.1 Langmuir (1 Langmuir = 10^{-6} Torr·s) and thus was a measure of surface coverage. The intensity and shape of the scattered helium peak as probes of surface conditions on Pt(111) were also studied by West and Somorjai [14]. They found large non-specular components due to surface roughness on chemically etched surfaces and on argon ion bombarded surfaces. Surfaces with disordered overlayers of C, CO and C_2H_2 gave very non-specular scattering while ordered CO overlayers were more specular although less so than clean smooth surfaces.

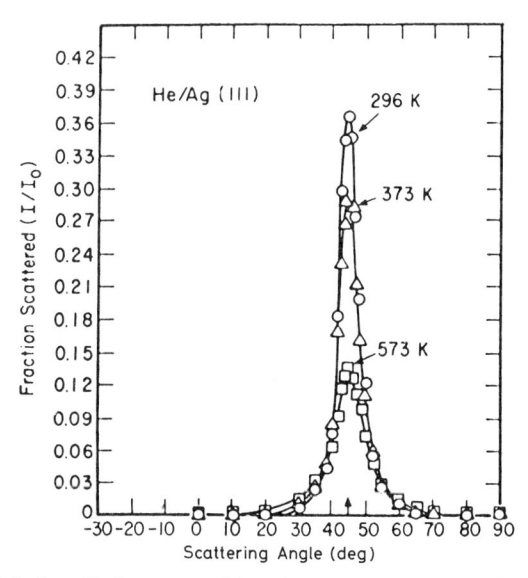

FIG. 34. Angular distribution of helium scattered from Ag(111) single crystal as a function of surface temperature [8].

In a study of helium scattering from single-crystal Ag(111), Sau and Merrill [8] found that the intensity at the specular angle decreased and the peak width increased as the surface temperature T_s increased. This is shown in Fig. 34. An increase in beam width is always associated with a decrease in intensity at the maximum. This broadening with increasing T_s is attributed to a 'thermal roughening' of the surface which increases as the mean square displacements of vibrating surface atoms increase with higher T_s. There is a decrease in specular intensity and an increase in peak width going from helium scattering from the close-packed fcc (111) face of silver to the more open bcc (110) plane of tungsten and to the still more open fcc (100) face of platinum [8]. These changes in surface structure are slight and microscopic but are still detectable by helium scattering. This technique has also allowed Sau and Merrill [8] to conclude that epitaxially grown silver films are more disordered than conventionally prepared clean single-crystal surfaces.

Ollis et al. [71] used beams of helium and neon as probes of the state of surfaces of molybdenum and rhenium during oxidation. They were able to obtain information on the coverage, binding states and binding energies of oxygen on their surfaces by measuring the intensity of gas atoms scattered specularly.

8. RESULTS OF INELASTIC SCATTERING STUDIES

8.1. Rare gas-metal systems

When a monatomic gas particle collides inelastically with a metal surface in the absence of chemical interactions, there can be energy transfer between the translational states of the atom and the vibrational states of the lattice. The incident atom can transfer energy into the lattice (phonon creation) or can absorb energy from it (phonon annihilation) and scatter at a higher energy than the incident particle, depending on the relative temperatures of the gas molecule and the surface.

In the same paper in which they reported on helium scattering studies, Sau and Merrill [8] also looked at inelastic scattering. Here the scattering is characterized by decreasing peak intensity with increasing θ_i and increasing T_s, as seen in Fig. 35 for krypton. Also obvious is the non-specular maximum of the scattered peak with the non-specularity increasing with increasing θ_i. For inelastic scattering there is no correlation between scattering intensity and microscopic surface roughness.

The authors also identified an inelastic scattering regime in which substantial trapping occurs [8]. Here, a significant number of incident atoms lose enough of their energy to become adsorbed. The regime is best illustrated by xenon scattering, as shown in Fig. 36, indicating large deviations from specularity and a substantial scattering contribution corresponding to a cosine distribution. This cosine component is from desorbed atoms which have equilibrated with the surface. Also typical of this regime is the increase in peak intensity with increasing T_s. This is due to the decreased trapping probability at high temperatures. The number of atoms trapped can be correlated well to the estimated depth of the attractive potential well for various gas-metal systems. It should be noted that these

LEED AND SURFACE TOPOGRAPHY 243

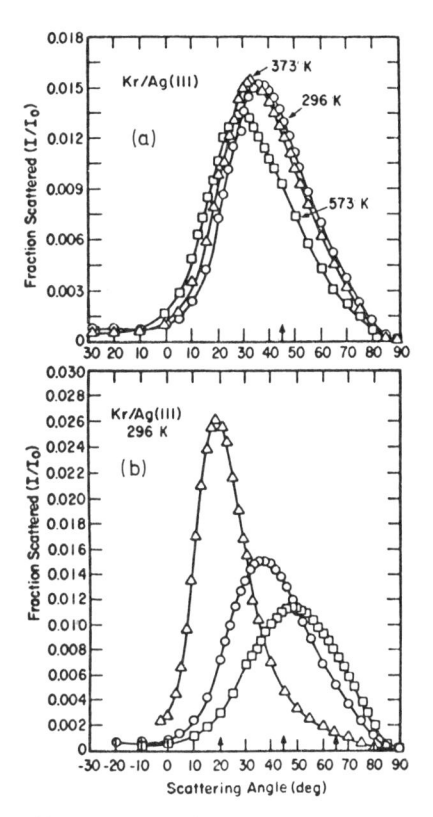

FIG. 35. Angular distribution of krypton scattered from Ag(111) single crystal as a function of (a) surface temperature and (b) incident angle [8].

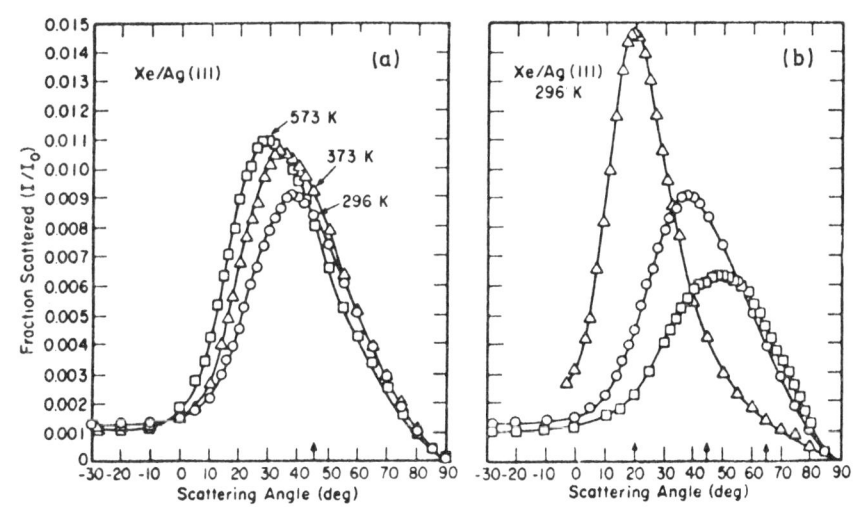

FIG. 36. Angular distribution of xenon scattered from Ag(111) single crystal as a function of (a) surface temperature and (b) incident angle [8].

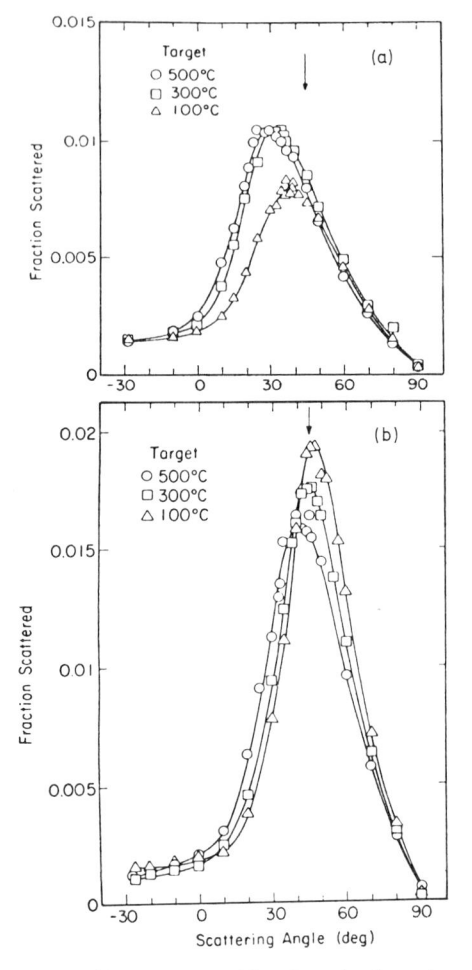

FIG. 37. Angular distribution of krypton scattered from Pt(111) at beam temperature of (a) 23°C and (b) 700°C [7].

observations of inelastic collisions are in qualitative agreement with the cube models.

Similar results were observed on scattering rare gases from Pt(111) [7] and W(110) [72]. In the Pt(111) experiments it was found that if an incident beam temperature of 700°C was used instead of a room temperature beam, the scattered peaks were closer to the specular angle and narrower. This is shown for krypton in Fig. 37. There is also a reversal in trend in intensity as a function of T_s. These trends are predicted by the three-dimensional lattice models, and the effect is attributed to a shorter collision time. A trend toward narrower angular distribution and a supraspecular shift in the peak maximum with increasing beam temperature were also observed earlier by Saltsburg and Smith [73] for heavy molecules scattered from Ag(111). Still another investigation by Romney and Anderson [12], using nearly monoenergetic argon nozzle beams with incident energies 0.05 - 5 eV, confirmed the supraspecular shift to about 0.3 eV but then showed a reversal in this trend at higher incident energies with the scattering maximum returning to the specular angle. This return to specular angle is not predicted by the cube models, which fail (not surprisingly) at the high incident energy limit. Miller and Subbarao [74] confirmed these findings and noted that the scattered peak width first decreases with incident energy and then increases again. The energy of the minimum peak width is a function of the gas scattered, being higher for heavier gases.

One of the best ways to analyse the translational energy transfer in an inelastic molecular beam-surface interaction is to measure the velocities of atoms or molecules scattered from a surface for an incident beam of known velocity. To date a few such measurements have been made for monatomic and diatomic gases. Early work by Fisher et al. [75] was carried out using polycrystalline nickel and stainless steel surfaces. Subsequently Yamamoto and Stickney [9] studied the average velocity of argon scattered from single-crystal W(110) using the phase-shift technique and a nearly monoenergetic nozzle beam source. Their results are shown in Fig. 38. They found that for a given incident velocity (5×10^4 cm · s^{-1}) the average velocity of the scattered gas increased as the angle of incidence increased. Also, the average energy decreased monotonically as a function of scattered angle away from the surface normal for all incident angles. The latter is predicted by the hard-cube model as shown by the dashed curves of Fig. 38. As noted, there is good qualitative but poor quantitative agreement. The experiment was also performed at various surface temperatures

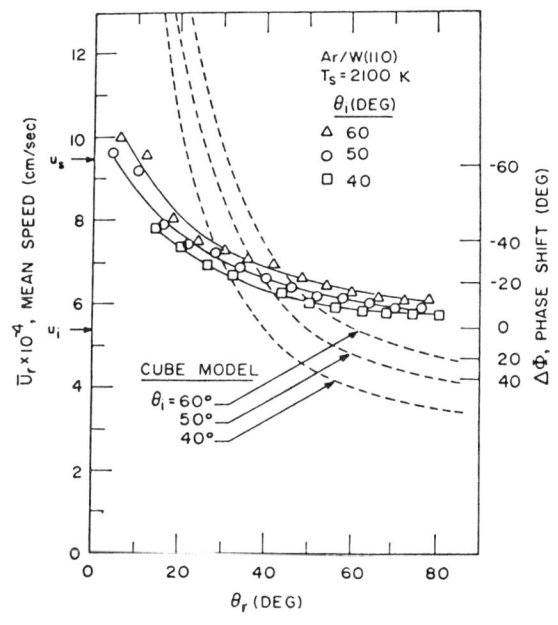

FIG. 38. Average velocity of argon atoms scattered from W(110) single crystal as a function of scattered angle for various incident angles. Arrows indicate U_i, the incident velocity, and U_s, the velocity the atoms would have if completely accommodated. Dashed lines show the predictions of the hard-cube model [9].

and it was found, not surprisingly, that the average velocity of scattered atoms increased as the surface temperature increased.

Recently Subbarao and Miller [20] studied the scattering of neon and argon from epitaxially grown Ag(111), using nozzle beams of various incident velocities. They found that at low energies (0.31 eV and lower) the mean velocities are highest at the surface normal and decrease monotonically towards the surface tangent, in agreement with Yamamoto and Stickney [9]. At high incident energy, 1.36 eV, the mean velocity is virtually constant with the scattering angle. Their measurements of the thermal spread in the scattered beam show nearly Maxwellian distributions at the surface normal and narrower distributions towards the tangent. At high incident energy the thermal spread appears constant. The scattered beams were all much more nearly Maxwellian than the incident beam. They also found that, at high incident beam energies, normal momentum transfer becomes less efficient and the scattering becomes more elastic. Also, there is an increase in tangential momentum transfer in agreement with the theory of McClure [48].

Siekhaus et al. [26] report a very interesting experiment in which they scattered various thermal beams from pyrolytic graphite at different temperatures and then measured the temperature of the scattered beam. They found that for O_2, D_2, xenon and krypton there was essentially no change in reflected beam temperature T_r for large variations in beam temperature T_b. Helium and neon beams showed a slight increase in T_r with increasing T_b. A most unusual result is that in several cases T_r was less than both T_b and the surface temperature T_s. As they point out, in a global exchange process this would violate the second law of thermodynamics. They also studied the effect of T_s on T_r for room-temperature beams, and these results are shown in Fig. 39 for the annealed basal plane of pyrolytic graphite. The data can be fitted to the empirical formula $T_r/T_c = 1 - \exp(-T_s/T_c)$, where T_c is a constant depending only on the incident gas and the nature of the solid surface. The authors suggest [26] that T_c, while only an empirical parameter, does have some characteristics of

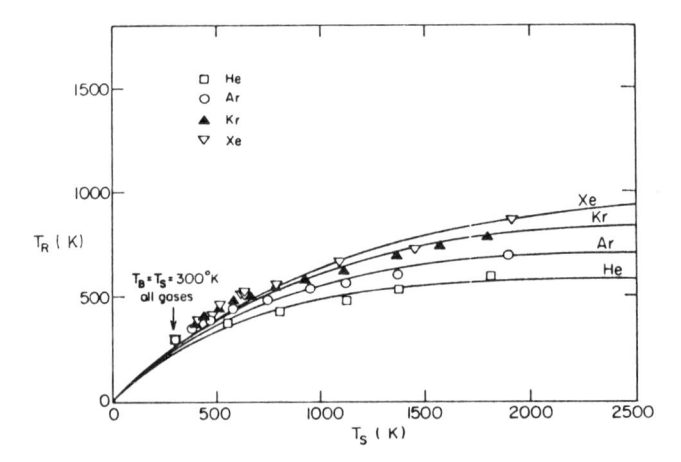

FIG. 39. Temperature of atoms scattered from the annealed basal plane of pyrolytic graphite as a function of surface temperature. Beam temperature is 27°C [26].

gas-surface binding energy and does increase with increasing atomic weight of the incident gas. They discuss their result of T_r being virtually independent of T_b and its dependence on T_s in terms of partial trapping, i.e. incident molecules are trapping long enough to uncouple the re-emission process from the incident energy but not long enough to completely equilibrate with the solid.

8.2. Scattering of diatomic and polyatomic molecules from metal surfaces

The case of scattering diatomic or polyatomic molecules from a metal surface is more complicated than for monatomic gases because of the additional possibility that the internal states of the molecule may interact with the vibrational states of the solid. The vibrational levels of most light molecules are at too high energy to be involved in a gas-solid interaction, but the rotational and translational states certainly are available for interaction with the solid phonons. In a case where the surface temperature is higher than the gas temperature, energy can be transferred from the lattice phonons to the translational or internal states of the molecule. If the gas temperature is higher than the surface temperature, the translational and internal energy levels of the molecule can interact with the lattice modes resulting in phonon creation.

Scattering of diatomic molecules was investigated by Palmer et al. [11] who first compared ^{3}He and ^{4}He scattering from epitaxial Ag(111) films and found them very similar in specular intensity and peak width. They then scattered H_2, D_2 and HD from the same surface and found that H_2 scattered in a manner very similar to that of helium but that HD and D_2 both gave a very much reduced specular intensity. The results for the hydrogenic molecules are shown in Fig.40. They [11] also looked at the scattering of HD and D_2 as a function of beam temperature, finding the distributions very broad for 80 K beams compared to 300 K and rather more specular at 1500 K than at 300 K. Their results are discussed in terms of the coupling of the lattice phonons with the rotational levels of the hydrogen molecules. For H_2 the lowest rotational transition energy is at 1032 cal/mole. For D_2 this is only 516 cal/mole. The estimated Debye phonon energy in silver is 450 cal/mole, rather close to the D_2 transition energy, while H_2 would certainly need a less probable multiphonon process to interact. There is also a transition at 516 cal/mole in HD. The conclusion drawn from this analysis [11] is that it should be much easier for D_2 and HD to couple with the lattice via their rotational states than for H_2. This question of rotational coupling was examined for D_2 on Pt(111) by Smith and Merrill [6]. They found that the fraction of diffuse D_2 scatter increased with increasing surface temperature. Since this is the reverse of the trend for trapping, they suggest that this increase in diffuse scatter is due to increased rotational coupling with the more highly excited lattice vibrations. Merrill and co-workers found similar results for D_2 scattered from W(110) [72] and most recently for scattering from single-crystal Ag(111) [8].

The scattering of diatomic and polyatomic molecules from Pt(100) was studied by West and Somorjai [5] for a variety of surface conditions. They found for NO, CO, N_2, O_2, H_2 and D_2 a broad scattered peak with a maximum at or near the specular angle. All the peaks were rather broad,

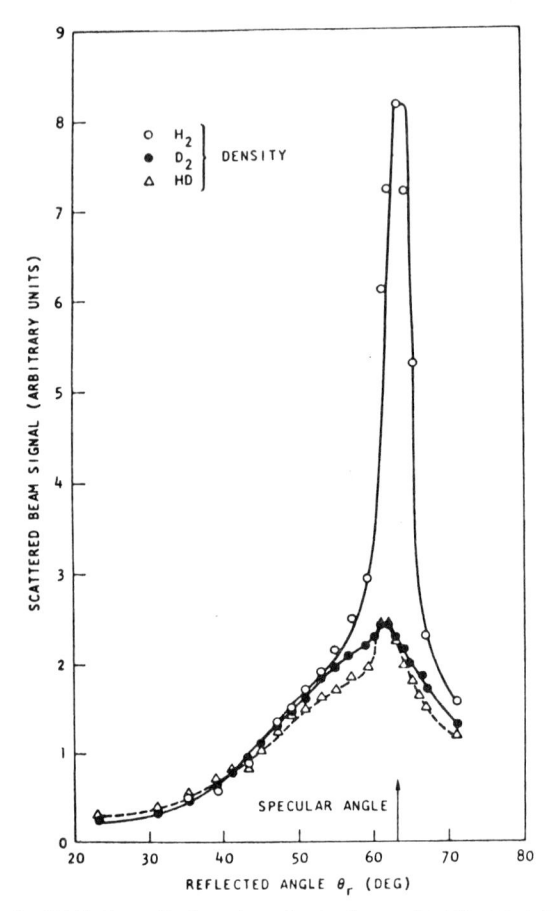

FIG. 40. Angular distribution of hydrogenic molecules from epitaxially grown Ag(111). Surface temperature = 570 K. Beam temperature = 300 K [11].

indicating substantial energy interaction or possibly surface roughness. They noted almost no difference between H_2 and D_2 scattering although broadening due to surface roughness may have hidden any changes due to rotational interactions. The polyatomic molecules CO_2, N_2O, NO_2, C_2H_2, NH_3 and methylenecyclobutane were also examined. For all except NH_3 there was a maximum in intensity very close to the specular angle and again the peaks were rather broad. For NH_3 the distribution was cosine, indicating complete accommodation. Some of the same molecules were scattered from a Pt(100) surface with a graphite overlayer. These results did not differ significantly from those on clean Pt(100).

The absence of complete accommodation for most molecules indicates inefficient energy transfer between the lattice modes and the incident translational energy even though these energies are of the same order of magnitude [5]. Indeed, it has been suggested by Beeby and Dobrzynski [76] that inefficient transfer is expected if translational and vibrational energies are nearly equal.

Both CO and acetylene chemisorb on Pt(100) to form rather well ordered overlayers. In the same series of experiments, West and Somorjai [5] also scattered CO from a Pt(100) surface with an ordered layer of CO and scattered C_2H_2 from an ordered layer of C_2H_2. The comparison between acetylene scattered from clean and acetylene-covered Pt(100) is shown in Fig. 41. For acetylene the distribution is very nearly cosine. For CO the distribution is very broad, and its maximum does not shift as the incident angle is changed. Both examples indicate nearly complete energy accommodation.

It is possible that when gases such as CO or acetylene are adsorbed, new vibrational modes exist which do allow efficient energy transfer with the incident CO and acetylene gas molecules [5]. It is also pointed out that for these rather heavy molecules with low-energy rotational transitions, it is likely that considerable interaction is taking place between the lattice phonons and the rotational states. There could also be transfer of rotational energy into translational energy of the gas molecule.

In other work on polyatomic molecular scattering, Saltsburg and Smith [73] looked at CH_4 and NH_3 scattered from Ag(111) films and found also that NH_3 has a cosine distribution and CH_4 is highly non-specular although non-cosine. Also, Saltsburg et al. [77] scattered CO_2, CO, O_2, N_2, H_2, D_2 and HD from Ag(111) films, finding broad peaks at or near the specular angle in all cases.

8.3. Direct observation of phonon interaction

In the papers in which high-resolution helium and neon diffraction was reported, Williams [66, 67] also reported small peaks in the vicinity of the specular peak, which he attributed to direct phonon interaction. These peaks are seen in Fig. 32 for neon. Both phonon absorption and emission are seen. From his data on helium and an assumption of surface phonons only, he was able to calculate a dispersion relationship. Fisher and Bledsoe [78] scattered 0.058 eV helium from LiF(001) and analysed the scattered velocity distributions, also finding peaks they assigned to single-phonon absorptions and emissions, with emissions dominating.

In a study of very low energy helium scattering from Ag(111) film, Subbarao and Miller [79] observed a peak in the angular distribution which they attribute to inelastic scattering from lattice phonons. The position of the peak maximum is a function of the incident beam velocity while the specular peak width remains constant. Recently, Subbarao and Miller [80] have expanded this work on helium on Ag(111) and come to the conclusion that the interaction involves bulk phonons only.

9. REACTIVE SCATTERING OF MOLECULAR BEAMS

How is the energy of a chemical surface reaction distributed among the reactants and the products? This is one of the fundamental questions of surface science, and it appears that the technique of reactive scattering of molecular beams can provide the answer. To demonstrate this let us consider a relatively simple surface reaction, the recombination of hydrogen

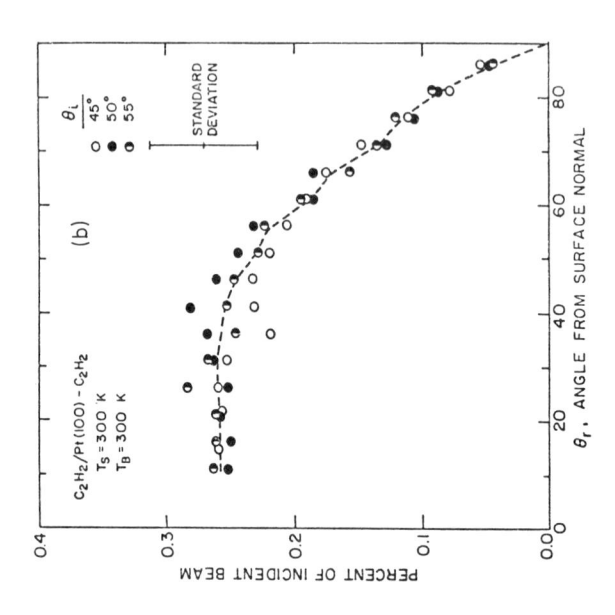

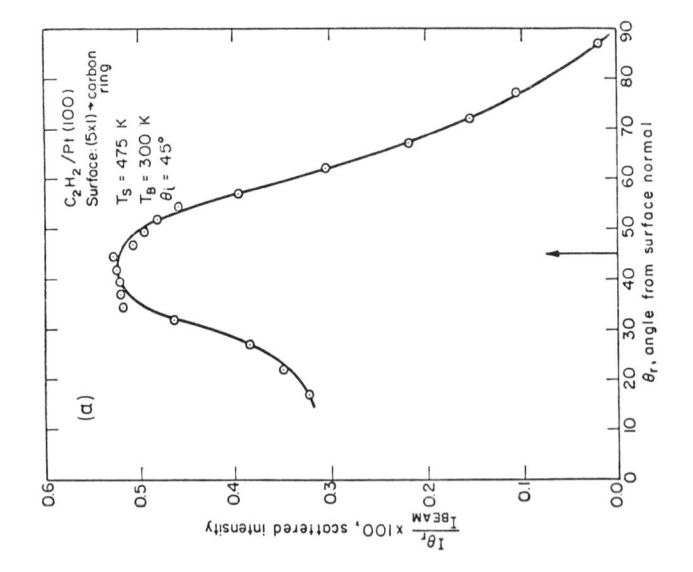

FIG. 41. Angular distribution of acetylene scattered from (a) clean Pt(100) and (b) acetylene-covered Pt(100) single crystals [5].

atoms on a metal surface. The hydrogen atoms emanate from a high-temperature source (greater than 2000 K) and have kinetic energy corresponding to the source temperature. As the hydrogen atoms impinge on the surface, which is held at a lower temperature, some of their translational energy is transferred to the solid through lattice vibrations (T-V energy transfer), and many of the atoms become trapped on the surface. As a result of surface diffusion, there is a high probability of recombination of the adsorbed atoms that is in general a highly exothermic reaction (over 100 kcal for hydrogen). The reaction energy is then partitioned between the solid and the desorbing molecules. By measuring the velocity of the desorbing molecules and their angular distribution, the distribution of chemical energy can be determined.

If the molecules stay on the surface long enough after the chemical reaction, they can come to equilibrium with the surface. In this circumstance, the desorbing molecular beam will have a cosine angular distribution and an average 'translational' temperature equal to the surface temperature. A non-cosine angular distribution and/or a higher translational energy than expected based on the surface temperature would indicate poor thermal equilibrium and that a portion of the chemical energy is converted to translational energy of the scattered molecules. If the desorbing molecules carry with them much of the energy of the chemical reaction, they may be in an excited rotational, vibrational or electronic state in addition to having increased kinetic energy. The partitioning of chemical energy between internal modes of the molecules and their translational energy is another important question that can be answered by molecular beam scattering studies utilizing measurements of the velocity of the reaction product and their angular distribution. So far, such detailed measurements of energy partitioning in surface chemical reactions have not been made, although the experimental techniques developed recently in various laboratories can provide such information. No doubt, investigations of this type will be carried out in the near future.

We shall now review the reactive scattering studies that have been carried out so far using molecular beams and clean solid surfaces. These reactive scattering studies may be divided into two classes: (a) chemical reactions where the surface acts as a catalyst for dissociation or recombination of the reactant molecules and (b) those in which the surface is one of the reactants.

9.1. Catalytic surface reactions

The hydrogen-deuterium exchange reaction $H_2 + D_2 = 2HD$ has been studied by Saltsburg et al. [81] on epitaxially grown nickel surfaces with (111) orientation. The angular distribution of HD was $\cos^3 \theta$ instead of $\cos \theta$, indicating incomplete accommodation of the reaction product with the surface. It was argued that HD molecules formed only when an H atom and a D atom were in close proximity and evaporated coherently to explain the angular dependence observed. The H_2-D_2 exchange reaction was studied by Bernasek et al. [34] using platinum single-crystal surfaces. No HD signal was detected on scattering a mixed hydrogen-deuterium beam from the (111) crystal face. However, when the diatomic molecules were scattered from a stepped high Miller index platinum surface, 5-10% HD product was

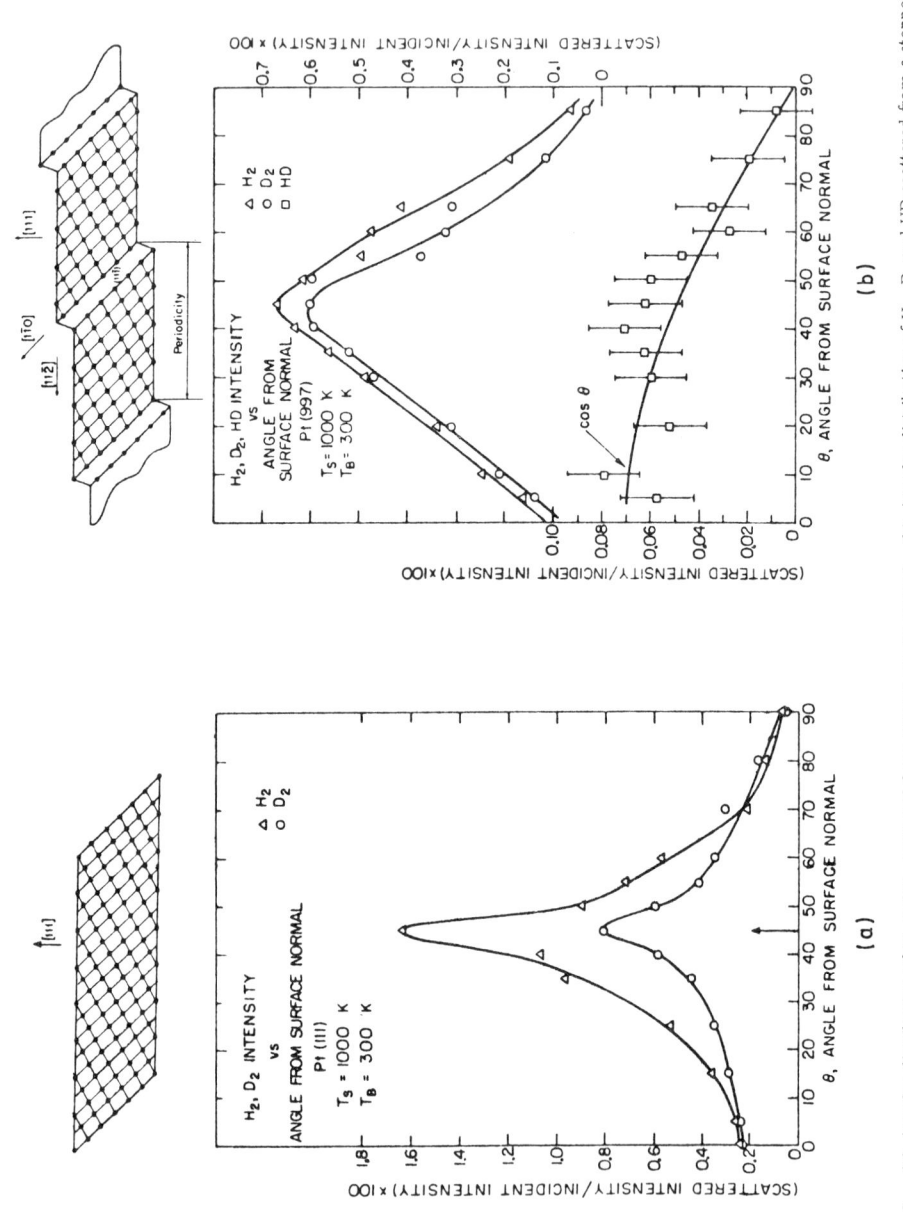

FIG. 42. (a) Angular distribution of H_2 and D_2 scattered from Pt(111) single crystal. (b) Angular distribution of H_2, D_2 and HD scattered from a stepped Pt single crystal. Schematic diagrams of the surfaces are shown above the curves. The HD distribution is shown with scale expanded [34].

detectable. High Miller index single-crystal surfaces are characterized by ordered arrangement of atomic steps one atom high separated by terraces of low Miller index orientation. In their study the terraces were 9 atoms wide and of (111) orientation. The observed H_2, D_2 and HD scattering distributions from the (111) and the stepped platinum surface are shown in Fig. 42. This study has shown that atomic steps play an all-important role in dissociating large binding energy diatomic molecules. Chopping the incident beam to generate an ac signal was done to obtain a better signal-to-noise ratio and to measure surface residence times by the phase-shift technique. Varying the chopping frequency has yielded HD residence times of 25 ms on a stepped platinum surface at a 700 K surface temperature. Such a long residence time should result in complete thermal equilibration between the surface and the reaction products. This was found by experiments as the desorbing HD beam exhibited cosine angular distribution.

Reactive scattering studies with two or more reactants can be carried out by either using a mixed molecular beam or pre-adsorbing one of the reactants on the surface and monitoring the reaction between one of the reactants in the molecular beam and the other reactant adsorbed. Although detailed comparison between the two modes of reactions have not yet been made, it is likely that the surface reaction kinetics may change as the reaction conditions are varied in this manner.

The dissociation of H_2 on tungsten surfaces was studied by Smith and Fite [82]. The reaction probability increased to 0.3 above 3000 K. The angular distribution of H atoms was cosine-like, indicating thermal equilibration between the hot surface and the desorbing H atoms. This was also corroborated by a long residence time, 280 μs, which was measured. The dissociation of H_2 on tantalum surfaces was studied by Krakowski and Olander [83] in the temperature range of 1100-2600 K. The reaction probability increased with increasing temperature just as for tungsten surfaces. The hydrogen atoms underwent complete thermal accommodation on the surface. By varying the kinetic energy ('beam temperature') of the incident beam they have determined an activation energy of 1.4 kcal/mole for the atomization surface reaction. Thus the atoms once formed cannot undergo recombination before desorption from the hot surface. The rate-limiting step in the reaction appears to be the desorption of H atoms, which has an activation energy of about 75 kcal/mole. The surface diffusion of H atoms does not require activation energy.

The dissociation of N_2O on hot tungsten has been investigated by Muschlitz et al. [84]. The decomposition probability approaches unity at high surface temperatures. The product N_2 was emitted with a cosine distribution while neither oxygen atoms nor oxygen molecules could be detected in the scattered beam. It appears that oxygen reacts with the tungsten surface during the decomposition of N_2O. Both N_2 and NO were found in the scattered beam and the ratio of the two species, N_2/NO, was approximately 12:1 at 2500 K. During the dissociation of N_2O on platinum surfaces studied by West and Somorjai [85], again both NO and N_2 were found, the NO concentration being larger than that found on desorption from tungsten surfaces. The angular distribution of the product NO molecule that formed by dissociation at the clean platinum surface was of the cosine type indicating complete accommodation of the NO molecule on the surface prior to re-emission. The angular distribution of NO product molecules is quite different, however, when they are emitted from carbon-covered platinum surfaces at 1125 K.

The angular distribution is certainly non-cosine, and it peaks at or near the specular angle. Such a peaked spatial distribution reflects a lack of energy accommodation during the surface dissociation reaction of N_2O on the carbon-covered Pt(100) surface and suggests a direct reactive scattering mechanism. N_2O may undergo a variety of chemical reactions on platinum and carbon-covered platinum surfaces. Those chemical reactions which can take place between the carbon on the platinum surface and N_2O are largely exothermic and yield CN, CO or CO_2 reaction products in addition to N_2 and NO on the clean platinum surface where the incident N_2O molecules can only undergo endothermic chemical reactions. In this case the dissociated species appear to be fully accommodated on the surface before re-emission, as indicated by the cosine angular distribution of the scattered beam. On the carbon-covered Pt(100) surface where surface reactions can be strongly exothermic there is evidence for direct reactive scattering, and NO molecules are emitted without complete energy accommodation between the incident beam and the surface, as indicated by the non-cosine angular distribution of the scattered beam. Direct scattering is commonly observed in studies of chemical reactions between crossed molecular beams that are exothermic and exoergic.

The chemical reaction between deuterium and oxygen to form D_2O occurring on the (111) platinum surface has been investigated by Smith and Palmer [86]. In these studies the molecular beam was D_2 and the oxygen was pre-adsorbed on the platinum surface which was maintained in an oxygen ambient. The production of D_2O was studied as a function of D_2 flux and temperature, oxygen pressure, platinum surface temperature, and angle of incidence of the D_2 beam. It was found that the reaction rate is proportional to the second power of the D_2 pressure and the 0.8 power of the oxygen pressure. The adsorption of D_2 requires an activation energy of 1.8 kcal/mole. It is proposed that in the adsorbed state the reaction proceeds upon the collision of four adsorbed deuterium atoms with an adsorbed oxygen molecule or activated oxygen complex.

Nutt and Kapur [87] reported on preliminary experiments involving the oxidation of ammonia using NH_3 and O_2 mixed molecular beams. The reaction products were N_2, H_2O and NO in the temperature range of 600-1200 K, and the reaction probability was about 0.1 - 0.2.

The oxidation of C_2H_4 on silver surfaces was studied by Smith et al. [88]. They found no evidence for the formation of ethylene oxide, CO_2 being the dominant reaction product. The surface temperature dependence of the CO_2 formation indicated an activation energy of 8 kcal/mole. The reaction probability was less than 10^{-2} at 820 K, and poisoning of the surface reaction by carbon which builds up on the silver surface was inferred from the experimental data. The hydrogenation of ethylene was studied using the (111) crystal face of platinum [89]. The reaction probability can be estimated to be about 10^{-4} in the range of 500-700°C. As a result, the formation of ethane could not be detected.

Olander et al. [29, 90] have studied the oxidation of both the basal plane and the prism plane of graphite. The product of the oxidation reaction is CO although a small CO_2 signal was also detectable during oxidation of the prism plane. The reaction rate was monitored as a function of surface temperature. From the chopping frequency dependence of the reaction probability they have concluded that there must be at least two parallel reactions, one slow and one faster, taking place on the graphite surface. For the basal plane the faster reaction is attributed to the migration of atomic oxygen over the surface to reaction sites where oxidation occurs. The rate constant k for this step is given by

$$k = 2.5 \times 10^7 \exp(-30 \text{ kcal}/RT) \text{ s}^{-1}$$

The slower reaction step is the desorption of CO, and its rate constant k_d is given by

$$k_d = 3 \times 10^{12} \exp(-50 \text{ kcal}/RT) \text{ s}^{-1}$$

There are two types of reaction sites postulated with surface concentrations of 10^{11} cm^{-2} and 10^8 cm^{-2}, respectively. Grain boundary and possible bulk diffusion of oxygen was found to be an important step in the oxidation of the prism face.

9.2. Reactions between gases and solid surfaces

The interaction between chlorine molecular beams and nickel surfaces has been studied by McKinley [91] and Smith and Fite [92]. The reaction probability is 0.8 at 1000 K, and both NiCl and NiCl$_2$ are detectable among the reaction products. The formation of NiCl predominates at low temperatures while NiCl$_2$ forms almost exclusively above 1400 K. The mechanism proposed involves a Ni$_2$Cl$_2$ surface intermediate. The formation of this dimer appears to be the rate-limiting step in the overall reaction while the desorption of NiCl and NiCl$_2$ are rapid-reaction steps. The residence time of the desorbing NiCl is 916 μs at 1150°C and 140 μs at 1300°C, and the NiCl residence times are even longer. The oxidation of germanium was studied by Anderson and Boudart [93] and Madix and Boudart [94]. The reaction probability was 0.04 and independent of temperature. However, the oxidation rate of the surface was dependent on the oxygen beam temperature, which indicates 100-200 cal/mole activation energy for the adsorption. It appears that the dissociative adsorption of oxygen is the rate-determining step in the reaction. The oxidation reaction using oxygen atoms instead of oxygen molecules was also investigated [95]. The reaction probabilities in this case are in the range 0.2-0.3, much higher than for oxygen molecules for surface temperatures in the range 830-1110 K. The difference in reactivities appears to be due to the requirement that both atoms in the oxygen molecule interact simultaneously with the surface atom. Thus the interaction probability depends on the orientation criterion.

Some of the surface reactions studied were endothermic (e.g. the decomposition of N$_2$O) while others were exothermic (oxidation reactions, etc.). It should be noted that only exothermic reactions are likely to yield reaction products with excess translational or internal energy because in this circumstance surplus chemical energy is available to the desorbing molecules. During endothermic reactions the reaction products are likely to equilibrate thermally with the surface and then desorb with a cosine spatial distribution and without excess translational or internal energy.

10. FUTURE TRENDS

The field of molecular beam-surface scattering is one of the frontier areas of surface science. The energy transfer that takes place in surface reactions between the gas and the surface atom is the key to understanding the reactivity of surfaces. Therefore, this field will grow rapidly over the next decade, we believe. There are many areas of development that can already be identified. Improvements in experimental techniques will permit the investigator to use single-crystal surfaces and to monitor the velocity of the scattered products as well as their angular distributions. Perhaps the velocity of the scattered beam is a more important experimental parameter in understanding energy transfer than the angular distribution of the emitted beam. The use of single-crystal surfaces of well defined structure permits studies of the correlation between gas-surface energy transfer and surface structure to explore the effects of atomic steps, the variation of atomic spacing and the structure of the adsorbed layer. Nozzle beam techniques will allow the use of more monochromatic molecular beams of higher intensity with corresponding improvements in measurements of residence times of adsorbed molecules, intensities of the emitted beams and their times of flight.

Various types of energy transfer processes can take place between the incident gas and the surface that involve T-V, R-V, V-V and other types of energy-transfer processes, as previously mentioned. Experiments can be devised that will separate these various processes and study the transition probabilities of each. The important question of how the total energy of the collision partners is distributed among the surface atom and the translational and internal energy modes of the emitted particles will then be answered.

One of the important areas of the molecular beam surface scattering field is the study of beams of condensable vapours as they interact with surfaces. Research in this area has been initiated recently in several laboratories, and there are many interesting problems that may be solved by this technique. Problems of crystal growth from the vapour, surface diffusion mechanisms, and energy transfer during condensation and nucleation can be studied this way.

Reactive scattering of molecular beams promises to shed light on the elementary steps of surface reactions. This field is just beginning to attract a great deal of interest and will grow rapidly during the next decade. Molecular beam surface scattering studies are likely to solve many important scientific and technical problems in the fields of surface science, heterogeneous catalysis, aerospace sciences and astrophysics.

REFERENCES (PART III)

[1] SALTSBURG, H., SMITH, J.N., ROGERS, M., Eds, Fundamentals of Gas-Surface Interactions, Academic Press, New York (1967).
[2] STICKNEY, R.E., in Advances in Atomic and Molecular Physics (BATES, D.R., ESTERMANN, I., Eds), Academic Press, New York (1967) 143.
[3] BERNASEK, S.L., SOMORJAI, G.A., Prog. Surf. Sci. 5 (1975) 377.
[4] MERRILL, R.P., Catal. Rev. 4 (1970) 115.
[5] WEST, L.A., SOMORJAI, G.A., J. Chem. Phys. 57 (1972) 5143.

[6] SMITH, D.L., MERRILL, R.P., J. Chem. Phys. 53 (1970) 3588.
[7] STOLL, A.G., SMITH, D.L., MERRILL, R.P., J. Chem. Phys. 54 (1971) 163.
[8] SAU, R., MERRILL, R.P., Surf. Sci. 34 (1973) 268.
[9] YAMAMOTO, S., STICKNEY, R.E., J. Chem. Phys. 53 (1970) 1594.
[10] SMITH, J.N., SALTSBURG, H., J. Chem. Phys. 40 (1964) 3584.
[11] PALMER, R.L., SALTSBURG, H., SMITH, J.N., J. Chem. Phys. 50 (1969) 4661.
[12] ROMNEY, M.J., ANDERSON, J.B., J. Chem. Phys. 51 (1969) 2490.
[13] PALMER, R.L., SMITH, J.N., SALTSBURG, H., O'KEEFE, D.R., J. Chem. Phys. 53 (1970) 1666.
[14] WEST, L.A., SOMORJAI, G.A., J. Chem. Phys. 54 (1971) 2864.
[15] SMITH, D.L., MERRILL, R.P., J. Chem. Phys. 52 (1970) 5861.
[16] ANDERSON, J.B., ANDRES, R.P., FENN, J.B., in Advances in Chemical Physics 10 (ROSS, J., Ed.), Wiley, New York (1966) 275.
[17] GORDON, R.J., LEE, Y.T., HERSCHBACH, D.R., J. Chem. Phys. 54 (1971) 2393.
[18] HAYS, W.J., RODGERS, W.E., KNUTH, E.L., J. Chem. Phys. 56 (1972) 1652.
[19] DYKE, T.R., TOMASEVICH, G.R., KLEMPERER, W., FALCONER, W.E., Surf. Sci. 57 (1972) 2277.
[20] SUBBARAO, R.B., MILLER, D.R., J. Chem. Phys. 58 (1973) 5247.
[21] JONES, R.H., OLANDER, D.R., KRUGER, V., J. Appl. Phys. 40 (1969) 4641.
[22] OLANDER, D.R., J. Appl. Phys. 40 (1969) 4650.
[23] SIEKHAUS, W.J., JONES, R.H., OLANDER, D.R., J. Appl. Phys. 41 (1970) 4392.
[24] OLANDER, D.R., KRUGER, V., J. Appl. Phys. 41 (1970) 2769.
[25] HARRISON, H., HUMMER, D.G., FITE, W.L., J. Chem. Phys. 41 (1964) 2567.
[26] SIEKHAUS, W.J., SCHWARZ, J.A., OLANDER, D.R., Surf. Sci. 33 (1972) 445.
[27] YAMAMOTO, S., STICKNEY, R.E., J. Chem. Phys. 47 (1967) 1091.
[28] JONES, R.H., OLANDER, D.R., SIEKHAUS, W.J., SCHWARZ, J.A., J. Vac. Sci. Technol. 9 (1972) 1429.
[29] OLANDER, D.R., SIEKHAUS, W.J., JONES, R., SCHWARZ, J.A., J. Chem. Phys. 57 (1972) 408.
[30] MADIX, R.J., SCHWARZ, J.A., Surf. Sci. 24 (1971) 264.
[31] DABIRI, A.E., LEE, T.J., STICKNEY, R.E., Surf. Sci. 26 (1971) 522.
[32] HAGENA, O.F., Appl. Phys. Lett. 9 (1966) 385.
[33] KODERA, K., et al., Jap. J. Appl. Phys. 10 (1971) 543.
[34] BERNASEK, S.L., SIEKHAUS, W.J., SOMORJAI, G.A., Phys. Rev. Lett. 30 (1973) 1202.
[35] GOODMAN, F.O., Surf. Sci. 26 (1971) 327.
[36] FEUER, P., J. Chem. Phys. 39 (1963) 1131.
[37] OMAN, R.A., J. Chem. Phys. 48 (1968) 3919.
[38] SMITH, J.N., SALTSBURG, H., PALMER, R.L., J. Chem. Phys. 49 (1968) 1287.
[39] LOGAN, R.M., STICKNEY, R.E., J. Chem. Phys. 44 (1966) 195.
[40] LOGAN, R.M., KECK, J.C., J. Chem. Phys. 49 (1968) 860.
[41] LORENZEN, J., RAFF, L.M., J. Chem. Phys. 49 (1968) 1165.
[42] LORENZEN, J., RAFF, L.M., J. Chem. Phys. 52 (1970) 1133.
[43] LORENZEN, J., RAFF, L.M., J. Chem. Phys. 52 (1970) 6134.
[44] LORENZEN, J., RAFF, L.M., J. Chem. Phys. 54 (1971) 674.
[45] McCLURE, J.D., J. Chem. Phys. 51 (1969) 1687.
[46] McCLURE, J.D., J. Chem. Phys. 57 (1972) 2810.
[47] McCLURE, J.D., J. Chem. Phys. 57 (1972) 2823.
[48] McCLURE, J.D., J. Chem. Phys. 52 (1970) 2712.
[49] GOODMAN, F.O., Surf. Sci. 7 (1967) 391.
[50] GOODMAN, F.O., Surf. Sci. 24 (1971) 667.
[51] GOODMAN, F.O., GILLERLAIN, J.D., J. Chem. Phys. 57 (1972) 3645.
[52] MANSON, R., CELLI, V., Surf. Sci. 24 (1971) 495.
[53] GOODMAN, F.O., Surf. Sci. 30 (1972) 1.
[54] BEEBY, J.L., J. Physics C 5 (1972) 3438.
[55] BEEBY, J.L., J. Physics C 5 (1972) 3457.
[56] TSUCHIDA, A., Surf. Sci. 14 (1969) 375.
[57] CABRERA, N., CELLI, V., GOODMAN, F.O., MANSON, R., Surf. Sci. 19 (1970) 67.
[58] CABRERA, N., CELLI, V., MANSON, R., Phys. Rev. Lett. 22 (1969) 346.
[59] BEEBY, J.L., J. Physics C 4 (1971) L395.
[60] WEINBERG, W.H., J. Physics C 5 (1972) 2098.
[61] WEINBERG, W.H., J. Chem. Phys. 57 (1972) 5463.
[62] MASON, B.F., WILLIAMS, B.R., J. Chem. Phys. 56 (1972) 1895.
[63] TENDULKAR, D.V., STICKNEY, R.E., Surf. Sci. 27 (1971) 516.

[64] ESTERMANN, I., STERN, O., Z. Phys. 61 (1930) 95.
[65] O'KEEFE, D.R., SMITH, J.N., PALMER, R.L., SALTSBURG, H., J. Chem. Phys. 52 (1970) 4447.
[66] WILLIAMS, B.R., J. Chem. Phys. 55 (1971) 1315.
[67] WILLIAMS, B.R., J. Chem. Phys. 55 (1971) 3220.
[68] O'KEEFE, D.R., SMITH, J.N., PALMER, R.L., SALTSBURG, H., Surf. Sci. 20 (1970) 27.
[69] HOINKES, H., NAHR, H., WILSCH, H., Surf. Sci. 30 (1972) 363.
[70] WEINBERG, W.H., MERRILL, R.P., J. Chem. Phys. 56 (1972) 2893.
[71] OLLIS, D.F., LINTZ, H.G., PENTENERO, A., CASSUTO, A., Surf. Sci. 26 (1971) 21.
[72] WEINBERG, W.H., MERRILL, R.P., J. Chem. Phys. 56 (1972) 2881.
[73] SALTSBURG, H., SMITH, J.N., J. Chem. Phys. 45 (1966) 2175.
[74] MILLER, D.R., SUBBARAO, R.B., J. Chem. Phys. 52 (1970) 425.
[75] FISHER, S.S., HAGENA, O.F., WILMOTH, R.G., J. Chem. Phys. 49 (1968) 1562.
[76] BEEBY, J.L., DOBRZYNSKI, L., J. Physics C 4 (1971) 1209.
[77] SALTSBURG, H., SMITH, J.N., PALMER, R.L., in The Structure and Chemistry of Solid Surfaces (SOMORJAI, G.A., Ed.), Wiley, New York (1969).
[78] FISHER, S.S., BLEDSOE, J.R., J. Vac. Sci. Technol. 9 (1972) 814.
[79] SUBBARAO, R.B., MILLER, D.R., J. Chem. Phys. 51 (1969) 4679.
[80] SUBBARAO, R.B., MILLER, D.R., J. Vac. Sci. Technol. 9 (1972) 808.
[81] SALTSBURG, H., SMITH, J.N., PALMER, R.L., Am. Vac. Soc. Symp. on Surface Science, Evaporation and Effusion, Los Alamos, 1969.
[82] SMITH, J.N., FITE, W.L., J. Chem. Phys. 37 (1972) 898.
[83] KRAKOWSKI, R.A., OLANDER, D.R., J. Chem. Phys. 49 (1968) 5027.
[84] COLTHARP, R.N., SCOTT, J.T., MUSCHLITZ, E.E., in Structure and Chemistry of Solid Surfaces (SOMORJAI, G.A., Ed.), Wiley, New York (1969).
[85] WEST, L.A., SOMORJAI, G.A., J. Vac. Sci. Technol 9 (1972) 668.
[86] SMITH, J.N., PALMER, R.L., J. Chem. Phys. 56 (1972) 13.
[87] NUTT, C.W., KAPUR, S., Nature (London) 220 (1968) 697.
[88] SMITH, J.N., PALMER, R.L., VROOM, D.A., J. Vac. Sci. Technol. 10 (1973) 373.
[89] SMITH, D.L., Ph. D. thesis, Dept. Chem. Eng., Univ. California, Berkeley (1969).
[90] OLANDER, D.R., JONES, R.H., SCHWARZ, J.A., SIEKHAUS, W.J., J. Chem. Phys. 57 (1972) 421.
[91] McKINLEY, J.D., J. Chem. Phys. 40 (1964) 120.
[92] SMITH, J.N., FITE, W.L., in 3rd Int. Symp. on Rarefied Gas Dynamics (LAURAMANN, J.A., Ed.), Academic Press, New York (1963) 430.
[93] ANDERSON, J., BOUDART, M., J. Catal. 13 (1964) 216.
[94] MADIX, R.J., BOUDART, M., J. Catal. 7 (1967) 240.
[95] MADIX, R.J., SASU, A.A., Surf. Sci. 20 (1970) 377.

PART IV

11. CATALYSIS BY SINGLE-CRYSTAL SURFACES AT LOW AND HIGH PRESSURES

The relationship between the structure of solid surfaces and their reactivity is one of the most important questions in modern heterogeneous catalysis. A great deal of experimental evidence has been accumulated [1-5] in recent years which indicates that the atomic structure of the surface has a marked influence on the nature and rate of surface reactions.

The further application of LEED as a tool to detect surface structures should allow advances in this area. Progress has, however, been retarded for two reasons. The first concerns the very small areas of single crystals which are used in LEED experiments (less than 1 cm^2). It was unclear whether the rather low rates of many catalytic processes would yield detectable quantities of products. Only recently has it been shown that catalytic reactions of low reaction probability, such as the n-heptane conversion to toluene by the dehydrocyclization reaction on platinum crystal surfaces, can readily be monitored as they take place from a single-crystal surface of 1 cm^2 in area by quadrupole mass spectrometry, and the surface structures that form as intermediate in this reaction have been monitored by LEED [6, 7].

The second problem involved is the simulation of the surface structure of small catalysis used in most industrial processes. These surfaces may have surface structures different from low-index crystal faces of low surface free energy, which were previously used in most LEED studies. This second difficulty has also been overcome by recent LEED studies, which revealed not only that low Miller index crystal faces show sufficient thermal stability to be considered important in catalytic studies, but also that certain high Miller index crystal faces exhibit ordered atomic steps [8, 9]. Since catalytic studies utilize small polydispersed particles likely to have both low-index and stepped surfaces of high Miller index, it would be important to study selectively the catalytic properties of surfaces differing by their atomic structures. This has been carried out in a series of investigations that studied the dehydrocyclization of n-heptane to toluene, as well as the accompanying side reactions of hydrogenolysis and isomerization of n-heptane on various platinum surfaces. The surface reaction has been studied at low pressures in the range of 10^{-5} - 10^{-4} Torr under static or steady-state conditions and the formation of toluene and other reaction products was monitored by a quadrupole mass spectrometer.

The surface structures of the platinum single crystals and those of the adsorbed gas were monitored by LEED. The surface composition was monitored by Auger electron spectroscopy (AES), which could detect the presence of unwanted impurities in concentrations of about 1% of a monolayer. The reader is referred to recent reviews for the detailed description of these techniques and their various applications [10, 11].

It was found that stepped platinum surfaces with terraces of (111) orientation were more reactive for dehydrocyclization than either the stepped platinum surfaces with terraces of (100) orientation or the Pt(111) low Miller index crystal face. These studies can also be carried out at high pressures (1 - 10 atm) using a gas chromatograph as a detector of the reaction products. Such studies are in progress; a description of the apparatus for the high-pressure experiments is given here, as well as the preliminary results of the high-pressure experiments using single-crystal surfaces, during which the ring-opening of cyclopropene has been studied.

The preparation of the various low-index and stepped crystal surfaces was described in the previous section. In this series of studies we have investigated the catalytic activity for various stepped platinum surfaces with (111) orientation terraces of various widths and atomic steps of monatomic height in the absence of adsorbed gases. One technique that revealed the importance of atomic steps in catalytic reaction studies is molecular beam scattering. Using a rotating mass spectrometer, one measures the angular distribution of the products scattered from one face of a single crystal about 1 cm^2 in size. By chopping the incident beam, the flight time and the residence time of the molecules on the surface can be measured. By chopping the scattered beam, the velocity of the scattered products can be determined. Thus using these techniques, energy transfer between the

FIG. 43. Schematic representation of chemical reaction of dehydrocyclization of n-heptane to toluene.

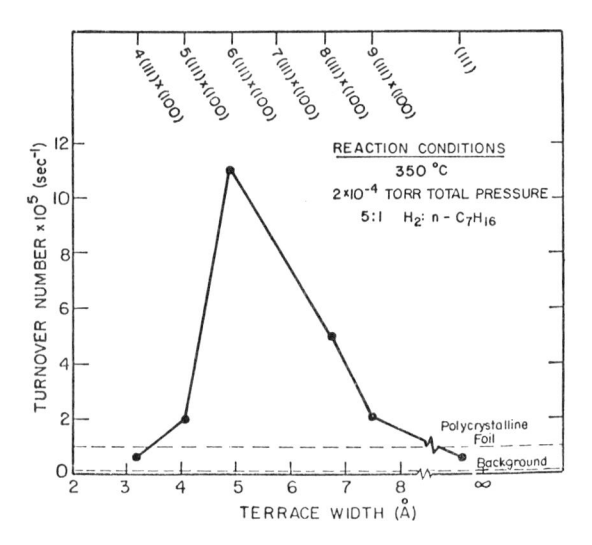

FIG. 44. Dehydrocyclization of n-heptane to toluene on platinum stepped surfaces.

incident gas atoms or molecules and the surface can be measured. Figure 43 shows the results of scattering a hydrogen and deuterium mixed molecular beam from a low Miller index (111) crystal face of platinum, as well as from a stepped platinum surface. No detectable amount of HD has formed on scattering from a Pt(111) crystal face. However, from a stepped surface that exhibits (111) orientation nine atom-wide terraces, about 10-15% HD forms [12].

Studies of this type indicate that atomic steps on surfaces play an all-important role in dissociating large binding-energy diatomic molecules, such as hydrogen, deuterium, oxygen, nitrogen and carbon monoxide. Clearly, the dissociation of diatomic molecules is one of the important reaction steps in many catalytic reactions. Figure 43 shows the schematic representation of the chemical reaction of the dehydrocyclization of n-heptane to toluene. The reaction conditions are also indicated. Even though four moles of hydrogen are produced during the formation of one mole of toluene from n-heptane, the reaction is carried out in industry in the presence of excess hydrogen, from 5 : 1 to 20 : 1. It was found that excess hydrogen was also necessary to obtain optimum yields for toluene from the various stepped surfaces. In Fig. 44 the turnover numbers, i.e. the number of toluene molecules formed per platinum atom per second, are plotted as a function of the surface structure of the various platinum surfaces utilized. Although both isomerization and hydrogenolysis reactions were taking place on all these crystal surfaces, there is a remarkable inactivity for the dehydrocyclization reaction on all but a few stepped platinum surfaces. It appears that the dehydrocyclization reaction takes place almost exclusively on crystal faces that are stepped and have terrace widths of about 5-7 atoms.

During this chemical reaction, the AES studies indicated that the surface is covered with a carbonaceous deposit. However, on those surfaces which showed high catalytic activity for dehydrocyclization, the carbonaceous layers were ordered. The carbon layer exhibited $(\sqrt{3} \times \sqrt{3})$ R - 30° surface structure, as well as several other structures, upon heating in vacuum to high temperatures. All these ordered structures appear to be active in forming the toluene products. Since only those crystal faces showed catalytic activity for the dehydrocyclization of n-heptane which had ordered carbonaceous overlayers, we believe that these overlayers play an all-important role in the complex catalytic reactions of hydrocarbons on platinum surfaces.

The surface structure sensitivity of many reactions has been predicted by Boudart, Balandin, and others, and our studies seem to confirm this. It appears that the more complex the chemical reaction, the more likely it is to be structure sensitive. It is clear that catalytic activity for structure-sensitive reactions can be optimized by creating an optimal atomic-surface structure. For dehydrocyclization of n-heptane, there is a factor of 50 increase in the rate of reaction on the optimal surface structure. Perhaps other crystal orientations not yet studied will be even better than those that have been studied.

The ordered carbon structures appear to play a unique role, at least at low pressures, on platinum surfaces in catalysing complex reactions. In many ways, catalysis involves carrying out two major functions by surfaces. First, the activation energy for a thermodynamically feasible surface reaction has to be lowered so that thermodynamic equilibrium can readily be achieved. Surfaces with atomic steps appear to help in this, probably because of the ease with which large binding-energy chemical bonds (HH, OH, CH, CC bonds) can readily break at atomic steps. Second, out of several competing thermodynamically feasible reactions, the surface must catalyse one while rejecting the others. It appears that the ordered-carbon structure formed catalyses the dehydrocyclization reaction of n-heptane while retarding the other competing reactions, all of which are thermodynamically feasible.

Thus, the ordered-surface structure appears to have enzyme-like activity. It forms a template for the formation of a desired product and the same template seems to poison the formation of other products considered to be less important in this particular reaction, such as products of hydrogenolysis and isomerization. Thus, the heterogeneous catalysis of complex reactions is similar to enzyme reactions, at least on platinum surfaces. It appears that other carbon structures which may have an even more complex structure will catalyse unique and complex catalytic reactions. This view of heterogeneous catalysis should be rather fruitful in searching for new unique catalysts for complex and novel surface reactions.

When working with a model system, such as one face of a single crystal, it is very important to try to extrapolate these results to conditions where

TABLE VI. COMPARISON OF INITIAL SPECIFIC RATE DATA FOR CYCLOPROPANE HYDROGENOLYSIS ON PLATINUM CATALYSTS

Type of catalyst	Calcd spec reaction rate @ $P_{CP}^0 = 135$ Torr and $T = 75°C$ (moles C_3H_8/ min·cm² Pt)	(molecules C_3H_8/min· Pt site)	Comments
Run 10A	2.1×10^{-6}		Rate on Pt(s)-[6(111) × (100)] single crystal based on $E^* = 12.2$ kcal/mole.
Run 12A	1.8×10^{-6}		
Run 15	1.8×10^{-6}		
Run 16	2.1×10^{-6}		
Av	1.95×10^{-6}	812[a]	
0.04 Wt% Pt on η-Al₂O₃	7.7×10^{-7} based on 100% Pt dispersion	410[b]	
0.3% and 2.0% Pt on η-Al₂O₃;	8.9×10^{-7}	480	$\eta_{CP} = 0.2$, $E^* = 8.5$ kcal/mole.
0.3% and 0.6% Pt on γ-Al₂O₃	2.5×10^{-6}	1340	$\eta_{CP} = 0.6$, $E^* = 8.5$ kcal/mole. (Dougharty reports $E^* = 8-9$ kcal/mole and $n = 0.2-0.6$)

[a] Value based upon 87% (111) orientation and 13% polycrystalline orientation.
[b] Based upon av Pt site density of 1.12×10^{15} atoms/cm². This value would be nearly equal to average of above values if dispersion was approximately 50%.

the prototype reactions and those carried out in industry can be compared. Thus it is important to carry out single-crystal surface reactions at high pressures, since much of chemical technology uses pressures of atmospheres, instead of the pressures used in the low-pressure studies of eight orders of magnitude lower. Catalytic reactions at atmospheric pressures have also been carried out in my laboratory, using a gas chromatograph as a detector. The cyclopropane ring opening to form propane has been studied and reaction rates within a factor of two, equal to those observed on highly dispersed catalysts, have been found. Activation energies of 12 kcal/mole were detected, which are also identical to the activation energies for the cyclopropane ring opening determined on polydispersed platinum catalyst surfaces used in the chemical industry. Thus it appears that the same rates and activation energies can be obtained from single-crystal surfaces as those found on polydispersed platinum catalysts.

Table VI compares the initial specific rates for cyclopropane hydrogenolysis on platinum catalysts with the results of the single-crystal studies.

Catalytic reaction may be studied in two ways. The more classical studies involve investigation of the catalytic activities of polydispersed systems identical to those used in industrial applications. For such systems, their surface structure, chemical composition and the all-important relationship between the reactivity and the atomic surface structure cannot be determined. The other type of study reported here involves studies of well defined single-crystal surfaces that are model systems of catalyst particles. These surfaces are well characterized and can readily be investigated by many of the modern techniques on an atomic scale (LEED, AES, molecular beam scattering, etc.).

It is, however, imperative to prove that these studies yield results which can describe the behaviour of industrial catalytic systems. Such studies can be carried out, as shown here, and quantitative agreement between the activities of single-crystal surfaces and polydispersed platinum catalysts can be obtained.

REFERENCES (PART IV)

[1] GWATHMEY, A.T., CUNNINGHAM, R.E., Adv. Catal. 10 (1958) 57.
[2] BOUDART, M., ALDAG, A.W., PTAK, L.D., BENSON, J.E., J. Catal. 11 (1968) 35.
[3] BALANDIN, A.A., Adv. Catal. 19 (1969) 1.
[4] BOUDART, M., Adv. Catal. 20 (1969) 153.
[5] ANDERSON, J.R., MacDONALD, R.J., SHIMOYAMA, Y., J. Catal. 20 (1971) 147.
[6] LANG, B., JOYNER, R.W., SOMORJAI, G.A., J. Catal. 27 (1972) 405.
[7] SOMORJAI, G.A., Catal. Rev. 7 (1972) 87.
[8] LANG, B., JOYNER, R.W., SOMORJAI, G.A., Surf. Sci. 30 (1972) 440.
[9] LANG, B., JOYNER, R.W., SOMORJAI, G.A., Surf. Sci. 30 (1972) 454.
[10] FARRELL, H.H., SOMORJAI, G.A., Adv. Chem. Phys. 20 (1971) 215.
[11] SZALKOWSKI, F.J., SOMORJAI, G.A., Adv. High Temp. Chem. 4 (1971) 137.
[12] BERNASEK, S.L., SOMORJAI, G.A., J. Chem. Phys. 62 (1975) 3149.

IAEA-SMR-15/11

NUCLEATION THEORY
AND CRYSTAL GROWTH

A.J. FORTY
Department of Physics,
University of Warwick,
Coventry,
United Kingdom

Abstract

NUCLEATION THEORY AND CRYSTAL GROWTH.
This paper discusses the basic ideas underlying the equilibrium structure of crystal surfaces in contact with the vapour, solution or liquid phases. The kinetics of crystal growth from the vapour or a dilute solution are examined and it is shown that growth cannot proceed at low supersaturations without the existence of dislocations in the growing crystal. The various possibilities of layer growth on a smooth interface, dislocation-assisted layer growth and continuous growth on a rough interface are examined for a crystal growing from its melt. The kinetics of growth and the quality of the grown crystal vary significantly according to the preferred mechanism of growth. Finally, the attempts to provide a unifying treatment of the structure of the solid/liquid interface are examined in terms of their usefulness for predicting the occurrence of layer or continuous growth.

1. INTRODUCTION

The continuing growth of a crystal occurs at its 'surface'. Consequently, crystal growth, surface structure and atomic processes in the vicinity of a surface are closely inter-related. Just as we can gain a better understanding of the mechanisms of crystal growth from a knowledge of surface structures, we can equally well gain insight into the surface microstructures by studying crystal growth. This will become evident during the discussions of the various forms of crystal growth throughout this paper.

The mode of growth, and hence the dynamic structure of the interface between the solid and parent phases, depends on the conditions under which growth proceeds. It depends on the 'driving force' for growth. There is clearly no single theory of crystal growth but, rather, several theories based on various simplified models which emphasize the dominant features of specific modes or 'regimes' of growth taking place in particular systems. We illustrate this by considering the following practical growth situations.

1.1. Growth of a crystal from its vapour

When a crystal is in equilibrium with its vapour (i.e. the vapour is just saturated) its surface is stable in the sense that there is a statistical balance between the rates of exchange of molecules by adsorption and evaporation. A departure from equilibrium caused by a lowering of the vapour pressure leads to overall evaporation of the crystal, whereas an increase of pressure in the vapour phase above saturation leads to growth. The driving force for growth from the vapour is therefore best described in terms of the supersaturation of the vapour, which we shall define for our purposes as

$$\frac{p - p_0}{p_0} = \frac{p}{p_0} - 1$$

p_0 being the equilibrium (or saturation) vapour pressure and p the ambient vapour pressure. The quantity p/p_0 is referred to as the saturation ratio α.

A theory of growth from the vapour phase must relate the rate of growth to the degree of supersaturation. In general the rate of growth increases steadily with supersaturation, although the detailed dependence changes from parabolic at low values to linear at higher supersaturations. We shall see in Section 3 how attempts to account for this variation led to the idea that crystal dislocations, and the influence of these on surface structure, play a dominant role in this type of growth [1].

1.2. The growth of thin films

This corresponds roughly to growth from the vapour at high degrees of supersaturation. In such circumstances the arrival rate of molecules at the substrate is high. Clusters of molecules form very easily and the growth of these leads to the formation of stable nuclei. The film spreads and thickens by the growth and coalescence of these nuclei. The kinetics of film growth depend on:

The energy of adsorption of a molecule at the substrate surface E_a;
The binding energy of a pair of molecules on the surface E_2;
The activation energy for diffusion of a molecule over the surface E_d;
The temperature of the substrate T; and
The rate of arrival of molecules from the evaporator R.

In this case, therefore, the driving force for growth cannot be so well defined. All we can do is attempt to characterize the various regimes of thin film growth in terms of all these parameters. Clearly this is a complicated process and it is still not clearly understood.

1.3. Growth of a crystal from solution

In the case of growth from the vapour it is supposed that after adsorption at a surface site the molecules diffuse over the surface and either re-enter the vapour phase (evaporate) or attach themselves to a surface step where they may remain strongly bound. In the case of growth from solution, however, the direct transport of molecules by diffusion through the solution will assume greater importance. This is because the presence of the solvent usually means that re-dissolution of molecules occurs more readily than re-evaporation; consequently surface diffusion lengths are more limited in the presence of the solution. In these circumstances, growth is controlled by diffusion of molecules through the boundary layer of solution in the vicinity of surface steps rather than adsorption and diffusion over the surface. Clearly growth from solution is a complex process and the growth rate depends on the hydrodynamic properties of the solution as well as its supersaturation.

1.4. The electrochemical deposition of thin films

This is a very important technological process. It will not be discussed further in this paper but some aspects of the problem are discussed in the paper by Parsons [2]. The rate of deposition depends on the temperature of the substrate, the concentration of the electrolyte and its viscosity, and also the surface potential.

1.5. Growth of crystals from their melts

Although technologically the most important method for growing crystals, growth from the melt is probably the least understood. Whereas in the modes of growth discussed in the preceding sections the crystal surface may be regarded as being atomically smooth, apart from monomolecular steps and isolated surface vacancies and adsorbed molecules (often described as a singular interface since it corresponds to a singularity in the γ-plot of surface energies), the interface between a crystal and its melt can be atomically rough. For such an interface there is no single plane at which the phases change from solid to liquid. The transition occurs over many atomic layers. Growth proceeds by a continuing process of molecular re-arrangement in the transition layer.

The macroscopic shape of the interface, as also the growth rate and the factors which control it, will depend on the type of interface; growth on a singular interface leads to a faceted crystal phase, whilst the rough interface produces a rounded shape, following the isotherms in the melt. In either case, the growth rate depends on the undercooling, ΔT, of the melt in the immediate vicinity of the interface below the solid/liquid equilibrium temperature.

2. THE EQUILIBRIUM STATE OF A CRYSTAL SURFACE IN CONTACT WITH ITS VAPOUR OR SOLUTION

As we have seen in the previous section, in many modes of growth the crystal surface may be regarded as atomically flat. To develop the theory of growth of such a surface we need to examine the state of such a surface and to consider in some detail the processes of adsorption, diffusion and re-evaporation of molecules that might be expected to take place. This is best done by considering first of all the structure of the surface when it is in equilibrium with its surroundings. In this section we consider only the simplest situation, a crystal surface in equilibrium with its vapour or a dilute solution. The ideas presented here were first discussed by Gibbs [3] and later developed by Volmer [4], Kossel [5], Stranski [6], Becker and Döring [7], Frenkel [8], and Burton et al. [9]. They form the basis of much of our present understanding of the growth of crystals.

The usual tacit assumption is that the faces of low crystallographic index of a perfect crystal in equilibrium with its vapour (or dilute solution) are flat. (Indeed, it can be argued that if any rough, high-index surfaces do occur they will grow so rapidly that they will eventually 'grow out', leaving only the flat surfaces in equilibrium.) Any quantitative discussion of the structure of such a surface can be greatly simplified by considering a Kossel crystal, one having a simple cubic structure and in which only nearest-neighbour interactions between the molecules are important.

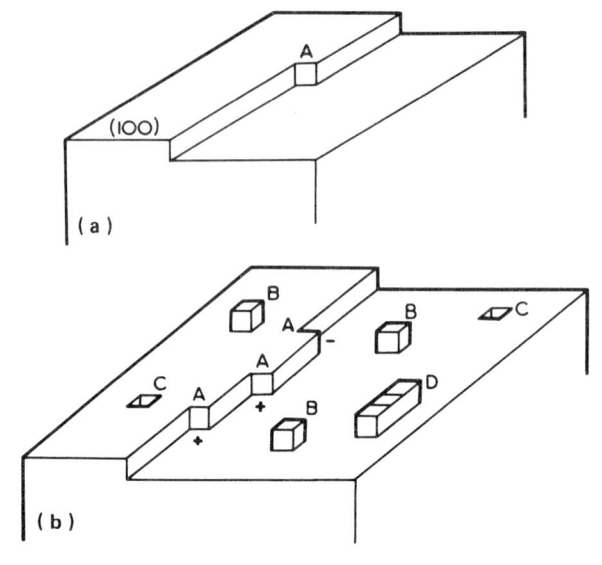

FIG.1. Surface of a simple crystal in equilibrium with its vapour (a) at low temperature and (b) at high temperature.

Such a surface may be represented schematically as in Fig.1. The surface will generally be divided into terraces by molecular steps and the steps will themselves contain molecular kinks (A) (or jogs). The density of steps will depend on the crystallographic orientations of the surface, on its previous history and on the nature of the environment. The density of kinks in a step depends on the orientation of the step relative to low index directions in the surface and also on the temperature of the crystal. At higher temperatures, the density of kinks increases and at still higher temperatures molecules will leave surface sites, creating surface vacancies (C), and adsorb onto the surface elsewhere, forming adatoms (B). This roughening of the surface increases as the temperature approaches the melting point of the crystal, resulting in an equilibrium distribution of steps, kinks, vacancies, adatoms and adatom clusters.

A valuable insight into the surface structure, and particularly the relative densities of the various forms of surface defect, can be obtained by considering a simple Kossel crystal and applying Boltzmann statistics as illustrated in the following:

2.1. Equilibrium density of kink sites in a surface step

We follow here the discussion given by Frenkel [8] and developed by Burton and Cabrera [10]. In thermal equilibrium the fraction of step sites which are positively kinked (N_+) and the fraction which are negatively kinked (N_-) are given by

$$N_+ \, N_- = \exp\left(-\frac{2E_k}{kT}\right) \tag{1}$$

where E_k is the energy of formation of a kink. This is the net energy required to take a molecule out of the step and place it on the step at some other site divided by the number of kinks so formed. In the Kossel crystal such a process requires a net amount of energy 2ϕ, where ϕ is the energy of interaction of a pair of nearest-neigbour molecules, and creates four new kink sites. Consequently E_k is approximately $\phi/2$. The density of kinks is also related to the orientation of the step relative to the nearest principal direction in the surface. In terms of the magnitude of this mis-orientation, θ,

$$N_+ - N_- = \tan \theta \qquad (2)$$

Solving Eqs (1) and (2) we find that the mean kink density is

$$N_+ - N_- = \left[4 \exp\left(-\frac{2E_k}{kT}\right) + \tan^2 \theta \right]^{\frac{1}{2}}$$

That this is usually quite a high density may be seen by considering the lowest temperature at which a crystal might be grown from its vapour to be one at which the vapour pressure is 10^{-10} atm (10^{-5} Nm^{-2}). Then E_{ev}/kT is 23 and, since the evaporation energy E_{ev} is 3ϕ, it follows that E_k/kT is about 4. Then, even in the closest packed direction ($\theta = 0$) a step will contain kinks only a few molecules apart. It is interesting to note that for finite values of θ the kink density can be very large and, since growth of the crystal proceeds by the addition of molecules to the step at kink sites, steps which are not parallel to crystallographic directions will grow very fast, eventually leaving only the slower-growing steps parallel to the principal directions. These are the steps normally observed on crystal surfaces.

Whilst the density of kink sites and, as we shall soon see, the density of adatoms and vacancies increases as the temperature of the crystal is raised, the number of surface steps does not increase. The entropy change by such an increase would not compensate for the very large internal energy required to create a step, and indeed there should be no steps on a low-index surface of a crystal in thermal equilibrium. Thus, once any initially present steps have grown out of a surface, the growth of the crystal is arrested; further growth requires new steps and these can be created only by a nucleation event from a sufficient number of adsorbed molecules (see Section 3).

2.2. Surface diffusion

Adsorbed molecules can diffuse very readily over the surface of a crystal. If a molecule arrives at a kink site before evaporating, it will be strongly bound there and will therefore contribute to the growth of the crystal. The rate of crystal growth thus depends on the surface migration distance. An estimate of this can be reached as follows.

If E_a is the energy of adsorption of a molecule at an individual site on the surface, the probability per second that such a molecule will evaporate is

$$\nu \exp\left(-\frac{E_a}{kT}\right)$$

ν being the natural frequency of vibration of a molecule at an adsorbed site. The probability per second that the same molecule will jump to a neighbouring adsorbed site is

$$\nu \exp\left(-\frac{E_d}{kT}\right)$$

where E_d is the activation energy for diffusion. It follows that the number of molecular jumps across the surface before the molecule evaporates will be

$$\exp(E_a - E_d/kT)$$

The migration of the molecule will be a random walk and hence the distance travelled before evaporation is

$$x_s = a \exp(E_a - E_d/2kT)$$

where a is the size of a molecular jump.

With suitable values for E_a and E_d, we find that this is several hundred jumps even at a moderate temperature. This is an important result because it follows that the rate of direct arrival of molecules from the vapour at a particular point on the crystal surface is generally small compared with the rate of indirect arrival by adsorption elsewhere followed by surface diffusion. Steps will advance over the surface by collecting molecules from a diffusion zone of width x_s.

2.3. The equilibrium densities of adsorbed molecules and surface vacancies

As the temperature is raised, a number of adsorbed molecules (B) (see Fig.1) appear on the surface and in thermal equilibrium with it. The activation energy for creating an adatom E_s, i.e. the energy to remove a molecule from a kink site in a step (A) to an adsorbed site (B), is 2ϕ for a Kossel crystal. Hence the fraction of the total number of surface sites occupied by such molecules is

$$\frac{N_s}{N_0} = \exp\left(-\frac{E_s}{kT}\right)$$

Since E_s is smaller than E_{ev}, the energy required to take a molecule from a kink site into the vapour 3ϕ, molecules will move onto adsorbed sites rather than evaporate completely.

Similarly the number of surface vacancies (C) is given by

$$\frac{N_v}{N_0} = \exp\left(-\frac{E_v}{kT}\right)$$

where E_v is the energy required to remove a molecule from within the surface layer and place it on a kink site. This again is equal to 2ϕ for a Kossel crystal and so we can expect the numbers of adatoms and surface vacancies to be about equal.

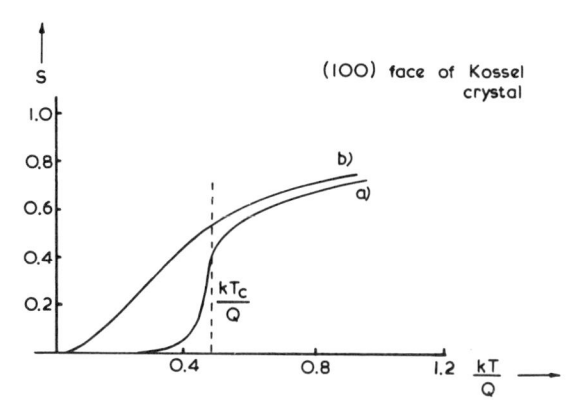

FIG.2. Variation of surface roughness S with temperature calculated by Burton and Cabrera for a Kossel crystal. S is defined as the number of unused molecular bonds parallel to the surface expressed as a fraction of the total number of molecules in the flat surface. Curve (a) shows the result when the surface roughening is treated as a co-operative phenomenon. Curve (b) is the result obtained if it is assumed that surface jumps occur independently.

2.4. Surface melting

For most purposes we can assume that the surface of a crystal between steps is perfectly flat, apart from the presence of isolated adatoms or surface vacancies. However, at a sufficiently high temperature it is thought possible that the surface in certain crystallographic orientations will develop a roughness on the atomic scale, as if the surface layer has melted. The surface is no longer flat but will contain discrete 'jumps' in surface level. This is an interesting idea because jumps in level provide a constant supply of sites for evaporation or condensation, and under these conditions the crystal will shrink or grow very rapidly. It should also mean that surface diffusion and related processes will be greatly enhanced above the surface melting temperature.

The problem is to estimate the number of surface jumps in thermal equilibrium as a function of temperature. This has been done by Burton and Cabrera [10]. They point out that jumps cannot occur independently. It is possible to have twice as many jumps as there are surface sites (i.e. a jump can be up or down) and therefore it is not possible to specify the existence of a jump at a given point on the surface without a knowledge of the state of the surface at every other point. In other words we are dealing with a co-operative phenomenon.

Whilst the surface can probably roughen over several molecular layers, Burton and Cabrera treated only a simplified two-level model of the rough surface. They were then able to employ the method developed by Onsager and others for the treatment of ferromagnetism, using the two-dimensional Ising model. Just as there is a critical temperature (Curie temperature) associated with an infinite specific heat in the case of the two-dimensional ferromagnet, so there is in the case of this two-dimensional crystal surface. Figure 2 shows the variation of surface roughness S with temperature calculated by Burton and Cabrera for a (100) face of a Kossel crystal. It is clear from this that there should be a critical temperature, T_c, below which the surface is smooth and above which it is rough. This is the so-called surface melting temperature. In the particular case of the (100) surface the computed melting temperature actually exceeds the bulk melting point. However, it is appreciably lower for higher index faces and these should therefore roughen and grow more rapidly at temperatures well below the bulk melting point.

The Burton and Cabrera treatment is greatly oversimplified but these conclusions are not significantly changed by a more sophisticated approach. The interesting possibility therefore exists that surfaces will 'melt' and therefore display extraordinary properties. There is no definite experimental information in support of this idea, but the work of Rhead and Henrion [11] on enhanced diffusion rates for crystal surfaces at high temperatures might provide indirect evidence.

3. GROWTH OF PERFECT CRYSTALS FROM THE VAPOUR

It can easily be shown that a low-index face of a perfect crystal will eventually become flat, without steps.

Consider the case where a step actually exists on the surface. Molecules will join the step by diffusing over the surface into the kink sites. Molecules will leave the step by detaching themselves from kink sites, diffusing over the surface as adsorbed molecules and then evaporating into the vapour phase. If the vapour is supersaturated, more molecules join the step than leave it and hence the step will advance across the surface. The actual rate of advance of the step depends on the rate of arrival of adsorbed molecules and this depends on the rate of surface diffusion. The latter can be obtained by solving the diffusion equation where the concentration of adsorbed molecules close to the step is near the equilibrium value ($\alpha = 1$), because the density of kinks in the step is always high and the arriving molecules can therefore always be accommodated, and is a maximum in between the steps. The rate of advance of the step is finite for any degree of supersaturation and hence the step will ultimately move to the edge of the crystal.

Thereafter, growth can only proceed by the creation of a new step. This requires a two-dimensional nucleation event, as was recognised as long ago as 1878 by Willard Gibbs.

3.1. Surface nucleation

In this section we consider the way a two-dimensional nucleus forms on a surface and grows by the aggregation of adsorbed molecules. Let us suppose for simplicity that we have a circular nucleus, radius ρ. This will grow by the addition of molecules to its periphery. It can also shrink by the evaporation of molecules. There will clearly be a critical size depending on whether the reduction in free energy associated with the condensation of adsorbed molecules is greater than the increase in free energy due to the creation of extra length of step at the periphery. The total excess free energy for a nucleus of radius ρ is

$$\Delta G = \frac{2\pi\rho}{a}\gamma - \frac{\pi\rho^2}{a^2}E_s$$

where γ is the edge energy per molecule ($\phi/2$ for a Kossel crystal). E_s is the energy of removal of a molecule from a step to an adsorbed site (2ϕ) and a is the molecular spacing in the crystal.

The critical size of nucleus is that for which ΔG begins to decrease with increasing size, giving

$$\rho_c = \frac{\gamma a}{E_s} \tag{3}$$

Now, since p is proportional to $\exp(-E_a/kT)$ and p_0 is proportional to $\exp(-E_{ev}/kT)$, and $E_{ev} - E_a = E_s$, it follows that the saturation ratio α is given by

$$\alpha = \frac{p}{p_0} = \exp\left(\frac{E_s}{kT}\right)$$

so that

$$E_s = kT \log \alpha$$

We have also noted that γ is $\phi/2$ for a Kossel crystal. Substituting in (3) we have therefore

$$\rho_c = \frac{a\phi}{2kT} \log \alpha \tag{4}$$

Thus the critical size of the nucleus depends on the strength of molecular binding and also the saturation ratio in the vapour.

A more exact calculation of ρ_c by Burton and Cabrera [10] takes into account the fact that the equilibrium shape of the nucleus need not be circular, but arrives at a similar value for the minimum radius of curvature for a stable nucleus.

3.2. Rate of formation of critical nuclei

The excess free energy of a nucleus of critical size is

$$E_n = \frac{\pi \phi^2}{4kT} \log \alpha$$

It follows that the probability of occurrence of a stable nucleus is proportional to $\exp(-E_n/kT)$ and the rate of formation of such nuclei will be

$$R \cdot S/a^2 \cdot \exp\left(-\frac{E_n}{kT}\right)$$

where R is the rate of arrival of molecules from the vapour phase at a particular surface molecular site and S is the area of the crystal surface. (S/a^2 is the number of surface sites.)

3.3. Rate of growth of a perfect crystal

Because surface steps are always densely kinked they will advance quickly over the surface. The rate of growth will therefore be limited only

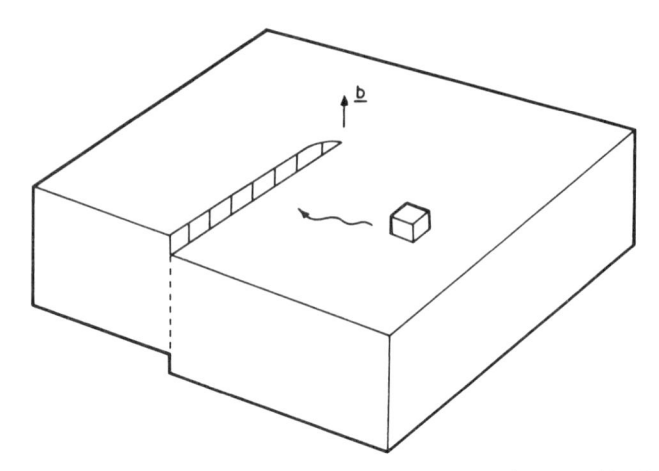

FIG.3. A crystal containing a screw dislocation. The dislocation raises a surface step of height equal to its Burgers vector.

by the rate of nucleation. It follows from the previous section that the rate of thickening of the crystal will be

$$J = R \cdot S/a^2 \cdot \exp\left(-\frac{\pi\phi^2}{4(kT)^2 \log \alpha}\right)$$

layers per second.

Now even in a dense vapour, $R < 10^{13}$ s^{-1}, and therefore for a crystal 1 mm in diameter, RS/a^2 must be $< 10^{27}$. Under normal growth conditions it is about 10^{22}. Thus, for a detectable growth rate (1 μm per month, say) $\log \alpha$ must be at least $(\phi kT)^2/90$ which, with a typical value of ϕ/kT, signifies a supersaturation ($\alpha-1$) of not less than 25%.

When we examine the expression for the nucleation rate more carefully we find that this is very sensitive to supersaturation. For a small change of supersaturation, at around 25% for a typical case, the rate of growth changes from being negligible to a value at which growth proceeds without hindrance by nucleation. This is quite contrary to the behaviour found in practice. For example, in their classical experiments on the rate of growth of iodine crystals from iodine vapour, Volmer and Schultze [12] observed growth at all measurable values of supersaturation, well below the critical value predicted by surface nucleation theory.

4. GROWTH OF IMPERFECT CRYSTALS

The discrepancy between the growth rates observed in practice and that predicted by nucleation theory was resolved by Frank [13]. He pointed out that real crystals are imperfect and, in particular, contain dislocations. If a dislocation emerges at a crystal surface and has a component of its Burgers vector perpendicular to the surface, then it must by definition raise a step in the surface. This is illustrated in Fig.3 for the special case of a screw dislocation. The height of the step is equal to the component of Burgers vector normal to the surface; in the example shown here it is equal to the full Burgers vector of the screw dislocation. The step will be in thermal equilibrium with the surface and the vapour and will therefore be kinked. The step will advance by the adsorption of molecules but, since it is terminated at one end by the dislocation it will be self-perpetuating. Consequently, growth of the crystal can proceed without the necessity for nucleation of new steps.

4.1. Spiral growth steps

In a dislocated crystal a step will extend from each dislocation to the edge of the crystal or to another dislocation of equal but opposite Burgers vector. If the vapour phase is supersaturated, the steps will advance because on average they receive more molecules than they lose by evaporation. All the molecules striking the surface from the vapour within a surface diffusion zone $\pm x_s$ about a step will eventually reach the step and condense at kink sites on it. Under a saturation ratio α, this will be $2\alpha Z x_s/a^2$ molecules per second per unit length of step, where Z is the rate of arrival of molecules at a lattice site from the just-saturated vapour (i.e. the number arriving per second or evaporating per second under equilibrium conditions). The rate at which molecules leave the step, diffuse over the zone $\pm x_s$ and eventually evaporate will be $2Z x_s/a^2$. Hence the net rate of gain of molecules by the step is

$$\frac{2(\alpha - 1)Z x_s}{a^2}$$

Since the molecular cross-section is a^2, the rate of advance of the step over the surface is

$$V_0 = 2(\alpha - 1) Z x_s$$

In estimating this velocity we should also take into account the probability β that a molecule will stick in the step having reached it. β is about unity in simple cases but can be considerably less in special cases where the kinks are not close together compared with x_s or the molecules have a complex shape.

The step attached to a dislocation will advance with this velocity, at a given supersaturation $(\alpha - 1)$, everywhere along its length except close to the dislocation. Here it must remain attached to the dislocation and consequently it will develop a curvature ρ. The radius of curvature cannot be less than the radius of the critical nucleus ρ_c, otherwise faster evaporation will occur to decrease the curvature. Burton et al. [9] have shown that the velocity of the curved step will be

$$V = V_0 \left(1 - \frac{\rho_c}{\rho}\right)$$

where V_0 is the velocity of the corresponding straight step $2(\alpha-1)Z x_s$.

Under conditions of steady growth the step winds up into a helicoidal growth hill centred on the dislocation. The spiral step will rotate as growth proceeds and have a stationary form given in polar co-ordinates by

$$\frac{r}{\rho_c} + \log\left(1 + \frac{r}{\sqrt{3}\,\rho_c}\right) = 2\left(1 + \frac{1}{\sqrt{3}}\right)\left[\theta - \frac{V_0 t}{2\rho_c\left(1 + \frac{1}{\sqrt{3}}\right)}\right] > 0$$

a good approximation being

$$\frac{r}{\rho_c} = 2\left(\theta - \frac{V_0 t}{2\rho_c}\right) > 0$$

4.2. Rate of growth of an imperfect crystal

If we assume that the growth of a particular crystal face is dominated by a spiral growth hill, we may estimate the rate of growth of the crystal from the number of turns of the spiral passing a fixed point per second and the height of the step. Thus rate of growth is

$$J = d\left[\frac{V_0}{4\pi\rho_c\left(1 + \frac{1}{\sqrt{3}}\right)}\right]$$

Now we have seen that V_0 is proportional to the supersaturation $(\alpha - 1)$. The critical radius of curvature ρ_c is proportional to $1/\log \alpha$, which is approximately $1/(\alpha - 1)$ for small values of α. We see therefore that for small supersaturations the growth rate is proportional to $(\alpha - 1)^2$.

At higher supersaturations, the turns of the spiral become so close together (spacing $< x_s$) that they begin to compete with each other for molecules from the vapour. At this stage a further increase in step density no longer contributes to the increase in growth rate and hence J is simply proportional to $(\alpha - 1)$.

Thus the theory of spiral growth predicts a rate of growth proportional to the square of the supersaturation at low degrees of supersaturation and a linear dependence at higher supersaturations. This is generally consistent with observations on growth of crystals from the vapour phase (cf. those of Volmer and Schultze) and is also found in studies of growth from dilute solutions. The theory has also been supported very extensively by the direct observation of spiral growth hills on the surfaces of a great many crystals [14]. There has not, however, been any real attempt to verify the theory in absolute quantitative terms.

5. GROWTH FROM THE MELT

Growth of a solid from its melt occurs if the melt is undercooled. Undercooling is measured as the difference in temperature between the actual solid/liquid interface and a virtual interface with approximately the same curvature but which is static and corresponds to the real interface in thermal equilibrium. This latter temperature corresponds to the true melting point. Thus the driving force for growth (the undercooling) is

$$\Delta T = T_m - T$$

Depending on the physical nature of the interface we can recognise three distinct modes of growth:

(a) Layer growth on an atomically flat and perfect interface (a so-called singular interface);
(b) Layer growth on an interface which is stepped by the presence of dislocations in the solidified crystal; and
(c) Continuous growth on a non-smooth (rough) interface.

We have already met (a) and (b) in the preceding sections in connection with the growth of a crystal from its vapour. There are, however, differences in the kinetics of growth because in growth from the melt we are dealing with two phases which are not greatly different thermodynamically.

In the following paragraphs these various modes of growth are considered more closely. In the final section attempts to find a unifying treatment are discussed.

5.1. Layer growth

On a perfect flat interface, growth can only proceed by the successive nucleation of island monolayers. In equilibrium there will be an exchange of molecules between the solid and the melt. Under a finite undercooling there will be an excess of molecules arriving from the melt. The first such molecules will have a small finite probability of sticking anywhere on the interface because all sites are equivalent and all involve the molecules being weakly bound because most of their nearest neighbours are still in the liquid phase. After a certain amount of condensation, however, additional molecules from the melt can either choose isolated sites similar to those occupied by their predecessors or they can take up adjacent positions. In this latter case the pairs of molecules so formed will be more strongly bound. As more molecules reach the cluster, so it will become more stable. Thus growth will proceed by the lateral spreading of these islands.

As in Section 3, we can estimate the critical size for a stable cluster by considering the excess free energy of a two-dimensional island containing n molecules:

$$\Delta G = -\alpha_1 n \Delta G_v + \alpha_2 \sigma_a \sqrt{n}$$

where α_1 and α_2 are factors (~ 1) depending on the shape of the molecules.

ΔG_v is the volume free energy change per molecule associated with the phase change (and may be taken to be $L\Delta T/T_m$ where L is the latent heat of melting) and σ_a is the interfacial free energy per molecule. The condition that further addition of molecules should produce a decrease in free energy leads to the result that the critical size for a stable cluster is n* molecules where

$$n^* = \left(\frac{\alpha_2}{\alpha_1} \frac{T_m \sigma_a}{2L\Delta T}\right)^2$$

For a low melting temperature T_m and large latent heat L this can be quite small, involving only a few molecules; a result which is markedly different from that for the critical nucleus for vapour growth, where several hundred molecules may be required to form a stable nucleus.

The rate of formation of stable nuclei I will be proportional to $\exp(-\Delta G^*/kT)$, where ΔG^*, the nucleation energy barrier, is derived from n^* and is

$$\Delta G^* = \frac{\alpha_2^2}{4\alpha_1} \frac{\sigma_a^2 T_m}{L\Delta T}$$

Since each nucleus accounts for the growth of the crystal by one molecular layer, the rate of growth is therefore $J = aI$ and this is proportional to

$$\exp\left(\frac{-\alpha_2^2}{4\alpha_1} \frac{\sigma_a^2 T_m}{LkT\Delta T}\right)$$

This means that at low undercoolings the growth rate will be negligibly small but will increase rapidly as ΔT increases. At large undercoolings the nucleation rate will be so high that the rate of arrival of molecules will be the limiting process. This is essentially the situation of continuous growth.

5.2. Screw dislocations and layer growth

There have been many attempts to observe the microstructure of solid/melt interfaces in order to establish the mechanism of crystal growth. All such attempts give ambiguous results because the interface is perturbed by observational methods employed; for example, the much used method of decanting the melt inevitably leads to artefacts arising from the rapid freezing of the residual liquid.

However, in the case of some materials (e.g. lead iodide [15] and ice [16]) there is strong evidence that the interface contains spiral growth hills, suggesting that screw dislocations in the grown crystal play an important part in the growth process.

It is possible therefore for layer growth to proceed by molecules from the melt becoming attached to spiral steps centred on dislocations. The spacing between steps on such a helicoidal surface is given approximately by $a\sqrt{n^*}$ (the radius of the critical nucleus). Then, if steps advance with a speed V, the growth rate will be

$$J = \frac{V}{\sqrt{n^*}} = \frac{V \, 2L\alpha_1 \Delta T}{\alpha_2 \sigma_a T_m}$$

Since V is proportional to the supercooling ΔT, it follows that the growth rate for a helicoidal interface is proportional to $(\Delta T)^2$.

5.3. Continuous growth

In this mode of growth it is considered that the interface is permanently rough so that all molecules arriving at it from the melt will stick and remain there for some time before melting again. A continuous exchange of molecules takes place at the interface so that the net rate of growth is

$$J = J_F - J_M$$

where J_F, the rate of flow of molecules into the interface, is of the form $J_F^0 \exp(-Q_F/kT)$ and J_M, the rate of flow of molecules back into the melt, is $J_M^0 \exp(-Q_M/kT)$. Since $Q_M - Q_F$ is the latent heat of melting (L) per molecule and because the net flow of molecules is zero when the interface is at the melting temperature (T_m), we have

$$\frac{J_F^0}{J_M^0} = \exp\left(\frac{-L}{kT_m}\right)$$

Thus,

$$J = J_F^0 \exp\left(-\frac{Q_F}{kT}\right)\left[1 - \exp\left(\frac{L}{kT_m} - \frac{L}{kT}\right)\right]$$

For a reasonably small undercooling ($\Delta T = T_m - T$)

$$\frac{L\Delta T}{kT_m T} \ll 1$$

Therefore,

$$J = J_F^0 \exp\left(-\frac{Q_F}{kT}\right) \frac{L\Delta T}{kT_m T}$$

and we see that the rate of growth is proportional to ΔT at small undercoolings.

It is interesting to note that since the escape of molecules from the interface is controlled by diffusion through the melt, the term $\exp(-Q_F/kT)$ is related to the coefficient of self-diffusion in the melt, which may be taken to be inversely proportional to the coefficient of viscosity η. Thus we may write

$$J \alpha \frac{1}{\eta}\left[1 - \exp\left(\frac{L\Delta T}{kT_m T}\right)\right]$$

for all undercoolings.

6. A UNIFYING THEORY OF GROWTH FROM THE MELT

It is evident from the foregoing that the kinetics of crystal growth depend strongly on the nature of the solid/liquid interface. There has been considerable interest in the problem of obtaining a proper description of the interface, and in particular in determining whether the interface should be rough or smooth. The most successful attempt is that by Jackson et al. [17]. A rather more fundamental approach to the problem has been made by Cahn [18] but this does not readily provide a criterion to distinguish between rough and smooth interfaces for particular substances.

TABLE I. JACKSON'S FACTOR χ CALCULATED FOR VARIOUS SUBSTANCES

Material	χ	Material	χ
Potassium	0.825	Lead	0.94
Copper	1.14	Silver	1.14
Mercury	1.16	Cadmium	1.22
Zinc	1.26	Aluminium	1.36
Tin	1.64	Gallium*	2.18
Bismuth*	2.36	Antimony	2.57
Germanium*	3.15	Silicon*	3.56
Water (Ice)*	2.62	CBr_4	1.27
Cyclohexanol	0.69	Succinonitride	1.40
Benzil*	6.3	Salol*	7.0

* Materials for which a faceted solid/liquid interface has been observed experimentally

6.1. The Jackson criterion

In this approach we start with a molecularly smooth interface and then consider the excess free energy associated with adding a certain number N_A of molecules to it disposed at random amongst the N possible sites for attachment. The extra energy involved is

$$\Delta G = -\Delta E_0 - \Delta E_1 + T\Delta S_0 + T\Delta S_1 - P\Delta V$$

where

ΔE_0 = change in internal energy associated with molecules being attached to the surface;

ΔE_1 = change in internal energy associated with the molecules on the interface interacting with other adsorbed molecules;

ΔS_0 = change in entropy associated with the adsorbed molecules passing to the solid phase from the liquid;

ΔS_1 = configurational entropy associated with the different possible sitings of the adsorbed molecules;

ΔV = change in volume due to molecules becoming associated with the new phase (negligible for the solid/liquid transition, but not for growth from the vapour).

Jackson examined all these contributions and, after using approximations appropriate to the solid/liquid interface, arrived at the relationship:

$$\frac{\Delta G}{NkT_m} = \chi N_A \left(\frac{N - N_A}{N^2}\right) - \log\left(\frac{N}{N - N_A}\right) - \frac{N_A}{N} \log\left(\frac{N - N_A}{N_A}\right)$$

where

$$\chi = \frac{L}{kT_m} \frac{\pi}{\nu}$$

π being the maximum possible number of nearest neighbours adsorbed molecules can have, and ν the bond energy per molecule.

If this quantity is evaluated for all possible values of N_A/N for various values of the parameter χ, it is found that the interface will be stable (ΔG a minimum) with half of the sites for adsorption occupied ($N_A/N = \frac{1}{2}$) if $\chi \leq 2$. This corresponds to a rough interface. If $\chi > 2$ the interface is stable if only a small fraction or nearly all the sites are occupied ($N_A/N \sim 0$ or 1). This is a smooth interface. This clearly offers a valuable criterion for determining the type of interface in terms of the material properties. Jackson's treatment is highly simplified, considering only two possible levels for molecules, in or above the interface. A more realistic model involving many possible levels (i.e. a more diffuse interface) has been used by Temkin [19]. However, this leads to results very similar to those of Jackson. This theory has had considerable success in predicting those materials which will develop a faceted interface, as can be seen from Table I. Quite apart from furthering our understanding of the nature of interfaces, this successful use of the χ factor has important technological application since the quality of a crystal grown from its melt is often dependent on securing a faceted interface so that layer growth can proceed.

REFERENCES

[1] FRANK, F.C., Adv. Phys. 1 (1952) 91.
[2] PARSONS, R., these Proceedings.
[3] GIBBS, W., Collected Works 1, Longmans Green, New York (1928) 325.
[4] VOLMER, M., Z. Phys. Chem. 119 (1926) 277.
[5] KOSSEL, W., Nachr. Ges. Wiss. Göttingen (1927) 135.
[6] STRANSKI, I.N., Z. Phys. Chem. 136 (1928) 259.
[7] BECKER, R., DÖRING, W., Ann. Phys. 24 (1935) 719.
[8] FRENKEL, J., Fiz. Zh. Moskva 9 (1945) 392.
[9] BURTON, W.K., CABRERA, N., FRANK, F.C., Philos. Trans. R. Soc. London Series A 248 (1951) 299.
[10] BURTON, W.K., CABRERA, N., Discuss. Faraday Soc. 5 (1949) 33.
[11] RHEAD, G.E., HENRION, J., in Diffusion Processes 1 (SHERWOOD, Ed.), Gordon and Breach, New York (1971).
[12] VOLMER, M., SCHULTZE, W., Z. Phys. Chem. A.156 (1931) 1.
[13] FRANK, F.C., Discuss. Faraday Soc. 5 (1949) 48.
[14] FORTY, A.J., Adv. Physics 3 (1954) 1.
[15] SEARS, G., J. Chem. Phys. Sol. 2 (1955) 37.
[16] KETCHAM, W.M., HOBBS, P.V., Philos. Mag. 18 (1968) 659.
[17] JACKSON, K.A., UHLMANN, D.R., HUNT, J.D., J. Cryst. Growth 1 (1967) 1.
[18] CAHN, J.W., Acta. Metall. 8 (1960) 554.
[19] TEMKIN, D.E., in Crystallisation Processes (SIROTA, GORSKII, VARIKASH, Eds), English transl. Consultants Bureau, Plenum, New York (1966).

IAEA-SMR-15/10

BRILLOUIN-WIGNER THEORY AND SOME SURFACE PROBLEMS*

R.O. JONES
Institut für Festkörperforschung der
  Kernforschungsanlage Jülich,
Jülich,
Federal Republic of Germany

Abstract

BRILLOUIN-WIGNER THEORY AND SOME SURFACE PROBLEMS.
Brillouin-Wigner perturbation theory, its relationship to the standard (Rayleigh-Schrödinger) theory, and some applications to surface problems are discussed.

1. INTRODUCTION

Brillouin-Wigner perturbation theory has proved to be of value in a number of surface problems. In view of this, and of its connection to self-consistent solutions in other branches of physics, it is useful to re-examine the theory and some of its applications. In spite of its customary title, Brillouin-Wigner (BW) theory is not strictly a perturbation theory, but yields an exactly equivalent expression for any finite secular determinant. Its structure is, however, very similar to the usual Rayleigh-Schrödinger (RS) perturbation theory. We begin by discussing the two approaches and their interrelationship.

2. PERTURBATION THEORY

The energy eigenvalues E and eigenfunctions ψ of a physical system with Hamiltonian $\hat{H}$ satisfy the Schrödinger equation:

$$\hat{H}\psi = E\psi \tag{1}$$

Expanding the wave function ψ in terms of any complete set of functions,

$$\psi = \sum_m c_m \phi_m \tag{2}$$

and substituting into (1) gives

$$\sum_m c_m \hat{H}\phi_m = \sum_m c_m E \phi_m \tag{3}$$

---

* Part of lecture series entitled "Electron States at Non-Metal Surfaces". For discussion of other topics see Ref.[8].

Multiplying both sides by ϕ_k^* and integrating over all space gives the secular equations:

$$(E - H_{mm})c_m = \sum_{k \neq m} H_{km} c_k \qquad (4)$$

where

$$H_{km} = \int d\tau\, \phi_k^* \hat{H} \phi_m$$

This discussion is perfectly general. Perturbation methods usually take for ϕ_m the eigenfunctions of the Hamiltonian $\hat{H}_0$ where

$$\hat{H} = \hat{H}_0 + \lambda \hat{V} \qquad (5)$$

The separation (5) is made so that $\hat{H}_0$ has a relatively simple structure and $\hat{V}$ is small, i.e. a 'perturbation'. The secular equations then become

$$(E - E_m^{(0)})c_m = \sum_{k \neq m} \lambda V_{mk} c_k \qquad (6)$$

where $E_m^{(0)}$ are the eigenvalues of the unperturbed system. The Rayleigh-Schrödinger technique for solving (6) is to assume that the eigenvalues E_m and the corresponding coefficients c_m can be expressed as a power series in an expansion parameter λ:

$$\begin{aligned} E_m &= E_m^{(0)} + \lambda E_m^{(1)} + \lambda^2 E_m^{(2)} + \ldots \\ c_m &= c_m^{(0)} + \lambda c_m^{(1)} + \lambda^2 c_m^{(2)} + \ldots \end{aligned} \qquad (7)$$

By substituting into the Schrödinger equation, collecting like powers of λ, and then setting $\lambda = 1$, we obtain

$$E_m = E_m^{(0)} + V_{mm} + \sum_{k \neq m} \frac{V_{mk} V_{km}}{E_m^{(0)} - E_k^{(0)}}$$

$$+ \sum_{k\ell \neq m} \left[\frac{V_{mk} V_{k\ell} V_{\ell m}}{(E_m^{(0)} - E_k^{(0)})(E_k^{(0)} - E_\ell^{(0)})} - \frac{V_{mk} V_{km} V_{mm}}{(E_m^{(0)} - E_k^{(0)})^2} \right] + \ldots \qquad (8)$$

$$\psi_m = \phi_m + \sum_{k \neq m} \frac{V_{km}}{(E_m^{(0)} - E_k^{(0)})} \phi_k$$

$$+ \sum_{k \neq m} \left[\frac{-V_{km} V_{mm}}{(E_m^{(0)} - E_k^{(0)})^2} + \sum_{\ell \neq m} \frac{V_{k\ell} V_{\ell m}}{(E_m^{(0)} - E_k^{(0)})(E_m^{(0)} - E_\ell^{(0)})} \right] \phi_k + \ldots \qquad (9)$$

This wave function is not normalized, though normalization may readily be carried out (see e.g. Ref. [1]). The method converges rapidly provided that $|V_{km}| \ll |E_m^{(0)} - E_k^{(0)}|$ for all k.

The Brillouin-Wigner technique, as modified by Feenberg [2], obtains expressions for the eigenvalues and coefficients in (4) having very similar structures to (8) and (9), but containing no matrix elements with repeated indices. As in the above case, we consider the mth state and write $c_m = 1$, so that

$$(E - H_{kk})c_k = H_{km} + \sum_{n \neq mk} H_{kn} c_n \tag{10}$$

The corresponding equations for other coefficients are

$$(E - H_{nn})c_n = H_{nm} + H_{nk}c_k + \sum_{p \neq nmk} H_{np} c_p \tag{11}$$

$$(E - H_{pp})c_p = H_{pm} + H_{pk}c_k + \sum_{q \neq pnmk} H_{pq} c_q \tag{12}$$

and so on. These equations are, of course, exact. We have written them in this way as we wish to obtain an expression free of the repeated index matrix elements which occur in (8) and (9). The solution proceeds by a method of successive approximations, the first-order result coming if we drop the summation in (10). Higher-order expressions come from including progressively more equations in the sequence (10) - (12). The final result [2] is an implicit formula for the energy and an explicit result (in terms of the corresponding eigenvalues) for each coefficient. Defining

$$\epsilon_{nmpq} = H_{qq} + \sum_{r \neq nm\ldots pq} \frac{H_{qr} H_{rq}}{(E_n - \epsilon_{nm\ldots pqr})}$$

$$+ \sum_{\substack{r \neq nm\ldots pq \\ s \neq nm\ldots pqr}} \frac{H_{qr} H_{rs} H_{sq}}{(E_n - \epsilon_{nm\ldots pqr})(E_n - \epsilon_{nm\ldots pqrs})} + \ldots \tag{13}$$

we obtain

$$c_k(E - \epsilon_{mk}) = H_{km} + \sum_{n \neq mk} \frac{H_{kn} H_{nm}}{(E - \epsilon_{mkn})}$$

$$+ \sum_{\substack{n \neq mk \\ p \neq mkn}} \frac{H_{kn} H_{np} H_{pm}}{(E - \epsilon_{mkn})(E - \epsilon_{mknp})} + \ldots \tag{14}$$

and, substituting in (4),

$$E = H_{mm} + \sum_{k \neq m} \frac{H_{mk} H_{km}}{(E - \epsilon_{mk})} + \sum_{\substack{k \neq m \\ n \neq mk}} \frac{H_{mk} H_{kn} H_{nm}}{(E - \epsilon_{mk})(E - \epsilon_{mkn})} + \ldots \qquad (15)$$

3. RELATIONSHIP BETWEEN BRILLOUIN-WIGNER AND RAYLEIGH-SCHRÖDINGER THEORIES

Feenberg [2] pointed out that a solution of the BW expansion could be obtained by using the geometric series expansion:

$$\frac{1}{(E^{(\mu)} - H_{kk})} = \frac{1}{(H_{mm} - H_{kk})} \sum_{\nu=0}^{\infty} \left(\frac{H_{mm} - E^{(\mu)}}{H_{mm} - H_{kk}} \right)^{\nu} \qquad (16)$$

and iterating from $E^{(1)} = H_{mm}$. This leads to an explicit formula for the energy in which the denominators on the right-hand side have been reduced to differences between diagonal matrix elements, as in RS theory. We take as an example the second-order BW result:

$$E = H_{mm} + \sum_{k \neq m} \frac{H_{mk} H_{km}}{(E - H_{kk})} \qquad (17)$$

Substituting the first-order result $E = H_{mm} = E_m^{(0)} + V_{mm}$ into the denominator and expanding as a geometric series gives

$$E = E_m^{(0)} + V_{mm} + \sum_{k \neq m} \frac{V_{mk} V_{km}}{(E_m^{(0)} - E_k^{(0)})} - \sum_{k \neq m} \frac{V_{mk} V_{km} V_{mm}}{(E_m^{(0)} - E_k^{(0)})^2} + \ldots \qquad (18)$$

The last term is the first repeated index term of the RS series (8). This is a general result. The series of terms with repeated indices in RS is implicitly summed in the BW expansion.

This derivation of BW theory shows clearly that it is an <u>exact</u> representation of the secular determinant and it is therefore free of the restriction $|H_{km}| \ll |H_{mm} - H_{kk}|$. The transcendental nature of the eigenvalue equation (15) means that an iterative solution is usually necessary. The implicit summation of <u>all</u> terms with repeated indices suggests, however, that the convergence of such a method will be more rapid than the RS approach.

An interesting aspect of this technique is its connection to self-consistent solutions in other branches of physics. In the above derivation, we have at each stage systematically included all terms of a given order.

The series is, in fact, a prototype self-consistent solution of which the
Dyson and Lippmann-Schwinger equations are further examples (see Ref.[3]
for a discussion of this point). The parallel with linked-cluster expansions
in many-body perturbation theory is also apparent.

4. APPLICATION TO SURFACE PROBLEMS

In spite of the above advantages, the BW equation (15) is an implicit
equation for E. In the case of a large secular determinant with numerous
non-zero off-diagonal matrix elements, it is therefore an inappropriate
basis for calculation. In certain cases, however, the BW approach has
provided insight into surface problems.

The first is in model calculations of surface states using 'bulk' potentials
of certain simple types, e.g. in a nearly free electron (NFE) calculation
assuming a bulk potential of the form:

$$V(x) = V_0 - W \cos gx$$

If the basis states are written

$$\psi_{\{{\rm even} \atop {\rm odd}\}} = \sum_{m=0}^{\infty} A_{2m+p} \left\{ {\cos \atop \sin} \right\} (2m+p)x \qquad (19)$$

the off-diagonal matrix elements are zero except those connecting adjacent
values of m. All third- and higher-order terms in (14) and (15) vanish,
and the result is the familiar Mathieu function continued fraction. Energy
levels for the surface states follow from a wave-matching calculation.

In the simplest tight-binding model, a linear chain with nearest-neighbour
overlaps, off-diagonal matrix elements,

$$\int dx \, \phi^*(x - x_\ell)[V(x) - U(x - x_m)] \phi(x - x_m) \qquad (20)$$

where $\phi(x - x_m)$, the atomic orbital, and $U(x - x_m)$, the potential for the mth
site, are zero unless $\ell = m \pm 1$. Once again, third- and higher-order terms
in the BW expressions vanish and the same continued fractions for the eigenvalues and coefficients result. The connection between the NFE and tight-
binding methods noted by Davison and Levine [4] is clear. The simplest
cases of each correspond to mathematically equivalent secular determinants.

The BW technique has also proved useful in examining surface effects
on the energy levels on the linear tight-binding chain in the presence of an
electric field F [5]. The unperturbed levels (in the absence of overlap)
form a 'Stark ladder', $E_n = \alpha + n\Delta$, where α is the energy of the atomic state
and $\Delta = eFa$, with a the interatomic spacing. The form of the resulting BW
expansion for $\delta E_n = E_n - \alpha - n\Delta$ is (β is the nearest-neighbour overlap, N
the number of atoms in the chain)

$$\delta E_n = \cfrac{\beta^2}{\delta E_n - \Delta - \cfrac{\beta^2}{\delta E_n - 2\Delta - \cfrac{\beta^2}{\ddots \cfrac{}{\delta E_n - (N-2-n)\Delta - \cfrac{\beta^2}{\delta E_n - (N-1-n)\Delta}}}}}$$

$$+ \cfrac{\beta^2}{\delta E_n + \Delta - \cfrac{\beta^2}{\delta E_n + 2\Delta - \cfrac{\beta}{\ddots \cfrac{}{\delta E_n - (n-1)\Delta - \cfrac{\beta^2}{\delta E_n + n\Delta}}}}} \qquad (21)$$

This equation has the solution $\delta E_n = 0$, not only in the trivial case $\beta = 0$ but also if both continued fractions contain the same number of terms. The latter is true for an infinite chain, so a Stark ladder results for all β. For a finite chain there is exact cancellation of the two continued fractions only for the central atom, but we see that an approximate ladder can exist, centred on the middle of the energy spectrum, if $N\Delta \gg \beta$. Bounds on the energy spectrum can be found by using theorems on continued fractions.

A more important application of BW theory is to the case where second-order terms are as important as the first-order terms in a perturbation expansion. This is particularly true if the unperturbed state is n-fold degenerate. The first step in the RS theory would give zero-order wave functions from the n × n secular determinant. Due to the large second-order terms, however, these wave functions are likely to be poor approximations to the correct solutions. It is clearly necessary to diagonalize the appropriate part of the secular matrix exactly or, equivalently, to use BW theory. This problem arises when there are successive perturbations of different types but of comparable magnitude [6] or in calculating the bulk pseudopotential of covalent semiconductors [7]. In the latter case, an examination of detailed pseudopotential calculations shows that the most striking feature of the diamond structure materials is the dominance of the $\vec{g} = (111)$ Fourier components $V_{\vec{g}}$. In the neighbourhood of the Fermi energy, second-order contributions from $(V_{111})^2$ are at least as important as other first-order contributions. Without a proper description of both effects, such as BW theory, resultant band-structure calculations are likely to be misleading. The strongly directional nature of the charge distribution in diamond structure materials is a direct consequence of the large second-order terms, and the approach has been made the basis of a theory of covalency. This model of the bulk band structure has formed the basis of a number of surface state calculations.

REFERENCES

[1] LANDAU, L.D., LIFSHITZ, E.M., Quantum Mechanics — Non-relativistic Theory, Pergamon Press, Oxford (1958), Section 38.
[2] FEENBERG, E., Phys. Rev. 74 (1948) 206.
[3] ZIMAN, J.M., Elements of Advanced Quantum Theory, Cambridge University Press (1969).
[4] DAVISON, S.G., LEVINE, J., Solid State Phys. 25 (1970) 1 (see p. 54 onwards).
[5] HEINRICHS, J., JONES, R.O., J. Phys. C 5 (1972) 2149.
[6] PRYCE, M.H.L., Proc. Phys. Soc. (London) 63 (1950) 25.
[7] HEINE, V., JONES, R.O., J. Phys. C 2 (1969) 719.
[8] JONES, R.O., in Surface Physics of Semiconductors and Phosphors, (SCOTT, C.G., REED, C.E., Eds), Academic Press, London (1975).

IAEA-SMR-15/9

LIQUID SURFACES

M.V. BERRY
H.H. Wills Physics Laboratory,
University of Bristol,
Bristol,
United Kingdom

Abstract

LIQUID SURFACES. 1. Liquid-vapour interfaces — surface tension: 1.1. Macroscopic mechanical theory. 1.2. Macroscopic thermodynamic theory. 1.3. Microscopic mechanical theory (exact). 1.4. Statistical-mechanical theory (exact). 1.5. Some approximate methods. 2. Liquid-vapour interfaces near solids — contact angle: 2.1. Exact macroscopic theories. 2.2. Approximate microscopic theories.

1. LIQUID-VAPOUR INTERFACES — SURFACE TENSION

1.1. Macroscopic mechanical theory

Liquids are distinguished from other phases by the fact that they possess a mobile surface. Solids possess a surface, but it is not mobile, and gases are mobile but need possess no free surface. Macroscopically the mobile liquid surface is characterized by a single mechanical quantity — the surface tension γ. To understand this concept we consider the manner in which two important fluid-mechanical quantities change as we approach the surface of a liquid. Let the z axis be perpendicular to the liquid surface, pointing out into the vapour or air (Fig.1).

First consider the mass density $\rho(z)$. This will be a constant ρ_L within the liquid, whereas outside it will take a very low value ρ_v. $\rho(z)$ does not change discontinuously but falls from ρ_L to ρ_v (Fig.2) over a distance of about two intermolecular spacings. (In considering $\rho(z)$ as varying smoothly over such small distances, we are imagining looking at the surface with a microscope over periods of time long compared with atomic collision times, so that we observe quantities that are time-averaged but not spatially averaged; we call such quantities 'macroscopic'.)

The second quantity we consider is the stress tensor $\sigma_{ij}(z)$; σ_{ij} is defined as the force in direction j acting across an imaginary interface in the fluid perpendicular to direction i, on the fluid with the smaller value of the i co-ordinate (Fig.3). For any material, if classical mechanics applies, Newton's second law may be applied to an elementary cube at point $\vec{r}$ and gives

$$F_i(\vec{r})\rho(\vec{r}) + \sum_{j=1}^{3} \frac{\partial \sigma_{ji}(\vec{r})}{\partial x_j} = \rho(\vec{r}) \frac{du_i(\vec{r})}{dt} \qquad (1)$$

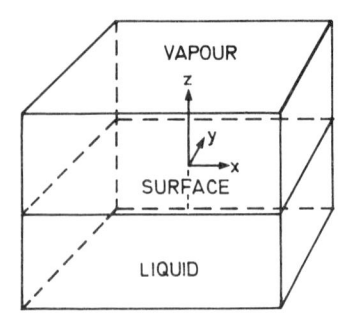

FIG.1. Liquid-vapour system in a box.

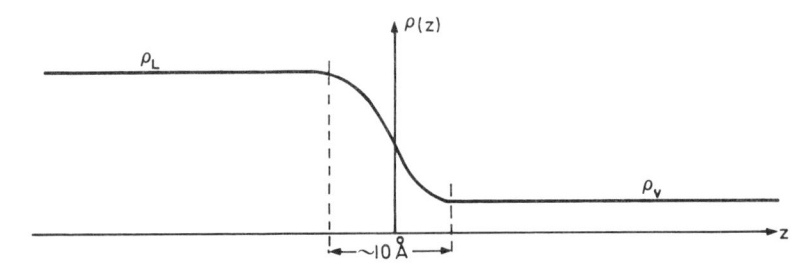

FIG.2. Density profile.

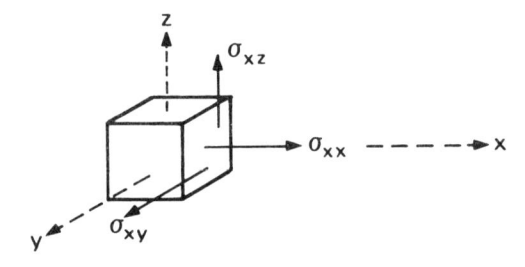

FIG.3. Definition of stress tensor.

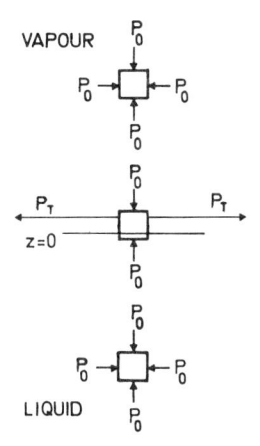

FIG.4. Forces on fluid elements at different heights.

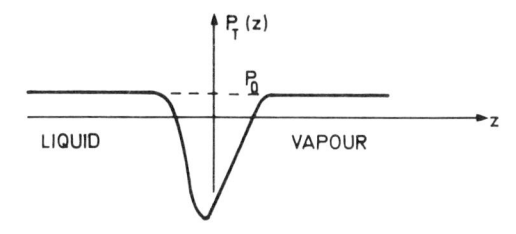

FIG.5. Variation of tangential pressure through the interface.

The first term involves $F_i(\vec{r})$, the body force per unit mass, and describes effects such as gravitation; these affect the macroscopic shapes of — e.g. — liquid droplets, but are totally negligible in the surface region. The right-hand side involves $u_i(\vec{r})$, the fluid velocity at $\vec{r}$; we are primarily interested in fluid statics, so we set $u_i = 0$. Thus Newton's law reduces to three equations of hydrostatic equilibrium:

$$\sum_{j=1}^{3} \frac{\partial \sigma_{ji}(\vec{r})}{\partial x_j} = 0 \qquad (2)$$

involving only the contact forces σ_{ij} between neighbouring elements of fluid.

Within the bulk liquid and vapour, the stress tensor is

$$\sigma_{ij}^{(bulk)} = \begin{pmatrix} -P_0 & 0 & 0 \\ 0 & -P_0 & 0 \\ 0 & 0 & -P_0 \end{pmatrix} \qquad (3)$$

where P_0 is the pressure (atmospheric, say). (If we had included the effect of gravity, we would have $P_{liq} = P_0 - \rho_L gz$, but the extra term is unimportant near the surface.) This tensor contains no shear stresses $\sigma_{i \neq j}$ because it describes a fluid at rest. Equation (3) satisfies the equilibrium condition (2), but is not the most general tensor that does so. How is the actual stress tensor $\sigma_{ij}(\vec{r})$ restricted by (2)? By symmetry, σ_{ij} can depend only on z, so (2) reads:

$$\frac{\partial \sigma_{31}(z)}{\partial z} = 0, \quad \frac{\partial \sigma_{32}(z)}{\partial z} = 0, \quad \frac{\partial \sigma_{33}(z)}{\partial z} = 0 \tag{4}$$

Thus the shear stresses σ_{xz} and σ_{yz} are always zero, and the normal pressure $P_N(z) \equiv -\sigma_{33}$ is constant through the interface zone and equal to P_0. A simple symmetry argument shows that the shear stresses σ_{xy} and σ_{yx} are always zero. This leaves only the possibility that the tangential pressure $P_T(z) \equiv -\sigma_{xx}(z) \equiv -\sigma_{yy}(z)$ could vary with z without upsetting hydrostatic equilibrium. Thus the most general tensor satisfying (2) is

$$\sigma_{ij}(z) = \begin{pmatrix} -P_T(z) & 0 & 0 \\ 0 & -P_T(z) & 0 \\ 0 & 0 & -P_N = -P_0 \end{pmatrix} \tag{5}$$

In particular, it is possible for $P_T(z)$ to decrease enormously in the surface layer, changing into a tension, and this is exactly what happens (Figs 4 and 5). The surface region is so thin that we can idealize this actual situation by imagining z = 0 to be a plane of discontinuity between liquid (ρ_L) and vapour (ρ_v), in which a tension acts, as follows: perpendicular to any line of unit length in the surface z = 0 (Fig.6) there acts a force γ with which the liquid surface on one side pulls the liquid surface on the other. To obtain γ it is necessary only to integrate the underpressure $P_0 - P_T(z)$ through the surface, to obtain the basic formula:

$$\gamma = \int_{-\infty}^{\infty} [P_N - P_T(z)] \, dz \tag{6}$$

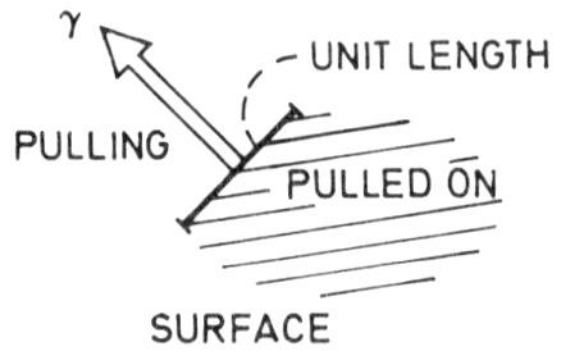

FIG.6. Definition of surface tension.

Of course γ is simply the surface tension of the liquid; this is the basic quantity characterizing the surface from the point of view of mechanics, and it determines the form taken by the surfaces of liquid masses acted on by other forces such as gravity and interactions with solid surfaces (see Section 2).

TOTAL FORCE ZERO

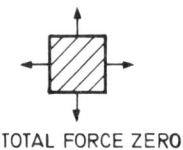

FIG. 7. A surface element in equilibrium.

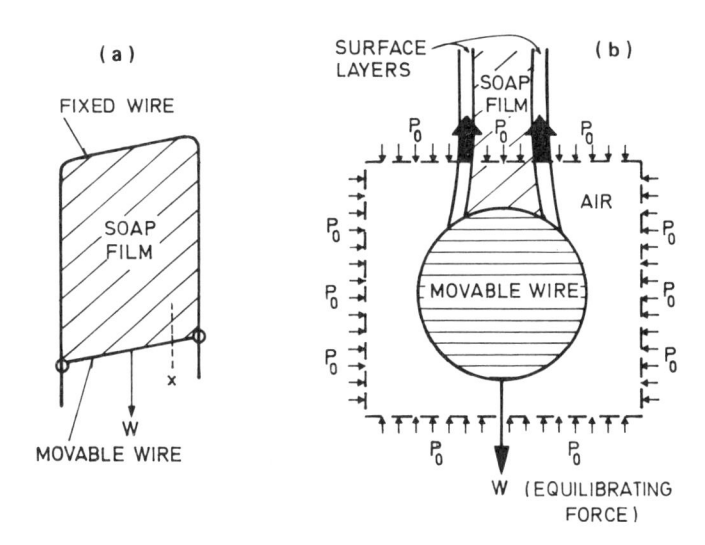

FIG. 8. Surface tension equilibrated by an external force: (a) experimental situation; (b) details of forces acting.

Provided that the radii of curvature of the surface are large compared with interatomic separations, and provided that any shear velocity differences between elements of the fluid separated by an interatomic distance are slow compared with the thermal velocities of the atoms, then γ may be treated as a constant for each liquid-vapour interface, depending only on the temperature T.

To detect the existence of γ there is no point in looking at a plane surface, since each element of it is in equilibrium (Fig.7). But we can show that γ exists by employing a situation involving unbalanced forces. Consider a soap film stretched over a wire frame with a movable bottom (Fig.8a), and look more closely at the cross-section at point x (Fig.8b). The film is many thousands of layers thick, so the two surfaces act independently. We do not need to consider yet exactly what happens at the wire-soap-air boundary (see Section 2), since we can consider the equilibrium of the matter contained within the volume defined by unit length of the movable wire and the area inside the dashed rectangle in Fig.8b. The bulk pressures P_0 act uniformly all round and cancel, while the surface tension gives a force 2γ acting to contract the film; to prevent this it is necessary to oppose the surface tension

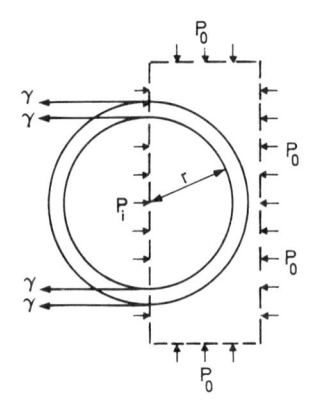

FIG. 9. Forces keeping a bubble in equilibrium.

with an equal and opposite force — say a weight of magnitude 2γ/unit length of wire. Such experiments are easily carried out and prove very directly that γ exists.

Another way of showing the existence of surface tension is to employ curved surfaces. The soap bubble (Fig.9) is a simple example; it has been blown up to a radius r and to an internal pressure P_i which exceeds the outside pressure P_0. Consider the equilibrium of the matter inside the dashed rectangle in Fig.9. This requires:

Force to right = force to left

i.e. $\pi r^2 (P_i - P_0) = 2(2\pi r \gamma)$

i.e. excess pressure $= P_i - P_0 = \dfrac{4\gamma}{r}$ (7)

This result can be generalized to give the excess pressure P_{ex} across a single surface (remember the soap bubble is a double surface) whose local radii of curvature are r_1 and r_2:

$$P_{ex} = \gamma \left(\frac{1}{r_1} + \frac{1}{r_2} \right) \qquad (8)$$

This is Laplace's equation, and is the basis for calculations of the forms of hanging, falling, spinning, vibrating or resting liquid masses.

Thus the properties of liquid surfaces differ from the bulk, and we can form the picture of a skin which is always stretched and trying to contract. This is not quite like an elastic skin (say a rubber balloon), because its tendency to contract (γ) is independent of how much it has already been

TABLE I. SURFACE TENSIONS FOR SOME FAMILIAR LIQUIDS

Substance	Temperature (K)	γ (Nm^{-1})
H_2	14	0.0028
N	26	0.0052
O_2	90	0.013
H_2O	293	0.073
	373	0.059
Soap solution	293	~0.025
Hg	235	0.498
Fe	1810	1.872
Os	3283	2.5

stretched, whereas for an elastic skin the contractile (restoring) force increases with the stretching (Hooke's law). Surface tension is a force per unit length, so that the units of γ are Nm^{-1}; in terms of cgs units we have

$$1 \text{ Nm}^{-1} = 10^3 \text{ dyne/cm} \tag{9}$$

Surface tensions for some familiar liquids are shown in Table I. It can be seen that liquid metals have easily the largest values of γ. In terms of tangential stresses, these surface tensions are enormous; if we assume the stresses act uniformly over the 10-Å thickness of the surface layer, then the stresses are given by

$$\text{stress} = -P_T \approx \frac{\gamma}{10 \text{ Å}} \sim 10^3 \text{ atm for water}$$

Although we employed soap films for illustrative purposes, their very existence depends on the fact that they consist of two components (soap and water), so that their composition can be, and is, different in the surface layer and the bulk. It is this capacity to adjust their composition that makes soap films stable against small disturbances. No single-component liquid can ever form a stable film — you cannot blow a water-bubble. It is the purpose of this paper to present the theoretical physics of surface tension in its simplest form, and from this point of view films represent a complication to be avoided; from now on, therefore, we consider only one-component ('pure') liquids.

1.2. Macroscopic thermodynamic theory

Now we show that the mechanical interpretation of surface tension is not the only one: γ is also a thermodynamic quantity characterizing the liquid surface. To see this, consider the system shown in Fig.1 and let the walls of

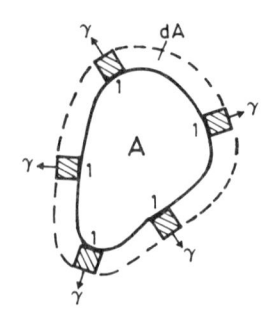

FIG.10. Expanding a surface does work.

the container be moved in such a way that the area A of the liquid surface increases by dA, while the volume of the system and its temperature T remain constant. How much work dW will be done on the system in this process? It is easy to see (Fig.10) that

$$dW = \gamma dA \tag{10}$$

What about the internal energy U of the system? This will change by dU, where

$$dU = TdS + dW \tag{11}$$

dS being the change in entropy of the system. At this stage we do not know the value of TdS; certainly it is not zero because the surface was expanded not adiabatically but isothermally. This suggests that we consider not U but the free energy F, defined by

$$F = U - TS \tag{12}$$

i.e.

$$dF = dU - TdS - SdT = -SdT + dW \tag{13}$$

In the process considered, dT was zero, so that (13) and (10) give $dF = \gamma dA$, i.e.

$$\gamma = \left(\frac{\partial F(T,V,A)}{\partial A}\right)_{T,V} \tag{14}$$

This gives a thermodynamic interpretation of γ, and may in fact be used to define this quantity instead of the mechanical equation (6). We emphasize, however, that both approaches are equally valid.

Since γ is independent of the area of the surface, we can integrate (14) to obtain

$$F(T,V,A) = \gamma(T)A + F_b(T,V) \equiv F_s(T,A) + F_b(T,V) \tag{15}$$

This expresses the total free energy as the sum of a contribution F_b from the bulk phases, and a contribution:

$$F_s(T,A) = \gamma(T)A \tag{16}$$

from the surface. This is equivalent to (14), and gives an interpretation of γ as surface free energy per unit area (there is no problem over units, since $Jm^{-2} = Nm^{-1}$ dimensionally). This interpretation easily explains the contractile tendency of liquid masses, and the spherical shape of falling droplets, for (13) states that for isothermal situations $dF = dW$, but dW is zero for a system in mechanical equilibrium (principle of virtual work), so that F is an extremum, which actually must be a minimum if the equilibrium is to be stable. For liquid masses V is essentially constant, so that only A changes and we have to minimize F_s; from Eq.(16) this implies that A itself is minimized at constant V, hence the sphericity of raindrops.

It is also possible to define other thermodynamic 'surface functions'. From (13), the surface entropy S_s is defined as

$$S_s \equiv -\left(\frac{\partial F_s}{\partial T}\right)_A = -\frac{d\gamma(T)A}{dT} \tag{17}$$

while from (12) the surface energy U_s is defined as

$$U_s = F_s + TS_s = \left[\gamma(T) - T\frac{d\gamma(T)}{dT}\right]A \tag{18}$$

We conclude that the single function $\gamma(T)$ provides a complete mechanical and thermodynamic description of liquid surfaces. What is the form of this function? Clearly it is never negative — if it were, stable droplets would have the maximum area for a given volume, corresponding to an infinitely puckered and convoluted surface, and this is never observed. Furthermore, we expect γ to vanish at the critical temperature T_c, where the distinction between liquid and vapour disappears. Thus $\gamma(T)$ should decrease to zero at T_c, and this is indeed observed. To compare results for different substances it is convenient to employ the reduced surface tension γ_r, a dimensionless quantity defined by

$$\gamma_r\left(\frac{T}{T_c}\right) \equiv \frac{\gamma(T)v_c^{2/3}}{kT_c} \tag{19}$$

where v_c is the critical volume per molecule and k is Boltzmann's constant. The results (Fig.11 and Table II) show that γ_r is almost a universal function of T/T_c, i.e. there is a 'principle of corresponding states' for surface tension, in terms of which $\gamma(T)$ for one liquid can be calculated quite accurately from measurements of $\gamma(T)$ on another liquid, provided v_c and T_c are known. The experimental data are well described over a large range by the empirical equation:

$$\gamma(T) = \gamma_0\left(1 - \frac{T}{T_c}\right)^I \tag{20}$$

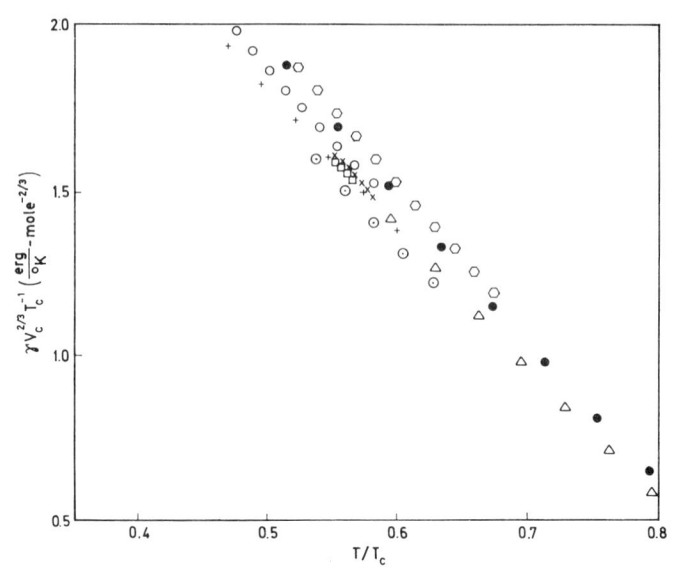

FIG.11. $\gamma(T) V_c^{2/3} T_c^{-1}$ plotted versus T/T_c. ⊙: neon; △: argon; ×: krypton; □: xenon; ●: nitrogen; ◯: carbon monoxide; +: methane; ○: oxygen. (From Buff and Lovett (1968).)

TABLE II. VALUES OF $\gamma_R = V_c^{2/3} T_c^{-1} \gamma(T)$ AT $T = 0.56 T_c$ (From Buff and Lovett (1968))

Substance	γ (dyne·cm^{-1})	γ_R (erg / K mole$^{-2/3}$)	γ_R/γ_R (Ar)
Ne	5.55	1.500	0.955
CH$_4$	13.790	1.548	0.986
Kr	16.081	1.568	1.000
Ar	13.28	1.570	1.000
Xe	18.98	1.583	1.008
O$_2$	14.04	1.605	1.022
N$_2$	10.42	1.662	0.059
CO	11.05	1.704	1.085

where the 'critical exponent' r is given by

$$r = 1.27 \pm 0.02 \tag{21}$$

For pure liquids, γ always decreases with T, but for liquids with more than one component it is possible for $d\gamma/dT$ to be positive in some temperature regions.

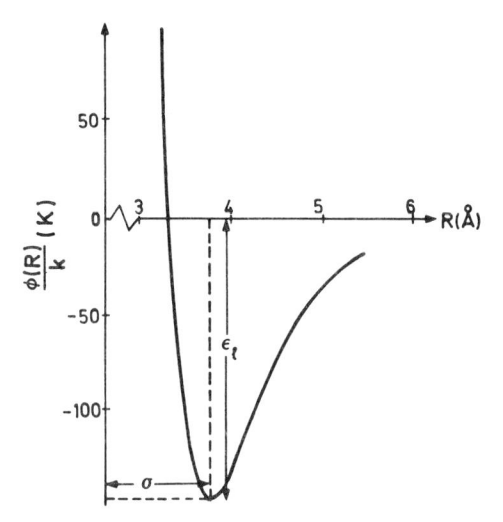

FIG.12. Typical interatomic potential (from Toxvaerd (1972 b)).

1.3. Microscopic mechanical theory (exact)

To calculate surface tension from the microscopic interactions between fluid molecules, we have to have a model. We choose the 'simple fluid' model, on which the molecules interact pairwise via a spherically symmetrical potential function $\phi(R)$, where $\vec{R}$ is the vector joining the centres of the two interacting molecules. For real fluids, of course, there are triplet and higher-order interactions (i.e. the force between two molecules is modified by others in the neighbourhood) and, further, the interaction depends on the orientation of $\vec{R}$ as well as its magnitude (i.e. on direction), and also on the relative velocity of the molecules (i.e. on temperature). If these complications are neglected, the resulting theory gives a very good description of bulk and interface properties of rare gases and liquids, and fluids with compact molecules, such as nitrogen, oxygen, carbon monoxide and methane; the theory does not work for liquid metals because of the electron gas, but it is possible to treat this separately. The potential $\phi(R)$ is attractive at large distances and repulsive at short distances, as shown in Fig.12 for argon; the attraction is responsible for the condensation of gases into liquids and solids, and the repulsion prevents interpenetration of molecules, so that in dense phases they act somewhat like 'hard spheres'. If we exclude the very lightest elements, we may use classical mechanics to describe the encounters between molecules, because their de Broglie wavelength is small in comparison with their size. Experimentally, $\phi(R)$ can be obtained from molecular scattering experiments or from the equation of state of nearly perfect gases.

In order to calculate γ from the mechanical definition (6) it is necessary to know the anisotropy $P_N - P_T(z)$ of the pressure in the interface. The pressure $P(\vec{\ell}, \vec{r})$ at a point $\vec{r}$ acting in a direction specified by a unit vector $\vec{\ell}$ is defined as follows: imagine a plane at $\vec{r}$ perpendicular to $\vec{\ell}$ (Fig.13), and consider an elementary area dS in this plane and containing $\vec{r}$. The plane divides the fluid

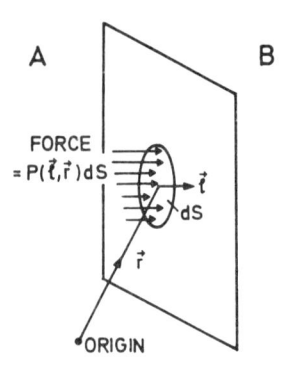

FIG.13. Definition of pressure.

into two regions, A and B, and the time average of the force exerted by the molecules in A on the molecules in B across dS is, by definition, $P(\vec{\ell}, \vec{r})\,dS$. Now, it is most important to realize that while from a macroscopic point of view A and B are distinct portions of fluid, on the microscopic level they are constantly exchanging particles. The average momentum per unit time transferred in this way from A to B across dS gives a kinetic contribution $P^K(\vec{\ell},\vec{r})\,dS$ to the pressure at $\vec{r}$. In addition, there is a static force component $P^f(\vec{\ell},\vec{r})\,dS$, equal to the average total force arising from $\phi(R)$ that the molecules in A exert on those in B which can be joined to them by a straight line intersecting dS. Thus we have

$$P(\vec{\ell}, \vec{r}) = P^K(\vec{\ell}, \vec{r}) + P^f(\vec{\ell}, \vec{r}) \tag{22}$$

We calculate P^K first. Temporarily we say that $\vec{\ell}$ points along the positive x axis. A molecule now in A, and moving with velocity $\vec{v}$ with mass m will carry normal momentum $m\,v_x$ across dS into B in unit time if it lies in a region (Fig.14) whose volume is $v_x\,dS$, and if v_x is positive. The average number of molecules in this volume is $n(\vec{r})v_x\,dS$, where $n(\vec{r})$ is the number density of molecules at $\vec{r}$, defined by

$$n(\vec{r}) \equiv \frac{\rho(\vec{r})}{m} \tag{23}$$

This includes molecules with all velocities $\vec{v}$; the force due to molecules moving from A to B across dS is the average of $n(\vec{r})v_x\,dS \times m v_x$ over the distribution of $\vec{v}$. This distribution is given by the Maxwell-Boltzmann law:

$$\mathscr{P}(\vec{v}) = \left(\frac{m}{2\pi kT}\right)^{3/2} \exp\left(-\frac{m v^2}{2kT}\right) \tag{24}$$

where $\mathscr{P}(\vec{v})\,d\vec{v}$ is the fraction of molecules whose velocities lie between $\vec{v}$ and $\vec{v}+d\vec{v}$. It is often thought, incorrectly, that this distribution only holds in an ideal gas. where molecular interactions embodied in $\phi(R)$ are infinitesimal,

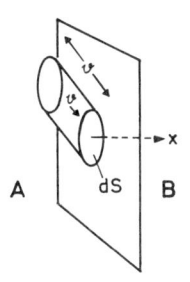

FIG.14. Momentum transfer across a surface.

but Eq.(24) actually holds rigorously for any potential that is not velocity dependent. This result is a consequence of the fact that the kinetic energy part of the Hamiltonian for each particle does not involve $\phi(R)$. Denoting averages by $\langle \rangle$, we require

$$\langle n(\vec{r}) m v_x^2 dS \rangle = n(\vec{r}) m \, dS \, \langle v_x^2 \rangle$$

$$= n(\vec{r}) m \, dS \int_{-\infty}^{\infty} dv_y \int_{-\infty}^{\infty} dv_z \int_0^{\infty} dv_x \left(\frac{m}{2\pi kT}\right)^{3/2} v_x^2 \exp\left(-\frac{mv^2}{2kT}\right)$$

$$= \frac{n(\vec{r}) kT \, dS}{2} \tag{25}$$

This is the force from A onto B due to molecules carrying positive x-momentum into B. However, there will be, on the average, an equal number of molecules carrying negative x-momentum out of B into A. Therefore, the total kinetic pressure of A on B is

$$P^K(\vec{\ell}, \vec{r}) = n(\vec{r}) kT \tag{26}$$

This depends only on the density at $\vec{r}$ and not on the direction $\vec{\ell}$ of the pressure. Therefore, P^K is isotropic, and cannot account for surface tension, which depends (Eq.6)) on the anisotropy $P_N - P_T$. If the potential $\phi(R)$ is negligible there is no static force component P^f, so that $P = P^K$ and (26) is the well-known equation of state for an ideal gas (to see this, observe that $nk = \mathcal{N}k/V = R/V$, where $\mathcal{N}$ is Avogadro's number, V is the molar volume, and R is the gas constant).

Now we must calculate $P^f(\vec{\ell}, \vec{r})$. The component along $\vec{\ell}$ of the force exerted by a molecule at $\vec{r}_A$ in A on the molecule at $\vec{r}_B$ in B is

$$-\nabla_{\vec{r}_B} \phi(|\vec{r}_B - \vec{r}_A|) \cdot \vec{\ell} = -\frac{\vec{R} \cdot \vec{\ell}}{R} \phi'(R) \tag{27}$$

where

$$\vec{R} \equiv \vec{r}_B - \vec{r}_A \tag{28}$$

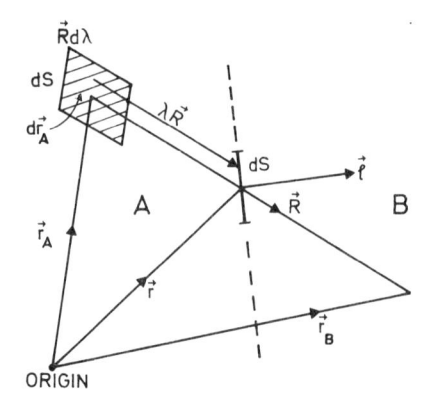

FIG.15. A group of A molecules that act on a B molecule.

and primes denote differentiation of a function with respect to its argument. For any two volume elements $d\vec{r}_A$, $d\vec{r}_B$, the average number of pairs of molecules, i.e. the number of molecules in $d\vec{r}_A$ multiplied by the number of molecules in $d\vec{r}_B$, is $n_2(\vec{r}_A, \vec{r}_B) d\vec{r}_A d\vec{r}_B$ where n_2 is the two-particle distribution function of the fluid. It is convenient to define the pair correlation function $g(\vec{r}_A, \vec{r}_B)$ by

$$n_2(\vec{r}_A, \vec{r}_B) \equiv n(\vec{r}_A) n(\vec{r}_B) g(\vec{r}_A, \vec{r}_B) \tag{29}$$

When $\vec{r}_A$ and $\vec{r}_B$ are well separated there is no correlation between the positions of the molecules, and $g(\vec{r}_A, \vec{r}_B)$ is unity. When $\vec{r}_A$ and $\vec{r}_B$ are very close together the exclusion effect arising from the repulsive part of $\phi(R)$ means that $g(\vec{r}_A, \vec{r}_B)$ is very small. For a bulk fluid phase $g(\vec{r}_A, \vec{r}_B)$ depends only on R, but we are particularly concerned with the interface, where the interparticle correlations are anisotropic.

To find P^f it is necessary to add up all the pair forces (27); the tricky part is to include only those pairs of molecules whose joining vector cuts dS. This only happens if the point $\vec{r}_A + \lambda \vec{R}$ lies on dS for some value of λ between zero and unity (Fig.15). Then if $\vec{R}$ is fixed the volume element in which the A molecules (say) must lie is

$$d\vec{r}_A = dS \, \vec{\ell} \cdot d\vec{R} \, d\lambda \tag{30}$$

(Fig.15). If we specify the positions of molecules by

$$\vec{r}_A = \vec{r} - \lambda \vec{R}, \quad \vec{r}_B = \vec{r} - \lambda \vec{R} + \vec{R} \tag{31}$$

then the number of contributing pairs is

$$n_2(\vec{r}_A, \vec{r}_B) d\vec{r}_A d\vec{r}_B = n_2(\vec{r} - \lambda \vec{R}, \vec{r} - (\lambda - 1)\vec{R}) d\vec{R} \, d\lambda \, dS \, \vec{\ell} \cdot \vec{R} \tag{32}$$

The total force along $\vec{\ell}$ across dS is found by multiplying this expression by either member of (27) and integrating over all λ between 0 and 1 and all $\vec{R}$ for which $\vec{\ell}\cdot\vec{R} > 0$; dropping the factor dS, we obtain

$$P^f(\vec{\ell},\vec{r}) = -\iiint_{\vec{R}\cdot\vec{\ell}>0} d\vec{R}\,\frac{(\vec{R}\cdot\vec{\ell})^2}{R}\phi'(R)\int_0^1 d\lambda\, n_2(\vec{r} - \lambda\vec{R},\, \vec{r} - (\lambda-1)\vec{R}) \qquad (33)$$

We can rewrite this as an integral over all $\vec{R}$ by realizing that (33) is unaltered if we reverse the $\vec{\ell}$ component of $\vec{R}$, replace λ by $1-\lambda$ and use the fact that n_2 is a symmetric function of $\vec{r}_A$ and $\vec{r}_B$. Thus we obtain

$$P^f(\vec{\ell},\vec{r}) = -\frac{1}{2}\iiint d\vec{R}\,\frac{(\vec{R}\cdot\vec{\ell})^2}{R}\phi'(R)\int_0^1 d\lambda\, n_2(\vec{r} - \lambda\vec{R},\, \vec{r} - (\lambda-1)\vec{R}) \qquad (34)$$

Combining this with (26) we find, for the pressure at $\vec{r}$ in direction $\vec{\ell}$,

$$P(\vec{\ell},\vec{r}) = n(\vec{r})kT - \frac{1}{2}\iiint d\vec{R}\,\frac{(\vec{R}\cdot\vec{\ell})^2}{R}\phi'(R)\int_0^1 d\lambda\, n_2(\vec{r} - \lambda\vec{R},\, \vec{r} - (\lambda-1)\vec{R}) \qquad (35)$$

For a bulk phase, where n is constant, I leave it as an exercise to show that this reduces to the more familiar expression:

$$P = nkT - \frac{2\pi n^2}{3}\int_0^\infty dR\, R^3 \phi'(R) g(R) \qquad (36)$$

In the bulk vapour, n is small and the second term is negligible. In the bulk liquid, n is large and the second term almost cancels the first (the attractive forces, where ϕ' is positive, dominate); P^f is negative, in fact.

To calculate γ using (6), we must specialize the general expression (35) to obtain $P_T(z)$ and $P_N(z)$ when there is a plane interface perpendicular to Z. For P_N, $\vec{\ell}$ is parallel to the z axis, and for P_T, $\vec{\ell}$ may be taken as any line perpendicular to Z, and we choose the x axis. Then we have, if $\vec{R} \equiv (X,Y,Z)$:

$$P_N(z) = n(z)kT - \frac{1}{2}\iiint d\vec{R}\,\frac{Z^2}{R}\phi'(R)\int_0^1 d\lambda\, n_2(z - \lambda Z, \vec{R})$$

$$P_T(z) = n(z)kT - \frac{1}{2}\iiint d\vec{R}\,\frac{X^2}{R}\phi'(R)\int_0^1 d\lambda\, n_2(z - \lambda Z, \vec{R}) \qquad (37)$$

Now we use (6), and notice that

$$\int_{-\infty}^\infty dz \int_0^1 d\lambda\, n_2(z - \lambda Z, \vec{R}) = \int_{-\infty}^\infty dz\, n_2(z, \vec{R}) \qquad (38)$$

we obtain, as the final result,

$$\gamma = \frac{1}{2} \int_{-\infty}^{\infty} dz \iiint \frac{d\vec{R}}{R} (X^2 - Z^2) \phi'(R) n_2(z, \vec{R}) \tag{39}$$

This rigorous microscopic formula for γ involves not only the intermolecular potential $\phi(R)$, but also the function $n_2(\vec{r}_A, \vec{r}_B)$, whose form in the interface region is unknown. From (29), it is clear that this function will involve the 'density profile' $n(z)$ (cf. Fig.2), and the anisotropic surface correlations embodied in $g(z, \vec{R})$. A relation between these quantities, which does not, however, determine them completely, can be obtained by recalling that for hydrostatic equilibrium (Eq.(4)) $P_N(z)$ must be constant through the interface. Differentiating the first equation in (37) and using the symmetry of $n_2(\vec{r}_A, \vec{r}_B)$, we obtain

$$kT \frac{dn(z)}{dz} = \iiint d\vec{R} \frac{Z}{R} \phi'(R) n_2(z, \vec{R}) \tag{40}$$

(it helps to replace λ by a new integration variable $z - \lambda Z$). This expresses the fact that the total force on any element of fluid is zero, on the average. By performing more delicate averages, with one or more molecules held fixed, it is possible to obtain a hierarchy of equations, of which (40) is the first; the second equation in the series relates $n_2(\vec{r}_A, \vec{r}_B)$ to the triplet distribution function $n_3(\vec{r}_A, \vec{r}_B, \vec{r})$, and successive equations relate to higher-order distribution functions. A major problem in microscopic mechanics is the solution of this infinite hierarchy; it seems unfair of nature to make the relatively low-order correlations embodied in n_2 depend on the complicated higher-order distributions. However, simple approximations are possible, as will be seen in Section 1.5.

To see qualitatively why γ is positive, we notice from (37) that it depends only on P_T^f and P_N^f. As we approach the interface from the liquid, $n(z)$ decreases and so, therefore, does the number of interacting molecular pairs. But this number decreases faster across the horizontal surface dS corresponding to P_N than for the vertical surface for P_T, so that $|P_N^f| < |P_T^f|$. But P^f is negative, so that, from (6), γ is positive.

1.4. Statistical-mechanical theory (exact)

Here we show that the exact microscopic formula (39) for γ can also be derived from the thermodynamic formula (14), employing the standard methods of Gibbsian statistical mechanics. Let the system (Fig.1) be closed, and contain N molecules with instantaneous positions $\{\vec{r}_i\} = \{\vec{r}_1 \cdots \vec{r}_N\}$ and momenta $\{\vec{P}_i\} = \{\vec{P}_1 \cdots \vec{P}_N\}$ in a volume V at temperature T. The Hamiltonian function H is just the instantaneous total energy:

$$H(\{\vec{r}_i\}, \{\vec{P}_i\}) = \sum_{i=1}^{N} \frac{P_i^2}{2m} + \frac{1}{2} \sum_{i \neq j = 1}^{N} \phi(|\vec{r}_i - \vec{r}_j|) \tag{41}$$

According to (14), we require the free energy F of the system: this is given by

$$F = -kT \log Z \tag{42}$$

where Z is the partition function of the system, defined as

$$Z = \iiint d\vec{P}_1 \cdots \iiint d\vec{P}_N \iiint_V d\vec{r}_1 \cdots \iiint_V d\vec{r}_N \exp\left[-\frac{H(\{\vec{r}_i\},\{\vec{P}_i\})}{kT}\right] \tag{43}$$

The probability distribution of the particles is

$$\Pi(\{\vec{r}_i\},\{\vec{P}_i\}) = \frac{\exp[-H(\{\vec{r}_i\},\{\vec{P}_i\})/kT]}{Z} \tag{44}$$

where $\Pi(\{\vec{r}_i\},\{\vec{P}_i\}) d\vec{r}_1 \cdots d\vec{r}_N \, d\vec{P}_1 \cdots d\vec{P}_N$ is the probability that any chosen instant molecule 1 will lie in the element $d\vec{r}_1$ about $\vec{r}_1$ with a momentum in the range $d\vec{P}_1$ about $\vec{P}_1$, molecule 2 will lie in $d\vec{r}_2$, $d\vec{P}_2$, etc. These formulae apply even when the equilibrium system is inhomogeneous, as is the case here; in particular, the form (41) for the Hamiltonian means that the $\vec{P}_i$ dependence factorizes out in (44), so that the particle velocities $\vec{v}_i = \vec{P}_i/M$ always have the Maxwell-Boltzmann distribution (25).

Using the relation:

$$\int_{-\infty}^{\infty} dp_x \exp\left(-\frac{P_x^2}{2mkT}\right) = \sqrt{2\pi mkT} \tag{45}$$

we can perform all the integrations over $\vec{P}_i$ in (43), and we get

$$Z = (2\pi mkT)^{3N/2} \iiint d\vec{r}_1 \cdots \iiint d\vec{r}_N \exp\left[-\frac{1}{2kT}\sum_{i \neq j=1}^{N}\sum \phi(|\vec{r}_i - \vec{r}_j|)\right] \tag{46}$$

According to (14) and (42) we must differentiate Z with respect to the area A of the interface, keeping V (and T) constant. Referring to Fig.1, we let the container have unit length along the y direction, and let its x dimension be A and its z dimension V/A. Then define new variables x_i' and z_i' by

$$\vec{r}_i \equiv \left(A x_i', y_i, \frac{V}{A} z_i'\right); \quad \iiint d\vec{r}_i = V \int_0^1 dx_i' \int_0^1 dy_i \int_0^1 dz_i' \tag{47}$$

With this transformation, the A dependence of Z occurs only in the arguments of the potential energy ϕ, since

$$|\vec{r}_i - \vec{r}_j| = \left[A^2(x_i' - x_j')^2 + (y_i - y_j)^2 + \frac{V^2}{A^2}(z_i' - z_j')^2\right]^{1/2} \tag{48}$$

so that

$$\left(\frac{\partial}{\partial A}|\vec{r}_i - \vec{r}_j|\right)_V = \frac{A(x_i' - x_j') - \frac{V^2}{A^3}(z_i' - z_j')^2}{|\vec{r}_i - \vec{r}_j|}$$

$$= \frac{(x_i - x_j)^2 - (z_i - z_j)^2}{A|\vec{r}_i - \vec{r}_j|} \tag{49}$$

Use of this result, together with (46), (42) and (14), gives for the surface tension

$$\gamma = -\frac{kT}{Z}\left(\frac{\partial Z}{\partial A}\right)_{V,T}$$

$$= \frac{\frac{1}{2}\iiint d\vec{r}_1 \cdots \iiint d\vec{r}_N \sum\sum_{i \neq j=1}^{N} \frac{\phi'(|\vec{r}_i - \vec{r}_j|)}{|\vec{r}_i - \vec{r}_j|}\left\{\frac{(x_i-x_j)^2 - (z_i-z_j)^2}{A}\right\} e^{-\frac{1}{2}\Sigma\Sigma\phi(|\vec{r}_i-\vec{r}_j|)}}{\iiint d\vec{r}_1 \cdots \iiint d\vec{r}_N\ e^{-\frac{1}{2}\Sigma\Sigma\phi(|\vec{r}_i-\vec{r}_j|)}} \tag{50}$$

To evaluate this expression, we can interchange the order of summation and integration, and realize that all of the N(N-1) terms in the sum are the same, because of the indistinguishability of the molecules. Thus we can write

$$\gamma = \frac{1}{2}\iiint d\vec{r}_1 \iiint d\vec{r}_2\ \frac{\phi'(|\vec{r}_1 - \vec{r}_2|)}{|\vec{r}_1 - \vec{r}_2|}\left\{\frac{(x_1 - x_2)^2 - (z_1 - z_2)^2}{A}\right\}$$

$$\times \left[\frac{N(N-1)\iiint d\vec{r}_3 \cdots \iiint d\vec{r}_N\ \exp\left(-\frac{1}{2}\sum\sum\phi(|\vec{r}_i - \vec{r}_j|)\right)}{\iiint d\vec{r}_1 \cdots \iiint d\vec{r}_N\ \exp\left(-\frac{1}{2}\sum\sum\phi(|\vec{r}_i - \vec{r}_j|)\right)}\right] \tag{51}$$

Now it can be seen from (44) that the quantity in square brackets in (51), multiplied by $d\vec{r}_1 d\vec{r}_2$, is N(N-1) times the probability that particle 1 will be found in $d\vec{r}_1$ and particle 2 in $d\vec{r}_2$, the position and momenta of the other particles being unspecified. However, there are N(N-1) ways of choosing particles which go into $d\vec{r}_1$ and $d\vec{r}_2$, so that the quantity in [ ] is simply the familiar two-particle distribution function $n_2(\vec{r}_1, \vec{r}_2)$, involving particles whose identity is not specified. Thus (51) becomes

$$\gamma = \frac{1}{2}\iiint \frac{d\vec{r}_1}{A} \iiint d\vec{r}_2\ \frac{\phi'(|\vec{r}_1 - \vec{r}_2|)}{|\vec{r}_1 - \vec{r}_2|}\left\{(x_1 - x_2)^2 - (z_1 - z_2)^2\right\} n_2(\vec{r}_1, \vec{r}_2) \tag{52}$$

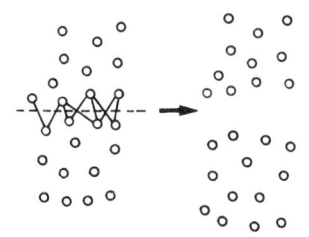

FIG.16. Surface tension is, approximately, work done in breaking 'bonds'.

Now we set $\vec{r}_2 - \vec{r}_1 \equiv \vec{R} \equiv (X,Y,Z)$ and integrate over $\vec{R}$ instead of $\vec{r}_2$. There is no dependence of the integrand on x_1 and y_1, because of the symmetry of the problem, and $\iint dx_1 dy_1$ cancels out the A in the denominator; we are left with

$$\gamma = \frac{1}{2} \int_{-\infty}^{\infty} dz_1 \iiint d\vec{R} \; \frac{\phi'(R)}{R} (X^2 - Z^2) n_2(z_1, \vec{R}) \qquad (53)$$

an expression identical with the result (39) obtained from microscopic mechanics. The regions of validity of the two derivations do not quite coincide, however: (39) holds even in non-equilibrium cases, but the derivation gives no method for calculating $n_2(\vec{r}_A, \vec{r}_B)$; (53) on the other hand, was derived using equilibrium statistical mechanics, but the derivation showed that in equilibrium $n_2(\vec{r}_A, \vec{r}_B)$ is given by the expression in [ ] in (51) — however, this involves N-fold integrations and so is hardly useful.

1.5. Some approximate methods

First we show that it is possible to obtain an order-of-magnitude estimate for γ without invoking any of the sophisticated theory of the last two sections. Imagine a column of liquid with cross-sectional area A. By breaking this column into two, new surface area 2A is created. If the operation is carried out isothermally, then (10) and (16) show that an amount of work $2\gamma A$ will be required. This work (Fig.16) comes from the breaking of molecular bonds in the interface. Most of these bonds are between pairs of molecules in each other's shell of nearest neighbours, where the bond strength is given by the depth ϵ_ℓ of the intermolecular potential $\phi(R)$ (Fig.12). Thus we have

$$\gamma \approx \frac{\epsilon_\ell}{2} \times \text{number of bonds/unit area} \qquad (54)$$

Each molecule is joined to Z_ℓ nearest neighbours, where Z_ℓ is the 'co-ordination number'. Of these Z_ℓ bonds from one molecule, $Z_\ell/2$ are broken during the creation of new liquid. The number of molecules per unit area is roughly the inverse square of the interparticle distance, i.e. about $n_L^{2/3}$. Thus our crude formula for γ is

$$\gamma \approx \frac{\epsilon_\ell Z_\ell n_L^{2/3}}{4} \tag{55}$$

For close-packed 'simple liquids', $Z_\ell \sim 8$. Values of γ calculated using this formula usually lie within a factor of two of the measured values. An order-of-magnitude value of ϵ — for use in emergencies only — is given by

$$\epsilon_\ell \sim kT_c \tag{56}$$

where T_c is the critical temperature. This formula is based on the idea that above T_c, where a close-packed phase cannot exist, the mean thermal energy kT exceeds ϵ_ℓ, all 'bonds' being broken. For all simple fluids (56) overestimates ϵ_ℓ by about 40%.

The simple result (55) takes no account of the effect on γ of (a) temperature, (b) the smooth change of density in the interface, (c) interactions between non-nearest-neighbours, (d) the existence of the vapour, and (e) correlations between molecular positions. Approximate methods taking some account of some or all of these effects fall into two groups. We start with the group of methods based on the exact formula (39). How can we find the two-body distribution function $n_2(\vec{r}_A, \vec{r}_B)$? According to (29), we need to know the density profile n(z), and the pair correlation function $g(z, \vec{R})$. The simplest approximation is to treat the vapour as a vacuum (this will not cause serious error except near the critical point), assume the density falls suddenly from n_L to zero, and assume $g(z, \vec{R})$ to be the liquid pair correlation function $g_L(R)$. Thus this first model postulates that

$$n_2(\vec{r}_A, \vec{r}_B) \approx n_L^2 g_L(R) \quad (z_A < 0 \text{ and } z_B < 0)$$
$$= 0 \text{ otherwise} \tag{57}$$

I leave it as an exercise to show that when used in (39) this leads to

$$\gamma \approx \frac{\pi n_L^2}{8} \int_0^\infty dR \, R^4 \phi'(R) g_L(R) \tag{58}$$

This simple formula was first derived (not from an exact theory) nearly forty years ago.

The result (58) is certainly an improvement on (55). It takes partial account of points (a), (c) and (e) at the beginning of the last paragraph, and can easily be modified to include point (d). It yields results that agree with experiment within about 20%. However, there are two points that must be made in this connection and they apply also to most of the later approximations we shall consider. The first is that $g_L(R)$ is not a microscopic quantity independent of $\phi(R)$; it can in principle be derived from $\phi(R)$ using bulk liquid statistical mechanics. But this is a vast subject and not strictly relevant here because we are concerned with those problems which specifically concern liquid surfaces. Therefore we shall treat $g_L(R)$ (and the corresponding

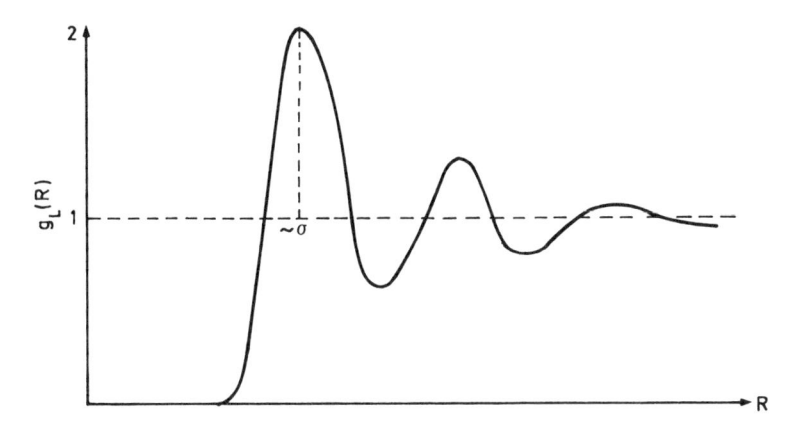

FIG.17. Typical liquid pair correlation function.

vapour correlation function $g_V(R)$) as capable of being measured experimentally by X-ray or neutron scattering, for example. Figure 17 shows the typical oscillatory form of $g_L(R)$ far from the critical point.

The second point is that the integral in (58) is very sensitive to the relative positions of the maximum of $\phi'(R)$ and the first maximum of $g_L(R)$, so that great care must be taken to use 'compatible data', i.e. $g_L(R)$ and $\phi(R)$ must either be measured under exactly the same conditions (temperature, pressure), or $g_L(R)$ must be calculated from $\phi(R)$ by one of the methods of bulk statistical mechanics. It is usually the case that $\phi(R)$ is not in fact measured precisely, but fitted to the Lennard-Jones expression:

$$\phi(R) = \epsilon_\ell \left(\left(\frac{\sigma}{R}\right)^{12} - 2\left(\frac{\sigma}{R}\right)^{6} \right) \qquad (59)$$

where ϵ_ℓ is the depth and σ the radius of the minimum of $\phi(R)$ (Fig.12).

The formula (58), based on (57), violates the equation (40) for hydrostatic equilibrium; in other words, if (57) is used to calculate the normal pressure $P_N(z)$(Eq.(37)), the result is not constant. This is because the simple form (57) does not take account of the smooth variation of the density through the interface. A better approximation for $n_2(\vec{r}_A, \vec{r}_B)$, which does take this into account, is

$$n_2(\vec{r}_A, \vec{r}_B) \approx n(\vec{r}_A) n(\vec{r}_B) g_L(R) \qquad (60)$$

The only error here is to assume that the correlations between molecules in the interface and the vapour are the same as in the liquid. This is not so serious as might at first appear, because: (a) $n_V \ll n_L$, so that the nature of the vapour hardly affects γ (provided $T \ll T_c$); (b) in the vapour, close molecular encounters are rare and, for large R, g_L and g_V both tend to unity; (c) in the interface most of the interactions, which cause the correlations, must occur with one molecule in bulk liquid. The approximation (60) leads to the following formula for γ:

$$\gamma \approx \frac{\pi}{2} \int_0^\infty dR\, g_L(R)\, \phi'(R) \left[- \int_{-R}^{R} dZ (R^2 Z - Z^3) \int_{-\infty}^{\infty} dz\, n(z)\, n(z+Z) \right] \qquad (61)$$

Thus the surface tension can be calculated if the density profile $n(z)$ is known. However, $n(z)$ can be calculated self-consistently from the hydrostatic equation (40), which becomes, on using (60),

$$kT \frac{dn(z)}{dz} = 2\pi n(z) \int_0^\infty dR\, \phi'(R)\, g_L(R) \int_{-R}^{R} dZ\, Zn(z+Z) \qquad (62)$$

The only unknown is $n(z)$. After a little transformation, (62) can be integrated once, to give

$$n(z) = n_L \exp\left[-\frac{4\pi n_L}{3kT} \int_0^\infty dR\, R^3 \phi'(R) g_L(R) \right.$$

$$\left. + \frac{\pi}{kT} \int_{-\infty}^{\infty} dZ\, n(z+Z) \int_{|z|}^{\infty} dR\, \phi'(R)\, g_L(R)\, (R^2 - Z^2) \right] \qquad (63)$$

This exponentially non-linear integral equation has been solved by a form of iteration, but the numerical difficulties are great because the exponentiation tends to cancel the stabilizing effect of the integration. After each iteration, the density profile shifts bodily, but its form is stable after about 40 iterations. What we need is an analytical understanding of equations like (63). However, one result that can be derived from (63) by letting z tend to $+\infty$ is a relation between n_L and n_V. For a given n_L, the value of n_V is several times too large; nevertheless, considerable insight into the nature of liquid-vapour coexistence (especially near T_c) can be obtained in this way. To calculate γ at any given temperature, then, we start with n_L for that temperature, use (63) to calculate n_V and $n(z)$, and then employ (61). The resulting values of γ are accurate to a few per cent, which is comparable with the error introduced by using simple model potentials such as (59).

However, an improvement on (60) is the simplest pair correlation function which changes smoothly from its liquid to its vapour forms:

$$n_2(\vec{r}_A, \vec{r}_B) \approx \frac{n(\vec{r}_A) n(\vec{r}_B)}{n_L - n_V} \left\{ \frac{[n(\vec{r}_A) + n(\vec{r}_B)]}{2} [g_L(R) - g_V(R)] + n_L g_V(R) - n_V g_L(R) \right\} \qquad (64)$$

This can be inserted into the hydrostatic equation (40), to give an equation similar to but more complicated than (63). The change from (60) to (64) is

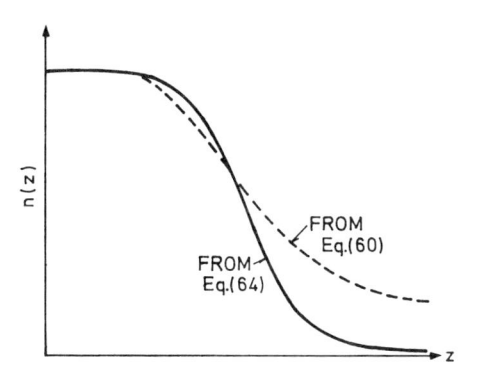

FIG.18. Calculated density profiles (from Toxvaerd (1972 a)).

said to affect the calculated value of γ by only a few parts in 10^5 (Toxvaerd, private communication), but this must surely be fortuitous. The density profiles differ considerably, however (Fig.18); that arising from (64) has the correct limiting values n_L and n_V.

For calculating γ it is probably unwise to use more sophisticated approximations of this type, because of the uncertainties in our knowledge of $\phi(R), g_V(R)$, and $g_L(R)$.

The second major group of approximate methods is not based on the exact microscopic formula (39), but on a different idea. This is quasithermodynamics, in which the notion that pressure can be defined at any point in the interface is generalized to cover extensive thermodynamic functions. These can be expressed as quantities 'per particle'. Thus the volume per particle $v(z)$ is simply

$$v(z) \equiv \frac{1}{n(z)} \qquad (65)$$

Likewise the free energy per particle, $f(z)$, exists; it can be defined rigorously and given exact statistical-mechanical expression in terms of $\phi(R)$ and $n_2(\vec{r}_A, \vec{r}_B)$.

Now, according to (16), the surface tension is simply the free energy associated with unit area of the interface, i.e. the free energy in a column of fluid of unit area extending from bulk liquid into bulk vapour, minus the free energy that the same column would contain if the liquid changed abruptly to vapour at $z = 0$. Thus

$$\gamma = \int_{-\infty}^{0} dz\, (n(z)f(z) - n_L f_L) + \int_{0}^{\infty} dz (n(z)f(z) - n_V f_V) \qquad (66)$$

where f_L and f_V are the free energies per particle in bulk liquid and vapour. Of course the actual column and the idealized discontinuous column must

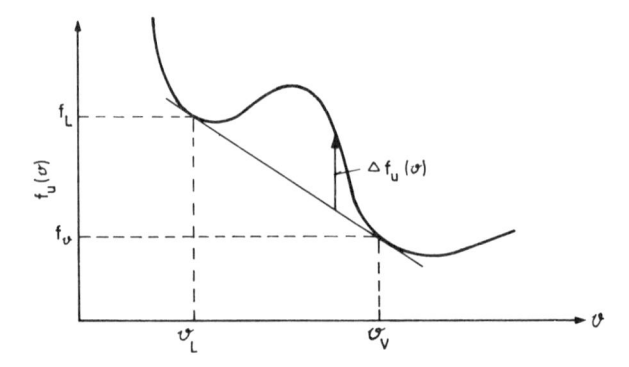

FIG.19. Analytic continuation of free energy between liquid and vapour phases.

contain the same number of particles, so that the location of the 'dividing surface' at z = 0 must satisfy

$$\int_{-\infty}^{0} dz\,(n(z) - n_L) + \int_{0}^{\infty} dz(n(z) - n_V) = 0 \tag{67}$$

After a little algebra, these equations can be shown to imply that

$$\gamma = \int_{-\infty}^{\infty} dz\,n(z)\Delta f(z) \tag{68}$$

where

$$\Delta f(z) \equiv f(z) - \left\{ \frac{f_V\left(\frac{n_L n_V}{n(z)} - n_V\right) + f_L\left(n_L - \frac{n_L n_V}{n(z)}\right)}{n_L - n_V} \right\} \tag{69}$$

In these quasithermodynamic methods, approximate formulae for f(z) are employed, which all involve n(z). Then the principle is used that the system will adopt the density profile that minimizes γ as given by Eq.(68). When the exact formula for f(z) is used, this method is identical to those discussed already: Eq.(68) becomes identical with Eq.(39), and the minimization condition simply gives again the condition (40) for hydrostatic equilibrium. (It is worth noting that an integrated form of (40) is sometimes used; this expresses the constancy of the 'chemical potential' across the surface.) When approximations are used for f(z), however, then the quasithermodynamic method and those based on (39) need not give the same results. Nevertheless, with sophisticated approximations, analogous to (64), the two groups of methods begin to converge.

Of greater interest is a simple approximation for f(z) which leads to a result that cannot obviously be obtained by any method based on (39). This is based on a major difference between f(z) in an inhomogeneous system and the free energies per particle f_L and f_V in the uniform bulk phases, namely that while f_L and f_V depend only on the densities n_L and n_V (and T), $f(z)$ cannot be expressed simply as a function of the local density $n(z)$, but depends on the density at all other levels in the fluid. An obvious approximation is to assume that $f(z)$ differs only slightly from $f_u(n(z))$, the free energy per particle of a hypothetical uniform fluid with density $n(z)$. Of course a uniform fluid with density lying between the coexistent densities n_L and n_V cannot exist, but we can obtain $f_u(n(z))$ by analytic continuation of the free energy $f(n)$ elsewhere on the same isotherm. The key assumption is that the difference between $f(z)$ and $f_u(n(z))$ depends only on $dn/dz\,(z)$, in the following way:

$$f(z) \approx f_u(n(z)) + K \left(\frac{dn(z)}{dz}\right)^2 \tag{70}$$

where K is a constant. Thus (68) becomes

$$\gamma \approx \int_{-\infty}^{\infty} dz \left[\Delta f_u(n(z)) + K \left(\frac{dn(z)}{dz}\right)^2\right] n(z) \tag{71}$$

where Δf_u is simply (69) with $f(z)$ replaced by f_u. (Symmetry forbids a term in (70) proportional to dn/dz, and any term proportional to d^2n/dz^2 would be equivalent to the term in $(dn/dz)^2$ so far as γ is concerned.)

Now (71) is an integral of the type which can be minimized by the methods of the calculus of variations: the unknown function is $n(z)$, and the integrand involves dn/dz. These methods give, for the condition on $n(z)$,

$$\left[\Delta f_u(n(z)) - K\left(\frac{dn(z)}{dz}\right)^2\right] n(z) = \text{const} \tag{72}$$

The 'constant' is zero, because dn/dz and Δf_u both vanish in the bulk phases as $|z| \to \infty$. Therefore, the density profile is determined by

$$\frac{dn(z)}{dz} = \sqrt{\frac{\Delta f_u(n(z))}{K}} \tag{73}$$

and γ can be expressed very simply by transforming (71) into an integral over n:

$$\gamma \approx 2\sqrt{K} \int_{n_V}^{n_L} dn\, n \sqrt{\Delta f_u(n)} \tag{74}$$

Thus to calculate γ we require the bulk function $\Delta f_u(n)$ and a value for the 'coupling constant' K.

Finding K requires another theory, but it is instructive to devise a simple model for $\Delta f_u(n)$. It is simpler to consider Δf_u and f_u as functions of 𝑣 rather than n. Then $f_u(v)$ for a given T must have the form shown in Fig.(19). This follows from basic thermodynamics: We must have

$$P_L = -\frac{\partial f_u(v_L)}{\partial v} = P_V = -\frac{\partial f_u(v_V)}{\partial v} \tag{75}$$

In addition, thermal and mechanical stability of the whole system implies that the Gibbs' function is extremal, i.e. that

$$dG \equiv d(F + PV) = VdP - SdT = 0 \tag{76}$$

If a fraction α is vapour, and $1-\alpha$ is liquid, then

$$G = N\alpha g_V + N(1-\alpha) g_L \tag{77}$$

For this to be stationary at all compositions α, $dG/d\alpha$ must vanish; therefore

$$g_V \equiv Pv_V + f_V = g_L \equiv Pv_L + f_L \tag{78}$$

i.e.

$$P = \frac{f_L - f_V}{v_V - v_L} \tag{79}$$

Taken together with (75), this implies the 'envelope' construction of Fig.(19). As $T \to T_c$, $v_L \to v_V$, and the curve representing Δf_u must become more and more symmetrical between its minima, both as a function of v and of n. Therefore, we may take

$$\Delta f_u(n) \xrightarrow[T \to T_c]{} \beta(n - n_V)^2 (n_L - n)^2 = \beta\left[\left(\frac{n_L - n_V}{2}\right)^2 - \left\{n - \frac{(n_L + n_V)}{2}\right\}^2\right] \tag{80}$$

where β is a constant.

Now we can find γ, using (74); we obtain

$$\gamma \approx \frac{\sqrt{\beta K} n_c (n_L - n_V)^3}{3} \tag{81}$$

For the density profile, (73) gives

$$n(z) \approx \frac{n_L + n_V}{2} - \left(\frac{n_L - n_V}{2}\right) \tanh\left[\frac{z(n_L - n_V)}{2}\sqrt{\frac{\beta}{K}}\right] \tag{82}$$

This profile is physically sensible, because its 'width' is proportional to $(n_L - n_V)^{-1}$, and so diverges as $T \to T_c$ as we would expect. Finally, by assuming that the simple fluid is 'analytic', i.e. that thermodynamic functions can be expanded in Taylor series about the critical point, we see that the dependence of (80) on $(n_L - n_V)^2$ suggests that

$$n_L - n_V \propto (T_c - T)^{1/2} \tag{83}$$

(van der Waal equation of state leads to the same result). Then (81) gives

$$\gamma \propto (T_c - T)^{3/2} \tag{84}$$

in fair agreement with the empirical results (21) and (22). Of course real fluids, even simple ones, are not analytic.

2. LIQUID-VAPOUR INTERFACES NEAR SOLIDS - CONTACT ANGLE

2.1. Exact macroscopic theories

A liquid-vapour interface meets a solid surface at what is, macroscopically, a line of three-phase contact. The angle θ_c (Fig.20) between the two surfaces, measured through the liquid and perpendicular to this contact line, is called the angle of contact. θ_c is the additional quantity needed in the application of Newtonian mechanics to the problem of determining the forms of the surfaces of liquid masses touching solids. For example, the shapes of droplets resting on solid surfaces can be calculated using Laplace's equation (8) to describe the joint action of gravity and surface tension in curving the liquid surface, but θ_c provides the necessary boundary condition at the three-phase line (Fig.21). It is conventional to call the case $\theta_c < \pi/2$ a wetting situation, and $\theta_c > \pi/2$ a non-wetting situation. If $\theta_c \leq 0$, we have a spreading situation; the liquid covers the whole solid, and there is no three-phase line. In practice, θ_c is found to be very sensitive to contamination of the solid surface, and even (if the droplet is moving) to whether the three-phase line is advancing or receding over the solid; this can be seen very clearly on rainswept window-panes (Fig.22). The concept of contact angle is central to a vast number of natural phenomena and technological processes involving wetting and waterproofing.

The contact angle is a basic mechanical and thermodynamic parameter of the three-phase line, and in Section 2.2 we shall show how it can be calculated directly from molecular interactions. We devote the remainder of this section to relating θ_c to other macroscopic parameters. First we adopt a mechanical approach. Consider the statical equilibrium of a mass of liquid, solid and vapour containing unit length of the three-phase line (Fig.20). What forces act? A hydrostatic pressure P_0 acts all round the volume and may be ignored (cf. Fig.8b). Then there are surface tension forces at the intersections of the three interfaces with the volume. The surface tension in the liquid-vapour interface we now call $\gamma_{\ell v}$, that in the solid-liquid interface we call $\gamma_{\ell s}$, and that in the solid-vapour interface we call γ_{sv}. The possibility of the existence of $\gamma_{\ell s}$ and γ_{sv} follows from theoretical arguments similar to those in Sections 1.1 and 1.2. Experimentally, it is difficult to detect effects of surface tensions involving solids, because solid surfaces are not mobile — the bulk static stress tensor can contain shear components giving rigidity to the solid. However, thin wires of gold near its melting point shrink unless a

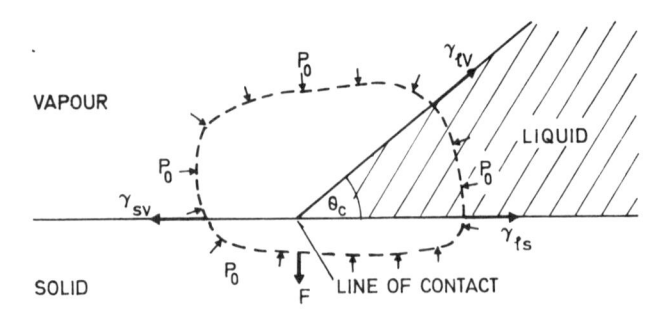

FIG. 20. Forces on matter near line of three-phase contact.

weight is applied (Fig.23), and X-ray diffraction indicates that the lattice spacing in tiny crystals is smaller than in large ones, suggesting that the material is under a high pressure similar in origin to that inside a bubble (Fig.9).

Now we consider the horizontal equilibrium of the forces on the matter shown in Fig.20; we must have

$$\gamma_{\ell v} \cos \theta_c + \gamma_{\ell s} = \gamma_{sv}$$

i.e.

$$\cos \theta_c = \frac{\gamma_{sv} - \gamma_{\ell s}}{\gamma_{\ell v}} \tag{85}$$

This is known as Young's equation, and expresses the three-phase parameter θ_c in terms of parameters characterizing the three surfaces separately. We must also consider the vertical equilibrium; this can only be achieved if there is a force F deep in the solid acting downwards (Fig.20), to counteract the tendency of the liquid surface tension to pull the solid upwards. We must have

$$F = \gamma_{\ell v} \sin \theta_c \tag{86}$$

For the solid to exert the force F, it cannot be perfectly rigid, but must strain a little near the line of contact; thus the perimeter of a droplet should lie on a ridge of solid, but the experimental evidence for this is slight and disputed. (For a liquid resting on a liquid, of course, the lack of rigidity leads to the formation of lens-shaped droplets, as with oil on soup.)

Now we adopt a thermodynamic approach. For the liquid mass to be in equilibrium, θ_c must have a value such that no work would be done in an infinitesimal translation of the line of three-phase contact; we are simply applying the principle of virtual work to the static liquid. Let the line of contact move a distance dx in two steps, as shown in Fig. 24. The first step, from (a) to (b), requires the creation of new liquid-vapour interface with area $dx(1 + \cos \theta_c)$; this requires work dW_{ab}, given by

$$dW_{ab} = \gamma_{\ell v}(1 + \cos \theta_c) dx \tag{87}$$

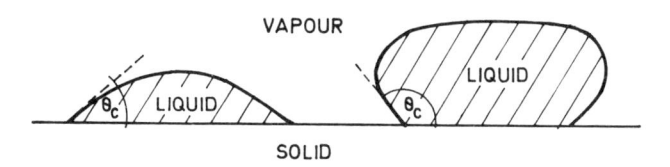

FIG. 21. Acute and obtuse contact angles.

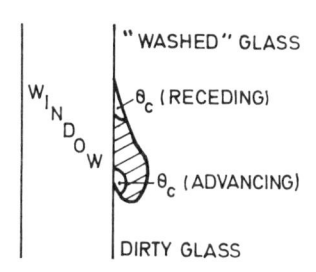

FIG. 22. Receding and advancing contact angles.

The step from (b) to (c) involves the destruction of solid-vapour area dx and liquid-vapour area dx and the creation of solid-liquid area dx, so that the required work dW_{bc} is

$$dW_{bc} = (-\gamma_{sv} - \gamma_{\ell v} + \gamma_{\ell s}) \, dx \tag{88}$$

The total work dW_{ac} is the sum of these quantities, and must be zero, so that

$$0 = \gamma_{\ell v}(1 + \cos \theta_c) - \gamma_{sv} - \gamma_{\ell v} + \gamma_{\ell s} \tag{89}$$

from which (85) follows, and, as expected, the thermodynamic and mechanical approaches yield the same result.

The quantity $-dW_{bc}/dx$ (Eq. 88) is called the work of adhesion, for obvious reasons. If we call it W_a, we have

$$W_a = \gamma_{sv} + \gamma_{\ell v} - \gamma_{\ell s} \tag{90}$$

It is a positive quantity, so that work is released in the process bc. Finally, combining (90) and (89), we obtain the useful relation:

$$1 + \cos \theta_c = 2 \cos^2 \frac{\theta_c}{2} = \frac{W_a}{\gamma_{\ell v}} \tag{91}$$

Because θ_c is so difficult to measure, there do not seem to be any graphs of its variation with temperature (cf. Fig. 11 for $\gamma_{\ell v}$). However, extensive measurements exist for particular solids of the variation of $\cos \theta_c$ with the

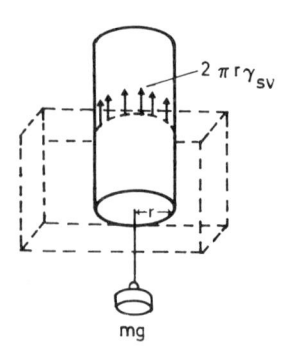

FIG.23. Surface tension in a solid wire equilibrated by a weight (cf. Fig.8).

surface tension of the liquid in contact with it (Fig.25). The resulting 'wettability curves' all have similar forms: $\cos\theta_c$ decreases approximately linearly as $\gamma_{\ell v}$ increases from a value γ_c — the critical surface tension of wetting — which corresponds to $\theta_c = 0$. The limiting tension γ_c is a characteristic of the solid — it is the surface tension of the liquid that just wets the solid. From (85), we have, when $\cos\theta_c = 1$,

$$\gamma_{sv} = \gamma_c + \gamma_{\ell s} \qquad (92)$$

where $\gamma_{\ell s}$ is the interfacial tension between the solid and this 'limiting liquid'. In the next section approximate microscopic arguments will be presented suggesting that in this case $\gamma_{\ell s}$ is zero; in actual cases, therefore, we might expect $\gamma_{\ell s}$ to be very small when $\theta_c = 0$, so that

$$\gamma_{sv} \approx \gamma_c \qquad (93)$$

but we know of no rigorous macroscopic argument that would convert this approximate equality into an exact equation.

2.2. Approximate microscopic theories

The process (Fig.24) used in defining contact angle involves no motion of molecules of the solid, so that θ_c cannot depend on the forces between solid molecules. However, the process does involve motion of the fluid molecules relative to one another and to the solid molecules. Therefore θ_c will depend not only on the potential $\phi(R)$ (Fig.12) between two fluid molecules, but also on the potential $U(R)$ between a fluid and a solid molecule. $U(R)$ is qualitatively similar to $\phi(R)$; in particular, it has an attractive long-range part whose depth we call $\epsilon_{\ell s}$.

The simplest theory for θ_c is based on the idea of 'breaking of bonds', as was the simplest theory for $\gamma_{\ell v}$ which led to Eq.(55). According to (91), we require the work of adhesion W_a, which, looked at in another way, is the work required to pull unit area of liquid away from the solid. By analogy with (54) we may write

$$W_a \approx \epsilon_{\ell s} \times \text{number of liquid-solid bonds/unit area} \qquad (94)$$

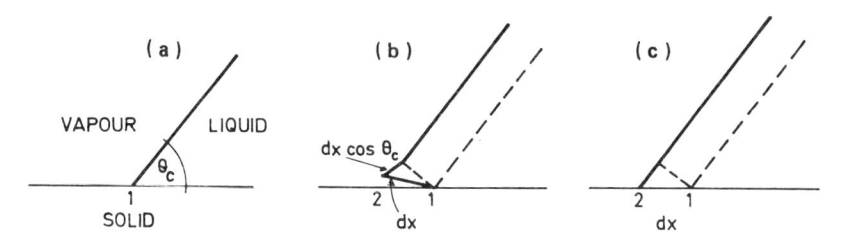

FIG. 24. Infinitesimal translation of contact line.

(there is no factor of $\frac{1}{2}$, because pulling liquid away from solid destroys only one liquid-solid interface, whereas breaking a column of liquid creates two liquid-vapour interfaces). If we now denote the solid co-ordination number and molecular number density by Z_s and n_s, then the number of bonds per unit area is about $\sqrt{Z_\ell Z_s}/2 \times \sqrt{n_s^{2/3} n_L^{2/3}}$, so that (94) becomes

$$W_a \approx \frac{\epsilon_{\ell s}}{2} (Z_\ell Z_s)^{1/2} (n_s n_L)^{1/3} \tag{95}$$

Combining this with (55) and using (91), we get

$$\cos^2 \frac{\theta_c}{2} \approx \frac{\epsilon_{\ell s}}{\epsilon_\ell} \left(\frac{Z_s}{Z_\ell}\right)^{1/2} \left(\frac{n_s}{n_L}\right)^{1/3} \approx \frac{\epsilon_{\ell s}}{\epsilon_\ell} \tag{96}$$

the last member involving approximations that for most liquid-solid systems are unimportant in comparison with those underlying (54) and (94).

This formula for θ_c is crude, admittedly, but it does capture the essential elements of the physics of wetting. If $\epsilon_{\ell s}/\epsilon_\ell > 1$ the solid attracts the liquid more than the liquid attracts itself and $\cos^2 \theta_c/2 > 1$, so that no contact angle exists and the liquid spreads over the solid. If $\frac{1}{2} < \epsilon_{\ell s}/\epsilon_\ell < 1$ then $0 < \theta_c < \pi/2$ and we have the case of wetting. Finally, if $0 < \epsilon_{\ell s}/\epsilon_\ell < \frac{1}{2}$, then $\pi/2 < \theta_c < \pi$ and we have the case of non-wetting, where liquid-liquid attraction considerably exceeds the liquid-solid attraction.

Now we simplify even further, and make a model which explains the qualitative form of the wettability curves of Fig.25. It is probable that $\epsilon_{\ell s}$ will lie somewhere between ϵ_ℓ and the depth ϵ_s of the solid-solid intermolecular potential function, i.e. that $\epsilon_{\ell s}$ is some mean value of ϵ_ℓ and ϵ_s. There is some reasoning, based on the detailed origin of the intermolecular forces, suggesting that it is correct to take the geometric mean, so that

$$\epsilon_{\ell s} \approx (\epsilon_\ell \epsilon_s)^{1/2} \tag{97}$$

and

$$\cos \theta_c \approx 2\sqrt{\frac{\epsilon_s}{\epsilon_\ell}} - 1 \approx 2\sqrt{\frac{\gamma_{sv}}{\gamma_{\ell v}}} - 1 \tag{98}$$

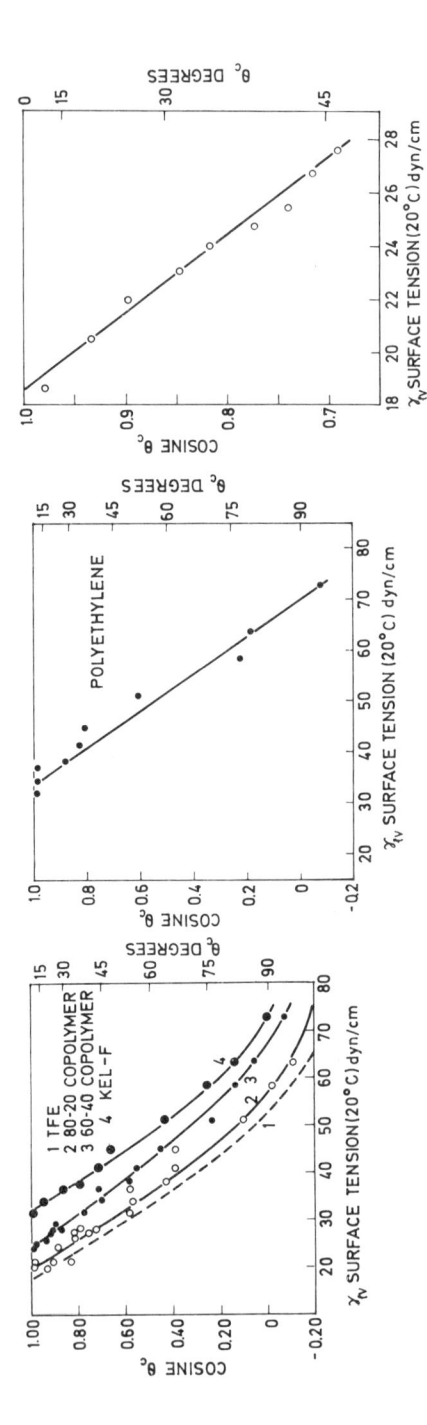

FIG. 25. (a) Wettability of copolymers of polytetrafluoroethylene and polychlorotrifluoroethylene. (b) Wettability of polyethylene; (c) Wettability of polytetrafluoroethylene by the n-alkanes. (From Zisman (1964).)

(In the last member we have used the fact that to this order of approximation γ_{sv} will be given by an approximation analogous to (54), but involving ϵ_s.) This formula does explain Fig.25, at least qualitatively, and moreover predicts that the critical surface tension of wetting γ_c satisfies (93) (any mean value of ϵ_ℓ and ϵ_s yields the same result). For larger values of γ_ℓ, $\cos\theta_c < 1$. However, the formula is quantitatively in error, because it predicts that the ratio of the value of γ_ℓ when $\theta_c = \pi/2$ to the value γ_c should be 4, whereas from Fig.25 this ratio is more like 2.5. Finally, we make a very crude final approximation and assume by analogy with (56) that

$$\epsilon_s \sim kT_m \tag{99}$$

where T_m is the melting point of the solid. This gives

$$\cos^2\frac{\theta_c}{2} \approx \left(\frac{T_m}{T_c}\right)^{1/2} \tag{100}$$

a formula which might give the general trend of θ_c across families of liquids and solids.

Now we turn to a method for calculating θ_c that is, in principle at least, much more firmly based on exact microscopic theory. We consider the statistical mechanics of the fluid molecules interacting via the potential $\phi(R)$ in an external field from all the solid molecules. If we consider the solid molecules as filling the half-space below the plane $z = 0$, then the potential $\psi(z)$ of this external field is

$$\psi(z) = 2\pi n_s \int_z^\infty dR(R-z)U(R) \tag{101}$$

where $U(R)$ is the fluid-solid molecular interaction potential, and we have averaged over the crystal structure of the solid — this is the 'flat smooth solid' approximation. For solids which are not flat or smooth, and may therefore undulate on all scales from the molecular to the macroscopic, and whose molecules may not be spherical, there will still exist a potential $\psi(\vec{r})$ describing the interaction of a single fluid molecule at $\vec{r}$ with the whole solid, but it will not depend simply on z. We require the molecular number density $n(\vec{r})$ of the liquid-vapour system in the presence of the potential $\psi(\vec{r})$ from the solid. The function $n(\vec{r})$ should show how the liquid-vapour interface recedes from the solid, and for large z the surfaces of constant n should be parallel planes making an angle θ_c with the solid surface, where, of course, θ_c is the contact angle.

To find $n(\vec{r})$ we start from an exact equation that is a generalization of Eq.(40):

$$kT\nabla n(\vec{r}) + n(\vec{r})\nabla\psi(\vec{r}) + \iiint d\vec{r}'\, n_2(\vec{r},\vec{r}')\nabla_{\vec{r}}\phi(|\vec{r}-\vec{r}'|) = 0 \tag{102}$$

This expresses the balance of forces on a volume element of a static inhomogeneous fluid in an external potential. However, we cannot find $n(\vec{r})$ from this

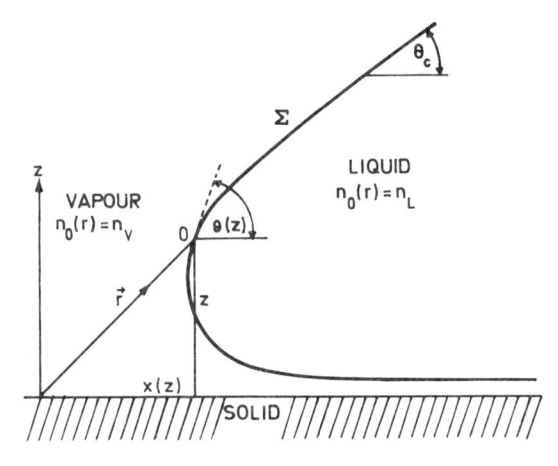

FIG. 26. Discontinuous approximate liquid-vapour interface curving near the solid.

equation, because we do not know the behaviour of $n_2(\vec{r}, \vec{r}')$ near the solid. Therefore we employ again the approximation (60), whereupon (102) gives, after some algebra, the following generalization of (63):

$$n(\vec{r}) \approx n_L \exp\left[-\frac{4\pi n_L}{3kT}\int_0^\infty dR\, R^3 \phi'(R) g_L(R) - \frac{\psi(\vec{r})}{kT} - \frac{1}{kT}\iiint d\vec{R}\, S(R)\, n(\vec{r}+\vec{R})\right] \quad (103)$$

where

$$S(R) \equiv -\int_R^\infty d\rho\, g(\rho)\phi'(\rho) \quad (104)$$

Thus we have a non-linear integral equation in which $n(\vec{r})$ is the only unknown function. Any attempt to solve (103) numerically would probably run into great difficulties, since even the simpler but similar equation (40) is computationally barely tractable. Therefore we proceed analytically and investigate how (103) behaves under iteration.

Any 'trial' solution $n_0(\vec{r})$ can be inserted into the exponent on the right-hand side of (103), and this should yield an improved approximation. Let us take for $n_0(\vec{r})$ a density function (Fig.26) changing discontinuously from n_L to n_V across a 'surface of tension' Σ whose form is initially unknown. After iteration, this will yield a 'smoothed' density function. Clearly the most stable iteration is one where the smoothed $n(\vec{r})$ most closely resembles $n_0(\vec{r})$. To achieve this, we choose the surface Σ in the following way: after iteration Σ must remain a surface on which n is constant. From (103) this implies that

$$\psi(\vec{r}) + \iiint d\vec{R}\, S(R)\, n_0(\vec{r}+\vec{R}) = \text{const for all } \vec{r} \text{ on } \Sigma \quad (105)$$

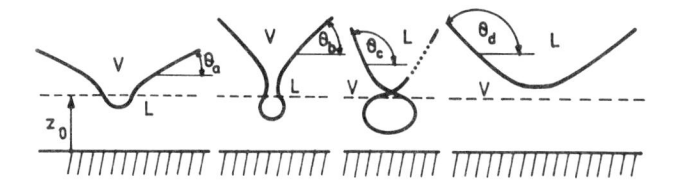

FIG.27. Impossible liquid-vapour interfaces resulting from incorrect choice of contact angle.

We can expand the integral for small R, since S(R) is a localized function (Eq.(104)). A careful analysis shows that only the lowest term is required; this involves the principal radii of curvature $r_1(\vec{r})$ and $r_2(\vec{r})$ of Σ at $\vec{r}$ and (105) becomes eventually

$$(n_L - n_V) \psi(\vec{r}) + \gamma_{\ell v} \left(\frac{1}{r_1(\vec{r})} + \frac{1}{r_2(\vec{r})} \right) = \text{const} \tag{106}$$

where $\gamma_{\ell v}$ is given by the approximation (58). This is a remarkable result, because it can be derived as follows, without using statistical mechanics: for statistical equilibrium, the pressure $P(\vec{r})$ must satisfy

$$\nabla P(\vec{r}) = \text{external force/unit volume of fluid} = -n_0(\vec{r}) \nabla \psi(\vec{r}) \tag{107}$$

If we assume that $P(\vec{r})$ is constant (e.g. atmospheric pressure) far from the solid, this predicts that the liquid pressure $P_\ell(\vec{r})$ and vapour pressure $P_V(\vec{r})$ are different at points $\vec{r}$ on Σ, the excess pressure across Σ being

$$P_{ex}(\vec{r}) = P_L(\vec{r}) - P_V(\vec{r}) = -(n_L - n_V) \psi(\vec{r}) \tag{108}$$

But this excess pressure is equilibrated by the surface tension $\gamma_{\ell v}$ acting via the curvature of Σ according to Laplace's equation (8) which when combined with (108) yields precisely (106).

To solve (106) and obtain a value for θ_c we treat a flat smooth solid, where $\psi(\vec{r}) = \psi(z)$. Let Σ be given by the curve $x(z)$ (Fig.26); then (106) becomes

$$(n_L - n_V) \psi(z) = -\frac{\gamma_{\ell v} x''(z)}{[1 + (x'(z))^2]^{3/2}} \tag{109}$$

This can be integrated once, to give

$$\cos \Theta(z) = \cos \theta_c + \frac{n_L - n_V}{\gamma_{\ell v}} \chi(z) \tag{110}$$

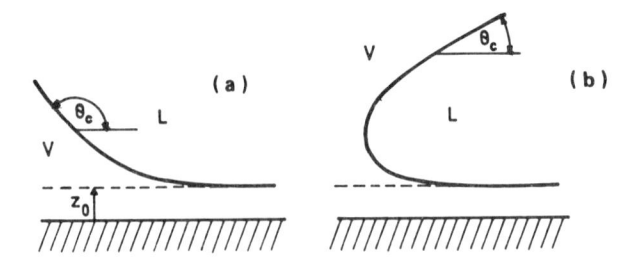

FIG.28. Form of stable liquid-vapour interface for (a) θ_c obtuse, (b) θ_c acute.

where

$$\tan \Theta(z) \equiv \frac{1}{x'(z)}, \quad \Theta(\infty) = \theta_c \tag{111}$$

(see Fig.26), and

$$\chi(z) \equiv \int_z^\infty d\lambda \psi(\lambda) \tag{112}$$

Starting from any given θ_c, Eq.(110) can be integrated in towards the solid to find the stable form of Σ.

So far θ_c is arbitrary. But it is found that all values of θ_c except one yield surfaces Σ that are either impossible or do not correspond to a liquid resting on a solid (Fig.27). The exceptional value of θ_c is of course the required contact angle. Analysis of Eq.(110) shows that it is given by

$$\cos \frac{\theta_c}{2} \approx \frac{-(n_L - n_V) \chi(z_0)}{2\gamma_{\ell v}} \approx \frac{n_s \int_{z_0}^\infty dR \, (R^2 - z_0^2)^2 \, U'(R)}{(n_L - n_V) \int_0^\infty dR \, R^4 g_L(R) \phi'(R)} \tag{113}$$

where z_0 is the height at which $\psi(z)$ vanishes. We can regard (113) as a rather sophisticated version of (96), just as (58) was a sophisticated version of (55). To obtain a similar approximation for the work of adhesion W_a, we use (113) and (91), together with (58), to obtain

$$W_a \approx -\pi n_s (n_L - n_V) \int_{z_0}^\infty dR \, R(R^2 - z_0^2) \, U(R) \tag{114}$$

which is a sophisticated version of (95). The stable forms of surface Σ are shown in Fig.28 for (a) the non-wetting case and (b) the wetting case. At first sight the behaviour near the solid seems peculiar, but it must be remembered that $z = 0$ is the plane containing the centres of the surface solid molecules, so we expect the density $n(\vec{r})$ of fluid atoms to be extremely small for $\vec{z}$ smaller than about the hard-core radius of the solid molecules, and this hard-core radius is roughly at $R = z_0$.

With the approximate equations (113) and (114) for θ_c and W_a, we have reached the end of the formal theoretical development. However, it should be pointed out that our approach leads not only to a formula for θ_c, but also to an understanding of the mechanism by which the liquid-vapour interface recedes from the solid while making a definite angle with its surface; in fact, in this microscopic theory, θ_c appears not as a boundary condition (as in macroscopic theory) but as an asymptotic condition applying at distances (microscopically) far from the solid.

BIBLIOGRAPHY: (ONLY AN OUTLINE, INTENDED TO GIVE AN INTRODUCTION TO THE LITERATURE)

Section 1: General reference: ONO, S., KONDO, S., Handb.Phys.10, Springer Verlag, Berlin (1960).
Experimental data: BUFF, F.P., LOVETT, R.A., in Simple Dense Fluids, Academic Press, New York (1968).
Papers dealing with approximate methods:
BERRY, M.V., DURRANS, R.F., EVANS, R., J.Phys.A 5 (1972)166.
CAHN, J.W., HILLIARD, J.E., J.Chem.Phys. 28 (1958) 258.
JOUANIN, C., C.R.Acad.Sci.Paris 268 (1968) 1597.
TOXVAERD, S., J.Chem.Phys. 55 (1971) 3116.
TOXVAERD, S., J.Chem.Phys. 57 (1972 (a)) 4092.
TOXVAERD, S., Prog.Surf.Sci. 3 (1972 (b)) 189.

Section 2: General references: GOOD, R.J., Adv.Chem. 43 (1964) 74.
ZISMAN, W.A., Adv.Chem. 43 (1964) 1.
ZISMAN, W.A., Wetting, Soc.Chem.Ind., London (1967), a useful collection of articles.
Theory: BERRY, M.V., J.Phys.A 7 (1974) 231.

Part II
PROPERTIES OF SURFACES

ELECTRONS AT METAL SURFACES

S. LUNDQVIST
Institute of Theoretical Physics,
Chalmers University of Technology,
Göteborg,
Sweden

Abstract

ELECTRONS AT METAL SURFACES.
1. Introduction. 2. Elementary aspects. 3. Band structure approach to electrons at metal surfaces. 4. The inhomogeneous electron gas at the metal surface. 5. Screening properties at a metal surface.

1. INTRODUCTION

This paper belongs to the introductory part of the Winter College on Surface Science and is meant to serve as an introduction and to give a background for many of the later papers. We wish to introduce important concepts, ideas and methods and discuss some of the basic physics involved, with no ambition, however, to give a full account of the theory. A certain amount of overlap with material presented in other papers cannot be avoided and may even be useful, particularly for those who study this field for the first time.

We start, in Section 2, with a description of the most basic aspects of the theory, based on the Sommerfeld model of free electrons. This serves as a basis to discuss some key features of the photoelectric effect, thermionic emission and field emission of electrons, and to introduce the concepts of work function, image potential, etc. The properties of the charge distribution near the surface will be discussed in a qualitative manner. The final part of Section 2 deals with the electromagnetic field at a metal surface and serves as an introduction of the surface plasmon and the surface polariton, which are discussed in much more depth in the papers by Professor Celli and others.

The electron states at metal surfaces are then briefly discussed in Section 3. A certain overlap with the paper by Dr. Jones is unavoidable. The elementary aspects of the band structure in one and three dimensions are given and a short discussion is also presented of some typical examples and comparison is made with experiments. The theoretical models are rather crude and the experimental comparisons and verifications very few at the time of writing. However, better theoretical calculations are in progress and experiments on photoemission and field emission will yield more precise experimental information.

Section 4 contains a presentation of the basic theory used to calculate surface properties. Much of the theoretical progress towards understanding the surface electronic structure in a more quantitative way has been made using the so-called density functional theory. A brief presentation of the theory is given. Exchange and correlation play an even more important role near the surface and are discussed; the approximations used are, however, fairly crude. The recent applications to the jellium model of a metal surface by Lang and Kohn are reviewed and a few comments made at the end about somewhat more realistic models where the ions are represented by pseudopotentials.

The last section deals with the response of a metal surface to external perturbations both static and dynamic. Many of these questions would carry us straight into the controversies of present research and a rather heavy mathematical apparatus. We avoid this by looking essentially at long-wavelength properties where a good approximate solution can be obtained for the non-local polarizability of the system. The asymptotic form of the dynamical image force and the long-wavelength contribution to the surface energy are then discussed, and a simple discussion given of the surface density oscillations in connection with a surface plasmon. Some brief remarks about results with more refined models conclude the work.

2. ELEMENTARY ASPECTS

2.1. The Sommerfeld model

The basic quantum properties of conduction electrons in a metal are given by the free-electron model. The complex field an electron feels from the interaction with all the other conduction electrons and with the metal ions is replaced by a constant potential. A potential step V_0 at the surface prevents the electrons from leaving the surface. The one-electron energy levels and wave functions for an electron in a box are easily found. According to the Pauli principle, no more than one electron is permitted to be in a given quantum state. The 'state' is characterized by a quantum label k; in this case the momentum and the spin of the electron. The average occupation number $\bar{n}_k$ in the state k is given by Fermi-Dirac statistics:

$$\bar{n}_k = \frac{1}{\exp[(\epsilon_k - \mu)/k_B T] + 1}$$

The energy eigenvalue is denoted by ϵ_k, k_B is the Boltzmann constant, T the temperature and the parameter μ is the chemical potential. When the temperature $T \to 0$, the distribution becomes the unit step function, $n_k = 1$ for all energy levels up to the Fermi energy $\epsilon_F = \mu$, and $n_k = 0$ for all energies larger than ϵ_F. At finite temperatures the step function varies smoothly around $\epsilon_k = \mu$ as indicated in Fig.1. The width of the transition region is of the order $k_B T$, which in a good metal is a very narrow energy interval compared with the Fermi energy.

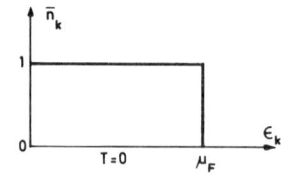

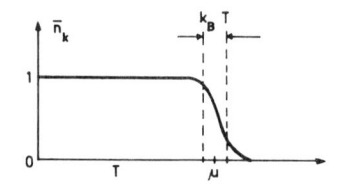

FIG.1. Fermi-Dirac distribution at T = 0 and T > 0.

In the Sommerfeld model there will be no dependence on the crystal or atomic structure. The only additional parameter needed to characterize the system is the density of conduction electrons $\rho = N/V$, where N is the total number of conduction electrons and V is the volume of the box. One usually characterizes the density by means of a dimensionless parameter r_s, being the radius in units of the Bohr radius a_0 of a sphere which an electron on the average occupies in the box; thus

$$\frac{4\pi(r_s a_0)^3}{3} = \frac{V}{N}$$

For metals, the values of r_s fall in the range 2-6.

This simple model for conduction electrons in a metal is sufficient to describe the key features of three important phenomena: the photoelectric effect, the thermionic emission and the field emission of electrons. We shall very briefly review the elements of these phenomena, and this provides the opportunity to introduce some important concepts, which will be returned to later.

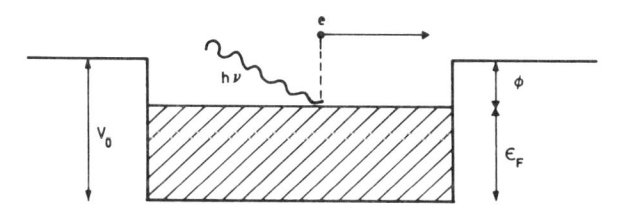

2.2. The photoelectric effect

In this basic experiment a light of frequency ν is shone on the metal and the photoelectric current is measured. One finds that the current is zero up to a limiting frequency ν_0. Above this frequency the photocurrent increases with increasing frequency ν. This phenomenon, the photoelectric effect, was explained by Einstein in 1905.

An electron can only leave the metal and contribute to the photoelectric current if it has absorbed a photon and been excited at least to an energy V_0 above the bottom of the conduction band. In this case we can with good approximation neglect the thermal smearing of n_k and regard the electron distribution as a step function, where all energy states up to ϵ_F are occupied. For an electron to leave the metal, we need at least an excitation energy

$$\phi = V_0 - \epsilon_F \tag{2.1}$$

The condition on ν for obtaining a photoelectric current is therefore $h\nu > \phi$. ϕ is called the work function for the metal (see Fig.2).

In Table I are given some typical values of the Fermi energy and the work function. We shall return later to a discussion of the work function.

FIG.2. The photoelectric effect.

TABLE I. TYPICAL VALUES OF THE
FERMI ENERGY AND WORK FUNCTION

Metal	Valency	ϵ_F (eV)	φ (eV)	r_s
Na	1	3.1	2.3	4.0
K	1	2.1	2.2	4.9
Cu	1	7.0	4.5	2.7
Ag	1	5.5	4.5	3.0
Mg	2	7.1	3.7	2.7
Al	3	11.7	4.2	2.1

2.3. Thermal emission of electrons

Thermionic emission refers to the flux of electrons which evaporate thermally from a heated metal, without applying any external electric field. To calculate the current we need first to calculate the flux of electrons up towards the surface and, second, to calculate the probability that the electron will actually penetrate the surface region.

Let us first consider the flux of electrons towards the surface, which is assumed to be the plane $x = 0$. The number of electrons per unit volume with the momentum in the element d^3p around $\vec{p}$, can be written as

$$dn = \frac{2d^3p}{h^3} \frac{1}{\exp\left[\left(\frac{p^2}{2m} - \mu\right)/k_B T\right] + 1} \tag{2.2}$$

The flux of electrons crossing the yz plane is given by the electron velocity in the x direction times the total number of electrons per unit area in the yz plane. Thus we obtain, using $v_x = p_x/m$:

$$dj = F(p_x) dp_x = \frac{2 p_x dp_x}{mh^3} \int \frac{dp_y\, dp_z}{\exp\left[\left(\frac{p^2}{2m} - \mu\right)/k_B T\right] + 1} \tag{2.3}$$

Introducing the new variable $u^2 = p_y^2 + p_z^2$, we can write $dp_y\, dp_z = 2\pi u\, du$ and obtain

$$F(p_x) dp_x = \frac{4\pi p_x dp_x}{mh^3} \int_0^\infty \frac{u\, du}{\exp\left[\left(\frac{p_x^2 + u^2}{2m} - \mu\right)/k_B T\right] + 1}$$

$$= \frac{4\pi m k_B T}{h^3} \ln\left\{1 + \exp\left[-\left(\frac{p_x^2}{2m} - \mu\right)/k_B T\right]\right\} \frac{p_x\, dp_x}{m} \tag{2.4}$$

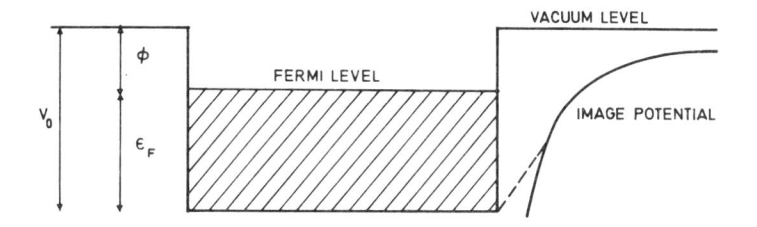

FIG.3. Image potential at a metal surface.

We now introduce the energy corresponding to the motion normal to the surface, $\epsilon_n = p_x^2/2m$, and obtain for the flux function the formula:

$$dj = F(\epsilon_n) d\epsilon_n = \frac{4\pi m k_B T}{h^3} \ln\left[1 + \exp\left(-\frac{\epsilon_n - \mu}{k_B T}\right)\right] d\epsilon_n \qquad (2.5)$$

The simple form for the incident flux function depends on the simple free electron energy formula $\epsilon(p) = p^2/2m$. We next turn to a discussion of the transmission function. To get a rough estimate the potential model described in Fig.3 could be used. However, this model is very crude and we take this opportunity to introduce a new important concept.

The abrupt change in the potential barrier at the metal-vacuum interface is not only an idealized model; it is also physically impossible to realize, because discontinuous changes in the potential imply infinite fields. In fact, the potential changes smoothly as a result of the image force. To understand the meaning of the image force, consider an electron approaching a plane metal surface. The electron will polarize the metal surface, which in turn exerts an attractive force K_i, the image force, on the electron with the magnitude $e^2/4x^2$. The corresponding potential V_i of the electron, measured with respect to the vacuum level, when it is at a distance x from the metal surface, is

$$V_i = \int_\infty^x K_i \, dx = -e^2/4x \qquad (2.6)$$

Thus the height of the potential step for $x > 0$ is no longer the constant value ϕ, but is rather a function of the distance x from the metal surface:

$$\phi(x) = \phi + V_i(x) \qquad (2.7)$$

However, Eq.(2.7) implies that the potential energy equals $-\infty$ at the metal surface. This anomaly is often circumvented by assuming that the image potential holds only for distances larger than some critical value of the order ~ 1 Å. For smaller distances one often assumes that it decreases linearly such that it coincides with the energy at the bottom of the band at $x = 0$.

The formula for the current density of electrons leaving the surface of the metal is

$$j = e \int F(\epsilon_n) D(\epsilon_n) d\epsilon_n \qquad (2.8)$$

where $D(\epsilon_n)$ is the probability that the electrons will penetrate the surface barriers. $D(\epsilon)$ is called the transmission function or barrier transmission coefficient.

The transmission coefficient can be related to the reflection coefficient for electrons coming from the outside and reflected from the metal. For elastic processes we have that $D(\epsilon) = 1 - R(\epsilon)$, where $R(\epsilon)$ is the reflection coefficient. This result follows from the quantum-mechanical theorem which states that the transmission coefficients for any potential barrier are necessarily the same for transmission in either direction.

Because of the infinite range of the image potential, only electrons with energies larger than V_0 (counted from the bottom of the band) can escape and thus we have

$$D(\epsilon) = \begin{cases} 0 & \epsilon < V_0 \\ 1 - R & \epsilon \geq V_0 = \mu + \phi \end{cases} \qquad (2.9)$$

For the total current we now obtain (treating R as constant)

$$j = \frac{4\pi e m k_B T}{h^3} (1 - R) \int_{\mu+\phi}^{\infty} \ln\left[1 + \exp\left(-\frac{\epsilon_n - \mu}{k_B T}\right)\right] d\epsilon_n \qquad (2.10)$$

Because $\epsilon_n - \mu \gg k_B T$ in the entire region of integration, we can write

$$\ln\left[1 + \exp\left(-\frac{\epsilon_n - \mu}{k_B T}\right)\right] \simeq \exp\left(-\frac{\epsilon_n - \mu}{k_B T}\right)$$

and obtain

$$j = \frac{4\pi e m k_B T}{h^3} (1 - R) \int_{\mu+\phi}^{\infty} \exp\left(-\frac{\epsilon_n - \mu}{k_B T}\right) d\epsilon_n$$

$$= \frac{4\pi m (k_B T)^2 e}{h^3} (1 - R) \exp\left(-\frac{\phi}{k_B T}\right) \qquad (2.11a)$$

or

$$j = A T^2 \exp\left(-\frac{\phi}{k_B T}\right)$$

$$A = \frac{4\pi m e k_B^2}{h^3} \qquad (2.11b)$$

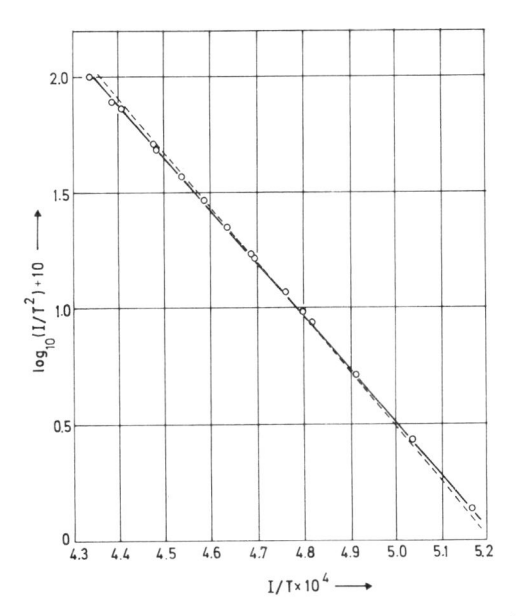

FIG. 4. Thermionic emission from tungsten. The data correspond to $\varphi = 4.5$ eV and $R = \frac{1}{2}$ (continuous line) and $R = 0$ (broken line). (From Fowler and Guggenheim [1].)

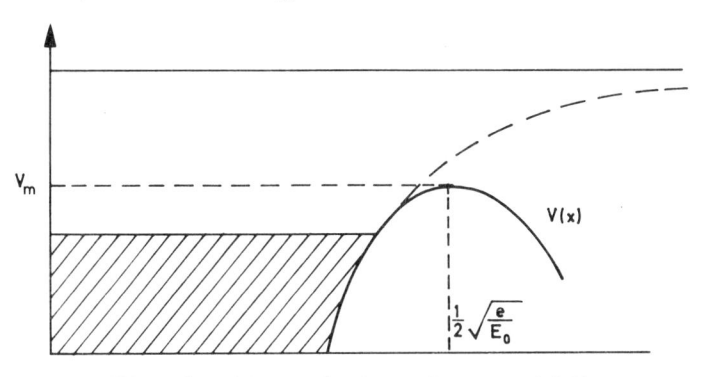

FIG. 5. Potential energy for electrons in an external field.

This is the Richardson-Dushman formula for thermal emission. The temperature dependence of this formula is in quite good agreement and it can be used to obtain the work function ϕ from experimental data. The results for tungsten are shown in Fig. 4.

2.4. Schottky emission

Let us now apply an electric field E_0 in addition to the temperature effect on the electron emission. In this case the surface barrier indicated in Fig. 5 will change and the image potential must now be replaced by

$$V(x) = -e^2/4x - eE_0 x \qquad (2.12)$$

The main effect will be to lower the potential barrier and thereby increase the thermal current of electrons. We seek the maximum of the potential and obtain

$$\frac{dV}{dx} = 0 = \frac{e^2}{4x_m^2} - eE_0$$

which gives

$$x_m = \frac{1}{2}\sqrt{\frac{e}{E_0}}$$

and

$$V_m = -e\sqrt{eE_0}$$

In comparison with the case of no field we see that the limit of integration must be changed to $\mu + \phi - e(eE_0)^{1/2}$ because of the lowering of the barrier. As a result, the electric field will give a strong increase in the yield and we obtain the formula

$$j = AT^2 \exp\left(-\frac{\phi - e\sqrt{eE_0}}{k_B T}\right) \tag{2.13}$$

showing that we get a substantial increase in the thermal current of electrons by applying an electric field.

2.5. Field emission from cold metals

In the presence of an electric field the earlier statement that electrons below the vacuum level are actually never free from the metal is no longer true. The barrier is of finite extension in the presence of the electric field and the electrons will be able to penetrate the barrier using the quantum-mechanical tunnel effect. As the external field E_0 is increased, the difference between V_m and μ decreases and the tunnel effect increases the flux of electrons. When the electric field becomes of the order of 10^7 V/cm the tunnel current becomes so large that the resulting current emission becomes independent of temperature. This phenomenon, known as cold emission, was first observed by Lilienfeld in 1922. Fowler and Nordheim [2] explained the phenomenon using a theoretical model sketched in Fig. 6.

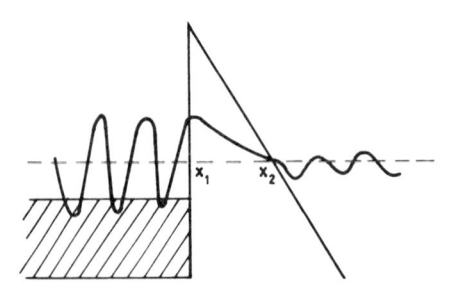

FIG.6. Fowler-Nordheim tunnelling.

Fowler and Nordheim considered a one-dimensional model for the tunnelling through the triangular barrier shown in Fig. 6. For a triangular barrier the transmission coefficient can be determined by wave function matching. To gain qualitative insight into the phenomenon, however, it is simpler to use the WKB method. The WKB transmission function for an electron passing the barrier from x_1 to x_2 will be (except for a prefactor of order unity)

$$D_{WKB}(\epsilon) \simeq \exp\left(-2 \int_{x_1}^{x_2} k(x)\, dx\right) \tag{2.14}$$

with

$$k(x) = \sqrt{\frac{2m}{\hbar^2} [V(x) - \epsilon]}$$

For the triangular barrier in field emission we have, choosing $x = 0$ at the potential step,

$$V(x) = \mu + \phi - eEx$$

and

$$x_1 = 0, \qquad x_2 = \frac{\mu + \phi - \epsilon}{eE}$$

The integral in Eq. (2.14) can be performed exactly and gives

$$D_{WKB}(\epsilon) = \exp\left[-\frac{4}{3eE}\left(\frac{2m}{\hbar^2}\right)^{1/2} (\mu + \phi - \epsilon)^{3/2}\right]$$

This procedure yields the well-known Fowler-Nordheim formula:

$$j = AE^2 \exp\left[-\frac{4(2m)^{1/2} \phi^{3/2}}{3\hbar E}\right] \tag{2.15a}$$

Typical values for the applied field are in the range $10^7 - 10^8$ V/cm and this gives a current density in the range $10^2 - 10^3$ A/cm^2.

As in the case of thermal emission, it is important to consider the influence of image forces. This will round off the surface potential and lead to a lowering of the effective work function, and consequently increase the tunnelling current.

Even when the image potential is taken into account, a result will be obtained of the form:

$$j = AE^2 \exp\left(-\frac{B\phi^{3/2}}{E}\right) \tag{2.15b}$$

but where A and B are no longer constants but slowly varying functions of $(e^3 E^{1/2})/\phi$. Because of the slow variation of A and B, the work function can be determined by plotting j/E^2 versus $1/E$ on semilog paper, which gives a straight line proportional to $\phi^{3/2}$. The temperature corrections to the Fowler-Nordheim equation are small at room temperature, giving a small increase in the current typically of the order a few per cent.

2.6. Energy distribution of field-emitted electrons

The field emission technique is a very important tool for studying the properties of electron states in the metal. We shall only present an elementary introduction to the subject and refer to the recent excellent review by Gadzuk and Plummer [3] for a detailed treatment.

To get a strong tunnel current a very large electrostatic field is needed, typically of the order 3×10^7 V/cm. To obtain such high fields the emitter (cathode) is usually prepared in the form of a very sharp point with a radius of the order 1000 Å. Then a voltage of the order of a few thousand volts applied to an anode will produce a sufficiently strong field at the emitter surface. The development of the field emission microscope by Müller has been essential for the experimental development. With an almost hemispherical tip he could project a greatly enlarged image onto a fluorescent screen of the electrons tunnelling from the emitter. There are three different sorts of data that can be obtained: (a) the total current, (b) micrographs, and (c) the energy distribution of electrons. The last is the one which gives information about electron states.

In the calculation of the tunnelling current in the previous section, as well as in the thermionic emission, we considered only the energy corresponding to the motion of electrons normal to the surface, which is an appropriate procedure for a planar surface. Because of the geometry of the field-emission microscope, however, one will not observe the energy distribution for electrons emitted in the normal direction only, but will observe the total energy distribution. Therefore the calculation of the current will have to be modified. When calculating the incident flux function we cannot, as before, integrate away the yz components of the motion, but must rather consider the flux of electrons having total energy ϵ and normal energy ϵ_n. For the electron gas model this is given as a product of the Fermi function $n(\epsilon)$, the group velocity $v_x = h^{-1} (\partial \epsilon / \partial k)$ and a density of states $\rho(\epsilon)$; thus,

$$F(\epsilon_1 \epsilon_n) = 2 e \overline{n}(\epsilon) v_x \rho(\epsilon) \tag{2.16}$$

The barrier transmission function $D(\epsilon_n)$ for the electron gas model is taken from the preceding section. We obtain the total current of electrons with energy ϵ by integrating over all normal energies ϵ_n consistent with ϵ; thus,

$$j_0' = \frac{dj_0}{d\epsilon} = \int_0^\epsilon F(\epsilon_1 \epsilon_n) D(\epsilon_n) d\epsilon_n \tag{2.17}$$

The label j_0' indicates that we are using the electron gas model.

Let us consider the low-temperature limit of Eq. (2.17). The Fermi function will cut off the energy distribution at the Fermi level ϵ_F and the exponential character of the transmission function will give an exponential decrease in the current with increasing distance from the Fermi level,

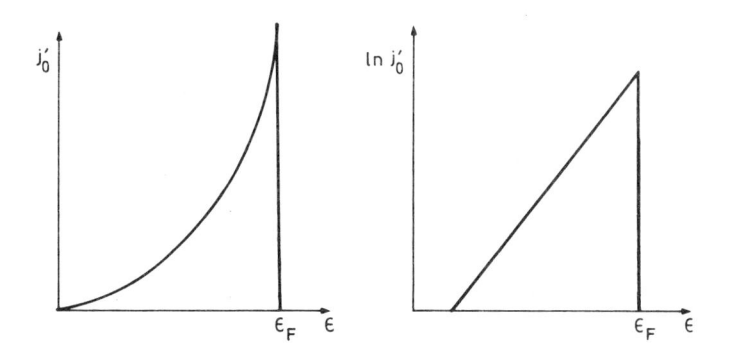

FIG.7. Schematic plot of the distribution of field-emitted electrons for an electron gas.

reflecting the increasing thickness of the tunnel barrier with decreasing energy. The form of the energy distribution is illustrated in Fig. 7. The logarithm of the current is usually plotted as a function of the energy.

The plot in Fig. 7 serves as a reference. The observed current j does not give a straight line for ln j and often also shows a high-energy tail above the Fermi level. One usually plots the radius $R(\epsilon) = j'(\epsilon)/j_0'(\epsilon)$, which gives an indication of the deviation from free electron behaviour because of band-structure effects, surface states, adsorbates or many-body effects. When band-structure effects are taken into account one needs to generalize the formula (2.17) for the current density. When calculating the transmission functions, we must now ask which dynamical quantities are conserved in the tunnelling process. If we assume that the tunnelling is elastic and that the surface is perfectly smooth, we should expect that energy ϵ and the transverse momenta $k_t = (k_y, k_z)$ are conserved, so that the transmission function will be of the form $D(\epsilon_1 k_t)$.

We can now proceed to calculate the number of electrons emitted with total energies in the interval $\epsilon \to \epsilon + d\epsilon$ and obtain

$$dj = \frac{dj}{d\epsilon} \, d\epsilon = 2e\bar{n}(\epsilon) \int \frac{d^3k}{(2\pi)^3} \frac{1}{\hbar} \frac{\partial \epsilon}{\partial k_x} D(\epsilon_1 \vec{k}_z) \qquad (2.18)$$

The integration should be taken in the volume between the energy surfaces ϵ and $\epsilon + d\epsilon$ in k space. The transformed volume element will be

$$\int d^3k \to \int \frac{d\epsilon \, dS}{|\nabla_k \epsilon|}$$

where the surface integral extends over the surface ϵ = constant. Thus we obtain

$$j'(\epsilon) = \frac{dj}{d\epsilon} = \frac{2e\bar{n}(\epsilon)}{\hbar(2\pi)^3} \int D(\epsilon_1 \vec{k}_t) \frac{\partial \epsilon/\partial k_x}{|\nabla_k \epsilon|} \, dS \qquad (2.19)$$

which is the appropriate generalization of the free-electron formula. If one wants to include many-body interactions, an even more complicated formula is obtained where the group velocity in Eq. (2.18) has to be replaced by a quantity derived from the single-electron spectral function.

2.7. Electron density at the surface

We shall discuss later, in more detail, the electron structure at the surface and, in particular, properties which require a knowledge of the electron distribution at the surface, such as the work function and the surface energy. In this introductory section we make only a few simple remarks. We consider the jellium model, i.e. the uniform electron gas where the actual field of the ions is replaced by that from a uniform positive background with a step-function behaviour at the surface (Fig. 8).

The electron wave functions have an exponential tail at the surface, which gives rise to a region of non-uniform density in the surface layer. The thickness of the surface layer is only of the order of 1 Å. In the surface layer the background charge is not compensated and we obtain a dipolar layer at the surface. The dipole layer in turn gives rise to a potential step V_{dip} which is an important contribution to the work function ϕ. The total potential felt by an electron is denoted by V_0. In Fig. 9 the behaviour of V_{dip} and V_0 close to the surface is shown schematically.

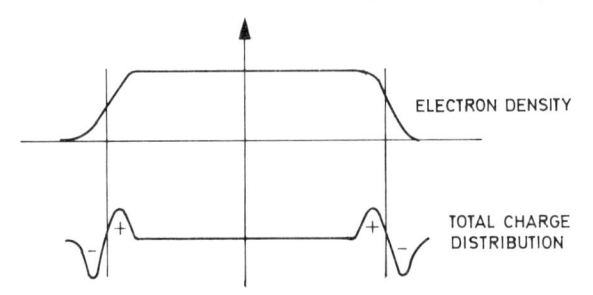

FIG. 8. Electron density and total charge distribution.

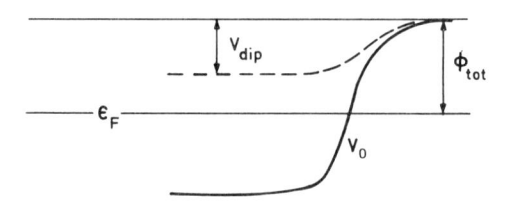

FIG. 9. Total barrier V_0 and dipolar part V_{dip}.

The major part of the surface potential comes from the exchange and correlation effects, which will be dealt with in detail later. The exchange effects arise from the Pauli principle, according to which electrons with parallel spins tend to stay apart from each other. Similar behaviour also holds for electrons with antiparallel spins because of their Coulomb repulsion. We say that an electron in an electron gas is surrounded by an exchange-correlation hole. The interaction with this hole gives a substantial contribution to V_0. When the electron leaves the metal, the hole will tend asymptotically into the polarization charge corresponding to the image potential. Thus the classical analogue to the exchange-correlation potential is the image potential.

Table II gives a few typical values of the work function ϕ, the barrier height V_0 and the dipolar part V_{dip}. It is a well-known fact that the work function is different for different orientations of the surface. This indicates that the atomic structure cannot be neglected. If we consider the surface (111), (100) and (110) of Ni (fcc), they are taken in order of increasing unevenness. On the (110) surface every second row of atoms is elevated. The positive ions now form an irregular background, as indicated in Fig. 10. The electrons will not exactly follow the positive background. This results in an additional contribution to the dipolar layer of opposite polarity and reduces the contribution to the work function. Thus the work function is smaller, as is shown in Table III.

TABLE II. TYPICAL VALUES OF ϕ, V_0 AND V_{dip}

Metal	ϕ (eV)	V_0 (eV)	V_{dip} (eV)
Na	2.35	4.85	0.79
Cu	4.4	11.4	2.91
Ca	2.80	5.9	1.55
Zn	4.24	15.2	4.45
Al	4.25	16.05	6.00
Pb	4.0	13.8	4.43
W	4.5	23.0	11.4

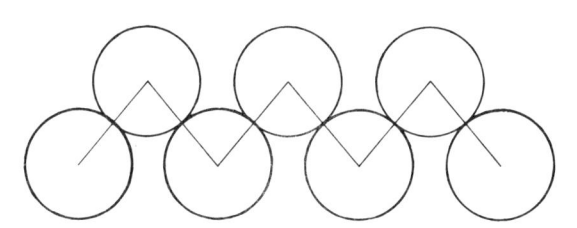

FIG. 10. Dipolar layer from surface ions. There will be an excess positive charge associated with the upper layer and a negative charge associated with the lower layer.

TABLE III. WORK FUNCTION ϕ FOR Ni AND W

Element	(100)	(110)	(111)
Ni (fcc)	4.75	4.2	4.68
W (bcc)	4.82	5.85	4.41

2.8. Collective electron modes: surface plasmons

We recall the motion of plasma oscillations in bulk using the jellium model. If we create a disturbance in the charge distribution in the system, an electric field will be set up, which will accelerate the electrons and cause a current to flow, which tends to eliminate the charge disturbance. Using the three classical equations: the Poisson equation, Newton's equation and the equation of continuity, an oscillation is found with the frequency

$$\omega_p = \sqrt{\frac{4\pi\rho e^2}{m}} \tag{2.20}$$

The quantized plasma oscillations are called plasmons. For typical metal densities, corresponding excitation energies $\hbar\omega_p$ fall into the range 3-20 eV.

We can, alternatively, look at the plasmons as a dielectric resonance. From the relation $D(\omega) = \epsilon(\omega) E(\omega)$, one finds that the condition for a self-oscillation is that an oscillation of the E field can be sustained even in the absence of an external driving field, i.e. $D(\omega) = 0$, and this gives the condition $\epsilon(\omega) = 0$. For the high-frequency oscillations of an electron gas at infinite wavelength we have that $\epsilon(\omega) = 1 - (\omega_p^2/\omega^2)$.

Similar simple arguments can be given for the long-wavelength surface plasmons [4]. We wish to calculate at what frequency will a charge fluctuation of infinite wavelength give rise to an infinite E field at the surface, i.e. we can have a self-oscillation in the E field. Let us consider the interface between two media having dielectric constants ϵ_1 and ϵ_2:

$$D_1 = \epsilon_1 E_1, \quad D_2 = \epsilon_2 E_2$$

Poisson's equation $\nabla D = 4\pi\rho$ gives

$$\epsilon_1 E_1 - \epsilon_2 E_2 = 4\pi\rho_s$$

where ρ_s is the surface charge density. If we have no external sources for the fields we have that $E_1 = -E_2 = E$ and we get

$$(\epsilon_1 + \epsilon_2)E = 4\pi\rho_s$$

Thus we can have a finite E for zero ρ_s only if

$$\epsilon_1 + \epsilon_2 = 0 \tag{2.21}$$

which is the condition for a surface oscillation. For the metal-vacuum interface $\epsilon_1 = 1 - (\omega_p^2/\omega^2)$ and $\epsilon_2 = 1$ and we obtain

$$\epsilon(\omega) + 1 = 0$$

$$1 - (\omega_p^2/\omega^2) + 1 = 0$$

which gives

$$\omega_s = \omega_p/\sqrt{2} \tag{2.22}$$

Thus the surface plasmons have a lower frequency than the bulk plasmons and are easily distinguished from the bulk oscillations.

It seems appropriate in this introductory paper to discuss in somewhat more detail the classical picture of surface modes. Some more refined details such as the dispersion and damping of surface plasmons require an advanced quantum-mechanical theory and will be treated later. However, the basic properties may be derived using classical arguments as was done by Ritchie [5, 6]. We refer to lectures by Mahan [7] for a more detailed discussion of the theory.

We start from Maxwell's equations:

$$\nabla \cdot \vec{D} = 4\pi\rho \tag{2.23}$$

$$\nabla \cdot \vec{B} = 0 \tag{2.24}$$

$$\nabla \times \vec{E} = -\frac{1}{c}\frac{\partial \vec{B}}{\partial t} \tag{2.25}$$

$$\nabla \times \vec{H} = \frac{4\pi \vec{J}}{c} + \frac{1}{c}\frac{\partial \vec{D}}{\partial t} \tag{2.26}$$

The surface modes generate electromagnetic fields at the surface. The fields are determined inside and outside the surface and the eigenvalue condition for the modes are obtained by matching the fields at the surface. Plasmons are non-magnetic modes so we set the magnetic permeability $\mu = 1$. We also assume that there is no charge density or current in the problem so that $\rho = 0$ and $J = 0$. The D and E fields are connected through the formula $D = \epsilon E$ and we shall assume that the dielectric function $\epsilon(\omega)$ depends only on the frequency. A more general theory should include a wave-number dependence as well; thus $\epsilon = \epsilon(q\omega)$. In the case of surface plasmons in metal, the neglect of the wave-number dependence is not important for the range of momenta to be considered. The dispersion of the surface plasmon having its origin in the quantum-mechanical treatment of the surface electrons sets in at wavelengths much shorter than the waves we shall study in the classical theory.

The modes are assumed to have the time dependence $e^{-i\omega t}$, where ω is the frequency to be determined. We first eliminate $\vec{H}$ and obtain

$$\nabla \times \nabla \times \vec{E} - \epsilon(\omega)\frac{\omega^2}{c^2}\vec{E} = 0 \tag{2.27}$$

Let us first consider bulk modes of the form $e^{i\vec{q}\cdot\vec{r}}$, where $\vec{q}$ is the wave vector:

$$\vec{q} \times (\vec{q} \times \vec{E}) + \frac{\omega^2}{c^2}\epsilon(\omega)\vec{E} = 0 \tag{2.28}$$

For longitudinal bulk modes $\vec{q}$ is parallel to $\vec{E}$ (or $\vec{q} \times \vec{E} = 0$), which gives the condition:

$$\epsilon(\omega) = 0 \tag{2.29}$$

which is the condition obtained earlier by a simple argument. For the
transverse modes we obtain the eigenvalue condition:

$$\epsilon(\omega) = \frac{q^2 c^2}{\omega^2} \tag{2.30}$$

The surface modes are obtained by solving Eq. (2.27) in the presence
of the surface. Let us assume a flat infinite surface between the solid
and the vacuum (Fig. 11). We consider a surface wave of the form e^{ikx}
moving in the x direction. All derivatives with respect to y will vanish.
From Eq. (2.28) we obtain

$$\left[\frac{\partial^2}{\partial z^2} - \gamma^2 \right] \vec{E} = 0$$

$$\gamma = \left(\frac{k^2 - \omega^2 \epsilon(\omega)}{c^2} \right)^{1/2} \tag{2.31}$$

which has the solution:

$$\vec{E} = \vec{E}_0 \, e^{-\gamma|z|} \tag{2.32}$$

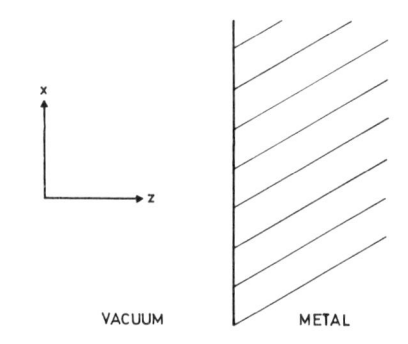

FIG.11. Interface between vacuum (z < 0) and solid (z > 0).

The $\vec{H}$ field also obeys Eq. (2.31) and therefore we also have

$$\vec{H} = \vec{H}_0 \, e^{-\gamma|z|} \tag{2.33}$$

We now have to match the fields inside the surface to those outside. To do
this we have to write out the Cartesian component of the Maxwell equations
(2.23) to (2.26) and obtain

$$\nabla \cdot \vec{E} = 0 \qquad \frac{\partial}{\partial z} E_z + ik E_x = 0 \tag{2.34}$$

$$\nabla \cdot \vec{H} = 0 \qquad \frac{\partial}{\partial z} H_z + ik H_x = 0 \tag{2.35}$$

$$\nabla \times \vec{E} = -\frac{1}{c}\frac{\partial \vec{H}}{\partial t} \quad \begin{cases} -\dfrac{\partial}{\partial z} E_y = \dfrac{i\omega}{c} H_x \\ \dfrac{\partial}{\partial z} E_x - ik E_z = \dfrac{i\omega}{c} H_y \\ ik E_y = +\dfrac{i\omega}{c} H_z \end{cases} \quad (2.36)$$

$$\nabla \times \vec{H} = \frac{1}{c}\frac{\partial \vec{E}}{\partial t} \quad \begin{cases} -\dfrac{\partial}{\partial z} H_y = -\dfrac{i\omega}{c} \epsilon E_x \\ \dfrac{\partial}{\partial z} H_x - ik H_z = -\dfrac{i\omega}{c} \epsilon E_y \\ ik H_y = -\dfrac{i\omega}{c} \epsilon E_z \end{cases} \quad (2.37)$$

These equations divide into two sets. The first set relates E_x, E_z and H_y. These modes have a non-vanishing field along the wave vector which gives the longitudinal character typical of plasmons and are the modes we are looking for. Technically they are the transverse magnetic modes. The remaining equations relate H_x, H_z and E_y, and describe transverse electric modes. There are no surface waves associated with these equations.

Inside the solid we have the dielectric function $\epsilon(\omega)$, and the exponential factor for the interior solution is

$$\gamma_i = \left[k^2 - \frac{\omega^2 \epsilon(\omega)}{c^2} \right]^{1/2}$$

In the vacuum we have $\epsilon = 1$ and the outside solution has the exponential factor:

$$\gamma_0 = \left[k^2 - \frac{\omega^2}{c^2} \right]^{1/2}$$

For the z component of E we obtain

$$\begin{aligned} E_z &= E_z^i e^{-\gamma_i z} e^{ikx} & z > 0 \\ E_z &= E_z^0 e^{\gamma_0 z} e^{ikz} & z < 0 \end{aligned} \quad (2.38)$$

The component E_x is obtained from Eq. (2.34):

$$E_x = \frac{i}{k}\frac{\partial}{\partial z} E_z = \begin{cases} -\dfrac{i}{k} \gamma_i E_z^i e^{-\gamma_i z} e^{ikx} & z > 0 \\ \dfrac{i}{k} \gamma_0 E_z^0 e^{\gamma_0 z} e^{ikx} & z < 0 \end{cases}$$

Matching E_x at the surface, $z = 0$ gives

$$\gamma_i E_z^i = -\gamma_0 E_z^0 \qquad (2.39)$$

The H_y component follows from Eq.(2.37) (the last component):

$$H_y = -\left(\frac{\omega}{kc}\right)\epsilon E_z = \begin{cases} \frac{\omega}{kc}\epsilon E_z^i e^{-\gamma_i z} e^{ikx} & z > 0 \\ \frac{\omega}{kc} E_z^0 e^{\gamma_0 z} e^{ikx} & z < 0 \end{cases}$$

and the matching condition gives

$$\epsilon E_z^i = E_z^0 \qquad (2.40)$$

which is seen to be the same as requiring the normal component of D to be continuous at the surface.

The eigenvalue equation for surface modes follows from Eqs (2.39) and (2.40):

$$\frac{\gamma_i}{\epsilon} = -\gamma_0 \qquad (2.41)$$

After squaring Eq.(2.40) and inserting the expressions for γ_i and γ_0 one obtains, after re-arranging the terms, the formula in standard form:

$$\left(\frac{ck}{\omega}\right)^2 = \frac{\epsilon(\omega)}{1 + \epsilon(\omega)} \qquad (2.42)$$

The dispersion relation $\omega(k)$ can be obtained by inverting the relation $k(\omega)$ implied by Eq.(2.42).

One important remark follows immediately from Eqs (2.41) and (2.42). Both γ_i and γ_0 are positive by definition and therefore surface plasmons can only exist at frequencies where $\epsilon(\omega) < 0$. Then Eq.(2.42) gives the even stronger condition that $\epsilon(\omega) < -1$. At large wave vectors $k \gg \omega/c$, both γ_i and γ_0 tend to the value k and in this case the eigenvalue condition becomes

$$\epsilon(\omega) + 1 = 0$$

which we obtained earlier by simple arguments, giving the result:

$$\omega_s = \omega_p/\sqrt{2}$$

Using the dielectric function:

$$\epsilon(\omega) = 1 - \omega_p^2/\omega^2$$

we can solve the eigenvalue equation. Introducing the notation $\omega_k = ck$, we obtain the equation:

$$0 = \omega^4 - \omega^2(2\omega_k^2 + \omega_p^2) + \omega_p^2 \omega_k^2$$

which has the solutions:

$$\omega_\pm^2 = \omega_k^2 + \frac{1}{2}\omega_p^2 \pm \frac{1}{2}\sqrt{\omega_p^2 + 4\omega_k^2} \qquad (2.43)$$

Only the solution ω_- represents surface plasmons. The root ω_+ has frequencies larger than ω_p and therefore violates the condition $\epsilon(\omega) < 0$. This unphysical root enters because we had squared the original eigenvalue equation in order to obtain Eq.(2.42) (see Fig.12). At large k, where $\omega_k \gg \omega_p$, the surface plasmon mode frequency approaches the value $\omega_s = \omega_p/\sqrt{2}$. The dispersive character enters at long wavelength because the 'pure' plasmon mode mixes with the photon modes ω_k. Because of the transverse electric field of the photons, they will couple to any other mode which has transverse components. The mixing is quite general and also occurs with bulk modes with transverse components, in which case the coupled modes are called polaritons. The coupled modes of photons and surface modes are called surface polaritons. In this case we have used the surface plasmons as the example, but the same type of coupled modes occurs with the coupling to optical surface phonons, as discussed by Dr. Andersson in these Proceedings.

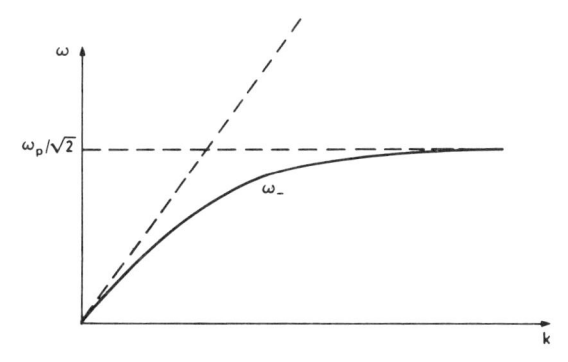

FIG.12. Dispersion relation for surface plasmon at long wavelengths.

3. BAND STRUCTURE APPROACH TO ELECTRONS AT METAL SURFACES

3.1. Introductory remarks

In this section a brief introduction is given to the band structure approach to the study of the electronic structure of metal surfaces. There have been rather few attempts to pursue the band structure approach to metal surfaces. The reason is partly that localized electron states at metal surfaces are not expected to show up as dramatically as the states localized at semiconductor surfaces.

Nevertheless surface states in metals may play an important role in chemical reactions at surfaces and therefore a knowledge from first principles of their properties may become important in chemisorption and catalysis. Furthermore, there are a number of recent experiments giving information about the electron structure of metal surfaces using field emission, photoemission and LEED, and where evidence of surface states appears.

A large number of research papers, review articles and monographs have dealt with the band structure approach to semiconductor surfaces, where the experimental material is very rich and where a large number of theoretical calculations have been published. These problems of the band structure at surfaces are dealt with extensively in the paper by Dr. Jones. In the present paper only an elementary introductory discussion of the surface state problem is given, to serve as a background to some remarks in connection with the rather few applications to metals.

The quantum theory of surface states has a long history. Tamm found in 1932 that surface states might occur in a modified Kronig-Penney potential. In this one-dimensional model the lattice potential is given by equidistant delta-function potentials. The solutions of the Schrödinger equation in a periodic potential can be written $\psi = u_k(x) e^{ikx}$ where $u_k(x)$ has the periodicity of the lattice, $u_k(x) = u_k(x+na)$. For real values of the wave number k, the solutions are ordinary Bloch waves extending throughout the lattice. Complex values, however, correspond to surface states, which decay exponentially when we move inwards from the surface. Such solutions have to be matched to exponential decaying solutions in the vacuum region.

Another early and often quoted discussion was given by Shockley in 1939. He considered a periodic potential with variable atomic distance a. He concluded that no surface states can occur for large interatomic separations. In this case the energy bands tend to the levels for free atoms. When a is decreased the atomic levels split into bands. For sufficiently small values of a, two discrete levels will be split off from the bands. They will be in the gap between the bands, and the corresponding wave functions are localized to the surface. The model calculations by Tamm and Shockley gave examples of how surface states may occur.

We shall give an elementary discussion of the nearly-free-electron (NFE) model and the tight-binding approach. We shall only mention the recently developed Green's function methods, which seem to have definite advantages in more realistic applications, and shall finally give a brief summary of the rather few calculations for metals and the experimental data on surface states in metals.

3.2. Surface states in one dimension: the nearly-free-electron model

In the infinite crystal the electron energies and wave functions are found by solving the Schrödinger equation (in atomic units):

$$[-\nabla^2 + V(\vec{r})] \psi_{n\vec{k}}(\vec{r}) = E_n(\vec{k}) \psi_{n\vec{k}}(\vec{r}) \tag{3.1}$$

In a periodic potential the solution has the well-known Bloch form:

$$\psi_{n\vec{k}} = e^{i\vec{k}\cdot\vec{r}} u_{n\vec{k}}(\vec{r}) \tag{3.2}$$

where $u_{n\vec{k}}(\vec{r})$ has the periodicity of the lattice. The vector $\vec{k}$ lies in the first Brillouin zone, and the energy eigenvalues form a set of bands, labelled n.

The introduction of a surface breaks the periodicity and lifts the restriction to Bloch states. To solve the Schrödinger equation for a given E we need a complete set of eigenfunctions. We have to add to the Bloch states those solutions with a complex wave vector $\vec{k}$ which decay exponentially in the direction normal to the surface. If these functions and their derivatives can be matched to the corresponding wave functions in vacuum with the same energy, we obtain new eigenstates which are localized to the surface region.

Let us now consider the Schrödinger equation in one dimension:

$$\left[-\frac{d^2}{dx^2} + V(x)\right]\psi = E\psi$$

where $V(x) = V(x+a)$, a being the lattice constant. We assume that the potential is weak so that we can expand around the average value V_0:

$$V(x) - V_0 = \sum_g V_g e^{igx}$$

where g are the reciprocal lattice points (multiples of $2\pi/a$).

Away from the zone boundary we can use ordinary perturbation theory and obtain (assuming that only one g contributes to the perturbation)

$$\psi_k = e^{ikx} + \frac{V_g e^{i(k-g)x}}{k^2 - (k-g)^2}$$

$$E_k = k^2 + \frac{|V_g|^2}{k^2 - (k-g)^2}$$

(3.3)

where we measure E_k relative to V_0. Near the zone boundary where $k \sim |k - g|$ these formulas break down and we have to use degenerate perturbation theory:

$$\psi_k = \alpha e^{ikx} + \beta e^{i(k-g)x} \quad (3.4)$$

For the coefficients we obtain

$$(k^2 - E)\alpha + V_g \beta = 0$$

$$V_g \alpha + [(k-g)^2 - E]\beta = 0$$

(3.5)

Equation (3.5) leads to

$$E = \frac{1}{2}\left[k^2 + (k-g)^2 \pm \sqrt{[k^2 - (k-g)^2]^2 + 4|V_g|^2}\right] \quad (3.6)$$

The wave function near the gap is given by

$$\psi_k = e^{ikx} + \frac{V_g}{E - (k-g)^2} e^{i(k-g)x}$$

$$= e^{ikx} + \frac{E - k^2}{V_g} e^{i(k-g)x} \qquad (3.7)$$

At the zone boundary $k = \frac{1}{2} g$, we obtain

$$E_\pm = (\tfrac{1}{2} g)^2 \pm |V|$$

$$\psi_\pm = e^{igx} \pm \frac{V}{|V|} e^{-igx} \qquad (3.8)$$

The wave functions are now standing waves, but their nature depends on the sign of V as shown in Table IV.

TABLE IV. WAVE FUNCTIONS AT THE BAND GAP

Energy	Wave function	
	V > 0	V < 0
E_+	$\cos \tfrac{1}{2} gx$	$\sin \tfrac{1}{2} gx$
E_-	$\sin \tfrac{1}{2} gx$	$\cos \tfrac{1}{2} gx$

This completes the résumé of the standard treatment of the linear NFE model, giving the solutions at and near the band gap.

To find new solutions let us analyse the energy formula (3.6) a little closer. Introducing the new wave number k' measured from the zone boundary (thus $k = \tfrac{1}{2} g + k'$), we obtain

$$E = E(k'^2) = (\tfrac{1}{2} g)^2 + k'^2 \pm \sqrt{V_g^2 + g^2 k'^2} \qquad (3.9)$$

For positive k'^2 this agrees with the standard result. But we have also found something new: the equation shows that we have also real solutions E for $k'^2 < 0$, i.e. for imaginary k' or complex k. These real solutions exist in the range $-V_g^2/g^2 < k'^2 < 0$ and indeed these solutions bridge the gap, as is indicated in Fig. 13.

This is an important result and shows that when we consider the energy as a function of k'^2, $E = E(k'^2)$, and permit negative values of k'^2 (complex wave numbers), there are no gaps in the spectrum; indeed $E(k'^2)$ is an analytic function of k'^2 throughout the gap except at the branch point at $k'^2 = -V_g^2/g^2$. (Note that we have to continue downwards to mid-gap from the upper branch and upwards to mid-gap from the lower branch.) This

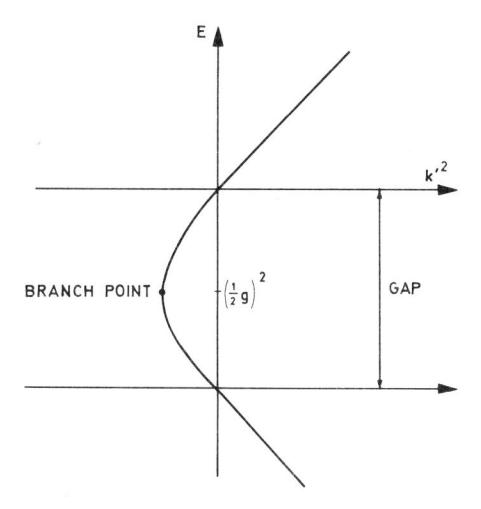

FIG.13. Energy band $E(k'^2)$ bridging the gap.

argument is valid at every gap, so by the continuation into negative k'^2 one single real function E is obtained. Indeed the real E lines never terminate but join one energy band to each other or go off to $E = -\infty$. The last situation appears at the bottom of the band. In our model we simply obtain $E = k'^2$ which tends to $-\infty$ when $k'^2 \to -\infty$. These results were first found by Heine [8]. Introducing instead the complex momentum $k = g/2 + i\kappa$, it simply means that we have to use the complex k plane and continue the gap-bridging branch as shown in Fig. 14.

We also need the wave functions for the solutions in the gap. We just insert Eq. (3.9) into Eq. (3.7) and find, using $k' = i\kappa$,

$$\psi_k = e^{-\kappa x} \cos\left(\frac{gx}{2} + \phi\right) \qquad (3.10)$$

where

$$e^{2i\delta} = \frac{E - k^2}{V_g} = \frac{E - \left(\frac{g}{2}\right)^2 + \kappa^2 - i\kappa g}{V_g}$$

$$\sin^2\delta = -\frac{g\kappa}{V_g} \; ; \quad \cos^2\delta = \frac{E - \left(\frac{g}{2}\right)^2 + \kappa^2}{V_g}$$

and

$$|\kappa|_{max} = \left|\frac{V_g}{g}\right|$$

We shall also need the logarithmic derivative:

$$\frac{\psi'}{\psi} = -\kappa - \left(\frac{g}{2}\right) \tan\left(\frac{gx}{2} + \phi\right) \qquad (3.11)$$

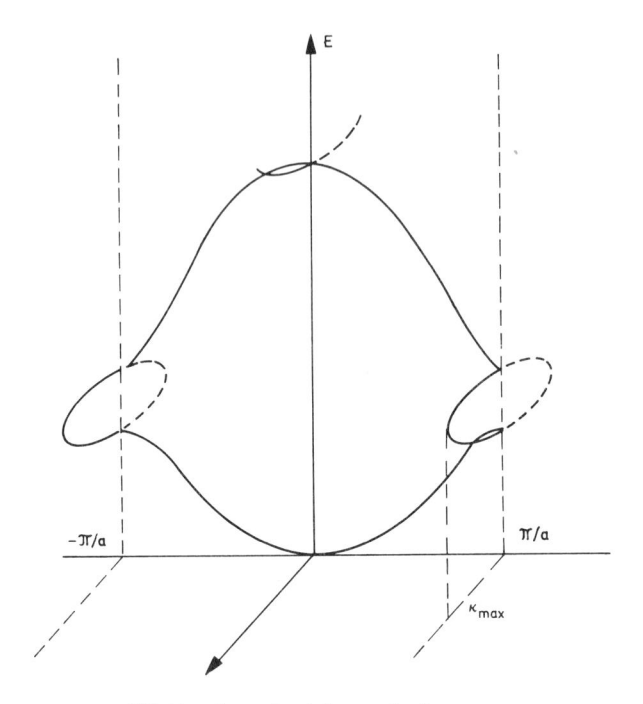

FIG. 14. Energy bands in complex k space.

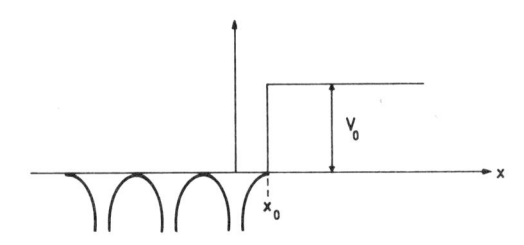

FIG. 15. Model of the surface potential.

Next we must determine if these solutions can be matched to vacuum states having the same energy. Let us consider the simple model shown in Fig. 15. For energies below the vacuum level, the exterior solution is

$$\psi(x) = e^{-qx}; \quad q = \sqrt{V_0 - E} \tag{3.12}$$

and the logarithmic derivative is

$$\frac{\psi'}{\psi} = -q = -\sqrt{V_0 - E}$$

We choose the origin at the centre of the surface atom; the matching coordinate will then be at $x_0 = a/2$ according to Fig. 15. Inside the crystal we must use the solution which decays for negative x, which means that $\kappa < 0$. The matching condition gives

$$\frac{\psi'}{\psi} = |\kappa| - \left(\frac{g}{2}\right) \tan\left(\frac{ga}{4} + \phi\right) = |\kappa| - \left(\frac{g}{2}\right) \tan\left(\frac{\pi}{2} u + \phi\right) = -\sqrt{V_0 - E} \quad (3.13)$$

If Eq. (3.13) is solved graphically a solution is found only in the case $V_g > 0$. Then we start with the sine function at the bottom of the band, and with increasing energy find the right slope at the boundary at a certain energy in the gap. This situation corresponds to the fact that we have the cosine (s state) at the bottom of the upper band and the sine (p state) at the top of the lower band.

3.3. The tight-binding model for a linear system

In this model we expand the solution of the Schrödinger equation in terms of atomic orbitals:

$$\psi(x) = \sum_\ell C_\ell \, \phi(x - \ell a) \quad (3.14)$$

where $\phi(x)$ is a solution of the Schrödinger equation for an isolated atom:

$$[-\nabla^2 + U(x)] \phi(x) = E_0 \phi(x)$$

We assume that the total lattice potential is the sum of atomic potentials, thus

$$V(x) = \sum_\ell V(x - \ell a)$$

For a finite linear chain of N atoms with nearest-neighbour interaction only, we have

$$\int dx \, \phi^*(x - \ell a) [V(x) - U(x)] \phi(x - ma) = \begin{cases} -\alpha & \ell = m \neq 1, N \\ -\alpha' & \ell = m = 1, N \\ -\beta & \ell = m \pm 1 \\ 0 & \text{otherwise} \end{cases}$$

$$\int dx \, \phi^*(x - \ell a) \phi(x - ma) = \begin{cases} 1 & \ell = m \\ \text{neglect otherwise} \end{cases} \quad (3.15)$$

Inserting Eq. (3.14) in the Schrödinger equation, we obtain the following set of equations for the coefficients C_l:

$$\left.\begin{array}{l} C_l(E - E_0 + \alpha) + (C_{l-1} + C_{l+1})\beta = 0 \quad l \neq 1, N \\ C_1(E - E_0 + \alpha') + C_2\beta = 0 \\ C_N(E - E_0 + \alpha') + C_{N-1}\beta = 0 \end{array}\right\} \quad (3.16)$$

The eigenvalues may be found from the secular equation:

$$\begin{vmatrix} E - E_0 + \alpha' & \beta & 0 & & \\ \beta & E - E_0 + \alpha & \beta & & \\ & & \ddots & & \\ & & \beta & E - E_0 + \alpha & \beta \\ & & & \beta & E - E_0 + \alpha' \end{vmatrix} = 0$$

We make the ansatz:

$$C_l = A e^{ilka} + B e^{-ilka}$$

and get

$$E = E_0 - \alpha + 2\beta \cos ka$$

The allowed values of k follow from the boundary conditions. Goodwin [9] found that there exists one surface state per surface provided that $|(\alpha-\alpha')/\beta| > 1$, with the energy

$$E_s = E_0 - \alpha + (\alpha - \alpha')\left[1 + \frac{\beta^2}{(\alpha - \alpha')^2}\right] \qquad (3.17)$$

All other states are extended. If $|(\alpha - \alpha')/\beta| < 1$ all states are extended.

3.4. Surface states in three dimensions

As was shown by the one-dimensional examples, the break in periodicity at the surface may introduce new solutions, because we no longer have the requirement that the component of momentum normal to the surface be real. Assuming an ideal flat surface, the momentum components in the surface plane will be real and can take all values inside a two-dimensional Brillouin zone. As a result we shall have bands of surface states in the three-dimensional case. We here extend the discussion given for the one-dimensional case and show how the complex momentum states may be used to calculate surface states. We return to the Schrödinger equation:

$$H\psi_{\vec{k}} = E(\vec{k})\psi_{\vec{k}}$$

with solutions of the form

$$\psi_{\vec{k}} = e^{i\vec{k}\cdot\vec{r}} U_{\vec{k}}(\vec{r})$$

To discuss the states for complex $\vec{k}$ it is more convenient to consider the equation for the function $U_{\vec{k}}(\vec{r})$, thus:

$$H(\vec{k}) U_{\vec{k}}(\vec{r}) = E(\vec{k}) U_{\vec{k}}(\vec{r})$$
$$H(\vec{k}) = H + 2\vec{k}\cdot\vec{p} + k^2 \qquad (3.18)$$

For real $\vec{k}$ and all bands these states form an orthonormal set in the crystal, the Hamiltonian is Hermitian and all eigenvalues are real. Energy bands in complex $\vec{k}$ space may be considered as analytic continuation from the domain of real $\vec{k}$, just generalizing the discussion from the one-dimensional case. The solutions for complex $\vec{k}$ will of course not be bounded at large distances. It is convenient to require that $U_{\vec{k}}(\vec{r})$ satisfies periodic boundary conditions also for complex $\vec{k}$. All new effects will then come from the exponential function.

From Eq.(3.18) we find the following symmetry properties (we note that $H(\vec{k})$ is no longer Hermitian):

$$\left.\begin{array}{rl} E(\vec{k}) &= E(\vec{k}+\vec{g}) \\ E(\vec{k}^*) &= E^*(\vec{k}) \\ E(-\vec{k}^*) &= E^*(\vec{k}) \\ E(\vec{k}) &= E(-\vec{k}) \end{array}\right\} \qquad (3.19)$$

We consider now the case of a planar surface parallel to the yz plane. The transverse momenta $k_y k_z$ can be regarded as fixed parameters, which we suppress in the notation. For given $k_x k_y$ the energy $E(k_x)$ can be considered as a function of the single complex variable k_x. The lines where E is real branch off symmetrically into the complex k_x plane, as was shown in the one-dimensional case discussed previously. It was shown by Blount [10] that $E(k_x)$ is an analytic function of k_x except at branch points off the real axis. These branch points are of simple square-root form:

$$E(k_x) - E_{bp} + \text{const}(k_{bp} - k_x)^{1/2} \qquad (3.20)$$

The real line in the band gap studied in Section 2 is a contour round the branch point at $k = \frac{1}{2} g + i\kappa_{max}$ leading from the lower band E_- to the upper band E_+.

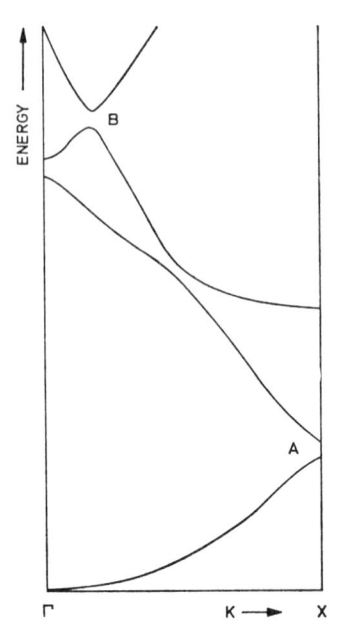

FIG.16. NFE band structure with gaps of kind A and B (from Ref.[11]).

3.5. Typical examples and comparison with experiments

We shall first give a typical example of surface states in a three-dimensional NFE system. Figure 16 shows a NFE band structure with the perpendicular component k of $\vec{k}$ along the line Γ - X in a fcc crystal. There are two different gaps, A and B. Gap A is the usual type occurring at the zone boundary and comes from the usual interaction of the free electron bands near the Bragg reflections. The gap of type B, however, may be anywhere inside the zone, depending on the momentum parallel to the boundary plane, $k_{\parallel} = k_x + k_y$.

The Schrödinger equation in atomic units ($\hbar^2/2m = 1$) is

$$[-\nabla^2 + V(\vec{r})] \psi_{\vec{k}} = E(\vec{k}) \psi_{\vec{k}}$$

and for weak potential $V(\vec{r})$ we use the plane wave expansion:

$$\psi_{\vec{k}} = \sum_{\vec{g}} \alpha_{\vec{k}-\vec{g}} \exp[i(\vec{k}-\vec{g})\cdot\vec{r}]$$

and obtain the equations:

$$[E^0_{\vec{k}-\vec{g}} - E(\vec{k})]\alpha_{\vec{k}-\vec{g}} + \sum_{\vec{g}'} V_{\vec{g}'-\vec{g}}\alpha_{\vec{k}-\vec{g}'} = 0$$

To investigate the case A we need only consider two waves $\vec{k}$ and $\vec{k} - \vec{g}$ mixed by the matrix element $V_{\vec{g}}$. Furthermore, k_x and k_y are just parameters, which we can suppress. We choose as an example the lowest gap at $\vec{g} = (002)$ in a fcc crystal and are back at the one-dimensional problem already discussed in Section 2. The secular equation is

$$\begin{vmatrix} k^2 - E & V_{200} \\ V_{200}^* & (k-g)^2 - E \end{vmatrix} = 0 \tag{3.21}$$

The solutions in the gap for complex $k = g/2 + i\kappa$ have the form

$$\psi_s(z) = A\, e^{\kappa z} \cos\left(\frac{g}{2} z + \phi\right) \tag{3.22}$$

and describe a wave localized to the surface region which decays into the solid when $z \to -\infty$. This wave has to be matched to an exponential wave outside the solid:

$$\psi_v(z) = B \exp\left(-\sqrt{V_0 - E'}\, z\right) \tag{3.23}$$

Precisely as in the one-dimensional case, the existence of a localized state in the gap A will depend on the sign of the matrix element V_g and the position of the boundary. We choose the co-ordinate $z = 0$ at an atom centre. The matching of the wave function is possible at some energy in the gap if $V_{200} > 0$ but not for $V_{200} < 0$.

$V_{200} < 0$ means that the electron lowers its energy when it has a high density at atom centres. This gives the cosine function at the lower band edge. If we raise the energy, the zeros of the wave function move inwards, and we get a derivative at the surface of opposite sign relative to the outside solution and therefore we have no surface state, as shown in Fig. 17.

If $V_{200} > 0$ we start with the sine function at the bottom; by moving the zeros inwards with increasing energy, a smooth matching is now obtained at some energy in the gap.

A gap of type B does not reduce to the one-dimensional model. The two-band approximation to the wave function will contain two different Fourier components parallel to the surface. This gives a more complicated matching problem and we refer to the original paper by Forstmann [11] for the details. He has shown that for small V_g there will always be a surface state independent of the sign of the potential.

Band gaps inside the Brillouin zone can also arise because of the hybridization which occurs when a free electron band crosses a tightly bound band or when two tightly bound bands cross each other.

The band structure of d-band metals shows typical band gaps due to crossover of the s band with the d band of the same symmetry as shown in Fig. 18.

The wave function near the gap shown in Fig. 18 contains quite a large number of Fourier components because the d state is mixed with an s-p state. Therefore several independent solutions have to be considered to achieve the matching both inside and outside the boundary. The complex

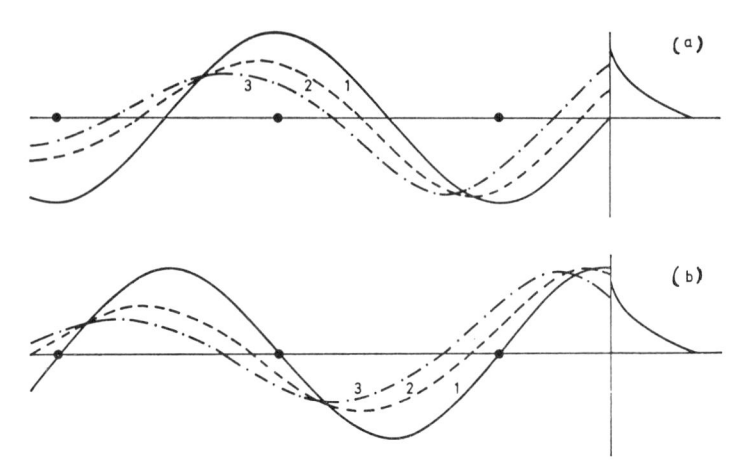

FIG.17. Wave functions near the surface for energies in the gap A. (a) $V_{200} < 0$; (b) $V_{200} > 0$. (From Ref.[11].)

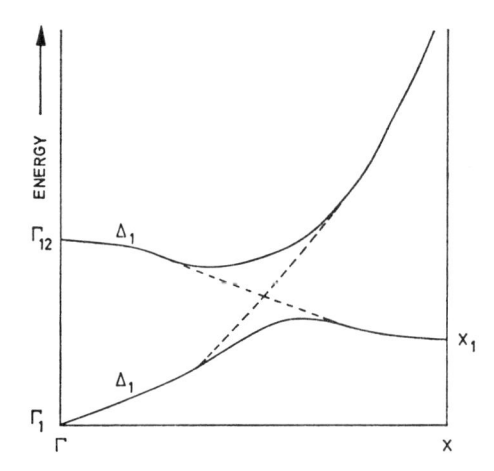

FIG.18. Typical band gap due to hybridization of the NFE band with the d band of the same symmetry (from Ref.[12]).

band structure for this case has been worked out by Forstmann and Pendry [12] We refer to their paper for details about the technique.

Here we shall only make some simple remarks of more hand-waving nature. Harrison [13] has shown that, near the point of degeneracy, the energy bands are given by the usual two-state hybridization formula:

$$E = \tfrac{1}{2}(E_s + E_d) \pm \tfrac{1}{2}\sqrt{(E_s - E_d)^2 + 4|\Delta_k|^2} \qquad (3.24)$$

The coupling term Δ_k is given by

$$\Delta_k = \langle k|\delta V|d\rangle - \langle k|d\rangle\langle d|\delta V|d\rangle$$

δV is the difference between the atomic potential and the true potential around a single ion, and the matrix elements are taken between plane waves and atomic d states.

Near the intersection at $k = k_c$ between the s and d band shown in Fig.18, one can expand the bands as

$$E_s = \epsilon_s + ak \quad , \quad E_d = \epsilon_d - bk$$

and obtain the parameters from a calculated band structure. For simplicity we choose $a = b = m$. We can now introduce complex k vectors and determine the exponential constant $\kappa = \kappa(E)$. A surface state is obtained of the form

$$\psi_s(z) = A_d(z) \, e^{\kappa z} \cos(k_c z + \delta) \tag{3.25}$$

The new thing is that $A_d(z)$ is now a function of z, since the hybridized wave functions may have a strong atomic component.

The treatment referred to here is due to Gadzuk [14]. A similar discussion for a spin-orbit split gap should be useful, just replacing Δ_k by ξ, the spin-orbit parameter.

There are rather few calculations for real metals. We just mentioned the work by Forstmann and Pendry [12]. They have done model calculations for fcc transition metals, particularly Cu and Ni. Their calculations show that for a (100) surface there exists a surface state at the lower end of the d-band region. The wave functions are confined to the two outer atomic layers. For Cu the state should occur around 5-6 eV below the Fermi energy and, for Ni, 4-5 eV below it. The energy position coincides with a maximum in the density of states seen in photoemission and in optical absorption. The comparison seems a little uncertain because of possible contamination in the experimental spectra. Another model calculation by Watts [15] gives a surface state at 2.75 eV but no surface state for Ni.

A study of surface states for simple metals with application to Al has been published by Boudreaux [16]. The monovalent metals have very few (if any) band gaps below the Fermi energy, and the existence of such gaps is necessary for obtaining intrinsic surface states. For divalent and trivalent metals, on the other hand, such gaps do exist. For the (111) face of Al, Boudreaux found a range of surface states between 0.6 Ry and 0.8 Ry (the Fermi energy is at 1.04 Ry). Also for the (001) surface he found surface states below the Fermi level but no states on the (011) surface. Similar results were obtained by Hoffstein [17], who also calculated the density of surface states. The surface density of states is shown in Fig.19.

The results show a rather pronounced peak at $E \sim 5.5$ eV, which is 1.5 eV below E_F (E refers to the vacuum level and the work function is around 4.0 eV). If the surface bands had been free electron bands the density of states would be constant. The peak is probably related to the matrix element V_{200} in the three-dimensional zone.

Experimental data for surface states on metal surfaces are still very few. We wish to draw attention to some exciting findings by Gadzuk and Plummer using field emission and photoemission from the (100) surface of W. We refer to their recent review [3] for a detailed discussion. In Fig.20 we have given the energy distribution $j'(\epsilon)$ for electrons emitted from the (100) plane, together with the free electron curve.

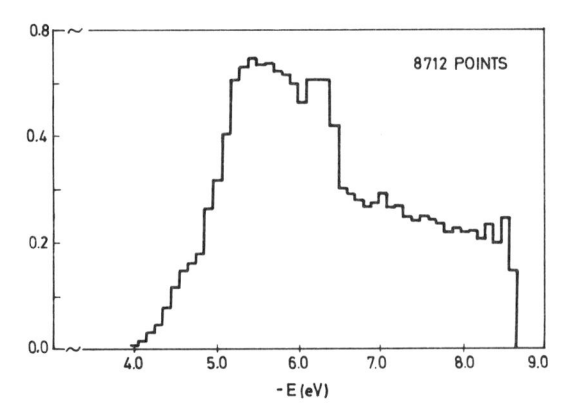

FIG.19. Calculated density of states for the (001) surface of Al (from Ref.[17]).

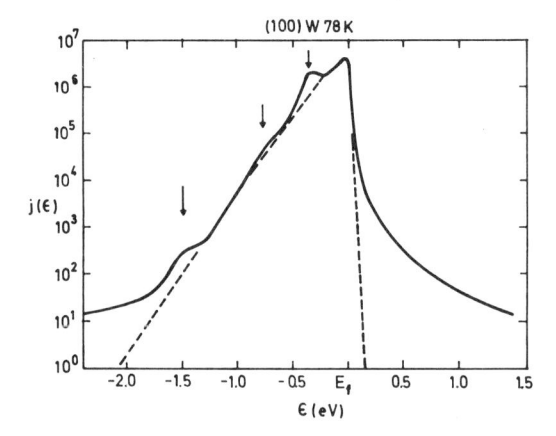

FIG.20. Energy distribution of electrons emitted from the (100) surface of W (from Ref.[3]).

There is a considerable amount of structure in the curve. This is even more clear if we plot the enhancement factor $R(\epsilon) = j'(\epsilon)/j'_0(\epsilon)$, as is shown in Fig.21 for several low-index planes of W. Plummer and Gadzuk interpret the extra emission as caused by surface states. The strong structure at -0.37 eV in the emission from the (100) plane seems to originate from a surface state in a spin-orbit split gap. In Fig.22 the relativistic energy bands are shown for the (100) direction calculated by Loucks [18].

Figure 22 shows that the strong maximum in $R(\epsilon)$ shows peaks at the energies of the spin-orbit gaps. The interpretation as surface states is given further support by the photoemission experiments by Waclawski and Plummer [19]. They have reproduced all the essential characteristics of the state observed in field emission.

This concludes the elementary introduction to surface states in metals. A number of topics have been left out, particularly the applications to LEED, which are treated in other papers. It should be emphasized that the techniques used in this paper are rather primitive. We wished to underline the intuitive

ELECTRONS AT METAL SURFACES 363

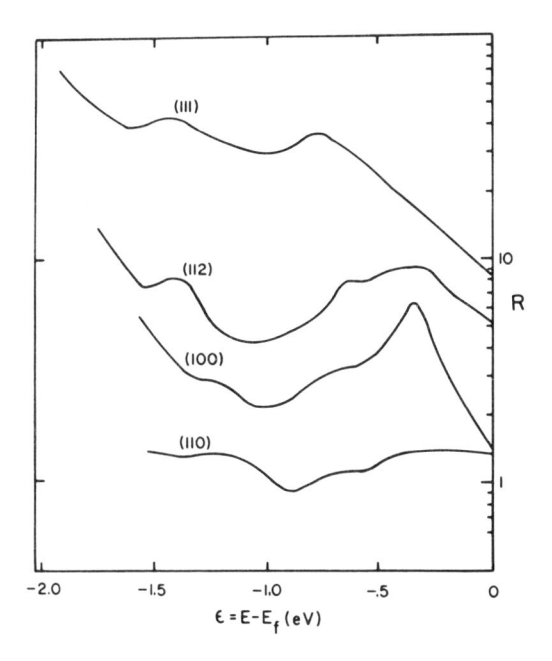

FIG.21. The enhancement factor $R(\epsilon)$ for several crystal orientations of W (from Ref.[3]).

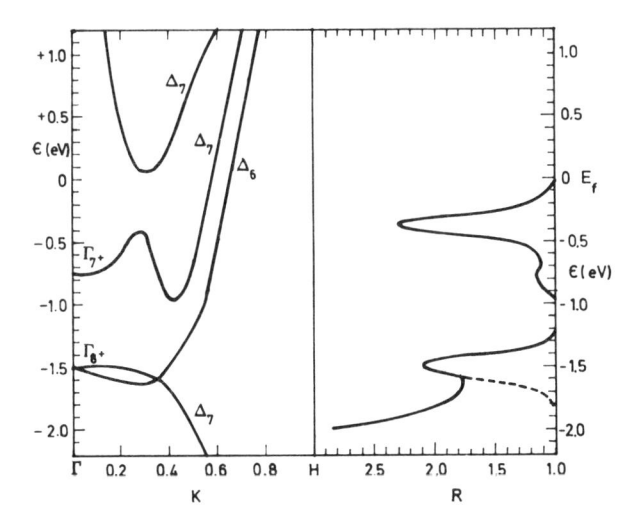

FIG.22. Relativistic energy bands for the (100) direction of W calculated by Loucks [18] together with the experimental enhancement factor $R(\epsilon)$.

physics involved rather than decribe tools for real-life applications. The
direct method of matching wave functions is simple and obvious in the
applications where we define our matching surface by a discrete step in the
potential. In real application with a continuous surface potential, the matching
procedure is less obvious. Similarly, for problems where distortions
from the ideal lattice positions occur - reconstruction of the surface or a
surface with impurities or adsorbed layers - more powerful methods must
be used. One also needs methods from which the surface density states
functions can be calculated with relative ease. The Green's function method
seems to be particularly well suited to this particular problem. We would
like in particular to mention the method developed by García-Moliner and
Rubio [20]. Their method is applicable to the general problem of two
systems A and B separated by a boundary, and the method determines the
Green's function for the combined system in terms of those for A and B.
The method seems very suitable to determine the surface states, but no
calculation has yet been made for a 'real' metal. It is only necessary to
know the wave functions and energy bands for real wave vectors k to
determine the Green's function. The generality and formal elegance of the
method suggest a wide range of future applications.

4. THE INHOMOGENEOUS ELECTRON GAS AT THE METAL SURFACE

4.1. The density functional theory

We now return to the question of the electron distribution at the surface
and discuss properties such as the charge density, the potential, the work
function, the surface energy, the image potential and even touch on the
problem of chemisorption. In all these discussions the electron density
$n(\vec{r})$ plays a central role.

We first introduce a theoretical framework from which the electron
density can be calculated. In these considerations we shall regard the
system as a fluid characterized by its local density $n(\vec{r})$. The standard
method to deal with an inhomogeneous electron fluid is the Thomas-Fermi
theory, which describes the gross features of atoms, molecules, solids,
impurities, etc. The method is, however, not sufficiently accurate to
describe properly the rapid density variations at the metal surface. One
needs in fact a much more accurate theory which can include the effects
of correlation and exchange in a satisfactory way. Such a theory has been
formulated by Kohn and co-workers [21, 22] and is known as the density
functional theory. Since practically all recent work on metal surfaces is
being based on this method, we shall give a brief résumé of the theory,
following the original papers closely.

We first formulate an exact variational principle for the ground-state
energy of a system in which variations with respect to the local density
$n(\vec{r})$ are considered. The electrons are moving under the influence of their
mutual Coulomb interaction in a given external potential. The Hamiltonian
has the form:

$$H = T + V + U$$

where T is the kinetic energy, V the energy in the external field and U the Coulomb interaction energy.

The ground-state electron density $n(\vec{r})$ is clearly a functional of the external potential $v(\vec{r})$. We wish to show that the converse is also true: that $v(\vec{r})$ is a unique functional of $n(\vec{r})$ (except for a constant). We assume that the ground state is non-degenerate.

To prove the theorem let us assume that a different potential $v'(\vec{r})$ would give rise to the same density $n(\vec{r})$. The ground state ψ' associated with $v'(\vec{r})$ cannot be the same as the ground state ψ belonging to $v(\vec{r})$ because ψ and ψ' satisfy Schrödinger equations with different potentials. We know that the expectation value of the energy has its minimum for the exact ground-state wave function and thus

$$E' = \langle \psi' | H' | \psi' \rangle < \langle \psi | H' | \psi \rangle = \langle \psi | H + V' - V | \psi \rangle$$

$$= E + \int [v'(\vec{r}) - v(\vec{r})] \, n(\vec{r}) \, d^3\vec{r}$$

However, we can just interchange primed and unprimed quantities and find

$$E < E' + \int [v(\vec{r}) - v'(\vec{r})] \, n'(\vec{r}) \, d^3\vec{r}$$

Thus our assumption that we can have two potentials $v(\vec{r})$ and $v'(\vec{r})$ with the same density $n(\vec{r})$ ends up with the inconsistency:

$$E + E' < E' + E$$

The conclusion is that $v(\vec{r})$ is (within a constant) a unique functional of $n(\vec{r})$. But because the choice of potential $v(\vec{r})$ determines the Hamiltonian H, we conclude that the full many-body ground state is a unique functional of $n(\vec{r})$.

Since the ground state ψ is a functional of $n(\vec{r})$, the same will hold for expectation values such as the kinetic energy and the interaction energy. We can therefore define

$$F[n(\vec{r})] = \langle \psi | T + U | \psi \rangle \tag{4.1}$$

where $F[n]$ is a universal functional, valid for any particle number N and any external potential. This functional plays the key role in the theory.

For a given external potential $v(\vec{r})$, we can now define the energy functional:

$$E_v[n] = \int d^3\vec{r} \, n(\vec{r}) v(\vec{r}) + F[n] \tag{4.2}$$

For the correct density $n(\vec{r})$, $E_v[n]$ gives the exact ground-state energy E.

The general result just presented is known as the Hohenberg-Kohn theorem.

It is straightforward to demonstrate that $E_v[n]$ has its minimum for the correct $n(\vec{r})$, with the obvious condition that

$$\int d^3\vec{r}\, n(\vec{r}) = N$$

for all trial functions $n(\vec{r})$. It follows that $n(\vec{r})$ can be determined from the equation:

$$\frac{\delta}{\delta n}[E_v[n] - \mu N] = 0$$

where μ is a Lagrange multiplier corresponding to the chemical potential of the system.

It is convenient to separate out the classical Coulomb interaction; thus

$$F[n] = \frac{1}{2}\int d^3\vec{r}\, d^3\vec{r}\,' \frac{n(\vec{r})\,n(\vec{r}\,')}{|\vec{r}-\vec{r}\,'|} + G[n]$$

and rewrite the energy functional as

$$E_v[n] = \int d^3\vec{r}\, n(\vec{r})\, v(\vec{r}) + \frac{1}{2}\int d^3\vec{r}\, d^3\vec{r}\,' \frac{n(\vec{r})\,n(\vec{r}\,')}{|\vec{r}-\vec{r}\,'|} + G[n] \qquad (4.3)$$

where $G[n]$ is a new universal functional of $n(\vec{r})$ containing the kinetic energy and the exchange and correlation parts of the interaction energy.

4.2. Self-consistent equations

We shall now discuss what kind of equations can be derived, starting from Eq. (4.3), to actually determine the density $n(\vec{r})$. There are two main available methods. One is to derive a set of equations similar to the Hartree equation; however, they include the full effect of many-body interactions on the charge density. The other way is to use Eq. (4.3) as the starting point to make systematic improvements of the Thomas-Fermi theory.

The starting point is Eq. (4.3). The functional $G[n]$ contains the total kinetic energy of the system and the exchange and correlation contributions to the interaction energy. Instead of dividing the functional as just indicated, the trick is to divide $G[n]$ into the following two parts:

$$G[n] \equiv T_0[n] + E_{xc}[n]$$

Here $T_0[n]$ is the kinetic energy of a system of non-interacting electrons having density $n(\vec{r})$. $E_{xc}[n]$ will thus be the total exchange and correlation contribution to the energy, which includes a substantial part from the kinetic energy of the system.

Next we apply the variational principle to the formula for the total energy, with the condition of constant number of particles, and obtain

$$\int d^3\vec{r}\,\delta n(\vec{r}) \left\{ \frac{\delta T_0[n]}{\delta n(\vec{r})} + v_{\text{eff}}(\vec{r}) \right\} = 0 \tag{4.4}$$

$$\int d^3\vec{r}\,\delta n(\vec{r}) = 0$$

Here,

$$v_{\text{eff}}(\vec{r}) = v(\vec{r}) + \int d^3\vec{r}\,'\,\frac{n(\vec{r}\,')}{|\vec{r}-\vec{r}\,'|} + v_{xc}(\vec{r}) \tag{4.5}$$

and

$$v_{xc} \equiv \frac{\delta E_{xc}[n]}{\delta n(\vec{r})} \tag{4.6}$$

$v_{\text{eff}}(\vec{r})$ is to be interpreted as an effective potential felt by an electron. The first two terms are the external potential and the average potential from the electrons in the system. The last term, v_{xc}, gives the contribution from exchange and correlation. If we were to set $E_{xc}[n] = 0$, the term v_{xc} would vanish and v_{eff} would reduce to the Hartree field, in which the electrons move independently of each other in the external field, plus the average field from the charge distribution $n(\vec{r})$. If we neglect v_{xc} we would retrieve the Hartree equations from Eqs (4.4) and (4.5).

The main point to note is that Eqs (4.4) and (4.5) have precisely the same form as for a system of N non-interacting electrons moving in a potential $v_{\text{eff}}(\vec{r})$. Therefore we can solve the problem as in the Hartree theory. We introduce single-particle wave functions $\psi_i(\vec{r})$ and express the density as

$$n(\vec{r}) = \sum_{i=1}^{N} |\psi_i(\vec{r})|^2 \tag{4.7}$$

The functions ψ_i satisfy the one-electron Schrödinger equation:

$$[-\tfrac{1}{2}\nabla^2 + v_{\text{eff}}(\vec{r})]\,\psi_i(\vec{r}) = \epsilon_i \psi_i(\vec{r}) \tag{4.8}$$

These equations must be solved by a self-consistent approach, because v_{eff} itself depends upon the ψ_i. It should be noted that only the density and ground-state properties that can be calculated from the density can be obtained by this method. The one-electron functions ψ_i are auxiliary quantities only to obtain the density and do not in a rigorous sense describe the physical one-electron properties.

The ground-state energy can now be written down in a form analogous to the Hartree theory and we obtain

$$E = \sum_{i=1}^{N} \epsilon_i - \frac{1}{2} \iint d^3\vec{r}\,d^3\vec{r}\,'\,\frac{n(\vec{r})\,n(\vec{r}\,')}{|\vec{r}-\vec{r}\,'|} + E_{xc}[n] - \int d^3\vec{r}\,v_{xc}(\vec{r})\,n(\vec{r}) \tag{4.9}$$

The units in this section correspond to taking e = m = ℏ = 1 so that the energy unit is 27.2 eV.

As already mentioned, there are also simpler ways to obtain approximate solutions. A very useful method described in the original paper of Hohenberg and Kohn [21] is to consider some simple approximation to the functional G[n] and minimize the expression for the total energy with respect to n($\vec{r}$). In practice one often considers a gradient expansion:

$$G[n] = \int d^3\vec{r} \, [g_0(n(\vec{r})) + g_2(n(\vec{r}))|\nabla n(\vec{r})|^2 + \ldots] \tag{4.10}$$

A whole hierarchy of approximate theories can be built up from this formula. For example, if we put $g_2 = 0$ and only consider the kinetic energy contribution we obtain the Thomas-Fermi theory. If we include the exchange energy we obtain the Thomas-Fermi-Dirac equation and we may obtain a further extension by adding the correlation energy. Analogous results are obtained for g_2 where we may include contributions from kinetic energy, exchange and correlation energy, respectively. The coefficient of the first gradient term is at present best given by the formula:

$$g_2(n) = \frac{1}{72n} - 0.0083 \frac{1}{n^{4/3}} + 0.0042 \frac{1}{n^{4/3}} \tag{4.11}$$

where the first term comes from the kinetic energy, the second from the exchange energy and the last from the correlation energy. The first two terms are exact whereas the last is valid in the high-density limit.

The presumed criterion for the validity of the gradient expansion is that $|\nabla n/n|$ is smaller than both the Fermi wave number k_F and the Thomas-Fermi screening wave number k_{TF} ($k_{TF} = (8k_F/3\pi)^{1/2}$). In the surface region of the jellium model one finds that $|\nabla n/n| \approx k_{TF}$ and therefore a density gradient approach is not well justified. Nevertheless Smith [23] has obtained reasonable results for some surface properties of jellium using this approach.

4.3. Exchange and correlation energy

To carry through a self-consistent calculation of the density, one has to specify the functional $E_{xc}[n]$, which describes the effects of exchange and correlation. The problem has not yet been solved in a satisfactory way for non-uniform densities, and in practice the zero-order contribution is often used in a gradient expansion; thus

$$E_{xc} = \int d^3\vec{r} \, g_{xc}(n(\vec{r})) \tag{4.12}$$

where

$$g_{xc}[n] = \epsilon_{xc}(n(\vec{r}))n(\vec{r})$$

and ϵ_{xc} is the exchange and correlation energy per electron in a uniform electron gas having density n($\vec{r}$).

Using this approximation, one obtains the formula:

$$v_{xc}(m(\vec{r})) = \frac{d}{dn}[\epsilon_{xc}(n)n] \qquad (4.13)$$

for the exchange and correlation potential, which in this approximation is equal to the exchange and correlation contribution μ_{xc} to the chemical potential of the corresponding uniform electron liquid.

The density functional scheme was originally developed as a method to include exchange. From the exchange energy functional E_x is found the corresponding potential v_x, given by the formula:

$$v_x = \mu_x = -\frac{1}{\pi}[3\pi^2 n(\vec{r})]^{1/3} \qquad (4.14)$$

This formula for the exchange potential was first derived by Dirac, and rederived by Kohn and Sham [22] using the method described here.

We comment briefly on the variety of exchange potentials used in practice. There is a different formula derived by Slater in which he proposed to approximate the non-local Hartree-Fock exchange potential by the average over the Fermi sea of a uniform electron gas, and he obtained the formula:

$$v_x^S = \frac{3}{2}\mu_x = -\frac{3}{2\pi}[3\pi^2 n(\vec{r})]^{1/3} \qquad (4.15)$$

There are also many applications where one introduces a variable parameter α and writes

$$v_{x\alpha} = \frac{3}{2}\alpha\mu_x \qquad (4.16)$$

It should be remarked that this so-called xα method, developed by Slater and co-workers, is not connected with the density functional approach. There are no free parameters in the density functional approach and there is a unique way of treating exchange and correlation in it.

The results for v_{xc} can be found from the calculations and interpolation formulas for an electron liquid, described, for example, in Ref.[24]. The result can be written in the form

$$v_{xc}(r_s) = \beta(r_s)v_x(r_s)$$

where $\beta(r_s)$ is a slowly varying function of the density parameter r_s, and $v_x(r_s)$ is the exchange potential defined by Eq. (4.14). In the high-density limit there is only exchange, so that $\beta(r_s)$ starts from the value 1 and increases monotonically towards values in the range 1.2 - 1.3 for metallic densities [25].

The extension of the density functional method to spin-polarized systems is of considerable interest for many surface problems. Some results for v_{xc} in the local density approximation have recently been published [26, 27].

The corresponding potential in the Hartree-Fock approximation has been used in many problems. The main effect of including correlation is to reduce the strong variations in the Hartree-Fock theory with spin polarization and to lower the potential over the entire range. Several successful applications of spin-dependent local potential are now being completed.

4.4. Application to metal surfaces: some definitions and formulas

We now discuss the application of the density functional theory to calculate some key properties of the surface, such as the density profile, the surface potential, the surface energy and work function, and the image potential.

A slight extension of the theory is needed so that we can consider the effect of adding or removing particles. At $T = 0$ we just replace the energy by $E - \mu N$ and consider variations in $n(\vec{r})$, keeping the chemical potential μ fixed. From Eq. (4.3) one obtains

$$\delta E - \mu \delta N = \int d^3\vec{r}\, \delta n(\vec{r}) \left\{ \varphi(\vec{r}) + \frac{\delta G[n(\vec{r})]}{\delta n(\vec{r})} \right\} - \mu \delta N = 0 \qquad (4.17)$$

where

$$\varphi(\vec{r}) = v(\vec{r}) + \int d^3\vec{r}'\, \frac{n(\vec{r}')}{|\vec{r} - \vec{r}'|} \qquad (4.18a)$$

is the total electrostatic potential in the system. Consider now a variation δn which does not conserve the number of particles. We see that Eq. (4.16) can only be fulfilled if

$$\varphi(\vec{r}) + \frac{\delta G[n]}{\delta n(\vec{r})} = \mu \qquad (4.18b)$$

One verifies directly that Eq. (4.18) must hold also for variations δn which conserve the number of electrons.

We shall use this result to obtain precise definitions of the various potentials. Let us consider a slab of metal whose surface consists essentially of two parallel faces with identical properties. We wish to find the work function for one of these surfaces and therefore remove one electron to a large distance outside the surface with the remaining $N-1$ electrons in the ground state E_{N-1}. The work function is by definition

$$\Phi = [\varphi(\infty) + E_{N-1}] - E_N \qquad (4.19)$$

Using $\mu = E_N - E_{N-1}$, we can write

$$\Phi = \varphi(\infty) - \mu$$

To obtain the standard form let us introduce the average interior potential $\overline{\varphi} = \varphi(-\infty)$, defined by

$$\overline{\varphi} \equiv \langle \varphi(\vec{r}) \rangle$$

i.e. the average electrostatic potential inside the system. We next define the chemical potential relative to $\bar{\varphi}$:

$$\bar{\mu} = \mu - \bar{\varphi} = \left\langle \frac{\delta G}{\delta n(\vec{r})} \right\rangle$$

and can finally write the alternative formulas:

$$\Phi = \varphi(\infty) - \mu = (\varphi(\infty) - \bar{\varphi}) - \bar{\mu} = \Delta\varphi - \bar{\mu} \qquad (4.20)$$

where

$$\Delta\varphi = \varphi(\infty) - \bar{\varphi} = \varphi(\infty) - \varphi(-\infty)$$

The last form of Eq. (4.20) is the standard form and expresses the work function in terms of the electrostatic potential difference for electrons far outside and deep inside the metal together with the chemical potential relative to the interior electrostatic potential.

We now specialize to the jellium model, i.e. an electron liquid in a compensating uniform background of positive charge extending to the plane $x = 0$.

The chemical potential μ is given by

$$\mu = \varphi(-\infty) + \bar{\mu}$$

where $\bar{\mu}$ is measured relative to the average electrostatic potential $\varphi(-\infty)$ deep inside the system. Recalling that $G = T_0 + E_{xc}$, we find that

$$\bar{\mu} = \tfrac{1}{2} k_F^2 + \frac{\delta E_{xc}}{\delta n} = \tfrac{1}{2} k_F^2 + \mu_{xc}(\bar{n}) \qquad (4.21)$$

In the interior of the metal, the effective potential will have the constant value:

$$v_{eff} = \bar{\varphi} + \mu_{xc}(\bar{n})$$

Parallel to the surface we have no forces and therefore the eigenfunctions can be written as

$$\psi_{k,k_y,k_z}(x,y,z) = \psi_k(x)\, e^{i(k_y y + k_z z)} \qquad (4.22)$$

Away from the boundary region we have

$$\psi_k(x) = \sin[k_x - \gamma(k)]$$

where the phase shift $\gamma(k)$ is a continuous function of k and depends on the self-consistent solution itself. The eigenvalues have the form:

$$\epsilon_{k,k_y,k_z} = \tfrac{1}{2}[k^2 + k_y^2 + k_z^2 - k_F^2] \qquad (4.23)$$

where we have chosen the Fermi energy as the zero level of energy, i.e.
$\mu = 0$.

The self-consistent scheme is defined by the set of equations:

$$\left\{ -\frac{1}{2} \frac{d^2}{dx^2} + \mathscr{v}_{\text{eff}}[n(x)] \right\} \psi_k(x) = \tfrac{1}{2}(k^2 - k_F^2) \psi_k(x) \tag{4.24}$$

with the condition that

$$\psi_k(x) \to \sin[k_x - \gamma(k)] \qquad x \to -\infty$$

The density profile is calculated from the formula:

$$n(x) = \frac{1}{\pi^2} \int_0^{k_F} (k_F^2 - k^2) [\psi_k(x)]^2 \, dk \tag{4.25}$$

4.5. Charge distribution and surface potential

The self-consistent equations were solved by Lang and Kohn [28] for bulk metallic densities in the range r_s = 2 - 6, using the jellium model. In Fig.23 we show the density profile for r_s = 2 and r_s = 5. We notice that there are characteristic oscillations at low density (r_s = 5). For the high average density corresponding to r_s = 2, however, a density profile is obtained similar to that obtained in the Thomas-Fermi theory.

The electrostatic potential and the total effective potential are shown in Fig.24. We notice that the inner electrostatic potential is quite small and shows oscillations.

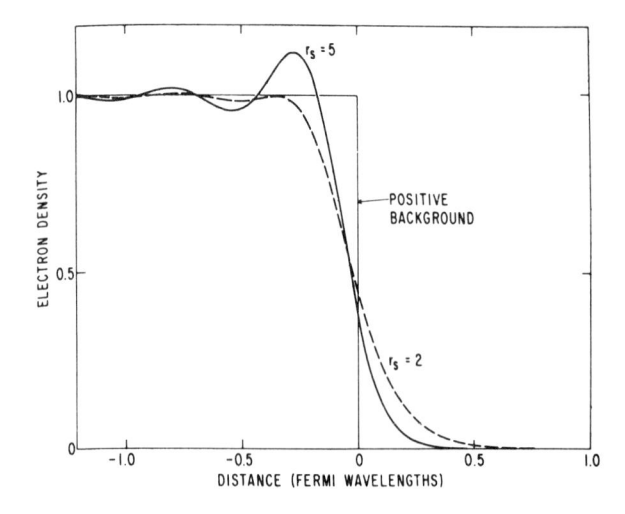

FIG.23. Self-consistent charge density near a metal surface at densities r_s = 2 and r_s = 5 (from Ref.[28]).

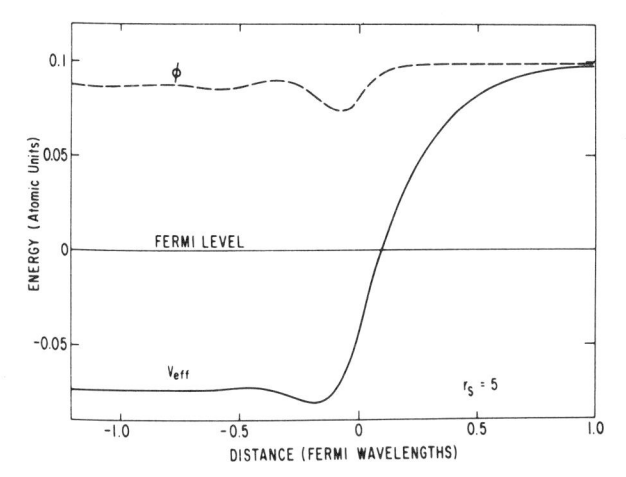

FIG.24. Electrostatic potential and total effective potential $V_{eff}(x)$ for $r_s = 5$ (from Ref.[28]).

An interesting exact relation between some parameters of the model has recently been derived by Budd and Vannimenus [29]. They show that the difference in the electrostatic potential at the surface and in the interior is given in terms of the pressure by the formula:

$$\varphi(0) - \varphi(-\infty) = -\frac{1}{\bar{n}} \frac{\partial E}{\partial \Omega} = \bar{n} \frac{\partial \epsilon}{\partial n} \qquad (4.26)$$

$\varphi(0)$ is the potential at the edge of the background charge, E is the total energy, Ω the volume and ϵ the average energy per electron. Eq.(4.26) is an exact result and therefore useful to check the accuracy of numerical calculation. Budd and Vannimenus calculated separately the left- and right-hand sides from the Lang and Kohn results and found excellent agreement at high densities and quite fair agreement even at the lowest densities.

4.6. Surface energy and work function

The surface energy σ of a crystal is defined as the energy required per unit area of surface formed when the crystal is split in two along a plane. The total energy of the crystal can be written as the sum of the three contributions:

$$E = T_0[n] + E_{xc}[n] + E_{es}[n] \qquad (4.27)$$

representing the kinetic energy for independent particles, the exchange-correlation energy and the total electrostatic energy.

Correspondingly we can write the surface energy of jellium as the sum of three terms:

$$\sigma_{jellium} = \sigma_0 + \sigma_{xc} + \sigma_{es} \qquad (4.28)$$

In Table V are listed some of the results of Lang and Kohn [28].

TABLE V. SURFACE ENERGY IN THE
JELLIUM MODEL IN UNITS OF erg/cm^2

r_s	σ_0	σ_{xc}	σ_{es}	$\sigma_{jellium}$
2	-5600	3260	1330	-1010
3	-720	750	170	200
4	-145	265	40	160
5	-30	115	15	100
6	-5	55	10	60

We observe that the kinetic energy part is negative. When we split the crystal the electrons spill over the edge of the positive background giving a region of lower electron density; hence the total energy is lowered. Table V shows clearly the enormous importance of exchange and correlation; indeed it is of the same order of magnitude as the total surface energy. We particularly note that $\sigma_{xc} \gg \sigma_{es}$ over the entire range of densities. Therefore the Thomas-Fermi or Hartree approximation would not be of much use for this problem.

The total surface energy becomes negative at densities higher than $r_s \simeq 2.7$, which is a failure of the jellium model. The reason is that the energy contributions are dominated by the reduction of the kinetic energy. Indeed the leading term in the high-density limit is of the form:

$$\sigma = -0.0763 \, r_s^{-9/2}$$

An entirely different approach to the surface energy in the jellium model was proposed by Schmidt and Lucas [30]. They considered the contributions to the surface energy from the zero-point energy of the collective modes, the plasmons. When the metal is cleaved, new modes are created - the surface plasmons - but at the same time the distribution of bulk modes is modified so that, for example, the f sum rule is satisfied also after the cleavage. Summing over the zero-point energies up to the cut-off momentum for plasmons, they arrived at the simple formula:

$$\sigma = C \, r_s^{-5/2}$$

The work by Schmidt and Lucas has stirred up an intense discussion about the surface energy problem and the controversy is still going on. It is clear that the plasmon theory is incomplete and gives, for example, the wrong high-density limit. On the other hand, their work showed that the collective modes seem to be important for the surface energy. A number of papers on this subject have appeared or are in print, but the situation is not at all clear at the moment. We shall comment on these problems later.

4.7. Interaction with external charges

Discussion of these problems requires the response function theory to be set up, which will be done in the final section. We shall therefore only report here some of the results obtained.

Lang and Kohn [31] applied the theory to a calculation of the profile of charge induced by a uniform electric field. A typical result is shown in Fig. 25.

It should be noted that the induced charge has its maximum at a distance x_0 outside the edge of the positive charge layer. The distance varies from 1.5 to 1.2 atomic units when r_s varies from 2 to 6.

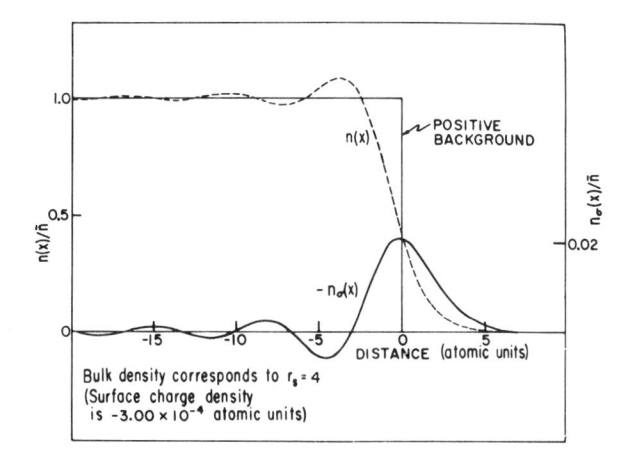

FIG.25. Typical profile of charge induced by a uniform electric field (from Ref.[31]).

Extending the calculation to the case of a point charge q far away from the surface, Lang and Kohn found that

$$v = -\frac{q^2}{4(x-x_0)}$$

where x_0 is the centre-of-charge co-ordinate just mentioned. Thus the idealized mathematical metal surface should be taken to pass through x_0.

An interesting application has recently been published by Smith, Ying and Kohn [32]. They consider the case of a proton outside a metal surface. They apply the theory by directly minimizing the energy functional $E_v[m(\vec{r})]$ with respect to the local density $n(\vec{r})$. The results corresponding to the adsorption of hydrogen on tungsten are shown in Figs 26 and 27. The results give a minimum about $x_m = 1$ a.u. with the corresponding absorption energy equal to E = 9 eV. This should be compared with the experimental value, which is 11 eV. Figure 27 shows their result for dipole moment. We see that the value is small and even changes sign near the minimum point for x_m.

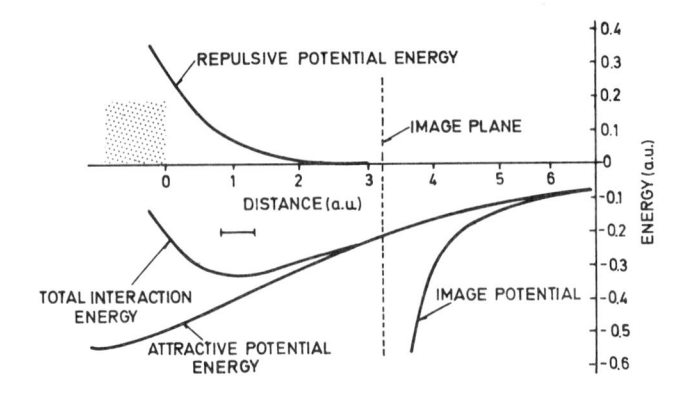

FIG.26. Proton-metal interaction energy versus separation distance (from Ref.[32]).

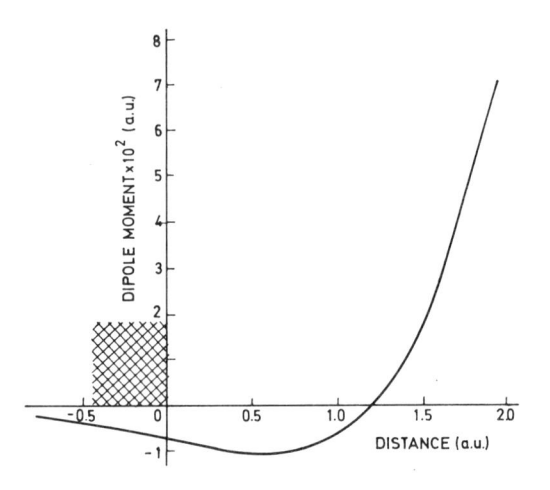

FIG.27. Hydrogen adatom dipole moment versus distance between proton and metal surface (from Ref.[32]).

It should be noted that approximations used in this work [32] are rather crude. The authors used the Thomas-Fermi type approach and linearized the equation with respect to the external charge. Work is in progress by several groups to find more accurate solutions.

4.8. Results for real metals

All the results discussed so far have been restricted to the jellium model. In the jellium model the effects of the ions in the lattice, which can be quite considerable, are completely ignored. Experimental data for the work function often show a strong dependence on the particular surface plane.

In recent calculations where the discrete ions have been included, the ions have usually been represented by some suitable model pseudopotential. Lang and Kohn [31] have used such an approach, treating the ion pseudopotential as a perturbation on the jellium results in order to discuss the anisotropy of the work function. In connection with their study of surface energies they point out [28] that there are additional large contributions to the surface energy coming from the ions. The presence of the ions will not in lowest order change the kinetic or exchange and correlation energy of the electrons at the surface but there will be a difference in the electrostatic energy of the system when the ions are included. After including these contributions they also obtain good agreement for the surface energy of high-density metals.

The self-consistent calculations for the jellium model have recently been extended to include the effects of the ions in a pseudopotential approach. Fully self-consistent calculations including a pseudopotential description of the ions have recently been performed by Applebaum and Hamann [33]. Their approach was first applied to the (100) surface of sodium and later to the (111) surface of silicon. In the case of sodium they find that the electron structure differs from that in the bulk only in the last atomic layer. As expected, the results for sodium do not differ in a qualitative way from those obtained by the jellium model.

The knowledge of the electron structure at the surface of simple metals is of course important from a methodological point of view. However, most of the interesting physics and chemistry is concerned with the electronic properties of transition metals such as nickel and tungsten. In the presence of d bands the pseudopotential approach has to be modified and concepts such as the renormalized atom picture seem to be useful [34]. The very first tentative applications to the electron properties of transition metal surfaces have recently appeared in the literature [35, 36]. Heavy numerical applications of self-consistent cluster methods are in progress for transition metal surfaces [37]. There is no doubt that the present unsatisfactory state of knowledge about transition metal surfaces is close to an end and that relevant theoretical information is just about to become available.

5. SCREENING PROPERTIES AT A METAL SURFACE

5.1. Key formulas in linear response theory

The dielectric response function for a system with a surface is a key quantity in the theory. It determines the induced charge density caused by a small external field. The poles in the response function as a function of frequency correspond to the physical resonances in the system, i.e. the elementary excitations of the system. They give information about surface excitations such as surface plasmons or particle-hole excitations. Surface excitation is an important topic that is dealt with in detail in the paper by Professor Celli, and only an introduction to this topic is given here. The dielectric response function or screening function is also a key quantity for our understanding of exchange and correlation at the metal surface and will therefore be of importance, for example, for discussion of the surface energy.

There are already numerous calculations of the screening properties of a metal surface using different assumptions about density profiles and potential barriers, the most recent ones even being based on self-consistent models. Rather than attempt an encyclopaedic review, we should like to emphasize the key model-independent features and try to avoid going into the merits and drawbacks of the different numerical approaches. Practically all results up to now have been obtained for the jellium model, which is discussed in Section 4.

We shall start by reviewing a few key formulas in linear response theory. We consider small perturbations around the unperturbed charge density ρ_0 and write

$$\rho(\vec{x}, t) = \rho_0(\vec{x}) + \rho_1(\vec{x}, t) \tag{5.1}$$

where $\rho_1(\vec{x}, t)$ is the induced charge caused by a small external potential $V_{ext}(\vec{x}, t)$. The relation between the two is given by

$$\rho_1(\vec{x}, t) = \int d^3x'\, dt'\, H(\vec{x}, \vec{x}', t-t')\, V_{ext}(\vec{x}', t') \tag{5.2}$$

where $H(\vec{x}, \vec{x}', t-t')$ is the retarded density-density response function, defined by

$$H(\vec{x}, \vec{x}', t-t') = -i\Theta(t-t')\langle [\hat{\rho}(\vec{x}, t), \hat{\rho}(\vec{x}', t')]\rangle \tag{5.3}$$

$\hat{\rho}(x, t)$ is the density operator, $\Theta(t)$ is the Heaviside step function and $\langle \cdots \rangle$ denotes the statistical average.

The effective electrostatic potential inside the system is given by

$$V_{eff}(\vec{x}, t) = \int d^3x'\, \varkappa(\vec{x}, \vec{x}')\, \rho_1(\vec{x}', t) + V_{ext}(\vec{x}, t)$$

where $\varkappa(\vec{x}, \vec{x}') = e^2/|\vec{x} - \vec{x}'|$ is the Coulomb potential. We now introduce the retarded response function $h(\vec{x}, \vec{x}', t-t')$ describing the response to $V_{eff}(\vec{x}, t)$; thus

$$\rho_1(\vec{x}, t) = \int d^3x'\, dt'\, h(\vec{x}, \vec{x}', t-t')\, V_{eff}(\vec{x}', t') \tag{5.4}$$

It is sometimes convenient to use a shorthand notation and consider all functions as matrices and vectors in the space and time variables; thus

$\rho_1 = H V_{ext}$

$\rho_1 = h V_{eff}$

$V_{eff} = \varkappa \rho_1 + V_{ext}$

from which it is straightforward to find that

$$H = h + h v H = (1 - h v)^{-1} h \qquad (5.5)$$

and

$$V_{eff} = (1 - h v)^{-1} V_{ext} \qquad (5.6)$$

which shows that $(1 - h v)^{-1}$ describes the screening of the external field and has the meaning of an inverse dielectric function.

In the lowest approximation, one takes for h the corresponding function h_0 for the non-interacting system, having the same ground-state density distribution $\rho_0(x)$; thus

$$h_0(\vec{x}, \vec{x}', t-t') = -i\Theta(t-t') \langle [\hat{\rho}(\vec{x}, t), \hat{\rho}(\vec{x}', t')] \rangle_0 \qquad (5.7)$$

This is the so-called random phase approximation (RPA). We shall not have reason to go beyond that approximation, although many of the formulas will have a more general validity. The Fourier transform of $h_0(\vec{x}, \vec{x}', t-t')$ is given by the formula

$$h_0(\vec{x}, \vec{x}', \omega) = \sum_n \left[\frac{\varphi_n(\vec{x}) \varphi_n^*(\vec{x}')}{\omega + i\epsilon - \omega_n} - \frac{\varphi_n(\vec{x}') \varphi_n^*(\vec{x})}{\omega + i\epsilon + \omega_n} \right] \qquad (5.8)$$

In Eq. (5.8), $\omega_n = \epsilon_p - \epsilon_h$ are the particle-hole energies of the non-interacting system and the corresponding amplitudes are $\varphi_n(\vec{x}) = \langle 0 | \hat{\rho}(\vec{x}) | n \rangle = \psi_p^*(\vec{x}) \psi_h(\vec{x})$, where $\psi_p(\vec{x})$ and $\psi_h(\vec{x})$ are the one-electron wave functions involved. The specialization of Eq. (5.8) to the jellium model is straightforward. Defining the x-y plane to be parallel to the surface, we introduce the two-dimensional wave vector $\vec{K}$ and notice that the wave function has the form

$$\psi_{\vec{K}, \nu}(x, y, z) = e^{i\vec{K} \cdot \vec{R}} \psi_\nu(\vec{x})$$

where $\vec{R} = (x, y)$ and $\psi_\nu(z)$ is the solution of the Schrödinger equation for the motions perpendicular to the surface. The one-electron energies are given by

$$\omega_{\vec{K}, \nu} = \frac{|\vec{K}|^2}{2m} + \omega_\nu$$

We shall only briefly indicate how the solution of the response equation:

$$\rho = h v \rho + h V_{ext} \qquad (5.9)$$

can be obtained using the Hilbert-Schmidt theory for inhomogeneous integral equations. We introduce the symmetrized kernel (letting ω be completely imaginary until the end of the calculations):

$$K = v^{\frac{1}{2}} h v^{\frac{1}{2}} \qquad (5.10)$$

and obtain the formal solution:

$$\rho = v^{-\frac{1}{2}} \frac{1}{1-K} K v^{-\frac{1}{2}} V_{ext} \tag{5.11}$$

An explicit representation can be given in terms of the eigensolutions of the homogeneous equation:

$$K\psi_\ell = \lambda_\ell \psi_\ell \; ; \quad \psi_\ell = v^{\frac{1}{2}} \rho_\ell \tag{5.12}$$

The functions ψ_ℓ form a complete set. They are orthogonal for different ℓ's and we can write

$$\langle \psi_\ell | \psi_{\ell'} \rangle = \delta_{\ell\ell'} N_\ell ; \quad 1 = \sum_\ell \frac{1}{N_\ell} | \psi_\ell \rangle \langle \psi_\ell |$$

where N_ℓ is a normalization constant. Using these relations, one obtains after a few elementary steps that the solution can be written as

$$H = v^{-\frac{1}{2}} \frac{1}{1-K} K v^{-\frac{1}{2}} = \sum_\ell \frac{\lambda_\ell}{1-\lambda_\ell} \frac{|\rho_\ell\rangle \langle \rho_\ell|}{\langle \rho_\ell | v | \rho_\ell \rangle} \tag{5.13}$$

where we can now relax the restriction on ω and let ω be real. For the jellium model,

$$H = H(q, z, z', \omega); \quad \lambda_\ell = \lambda_\ell(q, \omega) \; ; \quad \rho_\ell = \rho_\ell(q, z, \omega)$$

The poles of H correspond to the resonances of the system. The eigenmodes of the system are determined from the condition:

$$\lambda_\ell(q, \omega) = 1 \tag{5.14}$$

This gives the dispersion relation for the ℓ-th mode, and each eigenvalue is associated with a density oscillation amplitude $\rho_\ell(q, z, \omega)$ in space.

Some of the modes will be in the form of bulk plasmons, where the density oscillations form standing waves decaying in amplitude when we pass through the surface into vacuum. Other modes will be the surface plasmons, for which the density oscillation is localized to a region at the surface of the order of the width of the density profile. Other collective modes might be surface dipole modes and possibly also higher multipole modes. The existence of such higher modes seems possible only if the density profile is rather extended, and their existence from the experimental point of view does not yet seem settled. The electrostatic potential at a point outside the surface is strongest for the surface plasmon mode, is considerably weaker for the multipole modes and is practically zero for the bulk plasmon modes. There is, of course, also a continuous spectrum corresponding to individual screened particle-hole excitations.

5.2. Separation of the classical terms in the response

The response theory given in Section 5.1 is already in a form ready for applications. However, an important part of the response at high frequencies is of purely classical origin. This is true in particular for the long-wavelength limit of collective motions such as surface or bulk plasmons. In the formalism given, these classical effects enter the theory through the particular form of some limiting values of the density-density response function. For a general survey of the physical effects, it seems more convenient to recast the theory in such form that the classical terms occur explicitly in the equations and are kept separated from the terms that depend on the actual quantum dynamics of the system. This implies rewriting the theory in the spirit of the Thomas-Fermi method, but still retaining the full quantum properties. A general formulation along such lines was recently given by March and Tosi [38] dealing with the propagation of plasmons in periodic systems. Here we follow a similar approach developed by Mukhopadhyay and myself [39].

The idea is simply to separate the instantaneous classical response due to the electrostatic forces from the retarded response which depends on the dynamics of the system by rewriting the response equation in the form of a second-order differential equation in time. We therefore differentiate Eq. (5.4) twice with respect to time and obtain

$$\frac{\partial^2}{\partial t^2} \rho_1(\vec{x}, t) = \int d^3x' \left\{ -i \left\langle \left[\frac{\partial}{\partial t} \hat{\rho}(\vec{x}, t), \hat{\rho}(\vec{x}', t) \right] \right\rangle_0 \right\} V_{\text{eff}}(\vec{x}', t)$$

$$+ \int d^3x' dt' \left\{ -i \Theta(t - t') \left\langle \left[\frac{\partial^2}{\partial t^2} \hat{\rho}(\vec{x}, t), \hat{\rho}(\vec{x}', t') \right] \right\rangle_0 \right\} V_{\text{eff}}(\vec{x}', t') \quad (5.15)$$

From the equation of motion for $\hat{\rho}(\vec{x}, t)$ it follows that

$$-i \left\langle \left[\frac{\partial}{\partial t} \hat{\rho}(\vec{x}, t), \hat{\rho}(\vec{x}', t) \right] \right\rangle_0 = \rho_0(\vec{x}) \nabla_{\vec{x}}^2 \delta(\vec{x} - \vec{x}') + \nabla \rho_0(\vec{x}) \cdot \nabla_{\vec{x}} \delta(\vec{x} - \vec{x}') \quad (5.16)$$

Taking the Fourier transform, using the definition of V_{eff}, we obtain after some re-arrangements

$$\left\{ \omega^2 - \omega_{\text{pe}}^2(\vec{x}) + \nabla \rho_0(\vec{x}) \cdot \nabla_{\vec{x}} \mathcal{u} \right\} \rho_1 = -\widetilde{h} \mathcal{u} \rho_1 - \widetilde{h} V_{\text{ext}} \quad (5.17)$$

The classical terms all appear on the left-hand side. The first term on the right-hand side represents the retarded response in the system and the second term is the coupling to the external field. The dynamical response is given by $\widetilde{h}(\vec{x}, \vec{x}', \omega)$, which is related to the current-current response tensor through the formula:

$$\widetilde{h}(\vec{x}, \vec{x}', \omega) = -\sum_{\alpha\beta} \nabla_{\vec{x}}^\alpha \nabla_{\vec{x}'}^\beta C_{\alpha\beta}(\vec{x}, \vec{x}', \omega) \quad (5.18)$$

In the RPA, $C_{\alpha\beta}(\vec{x}, \vec{x}', \omega)$ is the Fourier transform of the unperturbed retarded current-current response function:

$$C_{\alpha\beta}(\vec{x}, \vec{x}', t-t') = -i\,\Theta(t-t')\,\langle[\hat{j}_\alpha(\vec{x}, t), \hat{j}_\beta(\vec{x}', t')]\rangle_0 \tag{5.19}$$

The coupling to the external field enters through the quantity:

$$\tilde{h}(\vec{x}, \vec{x}', \omega) = \ddot{h}(\vec{x}, \vec{x}', \omega) - \ddot{h}(\vec{x}, \vec{x}', 0) \tag{5.20}$$

The equation for the response function follows easily from Eq. (5.17). For the applications, it is more convenient to introduce a generalized polarizability function $\alpha(\vec{x}, \vec{x}', \omega)$ defined by

$$\alpha(\vec{x}, \vec{x}', \omega) = -\frac{1}{4\pi}\int d^3x''\, H(\vec{x}, \vec{x}'', \omega)\,\varkappa(\vec{x}'', \vec{x}') \tag{5.21}$$

The integral equation for $\alpha(\vec{x}, \vec{x}', \omega)$ is

$$\left\{\omega^2 - \omega_{pe}^2(\vec{x}) + \nabla\rho_0 \cdot \nabla\varkappa\right\}\alpha = -\ddot{h}\varkappa\alpha + \frac{1}{4\pi}\tilde{h}\varkappa \tag{5.22}$$

We note that the polarizability function is related to the inverse non-local dielectric function ϵ^{-1} through the formula:

$$\epsilon^{-1}(\vec{x}, \vec{x}', \omega) = \delta(\vec{x}-\vec{x}') - 4\pi\,\alpha(\vec{x}, \vec{x}', \omega) \tag{5.23}$$

The advantage of these equations is that the classical response appears explicitly and separated from the quantum terms. In problems where the collective behaviour plays a dominating role, one might treat the quantum dynamics in a simplified manner or even ignore them in limiting cases. In other problems where the quantum dynamics is of key importance there is no advantage in this formulation and it is probably more convenient to work directly with the formulas given in Section 5.1.

5.3. The jellium model once more

We consider again the semi-infinite jellium model with a uniform positive background extending through the half-space $z < 0$ and terminating at the plane $z = 0$. We shall then have a surface region around $z = 0$ in which the electron density decreases continuously from the bulk value ρ_0 towards zero. To simplify the equations we make use of the translational invariance in the z-y plane and introduce the two-dimensional Fourier transform:

$$f_K(z, z', \omega) = \int d^2r\, e^{i\vec{K}\cdot\vec{R}}\,f(\vec{R}, z, z', \omega) \tag{5.24}$$

It is also convenient to introduce the notations:

$$\chi = h\mathcal{u} \quad \text{and} \quad \ddot{\chi} = \ddot{h}\mathcal{u}$$

The key integral equation then becomes

$$\left\{\omega^2 - \omega_{pe}^2(\vec{x}) + \nabla\rho_0 \cdot \nabla\mathcal{u}\right\}\alpha(\omega) = -\ddot{\chi}(\omega)\alpha(\omega) + \frac{1}{4\pi}\{\ddot{\chi}(\omega) - \ddot{\chi}(0)\} \quad (5.25)$$

where we have used a matrix notation and only indicated the frequency dependence. For the jellium model we obtain explicitly

$$\{\omega^2 - \omega_{pe}^2(z)\}\alpha_K(z,z',\omega) + \frac{2\pi}{K}\frac{d\rho_0}{dz}\frac{d}{dz}\int dz'' \exp(-K|z-z''|)\alpha_K(z'',z',\omega)$$

$$= -\int dz'' \ddot{\chi}_K(z,z'',\omega)\alpha_K(z'',z',\omega) + \frac{1}{4\pi}\{\ddot{\chi}_K(z,z',\omega) - \ddot{\chi}_K(z,z',0)\} \quad (5.26)$$

where we have used the Fourier transform:

$$\mathcal{u}_K(z,z') = \frac{2\pi e^2}{K}\exp(-K|z-z'|) \quad (5.27)$$

The full solution requires a knowledge of the particle-hole spectrum of the model. For the present discussion we shall simply drop the terms containing $\ddot{\chi}_K(z,z',\omega)$. This is not related to any form of hydrodynamic approach, where the particle dynamics is only represented by the change in the static kinetic pressure. Since we are interested in high-frequency phenomena in the collisionless regime where hydrodynamics does not apply, we prefer to neglect the quantum dynamics altogether and look only at the response due to the classical terms. The response equation then reduces to

$$\alpha_K(z,z',\omega) = -\frac{1}{2}\cdot\frac{\exp(-K|z-z'|)}{\omega^2 - \omega_{pe}^2(z)}\frac{d}{dz}[\rho_0(z)\,\text{sgn}(z-z')]$$

$$+ \frac{2\pi}{\omega^2 - \omega_{pe}^2(z)}\frac{d\rho_0}{dz}\int dz'' \exp(-K|z-z''|)\,\text{sgn}(z-z'')\alpha_K(z'',z',\omega) \quad (5.28)$$

The solution requires that we specify the form of the density profile $\rho_0(z)$. However, in the limit $K = 0$ we can solve the equation and obtain

$$\alpha_0(z,z',\omega) = -\frac{\rho_0(z)}{\omega^2 - \omega_{pe}^2(z)}\delta(z-z')$$

$$-\frac{1}{2}\text{sgn}(z-z')\frac{d}{dz}\left[\frac{\rho_0(z)}{\omega^2 - \omega_{pe}^2(z)}\right] - \frac{1}{4}\frac{\rho_0\omega^2}{\omega^2 - \omega_s^2}\frac{d}{dz}\left[\frac{1}{\omega^2 - \omega_{pe}^2(z)}\right] \quad (5.29)$$

where ω_s is the surface plasma frequency.

To understand this result, let us place a unit oscillating test charge at $\vec{x}_0 = (0, 0, z_0)$, which will give rise to an external potential:

$$V_{ext}(\vec{x}, t) = \frac{e^{i\omega_t t}}{|\vec{x} - \vec{x}_0|}$$

Using the definition of α, we find for the Fourier component K of the induced charges the formula:

$$\rho_K(z, \omega) = -4\pi \, \alpha_K(z, z_0, \omega) \, 2\pi \, \delta(\omega - \omega_t) \qquad (5.30)$$

Using this relation between induced charge and the polarizability function, we see that the first term in Eq. (5.29) corresponds to the local response in the case of an infinite electron gas if the external charge is inside the medium. The response is local because we do not consider the contribution from the quantum dynamics which brings in spatial dispersion. Note that such a contribution is obtained throughout the entire density profile and does not only contribute in the interior. If we place the test charge for outside $z = 0$, then the first term vanishes, but the other terms do not. We may therefore regard the genuine surface contribution to α_0 as given by

$$\alpha_0^s(z, z_0, \omega) = \frac{1}{2} \frac{d}{dz}\left[\frac{\rho_0(z)}{\omega^2 - \omega_{pe}^2(z)}\right] - \frac{\rho_0 \omega^2}{4(\omega^2 - \omega_s^2)} \frac{d}{dz}\left[\frac{1}{\omega^2 - \omega_{pe}^2(z)}\right] \qquad (5.31)$$

We note the resonance at the surface plasmon frequency $\omega_s = \omega_0/\sqrt{2}$, where ω_0 is the bulk plasma frequency.

Although we have no general formula for $K \neq 0$ we note that for small K we can write

$$\alpha_K(z, z', \omega) \approx \alpha_0(z, z', \omega) \exp(-K|z - z'|) \qquad (5.32)$$

We can now derive some simple results. If we place the test charge sufficiently far from the surface, then only the $K = 0$ limit will contribute and we obtain for the total induced charge the result:

$$\int_{-\infty}^{+\infty} dz \, \rho_{K=0}(z, t) = \frac{\omega_s^2}{\omega_t^2 - \omega_s^2} e^{i\omega_t t} \qquad (5.33)$$

For a static test charge, i.e. $\omega_t = 0$, we obtain the result of electrostatics that the total induced charge is -1.

We can proceed to calculate the image potential by calculating the interaction energy between the test charge and the induced charge, i.e.

$$U(t) = \frac{1}{2} \int d^3x \, V_{ext}(\vec{x}, t) \rho_1(\vec{x}, t)$$

$$= -2\pi \exp(i2\omega_t t) \int_0^\infty dK \int_{-\infty}^{+\infty} dz \, \exp(-K|z - z_0|) \, \alpha_K(z, z_0, \omega_t) \qquad (5.34)$$

Now we have

$$\int_{-\infty}^{+\infty} dz \exp(-K|z-z_0|) \alpha_K(z, z_0, \omega_t) = \exp(-Kz_0) \int_{-\infty}^{z_0} dz \exp(Kz) \alpha_K(z, z_0, \omega_t)$$

$$+ \exp(Kz_0) \int_{z_0}^{\infty} dz \exp(-Kz) \alpha_K(z, z_0, \omega_t)$$

The second term vanishes if z_0 is far outside the surface. The first term contributes only for small K, where we can use Eq. (5.32) to obtain

$$\int_{-\infty}^{+\infty} dz \exp(-K|z-z_0|) \alpha_K(z, z_0, \omega_t) = -\frac{\exp(-2Kz_0)}{4\pi} \frac{\omega_s^2}{\omega_t^2 - \omega_s^2}$$

Therefore we obtain for the asymptotic limit of the dynamical image potential

$$U(t) = \frac{\exp(i 2\omega_t t)}{4z_0} \frac{\omega_s^2}{\omega_t^2 - \omega_s^2} \qquad (5.35)$$

This result has recently been obtained by Harris and Jones [40], using a different method. The contribution from surface plasmons has been studied by Šunjić et al. [41] and by Heinrichs [42]. The result in Eq. (5.35) is an exact result since it can be shown that the terms dependent on the dynamics of the electrons do not contribute to the asymptotic limit.

Another example is the van der Waals interaction between a molecule with polarizability $\alpha_a(\omega)$ and a metal surface, which has recently been treated applying the method given here by Mukhopadhyay and Mahanty [43]. For this problem one starts calculating the induced charge due to a unit dipole source at a point $\vec{x}_0$ far outside the metal surface:

$$\rho_{ext} = \hat{e} \cdot \nabla_x \delta(\vec{x} - \vec{x}_0) \qquad (5.36)$$

where $\hat{e}$ is a unit vector along the direction of the field. One next calculates the field due to the induced charge distribution; thus,

$$\vec{E}(\vec{x}, \omega) = -\nabla_x \int d^3x' \, \varkappa(\vec{x}, \vec{x}') \rho_{ind}(\vec{x}', \omega)$$

$$= 4\pi \nabla_x \int d^3x' \, d^3x'' \, \varkappa(\vec{x}, \vec{x}') \alpha(\vec{x}', \vec{x}'', \omega) \rho_{ext}(\vec{x}'', \omega)$$

$$= -4\pi \hat{e} \cdot \nabla_{x_0} \nabla_x \int d^3x' \, \varkappa(\vec{x}, \vec{x}') \alpha(\vec{x}', \vec{x}_0, \omega) \qquad (5.37)$$

This formula shows that, for a dipole far outside the surface, the leading contribution in z^{-1} will be determined by the limit $K \to 0$, and indeed the leading term in the field has the form:

$$\frac{1}{2z^3} \frac{\omega_s^2}{\omega^2 - \omega_s^2}$$

which is the analogue of the formula for the dynamical image potential. We refer to the original paper for details of the calculation.

The study of the screening properties for an external charge in the neighbourhood of the surface requires knowledge of the polarizability $\alpha_K(z, z', \omega)$ for finite K and requires the solution of the integral equation for the response, assuming a specific model for the surface potential and density profile $\rho_0(z)$. We refer to papers by Newns [44] and Beck and Celli [45] for a good discussion of linear response theory applied to the infinite barrier model. Many numerical illustrations of the induced potential and field at the surface are given in a paper by Beck et al. [46].

5.4. Surface plasmons and their dispersion

We have seen in the examples given in the last section that the surface plasmons play an essential role in the screening properties of a metal surface. They form the dominant mode of excitation at long wavelengths. In most applications of response theory we would need the eigenfrequencies $\omega_s(K)$ and often also the corresponding eigenfunctions. We shall first make a few remarks about some model-independent properties and make a few references to some recent calculations later.

We first consider the classical approach, where we neglect the quantum dynamics. For free oscillations of the system, the integral equation takes the form:

$$\rho(\vec{x}, \omega) = -\frac{1}{\omega^2 - \omega_{pe}^2(z)} \frac{d\rho_0}{dz} \frac{d}{dz} \int d^3x' \, v(\vec{x}, \vec{x}') \rho(\vec{x}', \omega) \tag{5.38}$$

or, in Fourier space,

$$\rho_q(z, \omega) = \frac{2\pi}{\omega^2 - \omega_{pe}^2(z)} \frac{d\rho_0}{dz} \int_{-\infty}^{+\infty} dz' \, \exp(-q|z - z'|) \, [\Theta(z - z') - \Theta(z' - z)] \tag{5.39}$$

For finite q, we have to specify the form of $\rho_0(z)$ to solve the equation; however, the limiting case $q = 0$ can be solved by inspection. The solution is of the form:

$$\rho_{q=0}(z, \omega) = A \frac{d}{dz} \frac{1}{\omega^2 - \omega_{pe}^2(z)} \tag{5.40}$$

and it satisfies the equation, provided that $\omega^2 = \omega_s^2$.

Thus the density oscillation corresponding to the surface plasmon is given by

$$\rho_{q=0}(z) = A \frac{d}{dz} \frac{1}{\omega_s^2 - \omega_{pe}^2(z)} \tag{5.41}$$

The eigenfunction has a width corresponding to the width of the surface region itself and the form is in the classical limit completely specified by $\rho_0(z)$. We note that there is a singularity at the value of z for which $\omega_{pe}(z) = \omega_s$. This singularity is a result of our neglect of the particle-hole spectrum. Even at q = 0 the surface plasmon is always degenerate with some particle-hole excitation. Therefore there is a decay of surface plasmons into particle-hole excitations, i.e. a damping of the collective oscillation. Including such a damping, the amplitude of the density oscillation will be finite everywhere.

We next turn to the dispersion of the surface plasmons. We note that irrespective of the surface profile the q = 0 limit is the classical surface plasmon frequency $\omega_s = \omega_0/\sqrt{2}$. The observation of the limiting frequency has therefore no information about the surface properties and one would have to study the dispersive properties. For small q we can write the dispersion relation as an expansion:

$$\omega_s(q) = \omega_s \left[1 + (A_1 + iA_2) \frac{q}{K_F} + (B_1 + iB_2) \left(\frac{q}{K_F}\right)^2 + \ldots \right] \tag{5.42}$$

The linear coefficient A_1 can be found from the result for q = 0 by using an exact result (within RPA), first found by Harris and Griffin [47], that

$$A_1 = \frac{\int_{-\infty}^{+\infty} dz\, z\, \rho_{q=0}(z)}{\int_{-\infty}^{+\infty} dz\, \rho_{q=0}(z)} \tag{5.43}$$

where $\rho_{q=0}(z)$ is the surface plasmon eigenfunction for q = 0.

Experimental data seem to indicate that the dispersion coefficient A_1 is negative [48]. Bennett [49] used a hydrodynamical model and was able to account for the experimental data for Mg. Later Beck and Celli [50], applying the RPA scheme to a model assuming a finite step at the metal surface, obtained results for A_1, A_2, B_1, B_2. They were in good agreement with experimental data for Al but less good for Mg. Recently Feibelman [51] has reported extensive calculations of A_1 and A_2, using as a basis the self-consistent potential and charge distribution by Lang and Kohn [28]. He also gives a survey of results for different magnitudes of the surface diffuseness and work function. All these calculations give a negative dispersion coefficient A_1; Feibelman also finds that A_2 is negative over the entire range of parameters studied.

The validity of the expansion in Eq. (5.42) is restricted to small momenta $q \ll K_F$. A calculation of the dispersion relation over a wide range of momenta has been made by Inglesfield and Wikborg [52]. They have solved the homogeneous integral equation in Eq. (5.12) in the RPA approximation and determined $\omega_s = \omega_s(q)$ from Eq. (5.14). They consider a step-potential

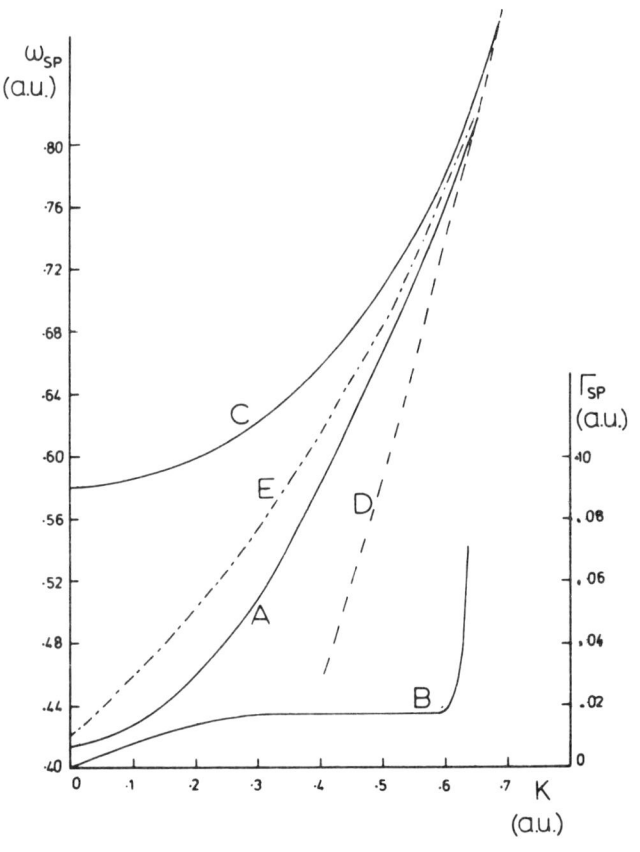

FIG.28. A: Surface plasmon frequency as function of momentum k. B: Damping Γ_{sp} of surface plasmon. C: Bulk plasmon branch. D: Landau cut-off energy ω_c. E: Surface plasmon branch in semiclassical approximation. (From Ref.[52].)

surface, and the bulk density corresponds to Al. Their result is illustrated in Fig. 28.

There is no indication of the negative dispersion at small momenta; on the contrary, the surprising feature is the very strong positive dispersion. We see that in fact the surface and bulk plasmon branches meet at a value of the momentum which is approximately equal to the plasmon cut-off momentum k_c, where the bulk plasmon branch meets the particle-hole continuum. We notice that there is a moderate damping of the surface plasmons at all momenta up to about the cut-off k_c, where the damping suddenly increases.

Inglesfield and Wikborg have also solved for the eigenfunctions and their results are illustrated in Fig. 29. We see that the long-wavelength plasmon amplitude is well concentrated at the surface region, but shows some oscillatory structure. For higher momenta the amplitude extends rather deep into the system.

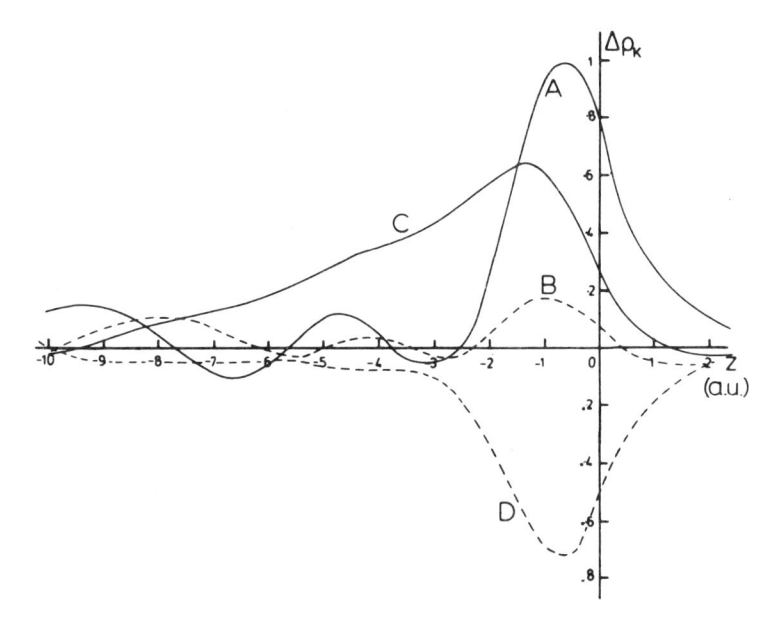

FIG.29. A: Re $\rho_q(z)$ for $q = 0.05$ a.u., $\omega = 0.42$ a.u.
B: Im $\rho_q(z)$ for the same values.
C and D: Re $\rho_q(z)$ and Im $\rho_q(z)$ for $q = 0.4$ a.u., $\omega = 0.58$ a.u.
Density corresponds to Al. (From Ref.[52].)

The question of the surface plasmon dispersion is not yet settled. On one hand, we have theoretical support that the dispersion should be negative at small momenta and, on the other, that the overall behaviour indicates that the surface plasmon branch shows a strong positive dispersion. From the experimental point of view there seems to be no accurate information about surface plasmon dispersion at present.

5.5. Remarks about the surface energy

Results from the previous sections will be used to illustrate some of the problems about the surface energy of free-electron-like metals. We have only derived the classical limit of the response function, and therefore can only comment on the long-wavelength contribution to the interaction part of the surface energy. However, it is precisely in this limit that the effects of the collective modes are most important.

The interaction part of the surface energy per unit area is defined as

$$E^s = \int_0^1 \frac{d\lambda}{\lambda} E^s(\lambda) \tag{5.44}$$

where

$$E^s(\lambda) = \frac{1}{2} \int d^3x_1 \, d^3x_2 \, v(\vec{x}_1, \vec{x}_2) \, \langle \hat{\rho}(\vec{x}_1 t) \hat{\rho}(\vec{x}_2 t) \rangle_\lambda^s$$

$$= 2 \int_0^\infty d\omega \, \text{Im} \int d^3x \, \alpha^s(\vec{x}, \vec{x}, \omega, \lambda)$$

$$= 2 \int \frac{d^2K}{(2\pi)^2} \int_0^\infty d\omega \, \text{Im} \int dz \, \alpha_K^s(z, z, \omega, \lambda) \tag{5.45}$$

The superscript s refers to the surface contribution and λ indicates that we have replaced e^2 by a variable coupling strength λe^2 in the formulas for $v(\vec{x}_1, \vec{x}_2)$ and $\alpha_K^s(z, z, \omega)$. Since we are using atomic units where $e = m = \hbar = 1$, the explicit effects of the transformation are not always obvious.

Using the translational invariance in the x-y plane, we can write

$$E^s = \int \frac{d^2K}{(2\pi)^2} E_K^s \tag{5.46}$$

with

$$E_K^s = 2 \int_0^1 \frac{d\lambda}{\lambda} \int_0^\infty d\omega \, \text{Im} \int dz \, \alpha_K^s(z, z, \omega, \lambda) \tag{5.47}$$

Here we can only make a remark about the long-wavelength contribution. Using our result for $K = 0$, Eq. (5.31), we first obtain

$$\int_0^\infty d\omega \, \text{Im} \int dz \, \alpha_0^s(z, z, \omega, \lambda) = \frac{\lambda^{\frac{1}{2}}}{8} \left[\omega_s - \frac{\omega_0}{2} \right]$$

and thus we find for the infinite-wavelength contribution,

$$E_0^s = 2 \int_0^1 \frac{d\lambda}{\lambda} \frac{\lambda^{\frac{1}{2}}}{8} \left(\omega_s - \frac{\omega_0}{2} \right) = \frac{1}{2} \left(\omega_s - \frac{\omega_0}{2} \right) \tag{5.48}$$

This is precisely the result obtained by Schmidt and Lucas [30] from rather classical considerations. They use the result in Eq. (5.48) for all momenta up to the plasmon cut-off momentum k_c.

It should be remarked that the long-wavelength contribution given by Eq. (5.48) only corresponds to the correlation energy contribution. In

TABLE VI. EXCHANGE AND CORRELATION CONTRIBUTIONS TO THE SURFACE ENERGY (in erg/cm^2)

r_s	Wikborg-Inglesfield	Lang-Kohn	Harris-Jones	Schmidt-Lucas
2.07 (Al)	1350	180	1200	1500
4	190	35	185	270
6	60	10	60	100

addition, one has to include the contribution from exchange, the decrease in the kinetic energy when a surface is created and also the electrostatic contribution from the ions. The theory by Schmidt and Lucas [30] ignores these effects and attributes the entire change in energy to the change in total energy of the collective modes. The theory of Lang and Kohn [28], on the other hand, seems to neglect the contributions from the surface modes but treats the change in the kinetic energy in an accurate way.

The vexed question about the surface energy of simple metals is not yet settled. On one hand, we have the results by Lang and Kohn [28] using the density functional scheme. At the other extreme, we have the plasmon theory by Schmidt and Lucas [30], attributing the entire effect to the collective modes. Several authors have tried to calculate the exchange and correlation contribution to the surface energy, using different models. We should like to refer particularly to work by Peuckert [53], Harris and Jones [52], Jonson and Srinivasan [55], Wikborg and Inglesfield [56], for further discussion. As an illustration, some results for Al are given in Table VI.

There seems to be a genuine discrepancy between the Lang and Kohn approach and the dielectric approach, where a substantial contribution from the surface collective modes comes in, as is seen from Table VI. No complete calculation has yet been presented where the kinetic energy and electrostatic contribution have been calculated with the same accuracy as in the work by Lang and Kohn and where within the same model the exchange and correlation has been accurately calculated from the dielectric approach.

REFERENCES

[1] FOWLER, R.H., GUGGENHEIM, E., Statistical Thermodynamics, Cambridge University Press (1960).
[2] FOWLER, R.H., NORDHEIM, L.W., Proc. R. Soc. (London) Ser. A 119 (1928) 173.
[3] GADZUK, J.W., PLUMMER, E.W., Rev. Mod. Phys. 45 (1973) 487.
[4] STERN, E.A., FERRELL, R.A., Phys. Rev. 120 (1960) 13.
[5] RITCHIE, R.H., Phys. Rev. 106 (1958) 874.
[6] RITCHIE, R.H., Surf. Sci. 34 (1973) 1.
[7] MAHAN, G.E., in Elementary Excitations in Solids, Molecules and Atoms (DEVREESE, J.T., KUNZ, A.B., COLLINS, T.C., Eds), Plenum, London and New York (1974).
[8] HEINE, V., Proc. Phys. Soc. 81 (1962) 300.
[9] GOODWIN, E.T., Proc. Camb. Philos. Soc. 35 (1939) 205.
[10] BLOUNT, E.I., Solid State Phys. 13 (1962) 362.
[11] FORSTMANN, F., Z. Phys. 235 (1970) 69.
[12] FORSTMANN, F., PENDRY, J.B., Z. Phys. 235 (1970) 75.

[13] HARRISON, W.A., Phys. Rev. 118 (1960) 1182.
[14] GADZUK, J.W., J. Vac. Sci. Technol. 9 (1972) 591.
[15] WATTS, C.M.K., J. Phys. F 1 (1971) 272.
[16] BOUDREAUX, D.S., Surf. Sci. 28 (1971) 344.
[17] HOFFSTEIN, V., Solid State Commun. 10 (1972) 605.
[18] LOUCKS, T.L., Phys. Rev. Lett. 14 (1965) 693.
[19] WACLAWSKI, B.J., PLUMMER, E.W., Phys. Rev. Lett. 29 (1972) 783.
[20] GARCIA-MOLINER, F., RUBIO, J., J. Phys. C 2 (1969) 1789.
[21] HOHENBERG, P., KOHN, W., Phys. Rev. 136B (1964) 864.
[22] KOHN, W., SHAM, L.J., Phys. Rev. 140A (1965) 1133.
[23] SMITH, J.R., Phys. Rev. 181 (1969) 522.
[24] HEDIN, L., LUNDQVIST, S., Solid State Phys. 23 (1969) 1.
[25] HEDIN, L., LUNDQVIST, B.I., LUNDQVIST, S., Solid State Commun. 9 (1971) 537.
[26] GUNNARSSON, O., LUNDQVIST, B.I., LUNDQVIST, S., Solid State Commun. 11 (1972) 149.
[27] VON BARTH, U., HEDIN, L., J. Phys. C 5 (1972) 1629.
[28] LANG, N.D., KOHN, W., Phys. Rev. B1 (1970) 4555.
[29] BUDD, H.F., VANNIMENUS, J., Phys. Rev. Lett. 51 (1973) 1218.
[30] SCHMIDT, J., LUCAS, A.A., Solid State Commun. 11 (1972) 415, 419.
[31] LANG, N.D., KOHN, W., Phys. Rev. B3 (1971) 1215.
[32] SMITH, J.R., YING, S.C., KOHN, W., Phys. Rev. Lett. 30 (1973) 610.
[33] APPLEBAUM, J.A., HAMANN, D.R., Phys. Rev. B6 (1972) 2166; Phys. Rev. Lett. 31 (1973) 106.
[34] HODGES, L., WATSON, R.E., EHRENREICH, H., Phys. Rev. B5 (1972) 3953.
[35] LEVIN, K., LIEBSCH, A., BENNEMAN, K.H., Phys. Rev. B7 (1973) 3066.
[36] FULDE, P., LUTHER, A., WATSON, R.E., Phys. Rev. B8 (1973) 440.
[37] JONES, R.O., private communication.
[38] MARCH, N.H., TOSI, M.P., Proc. R. Soc. (London) Ser. A 330 (1972) 373.
[39] MUKHOPADHYAY, G., LUNDQVIST, S., Nuovo Cim. 27B (1975) 1.
[40] HARRIS, J., JONES, R.O., J. Phys. C. 6 (1973) 3585.
[41] ŠUNJIĆ, M., TOULOUSE, G., LUCAS, A.A., Solid State Commun. 11 (1972) 1629.
[42] HEINRICHS, J., Phys. Rev. B7 (1973) 3478.
[43] MUKHOPADHYAY, G., MAHANTY, J., Solid State Commun. 16 (1975) 597.
[44] NEWNS, D.M., Phys. Rev. B1 (1970) 3304.
[45] BECK, D.E., CELLI, V., Phys. Rev. B2 (1970) 2955.
[46] BECK, D.E., CELLI, V., LO VECCHIO, G., MAGNATERRA, A., Nuovo Cim. B 68 (1970) 230.
[47] HARRIS, J., GRIFFIN, A., Phys. Lett. 34A (1971) 51; Can. J. Phys. 48 (1970) 2592.
[48] KUNZ, C., Z. Phys. 196 (1966) 311.
[49] BENNETT, A., Phys. Rev. B1 (1970) 203.
[50] BECK, D.E., CELLI, V., Phys. Rev. Lett. 28 (1972) 1124.
[51] FEIBELMAN, P., Ann. Phys. (in press).
[52] INGLESFIELD, J.E., WIKBORG, E., Solid State Commun. 14 (1974) 661.
[53] PEUCKERT, V., Z. Phys. 241 (1971) 191.
[54] HARRIS, J., JONES, R.O., J. Phys. F 4 (1974) 1170.
[55] JONSON, M., SRINIVASAN, G., Phys. Scr. 10 (1974) 262.
[56] WIKBORG, E., INGLESFIELD, J.E., Solid State Commun. 16 (1975) 335.

IAEA-SMR-15/8

ELECTROMAGNETIC SURFACE EXCITATIONS

V. CELLI*
International Centre for Theoretical Physics,
Trieste, Italy

Abstract

ELECTROMAGNETIC SURFACE EXCITATIONS.
 The general idea of elementary excitations. Surface excitations and bulk excitations. Macroscopic theory of electromagnetic excitations and long-wavelength limit. Bulk plasmons in metals, insulators and semiconductors. General considerations on surface plasmons. Dispersion relation for surface plasmons on a flat surface in the long-wavelength limit. Excitation of surface plasmons in diffraction gratings and Wood's anomalies. Reflectivity of rough surfaces. Electromagnetic excitations in thin films and layered structures: general theory in the long-wavelength limit. Example: frequency spectrum of surface excitations in a thin film in vacuo and in a film on a substrate. Experimental determination of surface plasmon frequency spectrum by the method of frustrated total reflection. Microscopic theory of surface plasmons at short wavelengths. Random phase approximation in an inhomogeneous electron gas. Hydrodynamic approximation. Dispersion relation of surface plasmons and the surface density profile.

1. INTRODUCTION

The concept of 'elementary excitations' has proved extremely useful in understanding the properties of large systems, especially in response to the action of external probes.

Classically, it arises from the concept of a 'normal mode' of the system for small oscillations around equilibrium. These oscillations can be mechanical, electrical, hydrodynamic, and so on, and are 'small' in the sense that they can be treated within a linear approximation, at least as a starting point. Deviations from linearity can be introduced as a refinement at a later stage. The response of the system to external fields is then described by the excitation of these normal modes with an amplitude proportional to the strength of the external fields. The thermal properties of the system can also be described, at least in part, in terms of the excitation of these normal modes. In general, however, the particles of the system are also capable of incoherent motions that contribute appreciably and even dominantly to the total free energy: we need only think of an ideal gas as an example. The presence of incoherent motions, as well as mode-mode interactions arising from non-linearities, gives rise to a decay of the amplitude of an excited normal mode with a characteristic relaxation time τ; thus the modal amplitude at time t, after the exciting field is switched off, will be

$$\text{Re}\,[A_0\,\exp[-i(\omega_R - i\Gamma)t]] \tag{1.1}$$

where A_0 is some complex function of the space variables, ω_R is the angular frequency of the mode and $\Gamma = 1/\tau$. Correspondingly, the complex amplitude

---
* Present address: Dept. of Physics, University of Virginia, Charlottesville, Va., USA

of the driven oscillations at frequency ω in the vicinity of an isolated modal frequency ω_R is proportional to

$$[\omega - (\omega_R - i\Gamma)]^{-1} \tag{1.2}$$

In other words, the complex linear response function has a pole at $\omega = \omega_R - i\Gamma$, in the lower-half complex frequency plane. Measurements of the response will show a Lorentzian peak of width Γ centred at ω_R.

Quantum mechanically, the amplitude (1.1) must be quantized in such a way that $|A_0|^2$ is proportional to $n + \frac{1}{2}$, where n is the number of quanta of energy $\hbar\omega_R$ present. Classical theory applies when $n \gg 1$. Depending on the phenomenon under consideration, such quanta are called phonons, plasmons, magnons, and so on. One quantum, whatever its nature, constitutes an 'elementary excitation' of the system. As long as the linear theory applies, arbitrary numbers of quanta can be excited in the various normal modes, or, in other words, the elementary excitations are independent of each other and behave as free particles. Refinements in the theory can be introduced through the interaction and scattering of the elementary excitations.

It is also possible to consider elementary excitations that do not correspond to a true classical mode: one-electron transitions from an occupied to an empty state are a prime example. In this case the response function still has the form (1.2), where $\hbar\omega_R$ is now the energy difference between the final and initial states. Of course, only one 'elementary excitation' can now exist for a given transition.

Thus the response of the system to external disturbances, including to some extent heating, is conveniently described if we know the 'excitation spectrum', i.e. the types of elementary excitations that can be present, with their frequency and damping.

The presence of surfaces will in general alter the whole excitation spectrum of the system. However, some excitations, known as 'bulk excitations' will only be affected marginally, in the sense that their frequency and damping is changed by terms of order d/L where L is a typical size of the system and d is the thickness of the surface region. The space-dependent amplitude A_0 in (1.1), for bulk excitations, is changed markedly in the surface region, which occupies a fraction d/L of the total volume, while through the bulk no essential change occurs up to order d/L. However, the surface can effectively mix excitations that have the same frequency in the bulk, as in the case of longitudinal and transverse phonons.

A fraction $\sim$ d/L of the excitations, known as 'surface excitations', are physically localized near the surface and have a frequency spectrum that is determined by the presence of the surface and by its structure. Examples of these excitations are the surface plasmons, surface phonons, surface polaritons and surface magnons.

2. PLASMONS

We shall discuss here the electronic excitations in a non-magnetic material, while the nuclei are held at their equilibrium positions. Further, we have in mind experiments such as optical absorption or inelastic electron scattering, where individual one-electron transitions are not resolved. The global effect of these transitions is described by the frequency-dependent dielectric tensor $\epsilon_{ij}(\vec{r}, \vec{r}'; \omega)$ that relates the electric displacement $D_i(\vec{r}, \omega)$ to the field $E_j(\vec{r}', \omega)$:

$$D_i(\vec{r}, \omega) = \int d^3 r' \sum_j \epsilon_{ij}(\vec{r}, \vec{r}'; \omega) E_j(\vec{r}', \omega) \qquad (2.1)$$

If ϵ is known, the fields are found by solving Maxwell's equations:

$$\nabla \cdot \vec{D} = 0 \qquad (2.2a)$$

$$\nabla \cdot \vec{B} = 0 \qquad (2.2b)$$

$$\nabla \times \vec{E} = i\omega \vec{B} \qquad (2.2c)$$

$$\nabla \times \vec{B} = -i\omega \vec{D} \qquad (2.2d)$$

We have assumed that there are no free charges and we have included in $-i\omega \vec{D}$ the current term $4\pi \vec{j}/c$. After a brief summary of the bulk solutions, we give a general discussion of the physically different regimes where interesting solutions occur in the presence of surfaces, and then consider particular cases.

2.1. Bulk plasmons

In an infinite medium, we neglect the variation of ϵ on the scale of the interatomic distance a when considering long-wavelength fields. Then ϵ depends only on the difference $\vec{r} - \vec{r}'$, and Fourier transforming (2.1) we have

$$D_i(\vec{k}, \omega) = \sum_j \epsilon_{ij}(\vec{k}, \omega) E_j(\vec{k}, \omega) \qquad (2.3)$$

In the following we assume for simplicity that the medium is isotropic or has cubic symmetry; then ϵ_{ij} must be of the form:

$$\epsilon_{ij} = \epsilon \delta_{ij} + \eta k_i k_j \qquad (2.4)$$

For transverse fields ($\vec{k} \cdot \vec{E} = 0$) we have then $\vec{D} = \epsilon \vec{E}$, while for longitudinal fields ($\vec{k} \times \vec{E} = 0$), $\vec{D} = (\epsilon + \eta k^2)\vec{E}$. The ratio $\eta k^2/\epsilon$ vanishes in the limit $k \to 0$, being of order $v^2 k^2/\omega^2$, where v is the average electron velocity. In the following we shall neglect this difference between longitudinal and transverse dielectric constants.

The solutions of (2.2) when $\vec{D}(\vec{k}, \omega) = \epsilon(\vec{k}, \omega) \vec{E}(\vec{k}, \omega)$ fall into two classes: longitudinal modes ($\vec{k} \times \vec{E} = 0$) with $\epsilon = 0$ and transverse modes ($\vec{k} \cdot \vec{E} = 0$) with $\epsilon = k^2 c^2/\omega^2$ [1]. For an electron gas, neglecting terms of order $(vk/\omega)^2$, we have

$$\epsilon(k, \omega) = 1 - \omega_p^2/\omega^2 \qquad (2.5)$$

where the plasma frequency ω_p is given in terms of the electron density ρ by

$$\omega_p^2 = 4\pi \rho e^2/m \qquad (2.6)$$

We find then longitudinal plasmons at $\omega = \omega_p$ and transverse plasmons at $\omega = (\omega_p^2 + k^2 c^2)^{1/2}$. In a real solid the condition Re $\epsilon(\omega) = 0$ may be satisfied for one or more values $\omega = \omega_R$. This is not enough to conclude that plasmons exist as well-defined excitations, because in general Im $\epsilon(\omega_R)$ need not be small. In the longitudinal case, $\vec{D}$ can be identified with the external field and the relevant response function is $E/D = 1/\epsilon$. Expanding near the point $\omega = \omega_R$ we have

$$\frac{1}{\epsilon} \simeq \frac{1}{(\omega - \omega_R)\epsilon_R' + i\epsilon_I} \tag{2.7}$$

where $\epsilon_R' = \partial \operatorname{Re} \epsilon/\partial \omega$ at $\omega = \omega_R$ and $\epsilon_I = \operatorname{Im} \epsilon(\omega_R)$. We find then that the response function has a pole at $\omega_R - i\Gamma$, where $\Gamma = \epsilon_I/\epsilon_R'$ gives the width of the plasmon resonance. This width arises from the decay of the plasmon in individual electron-hole excitations, which are to be regarded here as 'incoherent motions' according to the discussion given in the introduction. For $kv < \omega$ the width comes dominantly[1] from interband transitions, if any are present at $\omega = \omega_R$; when $kv > \omega$ there is a large contribution from intraband transitions, i.e. from the acceleration of electrons in partially filled bands, if any are present. For instance, in a doped semiconductor where the carrier density is ρ_c, the dielectric constant for $\hbar\omega < \Delta$ (Δ is the band gap) can be taken to be, at $k = 0$,

$$\epsilon(\omega) = \epsilon_0 - 4\pi \rho_c e^2/m^* \omega^2 \tag{2.8}$$

showing a plasma pole at $\omega_{pc}^2 = 4\pi \rho_c e^2 / \epsilon_0 m^*$ (where m^* is the effective mass of the carriers). On the other hand, for $\omega \gg \Delta$ we have $\epsilon \simeq 1 - 4\pi \rho e^2/m$ where ρ is the density of valence electrons. Typically, there will be a broad 'valence plasmon' with an energy of several eV, while the plasma resonance of the free carriers is in the range 1 - 50 meV. Generally, the frequency of a plasma resonance increases with the wave number k, according to $\omega_R(k) = \omega_p + \alpha(kv/\omega_p)^2$ where α is a constant of order unity.

2.2. Surface plasmons: general considerations

We have seen that the frequency and damping of bulk plasmons depend on the wave number k through the dimensionless ratio kv/ω_p, provided that ka can be taken as small. Here a is the lattice spacing and v the mean velocity of the electrons. For 'valence' plasmons v/ω_p is essentially the Thomas-Fermi screening length, which is not very different from a and from k_F^{-1} if $\hbar k_F$ is the Fermi momentum. Thus the excitation spectrum changes on the scale set by one length of microscopic character. In the transverse response, which is completely separate from the longitudinal, the dimensionless ratio kc/ω is also important.

The situation is far more involved when surfaces are present. A list of characteristic lengths of the problem includes for 'valence' resonances:

(a) Microscopic lengths: the 'thickness' d of the surface, in addition to v/ω, a, k_F^{-1};

---

[1] In a degenerate free-electron gas, at zero temperature, there is no damping until $k = k_c$, where the plasmon merges into the particle-hole continuum.

(b) Characteristic lengths of the excitation: the wavelength $2\pi/K$ in the direction parallel to the surface and the penetration depth γ^{-1}, in addition to c/ω;

(c) Other characteristic lengths of the medium: one (or more) geometrical size L (e.g. the thickness of a slab), the electron mean free path ℓ.

There are some restrictions on the relative size of these lengths: for instance, $a < (L, \ell)$ and $c/\omega > (K, v/\omega)$. Furthermore, some cases can be discarded because the excitations turn out to be overdamped ($\Gamma > \omega_R$). Even so, there are far too many possibilities [2] for us to cover systematically, especially if we want to consider superimposed thin films, each with its characteristic lengths. We discuss only some typical and important cases.

2.3. Semi-infinite medium at long wavelengths ($vK/\omega \ll 1$), flat surface

We assume that all the microscopic lengths of group (a) (in Section 2.2) and the electron mean free path ℓ are much smaller than the lengths of group (b), while $L \to \infty$. Then we have a simple local relation between $\vec{D}$ and $\vec{E}$:

In vacuum $\quad\quad \vec{D}(\vec{r}) = \vec{E}(\vec{r})$ (2.9)

In the medium $\quad\quad \vec{D}(\vec{r}) = \epsilon \vec{E}(\vec{r})$ (2.10)

The solution of the Maxwell equations (2.2) can now be obtained, subject to the appropriate boundary conditions at infinity and to the continuity of $\vec{B}$, $\vec{E}_\parallel$ and $\vec{D}_\perp$ across the boundary. There are 'bound-state' solutions with fields decaying exponentially at $\pm \infty$, and 'scattering-state' solutions, with a given incoming field at infinity. We shall look for the latter and compute the reflection amplitude f, defined as the ratio E_0/E_i, where E_i is the amplitude of the incoming electric field:

$$\vec{E}_i \exp[i(\vec{K} \cdot \vec{R} - k_0 z - \omega t)]$$

and E_0 is the amplitude of the reflected field:

$$\vec{E}_0 \exp[i(\vec{K} \cdot \vec{R} + k_0 z - \omega t)]$$

We can choose the $\vec{E}$ vector of the incoming wave to be either parallel to the surface (transverse electric or S waves) or in the plane of incidence (transverse-magnetic or P waves). It is physically clear that P waves can be expected to create an accumulation of electric charge at the surface, thus exciting surface plasmons. Thus the surface provides a coupling between transverse waves in the vacuum and charge accumulations on the surface which are longitudinal in character, coming from a discontinuity of E_z, i.e. a surface divergence. The reflection coefficient for P waves on a flat surface is worked out in text books (see also (2.22) below):

$$R = |f|^2 = \left| \frac{\epsilon(\omega) \cos\theta - [\epsilon(\omega) - \sin^2\theta]^{1/2}}{\epsilon(\omega) \cos\theta + [\epsilon(\omega) - \sin^2\theta]^{1/2}} \right|^2 \quad\quad (2.11)$$

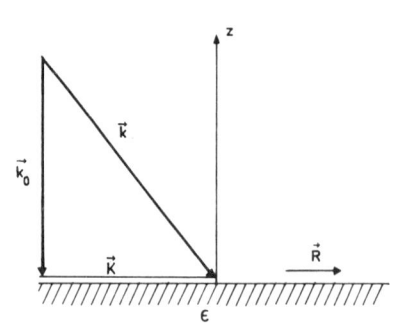

FIG.1. Surface scattering geometry.

where θ is the incidence angle (see Fig.1). The denominator cannot vanish for real values of ϵ and $0 < \cos\theta < 1$ (note that when $\epsilon < 1$ the square root is imaginary and $R = 1$, corresponding to total reflection).

We should now look for 'bound-state' solutions: these are properly the surface plasmons. The detailed treatment is given in Professor Lundqvist's paper and is not repeated here. We give the final result for the existence of normal modes:

$$\left(\frac{Kc}{\omega}\right)^2 = \frac{\epsilon(\omega)}{1 + \epsilon(\omega)} \qquad (2.12)$$

If we insert $\epsilon = 1 - \omega_p^2/\omega^2$, we find that the dispersion relation $\omega_R(K)$ always lies below the line $\omega_R = cK$. When Kc/ω_p is small, $\omega_R = cK$; when Kc/ω_p is large, ω_R approaches $\omega_p/\sqrt{2}$. There is no coupling between light incident on the surface and the surface plasmons if the surface is perfectly flat because, for a given K, the frequency of light $c\sqrt{K^2 + k_0^2}$ is always greater than $\omega_R(K)$. Nevertheless the reflection coefficient (2.11) contains in a hidden way the dispersion relation (2.12). Note that $\cos\theta = k_0/(k_0^2 + K^2)^{1/2}$ and continue formally the expression (2.11) to imaginary values of k_0. Since

$$k_0^2 + K^2 = \omega^2/c^2 \qquad (2.13)$$

the denominator of (2.11) becomes

$$i\left[\epsilon\sqrt{\frac{K^2c^2}{\omega^2} - 1} + \sqrt{\frac{K^2c^2}{\omega^2} - \epsilon}\right] \qquad (2.14)$$

which vanishes when (2.12) holds. Thus the bound-state frequencies appear as poles in the reflection amplitude for the scattering states (a well-known phenomenon in scattering theory!).

To observe surface plasmons by optical means, we can do one of two things: (a) create a very narrow gap between two adjacent surfaces; or (b) allow the admixture of higher values of K by using a grating or a rough surface. Gratings and rough surfaces are considered in the next two sections, while the narrow gap geometry will be treated together with the slab geometry.

Before proceeding, we write the resonance condition in the more general case of an interface between two media of dielectric constants ϵ_1 and ϵ_2. The quantities

$$\gamma_i = \sqrt{K^2 - \omega^2 \epsilon_i/c^2} \qquad (2.15)$$

are the inverse penetration lengths of transverse fields into the i-th medium (i.e. the fields are proportional to $\exp(-\gamma_i |z|)$). Then the reflection coefficient is singular when

$$\frac{\epsilon_1}{\gamma_1} + \frac{\epsilon_2}{\gamma_2} = 0 \qquad (2.16)$$

The solution ω_R of (2.16) starts along the $\omega = cK$ line (if $\epsilon_i \to 1$ for $K \to 0$) and approaches for large cK/ω the solution of $\epsilon_1 + \epsilon_2 = 0$. This 'electrostatic' equation is obtained by using only Poisson's equation instead of the full complement of Maxwell's equations or, as one often says, neglecting retardation effects. If $\epsilon_i = 1 - \omega_{pi}^2/\omega^2$, the 'interface plasmon' in the electrostatic limit is given by

$$\omega_R^2 = \tfrac{1}{2}(\omega_{p1}^2 + \omega_{p2}^2) \qquad (2.17)$$

2.4. Plasmon excitation in diffraction gratings

The conservation of parallel momentum can be frustrated in a neat way by making the surface into a grating. If b is the spacing of the grating, the n-th diffracted beam will have parallel momentum:

$$K_n = K + \frac{2\pi n}{b} = \frac{\omega}{c} \sin\theta + \frac{2\pi n}{b} \qquad (2.18)$$

where θ is the angle of incidence. Now, if K_n equals the surface plasmon wave number for the same frequency ω, as given by (2.12), a resonant matching occurs (see Fig.2) and the corresponding diffracted intensity shows a maximum (Fig.3). 'Anomalies' in the diffracted intensity from metal gratings were observed by Wood as far back as 1902, but the phenomenon was not used to study the surface plasmon dispersion relation until 1967 [3]. Ritchie et al. [4] used light with wavelengths from 3000 Å to 15000 Å to obtain the surface plasmon dispersion curves in aluminium and gold (Fig.4).

The formal theory of diffraction from a grating can be developed along the lines of the Rayleigh theory for sound diffraction [5, 6][2]. We restrict ourselves to the geometry of Fig.5, i.e. we consider only p-polarized light with the scattering plane perpendicular to the direction of the grooves (the y direction). This is an illustrative example that contains the interesting features of the problem; it can be shown that s-polarized light in the scattering geometry of Fig.5 does not couple to the surface plasmons, since the $\vec{E}$ field does not have a component perpendicular to the surface.

---

[2] This theory is not exact, since it uses expansions for the fields (see Eqs (2.19) and (2.21)) that are incomplete in the region of peaks and valleys of the surface profile. However, it leads to the correct results to lowest order in the surface roughness.

400   CELLI

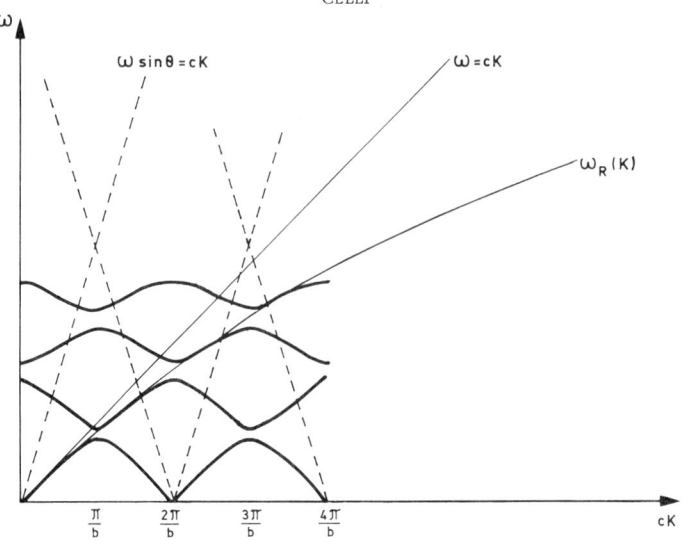

FIG. 2. Frequency spectrum of surface plasmons on a grating (schematic), also showing the frequency spectrum $\omega_R(K)$ for a flat surface (thin line). External light incident at an angle θ from the normal in the geometry of Fig. 3 gives rise to diffracted beams having a frequency spectrum shown by the broken lines. Resonant coupling to surface plasmons occurs where the broken lines cross the continuous curves.

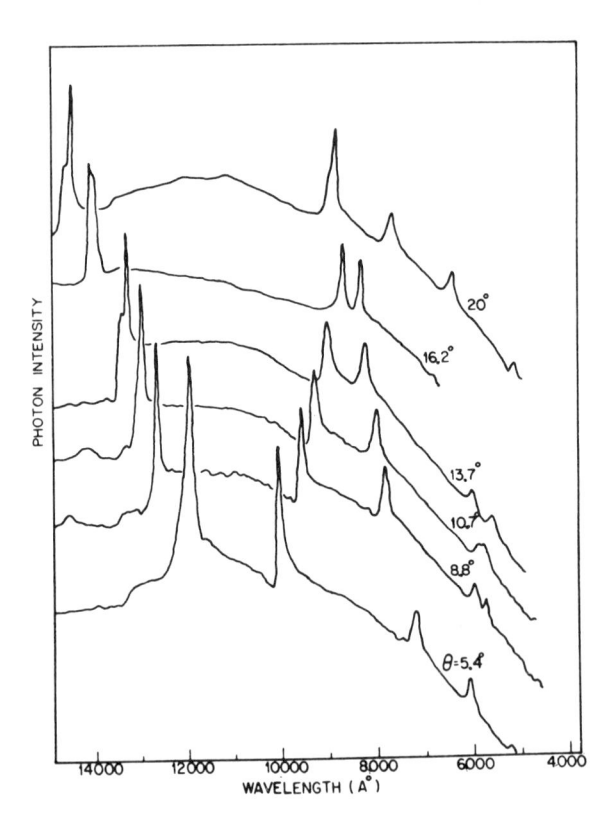

FIG. 3. p-polarized spectra of tungsten lamp diffracted by concave gratings for varying angles between entrance and exit slits (from Ref. [4]).

SURFACE EXCITATIONS

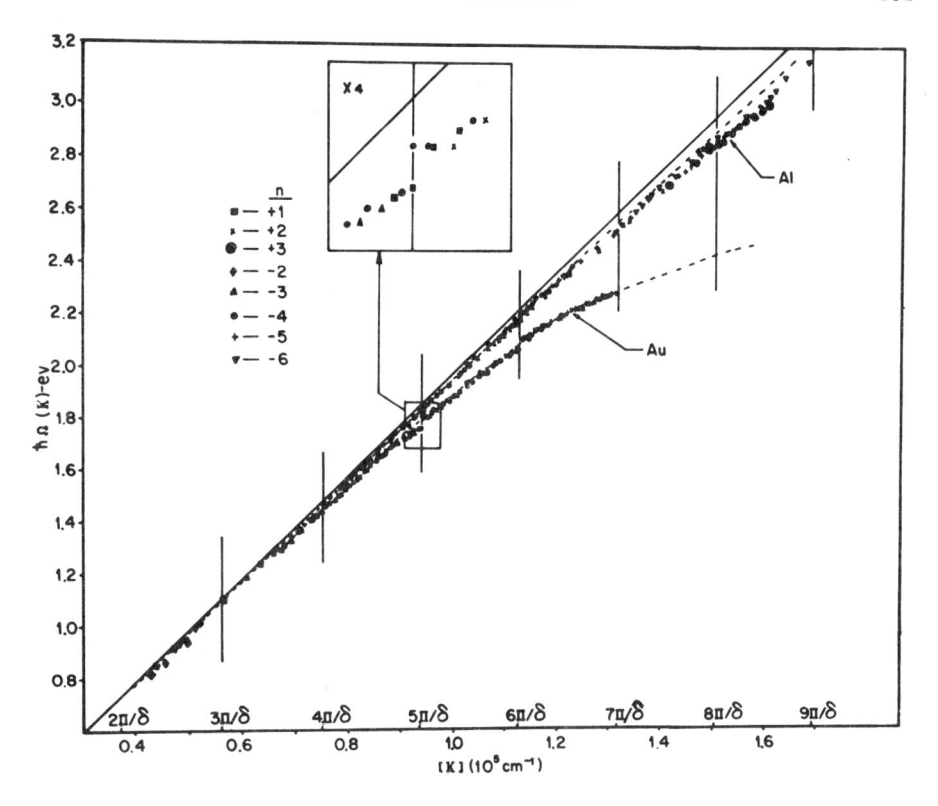

FIG.4. Dispersion curves of surface plasmons in Al and Au (from Ref.[4]).

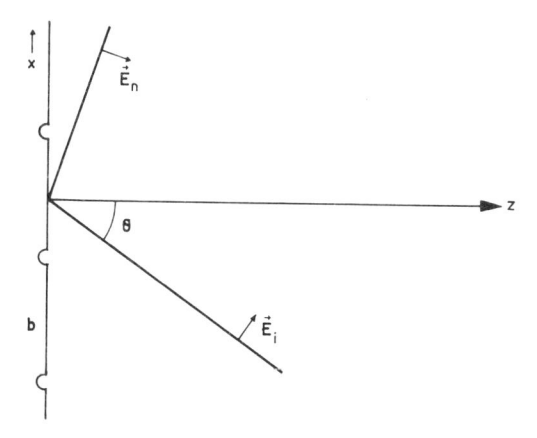

FIG.5. Scattering geometry for p-polarized waves incident on a grating when the plane of incidence is perpendicular to the direction of the grooves.

If the equation of the surface is $z = \zeta(x)$, for $z > \zeta$ the magnetic field in vacuum is given by

$$B_y = \exp[i(Kx-\omega t)] \left[B_i e^{-ik_0 z} + \sum_n B_n \exp[i(G_n x + k_n z)] \right] \quad (2.19)$$

where $G_n = 2\pi n/b$ and

$$k_n^2 + K_n^2 = k_0^2 + K^2 = \omega^2/c^2 \quad (2.20)$$

with $K_n = K + G_n$. Notice that k_n becomes imaginary for sufficiently large n. The beams corresponding to real values of k_n are true diffracted beams, the others are 'evanescent' waves that are confined to the region near the surface and not normally observed. However, these are the waves that can resonantly couple to the surface plasmons. The electric field for $z > \zeta$ has components E_x and E_z of the form (2.19). These can be obtained from the $\vec{B}$ field through the equation $-i\omega\vec{E}/c = \text{rot}\,\vec{B}$.

The magnetic field in the medium ($z<\zeta$) has component B'_y given by

$$B'_y = \exp[i(kx-\omega t)] \sum_n B'_n \exp[i(G_n x + \gamma'_n z)] \quad (2.21)$$

in the region where total reflection occurs, with

$$K_n^2 - \gamma_n'^2 = \omega^2 \epsilon/c^2 \quad (2.22)$$

The electric field for $z < \zeta$ is obtained from $-i\epsilon\omega\vec{E}'/c = \text{rot}\,\vec{B}$. The continuity conditions at the surface are

$$B_y = B'_y \quad ; \quad \frac{\partial B_y}{\partial n} = \frac{1}{\epsilon} \frac{\partial B'_y}{\partial n} \quad (2.23)$$

where $\hat{n}$ is the normal to the surface. Explicitly, the direction of $\hat{n}$ is given by $\nabla(z - \zeta(x))$, so that with $\zeta' = d\zeta/dx$

$$\frac{\partial}{\partial n} = (1 + \zeta'^2)^{-1/2} \left[\frac{\partial}{\partial z} - \zeta' \frac{\partial}{\partial x} \right] \quad (2.24)$$

Using the above equations, (2.23) becomes

$$B_i e^{-ik_0\zeta} + B_0 e^{ik_0\zeta} - B'_0 e^{\gamma'_0\zeta} + \sum_{n\neq 0} (B_n e^{ik_n\zeta} - B'_n e^{\gamma'_n\zeta}) e^{iG_n x} = 0 \quad (2.25)$$

$$-ik_0 B_i e^{-ik_0\zeta} + ik_0 B_0 e^{ik_0\zeta} - \epsilon^{-1} \gamma'_0 B'_0 e^{\gamma'_0\zeta}$$

$$+ \sum_{n\neq 0} (ik_n B_n e^{ik_n\zeta} - \epsilon^{-1}\gamma'_n B'_n e^{\gamma'_n\zeta}) e^{iG_n x} - \zeta'[iK(B_i e^{-ik_0\zeta}$$

$$+ B_0 e^{ik_0\zeta} - \epsilon^{-1} B'_0 e^{\gamma'_0\zeta}) + i\sum_{n\neq 0} K_n(B_n e^{ik_n\zeta} - \epsilon^{-1} B'_n e^{\gamma'_n\zeta}) e^{iG_n x}] = 0 \quad (2.26)$$

From (2.25) and (2.26) we can obtain a set of algebraic equations for the coefficients B_n, B'_n by taking Fourier components of both sides. We can always choose the origin so that the average value of $\zeta(x)$ is zero. If the 'surface wavity' $\zeta(x)$ is small, we can expand the exponentials containing ζ and find approximate equations in terms of the Fourier components:

$$\zeta_n = \frac{1}{b} \int_0^b \zeta(x) e^{-iG_n x} dx \qquad (2.27)$$

The resulting equations are:

$$[-ik_0(B_i - B_0) - \gamma'_0 B'_0] \zeta_n + B_n - B'_n = 0 \qquad (2.28)$$

$$B_i + B_0 - B'_0 + \sum_{n \neq 0} (ik_n B_n - \gamma'_n B'_n) \zeta_n^* = 0 \qquad (2.29)$$

$$[k_0^2(B_i + B_0) + \gamma_0^2 \epsilon^{-1} B'_0] \zeta_n - ik_n B_n + \gamma'_n \epsilon^{-1} B'_n$$
$$- G_n K(B_i + B_0 - \epsilon^{-1} B'_0) \zeta_n = 0 \qquad (2.30)$$

$$-ik_0(B_i - B_0) - \gamma'_0 \epsilon^{-1} B'_0 - \sum_{n \neq 0} \zeta_n^* [(k_n^2 + K_n G_n) B_n + \epsilon^{-1} (\gamma'^2_n - K_n G_n) B'_n] = 0 \qquad (2.31)$$

If we set $\zeta_n = 0$, we recover from (2.29) and (2.31) the result (2.11) for the reflection coefficient, while from (2.28) and (2.30) it follows that B_n and B'_n must vanish, unless the condition $i/k_n + \epsilon/\gamma'_n = 0$ is satisfied. This can happen only for imaginary values of k_n, $k_n = i\gamma_n$: it corresponds to the condition that one of the 'evanescent waves' is resonant with a surface plasmon. (See the discussion of (2.14).) In general, one can obtain B_n and B'_n from (2.28) and (2.30), insert the result in (2.29), (2.31) and solve explicitly for B_0 and B'_0. The reflection amplitude is then found to be

$$f = \frac{B_0}{B_i} = \frac{i\epsilon k_0 + \gamma'_0 - \sum_n \frac{|\zeta_n|^2}{D_n} (iA_n + B_n)}{i\epsilon k_0 - \gamma'_0 - \sum_n \frac{|\zeta_n|^2}{D_n} (iA_n - B_n)} \qquad (2.32)$$

where

$$D_n = \gamma'_n + \epsilon \gamma_n \qquad (2.33)$$

and A_n, B_n are lengthy expressions that simplify considerably when the resonance condition $D_n - 0$ is satisfied. Comparing with (2.14), we see that resonance occurs when a surface plasmon is excited. From the definitions (2.20) and (2.22) one finds that $D_n = 0$ is equivalent to

$$\gamma_n \gamma'_n = K_n^2 \qquad (2.34)$$

In general, the resonances are well separated and can be considered one at a time. We see that, if the plasmon damping is negligible, $|f|$ approaches unity at the plasmon pole, where $D_n = 0$. On the other hand, when D_n is of order K^{-1}, the terms in $|\zeta_n|^2$ are negligible and (2.11) is recovered. In the neighbourhood of the resonance, (2.32) has the characteristic Breit-Wigner form:

$$f \sim \frac{\omega - (\omega_R - i\Gamma_R) - i\Gamma_T}{\omega - (\omega_R - i\Gamma_R) + i\Gamma_T} \tag{2.35}$$

Whether f has a dip or a peak at resonance depends on the ratio Γ_R/Γ_T; note that Γ_R is the lifetime of a surface plasmon on a flat surface, Γ_T is the time spent by the photon as a virtual plasmon. The decrease in reflectivity $\Delta R = 1 - |f|^2$ is given by

$$\Delta R = \frac{2b \cos\theta_i \, a^3 (\omega/c)^3 |\zeta_n|^2}{(a^2-1)^2 (\sin^2\theta_i + a^2 \cos^2\theta_i)} [(1+a^2)\sin^2\theta_i - 2a \sin\theta_i$$

$$\times (\sin^2\theta_i + a^2)^{1/2} + a^2] \tag{2.36}$$

where θ_i is the incident angle ($\sin\theta_i = Kc/\omega$) and $a^2 = -\epsilon$.

In deriving (2.32), we have assumed that B_n does not mix appreciably with any other B_m; but this is certainly an oversimplification. We must recall that plasmons can undergo diffraction due to the presence of the grating, according to the general properties of waves in periodic media. We shall at first treat this phenomenon in the absence of an external field, with the expectation that the resulting resonant frequencies will appear in a formula of type (2.32) for the reflection amplitude when external fields are included. From (2.25) and (2.26) we find coupled homogeneous equations for the fields $B_n B'_n B_m B'_m$. When the determinant of the coefficients vanishes, there are self-sustained solutions that describe the surface plasmons on the rough surface. The explicit equation, putting for brevity $\zeta = \zeta_{n-m}$, is

$$\begin{vmatrix} 1 & -\epsilon & -\zeta\gamma_m & -\epsilon\gamma'_m\zeta \\ -\gamma_n\zeta^* & -\epsilon\gamma'_n\zeta^* & 1 & -\epsilon \\ -\gamma_n & -\gamma'_n & [\gamma_m^2 + (G_n - G_m)K_m]\zeta & -[\gamma'^2_m + (G_n - G_m)K_m]\zeta \\ [\gamma_n^2 - (G_n - G_m)K_n]\zeta^* & -[\gamma'^2_n - (G_n - G_m)K_n]\zeta^* & -\gamma_m & -\gamma'_m \end{vmatrix} = 0 \tag{2.37}$$

Solving this equation, we find, as expected, that a gap arises in the plasmon spectrum when R_m and R_n vanish simultaneously, i.e. when $K_m = -K_n$. In this case, the two roots of (2.37), to lowest order in ζ, are given by

$$\pm 2(\epsilon-1)|\zeta| = \frac{1}{\gamma_n} + \frac{\epsilon}{\gamma'_n} \tag{2.38}$$

It will be recognized that the argument leading to (2.37) and (2.38) is entirely analogous to the treatment of 'almost free electrons' in the band theory of solids; thus the plasmon spectrum must now be regarded as a many-valued periodic function. It is seen from Fig.2 that plasmon resonances will show up not only in the specular beam (corresponding to the line $\omega = cK/\sin\theta$), but also in the diffracted beams (the m-th diffracted beam corresponds to the portion of the line $2\pi n/b + cK/\sin\theta$ that lies above the 'light cutoff' $\omega = cK$). In other words, the amplitude f_m of the m-th diffracted beam will contain not only terms proportional to ζ_m but also terms proportional to $\zeta_{m-n}\zeta_n$:

$$f_m \sim c_1 \zeta_m + c_2 \frac{\zeta_{m-n}\zeta_n}{\omega - (\omega_R - i\Gamma_R) + i\Gamma_T} \tag{2.39}$$

The first term represents the direct transitions from the incoming beam to the m-th beam; the second term describes indirect transitions passing through the n-th beam, which is an evanescent wave resonant with the surface plasmon.

Since external light can couple to surface plasmons in a grating, it must also be possible to cause a grating to emit light by exciting surface plasmons, e.g. through electron bombardment. Such emitted light has been observed and its frequency ω has been monitored as a function of the angle from the surface normal. Since $K = \omega\sin\theta/c$, the measurement of ω and θ gives the momentum of the emitting plasmon, up to a multiple of $2\pi/b$. Teng and Stern [3] reflected 5400 Å light from an Al grating and obtained excellent correlation between 'Wood's anomalies' in the reflected intensity and the peaks in the angular distribution of light emitted at 5400 Å during electron bombardment from the same grating. The polarization of the emitted light indicates that its electric field is always perpendicular to the direction of the grating rulings. This is easy to understand, since charge accumulations at the rulings must be the source of the radiation.

2.5. Plasmons on rough surfaces

A rough surface provides an 'uncontrolled' coupling between the plasmons and the radiation field. The physical source of the coupling is the same as in the case of gratings except of course that, instead of a controlled 'wavity' $\zeta(x)$ with Fourier coefficients ζ_n, we now have a stochastic 'roughness function' $\zeta(x,y) \equiv \zeta(\vec{R})$ with a Fourier spectrum:

$$\zeta(\vec{Q}) = \int d^2\vec{R}\, e^{-i\vec{Q}\cdot\vec{R}}\, \zeta(\vec{R}) \tag{2.40}$$

Now all first-order scattering effects, such as the amount of diffusely reflected light, are proportional to $|\zeta(Q)|^2$, which is to be statistically averaged over the surface. We note that $|\zeta(Q)|^2$ equals $g(Q)$, the Fourier transform of the autocorrelation function for surface roughness:

$$g(\vec{R}) = \int d^2\vec{R}\, \zeta(\vec{R} - \vec{R'})\,\zeta(\vec{R'}) \tag{2.41}$$

In 'ordinary' or 'elastic' light scattering from surface roughness, the momentum transfer $\hbar\vec{Q}$ is such that Q is of order ω/c, where ω is the frequency of the light.

On the other hand, we can consider that 'resonant' scattering through the excitation of a surface plasmon is also a first-order process, in the limit where many channels are available for plasmon decay (cf. the discussion of (2.35)). Effectively, once the incoming photon is converted into a surface plasmon, it is lost to the specular beam. The decrease in specular reflectivity ΔR is given by $1 - |f|^2$, where f has essentially the form (2.32) or (2.35). We can conclude then that ΔR is proportional to $g(Q)/|D_Q|^2$, where D_Q, defined by analogy with D_n, has a sharp resonance when $\omega = \omega_R(\vec{K}+\vec{Q})$. To find ΔR, we must integrate over all $\vec{Q}$'s; however, $|D_Q|^{-2}$ has a sharp peak at $\vec{Q}_R$ such that

$$\omega_R(\vec{K}+\vec{Q}_R) = \omega \tag{2.42}$$

Therefore the decrease in specular reflectivity due to the surface plasmons is roughly proportional to $g(Q_R)$. If $g(Q)$ has a peak for $Q \sim 2\pi/\bar{a}$, the specular reflectivity will show a plasmon-induced dip at a frequency $\omega_R(2\pi/\bar{a})$, supposing $K\bar{a} \ll 1$. Such a correlation of reflectivity dips with surface roughness has been observed (e.g. Ref.[7]). Note that $\bar{a}$ is the correlation length for surface roughness, i.e. essentially the typical scale of surface irregularities.

As in the case of gratings, surface plasmons excited by other means on a rough surface will decay with the emission of light of frequencies up to the maximum value of $\omega_R(K)$, which is $\omega_p/\sqrt{2}$ in simple cases. The amount of light emitted in a given frequency range depends on the efficiency of plasmon excitation at that frequency, on the surface roughness function (being proportional to $|\zeta(Q)|^2$ to lowest order), and on the density of surface plasmon states in that range. Because ω_R flattens out for large K, the density of states for surface plasmons is singular for $\omega \sim \omega_p/\sqrt{2}$, and therefore strong light emission will be observed in the vicinity of this frequency if sufficient surface roughness is present.

The discussion of electromagnetic excitations in gratings and on rough surfaces has been based entirely on a classical solution of the boundary value problem leading to (2.25) and (2.26). An alternative approach is to carry out a transformation that replaces the boundary conditions with additional terms in the wave equations, or in the Hamiltonian corresponding to these equations. Such a transformation [8] consists of introducing new variables $u_1 = x_1$, $u_2 = y$, $u_3 = z - \zeta(x,y)$. The surface has now been flattened out into the plane $u_3 = 0$ and the boundary conditions can be handled without difficulty. On the other hand, additional terms arise when $\partial/\partial x$ is replaced by

$$\frac{\partial}{\partial u_1} - \frac{\partial \zeta}{\partial x}\frac{\partial}{\partial u_3} \tag{2.43}$$

and similarly for $\partial/\partial y$. The terms proportional to $\nabla \zeta$ are then treated by standard perturbation theory or resonant scattering theory. A comprehensive theory of the electromagnetic properties of uneven surfaces has been developed in this fashion by Ritchie and Elson [9]. The advantage of this formulation (at least from the physicist's point of view) is that surface plasmons and photons are each regarded as 'free particles', with a suitable 'interaction Hamiltonian' causing their mixing and scattering, due to the uneven surface. To lowest order in perturbation theory, the scattered intensity is proportional to $|\zeta_G|^2$ and the proportionality coefficient should be the same as in the Rayleigh-Fano theory.

2.6. Electromagnetic excitations in films and layered structures

All the results to be given in this section are derived within the classical long-wavelength approximation, where the local relation $\vec{D}(\vec{r}) = \epsilon \vec{E}(\vec{r})$ holds, with a frequency-dependent dielectric constant $\epsilon = \epsilon(\omega)$. This means that the geometrical sizes L and the penetration depths γ should be longer than the microscopic lengths and the effective electron mean free path $v/|\omega + i/\tau|$. Nevertheless these results have been applied to the discussion of atomic monolayers, with the hope that the classical theory retains some qualitative validity.

The solution of Maxwell's equations in a film geometry presents no difficulty. In the i-th medium, the field components will result in general from the superposition of an incident (+) wave, proportional to

$$\exp[i(Kx + k_i z - \omega t)] \tag{2.44}$$

and a reflected (-) wave, proportional to

$$\exp[i(Kx - k_i z - \omega t)] \tag{2.45}$$

with

$$K^2 + k_i^2 = \epsilon_i \frac{\omega^2}{c^2} \tag{2.46}$$

Note that k_i can be imaginary, in which case we shall put $k_i = i\gamma_i$ where $1/\gamma_i$ is the penetration depth in the i-th medium. Once again, we restrict ourselves to the case of p-polarized waves, with the magnetic field parallel to the interfaces. Let B_i be the magnitude of the field on the left of the i-th interface; the continuity conditions, as we have seen in Section 1, require that B and $\epsilon^{-1}(\partial B/\partial z)$ be continuous across the interface. The fields on the right of the interface are easily found from these conditions and from (2.44)-(2.46). To express the results in a concise form, we introduce a column vector $\vec{b}_i$ having components $B_i^{(+)}$ and $B_i^{(-)}$. Then the fields on the right of the (i-1)-th interface are described by the vector $\vec{t}_i \vec{b}_{i-1}$, where the matrix $\vec{t}_i$ is given by

$$\begin{pmatrix} \frac{1}{2}(1+r_i) & \frac{1}{2}(1-r_i) \\ \frac{1}{2}(1-r_i) & \frac{1}{2}(1+r_i) \end{pmatrix} \tag{2.47}$$

with

$$r_i = \frac{k_i \epsilon_{i-1}}{k_{i-1} \epsilon_i} \tag{2.48}$$

If the thickness of the i-th medium is L_i, the fields at the interface with the (i+1)-th medium are $\vec{b}_i = \vec{p}_i \vec{t}_i \vec{b}_{i-1}$, where

$$\vec{p}_i = \begin{pmatrix} e^{ik_i L_i} & 0 \\ 0 & e^{-ik_i L_i} \end{pmatrix} \tag{2.49}$$

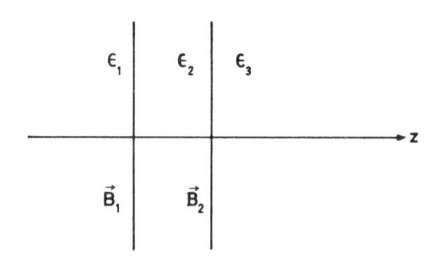

FIG. 6. Superimposed layers of dielectric constants ϵ_1, ϵ_2, ϵ_3. $\vec{B}_1$ is the magnetic field in medium 1 at the interface with medium 2, etc.

We can call $\vec{t}_i$ transmission matrices and $\vec{p}_i$ propagation matrices. Proceeding in this fashion, we find in the case of a film of thickness L_2 and dielectric constant ϵ_2, sandwiched between two media of dielectric constants ϵ_1 and ϵ_3 (see Fig.6), the following expression for the fields in medium 3 at the interface:

$$\vec{t}_3 \vec{p}_2 \vec{t}_2 \vec{b}_1 \tag{2.50}$$

where $\vec{t}_i$ and $\vec{p}_i$ are defined in (2.47) and (2.49).

We consider a field incident from medium 3. Then only a transmitted field $\vec{B}_1^{(-)}$ exists in medium 1. One finds

$$\vec{t}_3 \vec{p}_2 \vec{t}_2 \vec{\ell}_1 = \binom{N}{D} B_1^{(-)} \tag{2.51}$$

where

$$N = \frac{1+r_3}{2} \frac{1-r_2}{2} e^{ik_2 L_2} + \frac{1-r_3}{2} \frac{1+r_2}{2} e^{-ik_2 L_2} \tag{2.52}$$

$$D = \frac{1-r_3}{2} \frac{1-r_2}{2} e^{ik_2 L_2} + \frac{1+r_3}{2} \frac{1+r_2}{2} e^{-ik_2 L_2} \tag{2.53}$$

The reflection amplitude is $f = N/D$. The condition for the existence of surface excitations is $D = 0$. There are several particular cases of this general relation that are of special interest:

(a) Thin film in vacuum

In this case, $\epsilon_1 = \epsilon_3 = 1$ and $r_3 = 1/r_2 = \gamma_1 \epsilon_2 / \gamma_2$. Then the condition $D = 0$ becomes

$$\frac{1+r_2}{1-r_2} = \pm e^{-\gamma_2 L_2} \tag{2.54a}$$

or

$$r_2^{(+)} = -\tanh \tfrac{1}{2} \gamma_2 L_2 \tag{2.54b}$$

$$r_2^{(-)} = -\coth \tfrac{1}{2} \gamma_2 L_2 \tag{2.54c}$$

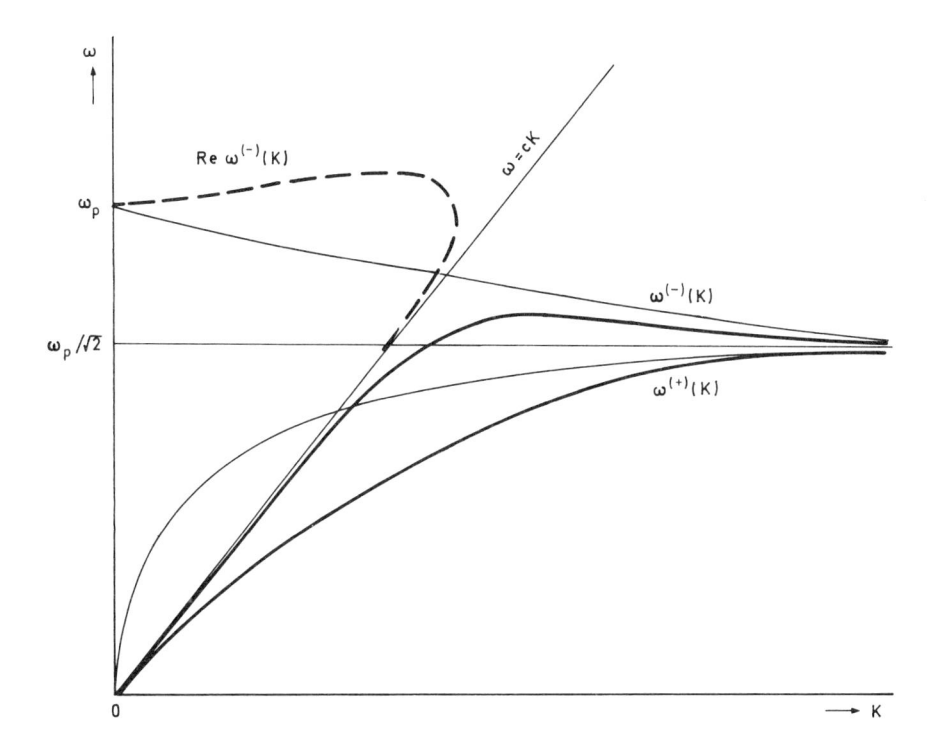

FIG. 7. Frequency spectrum of surface excitations in a thin film, with retardation (thick lines) and without retardation (thin lines). (From Ref. [2].)

where the plus and minus signs correspond to a symmetric and antisymmetric solution. In the electrostatic limit, $\gamma_2 = \gamma_1 = K$ and $r_2 = 1/\epsilon_2$, so that we obtain

$$\frac{\epsilon_2 + 1}{\epsilon_2 - 1} = \pm e^{-KL_2} \tag{2.55}$$

If $\epsilon_2 = 1 - \omega_p^2/\omega^2$, the following solutions of (2.53) are found:

$$\omega^{\pm}(K) = \frac{\omega_p}{\sqrt{2}} (1 \mp e^{-KL_2})^{1/2} \tag{2.56}$$

These are shown schematically as thin lines in Fig. 7. It can be seen that for $KL_2 \gg 1$ the two modes become degenerate at the frequency $\omega_p/\sqrt{2}$: this corresponds to independent oscillations on the two surfaces of the film. For smaller values of KL_2, the fields on the two surfaces begin to couple. The mode $\omega^{(+)}$ approaches zero for $K \to 0$ even in the electrostatic approximation (2.56); when the retardation effects are introduced according to (2.52) the mode is shifted to lower frequency, so that it lies below the $\omega = cK$ line, as shown by the thick lines in Fig.7. The mode $\omega^{(-)}$, which approaches ω_p in the electrostatic approximation (2.56), is affected more drastically by retardation. It is found that (2.54c), with $\epsilon_2 = 1 - \omega_p^2/\omega^2$, has a real solution that does not cross the line $\omega = cK$ and approaches zero as K vanishes. For this mode, as well as the $\omega^{(+)}$ mode, γ_2 tends to ω_p/c in the limit $K \to 0$. However, it is also possible to have a solution of (2.54c) with γ_2 vanishing in the limit $K \to 0$. In this case (2.54c) can be written

$$i\sqrt{\frac{\omega^2}{c^2} - K^2} = -\frac{L}{2}\left(K^2 - \epsilon\frac{\omega^2}{c^2}\right)$$

The corresponding frequency is in the neighbourhood of ω_p (see dashed line in Fig.7) and has a finite imaginary part, indicating that the mode decays by coupling to outgoing waves. This mode [10] is commonly referred to as the 'radiative plasmon in a thin film'.

(b) Film on substrate

In this case medium 3 is the vacuum; medium 2 is a film deposited on medium 1. Hence $r_2 = \gamma_2 \epsilon_1/\gamma_1 \epsilon_2$ and $r_3 = K\epsilon_2/\gamma_2$. From (2.53), the condition for existence of surface excitations is

$$(1 - r_2)(1 - r_3) e^{-\gamma_2 L_2} + (1 + r_2)(1 + r_3) e^{\gamma_2 L_2} = 0 \qquad (2.57)$$

To understand the character of the excitations, let us consider first the electrostatic limit when $r_2 = \epsilon_1/\epsilon_2$ and $r_3 = \epsilon_2$. Then we have

$$\frac{(\epsilon_2 - 1)(\epsilon_2 - \epsilon_1)}{(\epsilon_2 + 1)(\epsilon_2 + \epsilon_1)} = e^{2\gamma_2 L_2} \qquad (2.58)$$

As $L_2 \to \infty$, we obtain the surface plasmon of medium 2:

$$\epsilon_2 + 1 = 0 \qquad (2.59)$$

and the 'interface plasmon' at the boundary between 1 and 2:

$$\epsilon_1 + \epsilon_2 = 0 \qquad (2.60)$$

On the other hand, as $L_2 \to 0$, we see that (2.58) reduces to

$$\epsilon_2(\epsilon_2 + 1) = 0 \qquad (2.61)$$

This has roots at the bulk plasma frequency ω_2 of medium 2:

$$\epsilon_2 = 0 \qquad (2.62)$$

and at the surface plasmon frequency of medium 1:

$$\epsilon_1 + 1 = 0 \qquad (2.63)$$

A schematic plot of the solutions of (2.58) is given by the thick lines in Fig.8 [11, 12]. We have assumed that $\epsilon_i = 1 - \omega_i^2/\omega^2$ for $i = 1, 2$.

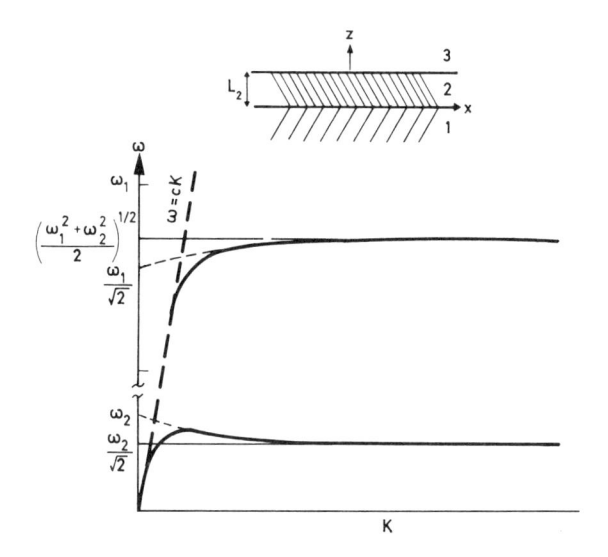

FIG. 8. Dispersion relation of surface excitations in a film deposited on a metal substrate (from Ref. [2]).

The thin lines in Fig. 8 give the solution of (2.57), which includes retardation effects. As usual, retardation effects become important when $cK \approx \omega_p$, causing the dispersion curves to bend down so as to lie below the $\omega = cK$ line.

The results described here have been used to discuss the effect of adsorbed layers on the excitation spectrum of a metal surface [13]. For instance, in electron energy loss experiments from tungsten, the loss peak associated with the surface plasmon in tungsten shifts to lower frequencies when caesium is adsorbed on the surface. This is in qualitative agreement with the (simple) prediction that the observed frequency should correspond to the surface plasmon frequency for caesium in the limit when the caesium layer is thicker than the typical wavelength of the excitations responsible for the loss. In electron scattering, this wavelength should be of the order of several lattice spacings, but much smaller than c/ω_p.

It is clear that the simple theory outlined here is not applicable to caesium layers of atomic thickness; however, Newns [14] has shown that in a highly simplified microscopic model an excitation mode is also obtained at the bulk plasmon frequency of caesium, as predicted by (2.62) in the limit of small KL_2.

(c) <u>A thin air gap: the method of frustrated total reflection</u>

Frustrated total reflection permits the observation of plasmons on a flat surface (the experimental results quoted here are taken from Ref. [15]).

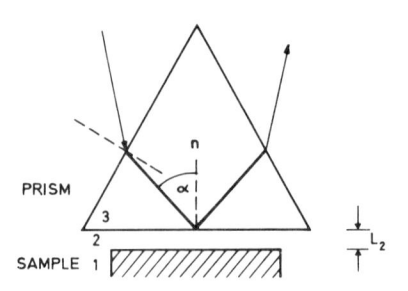

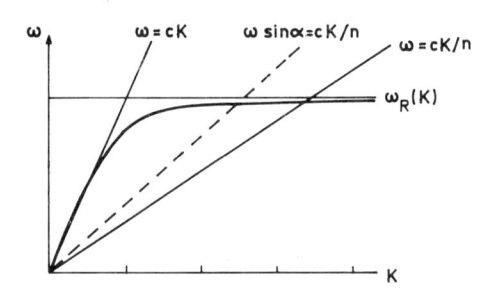

FIG.9. Frustrated total reflection of light. Surface plasmons can be excited in a sample separated by a thin air gap from the base of the prism of dielectric constant n^2. The graph shows the crossing of the frequency curve for surface plasmons $\omega_R(K)$ with the broken line giving the frequency of light incident at an angle α. (From Ref.[15].)

The experimental set-up is shown schematically in Fig.9. The prism can be regarded as an infinite medium of real dielectric constant $\epsilon_3 = n^2$ (n is the index of refraction). We have then

$$K = \frac{n\omega}{c} \sin \alpha \qquad (2.64)$$

and

$$k_3 = \frac{n\omega}{c} \cos \alpha \qquad (2.65)$$

where α is the angle of incidence, as shown in Fig.9. The air gap between the prism and the sample must have thickness such that $\exp(-\gamma_2 L_2)$ is small but not totally negligible. Here

$$\gamma_2 = \sqrt{K^2 - \frac{\omega^2}{c^2}} = \frac{\omega}{c}\sqrt{n^2 \sin^2 \alpha - 1} \qquad (2.66)$$

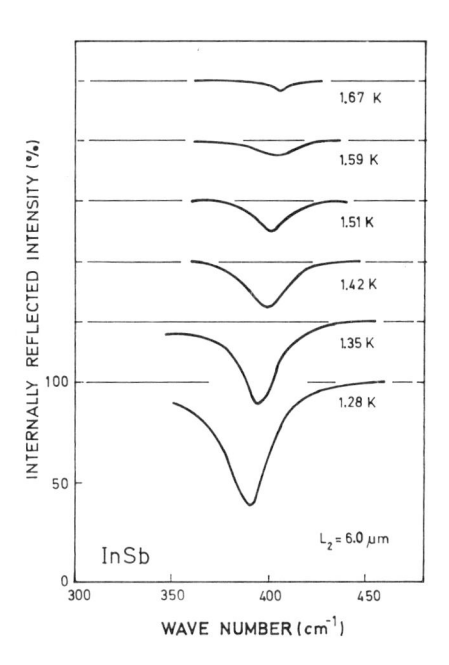

FIG.10. Reflection coefficient versus wave number ($\omega/2\pi c$) in the experimental set-up of Fig.9 for different values of $n \sin \alpha$. The sample was InSb with a bulk plasma frequency ω_p (conduction plasmon), corresponding to a wave number of 426.5 cm^{-1}. (From Ref.[15].)

In this case, $r_3 = k_3/i\gamma_2\epsilon_3$ is purely imaginary, hence $(1-r_3)/(1+r_3)$ is of the form $e^{i\eta}$ (this corresponds to total reflection at the lower face of the prism, in the absence of the sample). The reflection amplitude f can be written, from (2.52) and (2.53):

$$f = \frac{N}{D} = \frac{f_2 + e^{i\eta}}{f_2 e^{i\eta} + 1} \tag{2.67}$$

with

$$f_2 = \frac{1-r_2}{1+r_2} e^{-2\gamma_2 L_2} \tag{2.68}$$

We see from (2.67) that $|f|$ is unity, except when $\text{Im} f_2 \neq 0$. The surface plasmon of medium 1 is precisely at the frequency where $\text{Re}(1+r_2) = 0$ and $\text{Im} f_2$ has a sharp peak (which becomes a δ function if the lifetime of the surface plasmon is infinite). Thus we expect a dip in the reflection coefficient when the conditions for excitation of a surface plasmon are met: this is clearly seen in the experimental data reproduced in Fig.10. The crucial point, according to the discussion given after Eq.(2.14), is that the value of k_2 is complex, so that the value of K exceeds ω/c: in fact, as seen from (2.64), K can reach the maximum value $n\omega/c$. As shown in Fig.10,

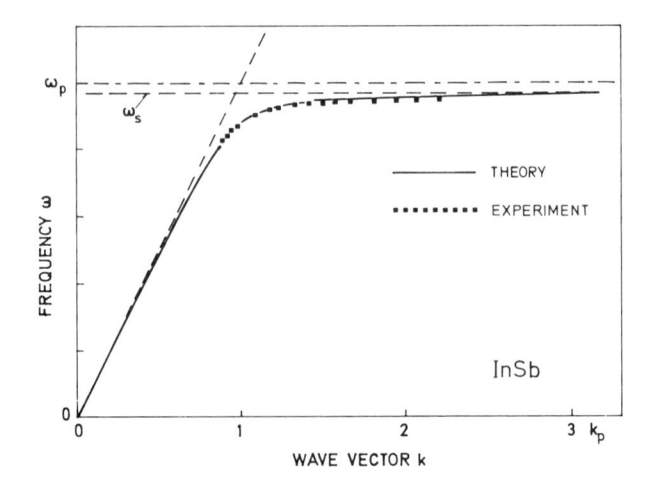

FIG.11. Frequency spectrum for surface plasmons in InSb determined by the method of frustrated total reflection, as shown in Figs 9 and 10 (from Ref.[15]).

the line $n\omega \sin\alpha/c$ intersects the surface plasmon dispersion relation for all values of $\sin\alpha > n^{-1}$. A plot of the dispersion relation obtained by Fischer et al. [15] for InSb is shown in Fig.11; notice that the surface plasmon frequency approaches the value $[\omega_c \epsilon_0 /(\epsilon_0 + 1)]^{1/2}$, where ω_c is the plasmon frequency of the carriers (see the discussion of Eq.(2.8)).

3. MICROSCOPIC THEORY

The extension of the theory of electronic surface excitations to microscopic wavelengths presents considerable difficulties. A summary is given here of some results that have been obtained within the random phase approximation (RPA); in fact much of this section is a general review of the RPA and related approximations in an inhomogeneous system.

The general RPA expression for the longitudinal dielectric constant can be obtained quite simply by considering that each electron moves in the average potential $V_0(\vec{r}) + V_1(\vec{r})$. Here $V_0(\vec{r})$ is the effective one-electron potential in the absence of external fields, and

$$V_1(\vec{r}) = V_{ext}(\vec{r}) + e \int d^3r' \rho_1(\vec{r}')/|\vec{r} - \vec{r}'| \qquad (3.1)$$

where V_{ext} is the external potential and $\rho_1(\vec{r})$ is the excess charge induced in the system by V_{ext}. Hence $V_1(\vec{r})$ is the total one-electron potential due to the external field.

To find $\rho_1(\vec{r})$, we solve the linearized Liouville equation for the density matrix:

$$i\hbar \frac{\partial \hat{\rho}_1}{\partial t} = [H_0, \hat{\rho}_1] + e[V_1, \hat{\rho}_0] \qquad (3.2)$$

where H_0 is the one-electron Hamiltonian in the absence of V_1 and $\hat{\rho}_0$ is the corresponding density matrix. Let ψ_i be the eigenfunctions of H_0:

$$H_0 \psi_i = E_i \psi_i \tag{3.3}$$

and consider separately each Fourier component of V_1 so that all quantities are proportional to $e^{-i\omega t}$. Then (3.2) can be solved by taking matrix elements between ψ_i^* and ψ_j, using the fact that $\hat{\rho}_0$ is diagonal in these states:

$$(\hat{\rho}_0)_{ij} = f_i \delta_{ij} \tag{3.4}$$

where $f_i = f(E_i)$ is the Fermi distribution. We find

$$[\hbar\omega - (E_i - E_j)](\hat{\rho}_1)_{ij} = e(f_j - f_i) V_{ij} \tag{3.5}$$

where

$$V_{ij} = \int d^3r \, \psi_i^*(\vec{r}) V_1(\vec{r}) \psi_j(\vec{r}) \tag{3.6}$$

Now the density $\rho_1(\vec{r})$ is

$$\rho_1(\vec{r}) = e \sum_{ij} (\hat{\rho}_1)_{ij} \psi_i(\vec{r}) \psi_j^*(\vec{r}) \tag{3.7}$$

From these equations we find

$$\rho_1(\vec{r}) = e \int d^3r' L(\vec{r}, \vec{r}') V(\vec{r}') \tag{3.8}$$

where

$$L(\vec{r}, \vec{r}') = \sum_{ij} \frac{f_i - f_j}{E_i - E_j - \hbar\omega} \psi_i^*(\vec{r}) \psi_j(\vec{r}') \psi_i(\vec{r}) \psi_j^*(\vec{r}') \tag{3.9}$$

is the linear response function of the density to the total field $V_1(\vec{r})$.

To find the relation between V_1 and the external potential V_{ext} we insert (3.8) in (3.1). We find

$$V_{ext}(\vec{r}) = \int d\vec{r}' \, \epsilon(\vec{r}, \vec{r}') V_1(\vec{r}') \tag{3.10}$$

where

$$\epsilon(\vec{r}, \vec{r}') = \delta(\vec{r} - \vec{r}') - e^2 \int \frac{d^3r''}{|\vec{r} - \vec{r}''|} L(\vec{r}'', \vec{r}') \tag{3.11}$$

is the longitudinal dielectric function.

Clearly, $\epsilon(\vec{r}, \vec{r}')$ is not simply proportional to $\delta(\vec{r} - \vec{r}')$, as assumed in the macroscopic theory. In a homogeneous system, ϵ is a function only of $\vec{r} - \vec{r}'$, with a range of order of the Thomas-Fermi screening length. Within a few screening lengths from the surface, ϵ is altered in a drastic way, falling to zero if either $\vec{r}$ or $\vec{r}'$ is outside the medium.

We recall that retardation effects are important on the scale set by the length c/ω, which is much longer than the screening length v/ω_p (where v is the average particle velocity) or the interparticle spacing. Therefore it is sufficient to give a microscopic treatment within the electrostatic limit, as was already done in deriving (3.11). Even so, the problem of computing $\epsilon(\vec{r}, \vec{r}')$ and then solving (3.10) with $V_{ext} = 0$ to find the self-sustained excitations is by no means simple. Detailed calculations have been carried out taking a potential V_0 that depends only on the z co-ordinate, perpendicular to the surface. In this case the wave functions ψ_i have the form

$$e^{i\vec{k}\cdot\vec{R}} \varphi_n(z) \tag{3.12}$$

and the function $\epsilon(\vec{r}, \vec{r}')$ can be Fourier-transformed into $\epsilon(K; z, z')$ where $\vec{K}$ is the wave vector of the disturbance. The solution of

$$\int \epsilon(K; z, z') V_1(K, z') dz' = 0 \tag{3.13}$$

for a given K and ω yields the dispersion relation $\omega(K)$. This is usually written, for small K,

$$\omega = \omega_0 \left(1 + A \frac{K}{k_F} + B \left(\frac{K}{k_F}\right)^2 + \ldots\right) \tag{3.14}$$

where ω_0 (as will be shown) is the surface plasmon frequency from $\epsilon_{bulk} + 1 = 0$ and k_F is the Fermi wave vector. Much interest has centred on the sign and size of the linear coefficient A, which from some experiments appears to be negative.

Several simplified treatments of the problem have been proposed in the hope of obtaining the coefficients A and B without all the labour. We can see how great simplifications can arise if we assume that only terms with $|E_i - E_j| \ll \hbar\omega$ contribute significantly to the sum in (3.9). We find then, to leading non-vanishing order,

$$\int L(\vec{r}, \vec{r}') V_1(\vec{r}') d^3r' = \frac{1}{\hbar^2\omega^2} \sum_{ij} (f_i - f_j)(E_j - E_i) V_{ij} \psi_i(\vec{r}) \psi_j(\vec{r}) \tag{3.15}$$

Now

$$\hbar^{-2}(E_j - E_i) V_{ij} = \int \psi_i^* (V_1 H_0 - H_0 V_1) \psi_j d^3r'$$

$$= \frac{1}{m} \int (\nabla \psi_i^*) \cdot (\nabla V_1) \psi_j d^3r' \tag{3.16}$$

hence

$$\sum_j f_i (E_j - E_i) V_{ij} \psi_i(\vec{r}) \psi_j^*(\vec{r}) = -\frac{1}{m} f_i \nabla \psi_i^*(\vec{r}) \cdot \nabla V_1(\vec{r}) \psi_i(\vec{r}) \tag{3.17}$$

This is inserted in (3.15) and the term containing f_j is similarly transformed. Recombining the two terms we find

$$\rho(\vec{r}) = e \int L(\vec{r},\vec{r}') V(\vec{r}') d^3r' = -\frac{e}{m\omega^2} \nabla(\rho_0(\vec{r}) \nabla V_1(\vec{r})) \qquad (3.18)$$

where

$$\rho_0(\vec{r}) = \sum_i f_i |\psi_i(\vec{r})|^2 \qquad (3.19)$$

is the unperturbed density of the medium. Equation (3.18) is appealingly simple. In the case of a homogeneous medium, where ρ_0 is a constant, it leads to $\omega^2 = 4\pi \rho_0 e^2/m$ when taken in conjunction with Poisson's equation, $\nabla^2 V_1 = -4\pi e \rho_1$. If ρ_0 is a function of z alone, with $\rho_0(-\infty) = 0$ and $\rho_0(+\infty) = \overline{\rho}_0$, one finds that $\omega^2 = 2\pi\overline{\rho}_0 e^2/m$, as expected. However, it does not tell us anything reliable about the dispersion coefficients A and B of (3.14).

One can think of expanding $(E_i - E_j - \hbar\omega)^{-1}$ to higher orders. This procedure is correct for free electrons in the bulk, where

$$E_i - E_j = E_{\vec{p}+\vec{k}} - E_{\vec{p}} = \hbar^2\left(\frac{\vec{p}\cdot\vec{k}}{m} + \frac{k^2}{2m}\right) \qquad (3.20)$$

since the size of p is restricted to being no greater than k_F by the Fermi functions that appear in (3.9). If $k \ll k_F$, the term $k^2/2m$ can be neglected in (3.20) and the condition $|E_i - E_j| < \hbar\omega$ becomes simply $v_F k < \omega$. For free electrons in the bulk, k is the momentum of the excitation, so that the procedure is exact for long-wavelength excitations. In the presence of a surface, however, only the parallel component $\vec{K}$ is fixed by momentum conservation for a smooth surface, while the perpendicular component k_z can assume arbitrary values (here $\vec{k} = (\vec{K}, k_z)$). In particular, for some values of k_z the denominator in (3.9) will vanish, and consequently ϵ will acquire an imaginary part, according to the well-known prescription:

$$\frac{1}{\hbar\omega + i\eta - (E_i - E_j)} = \frac{P}{\hbar\omega - (E_i - E_j)} - i\pi \delta(\hbar\omega - (E_i - E_j)) \qquad (3.21)$$

where η is a positive infinitesimal and P denotes that the principal part is to be taken. Of course, in a real solid there is also the possibility of interband transitions, i.e. E_i and E_j can be the energies of electrons in two different bands. These transitions give rise to an imaginary part of ϵ, corresponding to real absorption processes, even in the bulk at infinite wavelengths, whenever ω exceeds the relevant energy gap Δ.

Thus an expansion in powers of $(E_i - E_j)/\hbar\omega$ is approximate at best, even in the bulk.

An alternative method of obtaining equations for the density fluctuations is to relate ρ_1 to the current $\vec{j}$ through the equation of continuity:

$$\frac{\partial \rho_1}{\partial t} + \nabla \cdot \vec{j} = 0 \qquad (3.22)$$

and then obtain an equation for j. (This method has been used by March and Tosi [16].) This procedure is equivalent to writing a second-order equation for $\hat{\rho}_1$, i.e. computing $\partial^2 \rho_1/\partial t^2$ as is shown below. We find, differentiating (3.2),

$$-\hbar^2 \frac{\partial^2 \hat{\rho}_1}{\partial t^2} = [H_0, [H_0, \hat{\rho}_1]] + [H_0, [V, \hat{\rho}_0]] \qquad (3.23)$$

Taking matrix elements, and Fourier transforming,

$$\hbar^2 \omega^2 (\hat{\rho}_1)_{ij} = (E_i - E_j)^2 (\hat{\rho}_1)_{ij} + (E_i - E_j)(f_i - f_j) V_{ij} \qquad (3.24)$$

When we compute $\rho_1(\vec{r})$ according to (3.7), we see that the last term of (3.24) is precisely the same as was considered in (3.15), leading to the result (3.18). The term $(E_i - E_j)^2 (\hat{\rho}_1)_{ij}$ can be handled in a similar fashion; one finds that it leads to two contributions, the first of which is

$$-\frac{e}{m\omega^2} \nabla \cdot (\rho_1 \nabla V_0) \qquad (3.25)$$

This added to (3.18) gives, with $\rho = \rho_0 + \rho_1$, $V = V_0 + V_1$,

$$-m\omega^2 \rho_1 = e\nabla \cdot (\rho \nabla V) - e\nabla \cdot (\rho_0 \nabla V_0) \qquad (3.26)$$

which can be understood very simply in terms of the continuity equation (3.22) and of Newton's equations of motion, since $-\nabla V$ is just the total force acting on the electrons (the term $\rho_0 \nabla V_0$ must be subtracted because it is balanced by the equilibrium pressure of the electron gas).

The second contribution coming from $(E_i - E_j)^2 (\hat{\rho}_1)_{ij}$ can be written as the divergence of the stress tensor:

$$\pi_{\alpha\beta}(\vec{r}) = e \int \frac{d^3 p}{(2\pi)^3} p_\alpha p_\beta f(\vec{p}, \vec{r}) \qquad (3.27)$$

where $f(\vec{p}, \vec{r})$ is the (first-order) part of the Wigner joint distribution function due to V_1. Precisely, if we introduce

$$\rho_1(\vec{r}_1 \vec{r}_2) = \sum_{ij} (\hat{\rho}_1)_{ij} \psi_i(\vec{r}_1) \psi_j^*(\vec{r}_2) \qquad (3.28)$$

then

$$f(\vec{p}, \vec{r}) = \int d^3 r' \, e^{-i\vec{p}\cdot\vec{r}'} \rho_1\left(\vec{r} + \frac{\vec{r}'}{2}, \vec{r} - \frac{\vec{r}'}{2}\right) \qquad (3.29)$$

Classically, $f(\vec{p}, \vec{r})$ is the excess density of particles with momentum $\hbar\vec{p}$, at point $\vec{r}$, due to the action of V_1. It is worth noting that Eq.(3.16), with the addition of $\nabla \cdot \vec{\pi}$ on the right-hand side, can be derived classically from the collisionless Boltzmann equation for $f(\vec{p}, \vec{r})$. Of course, the catch in this formulation is that $\pi_{\alpha\beta}$ cannot be expressed in terms of $\rho_1(\vec{r})$ alone, unless a drastic approximation is introduced as follows: it is assumed that $\pi_{\alpha\beta}$ is diagonal, $\pi_{\alpha\beta} = -P_1 \delta_{\alpha\beta}$, and that the excess pressure p_1 is linearly related to ρ_1 through the local equilibrium compressibility of the electron gas. This leads to a closed set of equations for ρ_1 and V_1 (supposing that ρ_0 and V_0 are known), which have been used by Bennett [17] to calculate the spectrum of surface excitations, for a simple model of the surface density profile $\rho_0(z)$.

Having described the basic microscopic approach, we shall now derive a general relation due to Harris and Griffin [18] for the linear coefficient A of the surface plasmon dispersion relation (3.14). We start from (3.8) and (3.9) in the case where ρ_0 depends only on z (jellium model with a flat surface). Then $L(\vec{r}, \vec{r}')$ depends only on z, z' and on the difference $\vec{R} - \vec{R}'$ of the co-ordinates parallel to the surface; Fourier transforming, we arrive at the following equation, where $\varphi_m(z)$ is a wave function for the perpendicular motion:

$$\rho_1(K, z) = \sum_{m, n, \vec{P}} \frac{f_{\vec{P}+\vec{K}, m} - f_{\vec{P}, n}}{E_{\vec{P}+\vec{K}, m} - E_{\vec{P}, n} - \hbar\omega} \varphi_m(z) \varphi_n^*(z)$$

$$\times \int dz' \varphi_m^*(z') V(K, z') \varphi_n(z') \qquad (3.30)$$

The potential $V_1(K, z)$ is related to $\rho_1(K, z)$ by Poisson's equation:

$$\left(K^2 - \frac{d^2}{dz^2} \right) V_1(K, z) = 4\pi e \rho_1(K, z) \qquad (3.31)$$

which has the solution:

$$V_1(K, z) = \frac{2\pi e}{K} \int dz' e^{-K|z-z'|} \rho_1(K, z') \qquad (3.32)$$

We now integrate both sides of (3.30) over all z. Because of the orthogonality properties of the φ's, only the term m = n remains in the sum; that means that the denominator is simply

$$E_{\vec{P}+\vec{K}} - E_{\vec{P}} - \hbar\omega = \frac{\hbar^2}{2m}(2\vec{K}\cdot\vec{P} + K^2) - \hbar\omega \qquad (3.33)$$

It is now rigorously possible to expand in powers of ω^{-1} for small values of K; with the result that $(E_{\vec{P}+\vec{K}} - E_{\vec{P}} - \hbar\omega)^{-1}$ can be replaced by $\vec{K}\cdot\vec{P}/m\omega^2$ to leading order, since the term $(\hbar\omega)^{-1}$ and the term $K^2/2m\omega^2$ vanish when the sum over $\vec{P}$ is carried out. The result is then simply

$$\int \rho_1 dz = \frac{2\pi e^2}{m\omega^2} K \int dz\, dz'\, \rho_0(z) e^{-K|z-z'|} \rho_1(K, z') \qquad (3.34)$$

where $\rho_0(z)$ is the unperturbed density. In the right-hand side we cannot put $\exp(-K|z-z'|) = 1$ because then the integral of $\rho_0(z)$ diverges. Instead, we put first

$$\rho_0(z) = \bar{\rho}_0[\theta(z) + \tilde{n}(z)] \qquad (3.35)$$

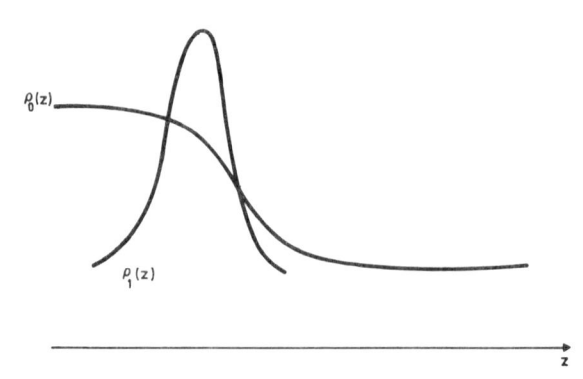

FIG. 12. Schematic drawing of the static surface density profile $\rho_0(z)$ and of the fluctuating density of the surface plasmon $\rho_1(z)$.

where $\bar{\rho}_0$ is the average density in the metal, $\theta(z)$ is a step function that vanishes outside the metal and $\tilde{n}(z)$ is strongly localized near the surface. Now the term containing $\theta(z)$ can be evaluated exactly, while in the term containing $\tilde{n}(z)$ we can take $\exp(-K|z-z'|) = 1$ without difficulties. The result is

$$\int \rho_1 \, dz = \frac{\omega_p^2}{2\omega^2} \left\{ \int \rho_1 \, dz \left[1 + K \int \tilde{n}(z') \, dz' \right] + K \int dz' z' \rho_1(z') \right\} \qquad (3.36)$$

with $\omega_p^2 = 4\pi \bar{\rho}_0 e^2/m$. We can choose the origin so that

$$\int \tilde{n}(z') \, dz' = 0 \qquad (3.37)$$

and find then the dispersion relation:

$$\omega^2 = \tfrac{1}{2} \omega_p^2 \left[1 + K \int dz' z' \rho_1(z') \Big/ \int dz \, \rho_1(z) \right] \qquad (3.38)$$

This is the Harris-Griffin relation [18]. It shows that the linear coefficient A of the surface plasmon dispersion relation is proportional to the first moment of the fluctuating density $\rho_1(z)$: the sign of this depends on where $\rho_1(z)$, which is strongly peaked near the surface, is located with respect to the unperturbed density profile $\rho_0(z)$ (see Fig.12).

Experimentally, the dispersion relation appears to be remarkably flat, within a rather large uncertainty [19]. Theoretically, there are calculations by Bennett [17] using the hydrodynamic approach described above and a density profile of the type shown in Fig.13. The linear coefficient A in (3.14) is found to decrease with increasing d, passing through zero when d is of the order of a few Å. Calculations based on the complete RPA response function also show the trend towards decreasing A for smoother surface profiles [20-22]. The effect of surface roughness and of absorbed layers has also been considered, but there are considerable difficulties in obtaining reliable results for realistic models.

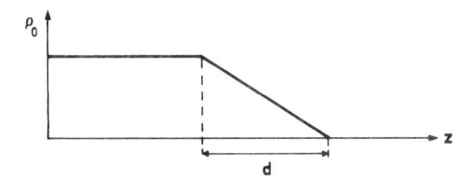

FIG.13. Surface density profile assumed by Bennett (Ref.[17]).

For larger K values the expansion (3.14) breaks down. An explicit relation for ω as a function of K has been obtained by Ritchie and Marusak [23] for a medium of constant density up to the surface (this corresponds to taking d = 0 in Fig.13 and is not equivalent to taking an infinitely high potential step at the surface). The Ritchie-Marusak relation is

$$1 + \frac{K}{\pi} \int \frac{dk_z}{K^2 + k_z^2} \frac{1}{\epsilon(K, k_z; \omega)} = 0 \qquad (3.39)$$

where $\epsilon(K, k_z; \omega)$ is the bulk dielectric function for frequency ω and momentum $\vec{k}$ with components $\vec{K}$ parallel to the surface and k_z in the perpendicular direction. This leads to a dispersion relation starting from $\omega_p/\sqrt{2}$ and increasing rapidly, so that at $K = k_c$ it crosses the bulk plasmon dispersion relation and both merge in the bulk particle-hole continuum. While the behaviour at small K is dependent on the details of the surface profile, as already discussed, the subsequent increase and the crossing with the bulk plasmon at k_c appear to be model-independent, as shown by the calculations [24, 25] that have been performed for a surface described by a potential step of finite height.

ACKNOWLEDGEMENTS

I wish to thank M. Berry, K. Rangarajan and other participants in the ICTP Winter College for pointing out errors in the first draft of these lecture notes and suggesting improvements. I am especially indebted to A. Marvin, who supplied the explicit formulas of Section 2.4 for the scattered intensity.

REFERENCES

[1] LINDHARD, J., K. Dan. Vidensk. Selsk., Mat.-Fys. Medd. 28 (1954) 8.
[2] ECONOMOU, E.N., NGAI, K.L., Adv. Chem. Phys. 27 (1974) 265.
[3] TENG, Y.T., STERN, E.A., Phys. Rev. Lett. 19 (1967) 511.
[4] RITCHIE, R.H., ARAKAWA, E.T., COWAN, J.J., HAMM, R.N., Phys. Rev. Lett. 21 (1968) 1530.
[5] LORD RAYLEIGH, Theory of Sound 2, 2nd Edn, Dover, New York (1945) 89.
[6] FANO, U., J. Opt. Soc. Am. 31 (1941) 213. (See footnote p. 399.)
[7] ENDRIZ, J.E., SPICER, W.E., Phys. Rev. B4 (1971) 4144.
[8] FEDDERS, P.A., Phys. Rev. 165 (1968) 580.
[9] ELSON, J.M., RITCHIE, R.H., Phys. Status Solidi B62 (1974) 461.
[10] KLIEVER, K.L., FUCHS, R., Phys. Rev. 153 (1967) 498.

[11] NGAI, K.L., ECONOMOU, E.N., Phys. Rev. B4 (1971) 2132.
[12] NGAI, K.L., ECONOMOU, E.N., COHEN, M.H., Phys. Rev. Lett. 22 (1969) 1375.
[13] GADZUK, J.W., Phys. Rev. B1 (1970) 1267.
[14] NEWNS, D.M., Phys. Lett. 38A (1972) 371.
[15] FISCHER, B., MARSCHALL, N., QUEISSAR, H.J., Surf. Sci. 34 (1973) 50.
[16] MARCH, N.H., TOSI, M.P., Proc. R. Soc. (London), Ser. A 330 (1972) 373.
[17] BENNETT, A.J., Phys. Rev. B1 (1970) 203.
[18] HARRIS, J., GRIFFIN, A., Can. J. Phys. 48 (1970) 2592; Phys. Lett. 34A (1971) 51.
[19] KLOOS, T., RAETHER, H., Phys. Lett. 3 (1973) 157.
[20] BECK, D.E., Phys. Rev. B4 (1971) 155.
[21] BECK, D.E., CELLI, V., Phys. Rev. Lett. 28 (1972) 1124.
[22] FEIBELMAN, P.J., Phys. Rev. Lett. 30 (1973) 925; Phys. Rev. B9 (1974) 5077.
[23] RITCHIE, R.H., MARUSAK, A.L., Surf. Sci. 4 (1968) 234.
[24] WIKBORG, E., INGLESFIELD, J., Solid State Commun. 14 (1974) 661.
[25] GRIFFIN, A., KRANZ, H., HARRIS, J., to be published.

SURFACE SPECTROSCOPY

G. CHIAROTTI
Institute of Physics,
University of Rome,
Rome, Italy

Abstract

SURFACE SPECTROSCOPY.
1. Introduction. 2. Surface states. 3. Surface-charge region. 4. Optical methods. 5. Electron energy losses. 6. Conclusions.

1. INTRODUCTION

The aim of this work is to review the experimental evidence of the existence of electronic surface states in semiconductors and insulators, the methods by which their energy distribution can be obtained and the existing information on their 'spectrum'. Emphasis will be placed on 'clean' surfaces in order to allow a comparison (at least qualitative) with the theory, although results on 'real' surfaces will occasionally be presented.

Various methods have been used to obtain the energy distribution of electronic surface states, mainly:

(a) The study of the transport properties of carriers in the space-charge region at the surface of a semiconductor;
(b) Optical absorption (or, more generally, optical properties) of the surface region;
(c) Surface photoconductivity;
(d) Surface photoemission; and
(e) Energy-loss spectra.

Many other methods of great potentiality (e.g. Auger spectroscopy, ESCA, appearance-potential spectroscopy, neutralization spectroscopy, etc.) have been used so far, mainly for studying chemical impurities at the surface, and are not dealt with here.

The transport properties of the space-charge layer, on the other hand, are the subject of a paper in these Proceedings by Professor Many [1] and are considered here briefly to introduce the characteristic parameters of the space-charge region and to describe some of the earliest experiments on the energy distribution of surface states on clean surfaces.

2. SURFACE STATES

Electronic states localized at the surface of a semiconductor (or insulator) arise for two different reasons: the abrupt termination of the periodic potential at the surface (or its modifications) and/or the presence of impurities or defects.

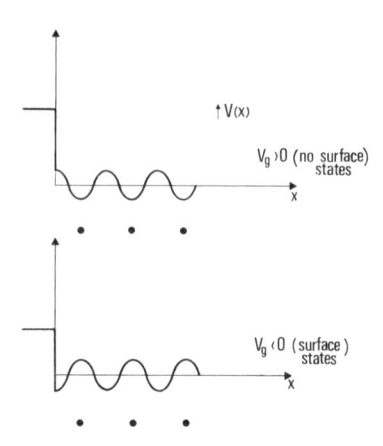

FIG.1. Schematic one-dimensional representation of the potential at the surface for the two cases $V_g > 0$ and $V_g < 0$. The position of the atoms is shown by the dots.

The theory of the intrinsic states (due to the termination of the potential) is presented in the paper by Professor Jones in these Proceedings [2].

Intrinsic states are characterized by an imaginary component of the k vector in the direction normal to the surface. Their energy corresponds in general (but not necessarily) to that of the forbidden gaps. Their existence depends upon the way the potential is terminated at the surface and therefore upon surface reconstruction.

In a one-dimensional model of nearly free electrons, with periodicity extending up to the surface plane, surface states exist only when the Fourier-transform V_g of the bulk pseudo-potential is negative (see Fig. 1 and Refs [1] and [3]). In Ref.[3], notice a different choice of the origin which introduces a change of sign in V_g.

In the same model, surface states can exist at the interface between two materials (say a semiconductor and its oxide layer) only if V_g has opposite signs on the two sides of the interface. If this is not true, the wave functions cannot be matched at the interface, as shown in Fig. 2. In a number of cases then, the oxidation of the surface may cause the surface states to disappear. This result is used in various ways in experiments on clean surfaces, where surface properties are compared before and after oxidation of the surface.

From the point of view of the theory of the chemical bonds, the condition $V_g < 0$ is that appropriate to a covalent solid, where the electrons are localized between the nuclei (see Fig. 1). Surface states are then associated with the unsaturated 'dangling' bonds at the surface. Saturation of the dangling bonds during oxidation will cause the disappearance of the surface states.

In ionic solids the occurrence of surface states is mainly due to the lowering of the Madelung energy at surface positions. This in turn lowers the energy required for transferring an electron from a surface anion to a surface cation, thus introducing surface levels in the gap.

Extrinsic surface states are due to impurities, adsorbed or chemisorbed at the surface of the solid or at the interface between a solid and its oxide or

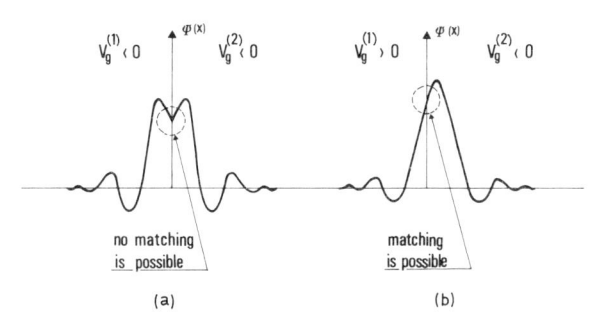

FIG.2. Schematic representation of the matching of the wave function at the interface between two crystals (a) when V_g has the same sign and (b) when V_g has opposite signs on the two sides of the interface. In case (a) no matching is possible.

adsorbed layer. They provide energy levels similar to those of donors and acceptors in semiconductors.

Though extrinsic surface states are of great importance in chemisorption and catalysis and in field-effect devices, our main concern here will be the study of intrinsic states that are expected on atomically 'clean' surfaces obtained from cleavage in ultra-high vacuum (10^{-10} Torr or better) or from 'cleaning' a 'real' surface by means of argon bombardment followed by annealing at high temperatures (for the preparation of surfaces see Chapter 5 of Ref.[4]).

3. SURFACE-CHARGE REGION

Both intrinsic and extrinsic surface states act as donors or acceptors and modify the position of the Fermi level at the surface. In other words, the surface acquires a potential difference with respect to the bulk or, if one prefers, the energy bands bend at the surface. The situation is sketched in Fig.3. In Fig.3(a) the surface has been considered as a separate phase describable by its own Fermi level and with a given distribution of states in the gap. Equilibrium conditions are shown in Fig.3(b) with the bands bending downwards and the formation of a space-charge layer.

The physical reason for bending is that some of the electrons in surface states find lower available energy levels in acceptor states (in Fig.3 the bulk is p type). At $T = 0$ the space charge is simply made up of negatively charged 'ionized' acceptors'. At $T \neq 0$, part of the negative space charge is due to an excess of electrons in the conduction band. In both cases the negative space charge is neutralized by a 'fixed' positive charge in the surface states.

The main problem of the theory of the space-charge region is: with a given distribution of surface states, find the band bending at the surface and calculate the variation of the important physical quantities, like conductivity, that depend on it.

The theory of the space-charge region is developed in several books. In the following, reference is made to the book by Many et al. [5] (MGG hereafter) to which the reader is referred for the derivation of the main results.

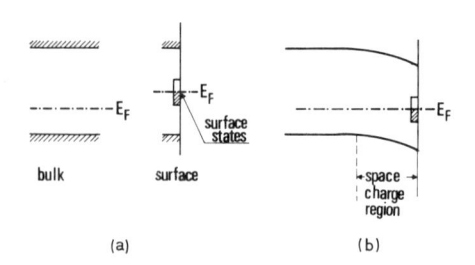

FIG. 3. Band bending at the surface of a semiconductor. (a) Scheme of energy levels in the bulk (acceptor levels have been omitted) and at the surface when the two phases are not in equilibrium. (b) Equilibrium condition.

With reference to Fig. 4, let us define the intrinsic level E_i that is parallel to the band edges and that, in the bulk, coincides with the Fermi level of the intrinsic material, the electrostatic potential Φ:

$$q\Phi = E_F - E_i \tag{1}$$

and the potential difference V:

$$V = \Phi - \Phi_b \tag{2}$$

Φ_b being the bulk potential. More often, adimensional quantities called the reduced potentials are introduced:

$$u = \frac{q\Phi}{kT}, \quad v = \frac{qV}{kT} = u - u_b \tag{3}$$

$v_s > 0$ and $v_s < 0$ ($v_s = v$ at the surface) correspond to downward and upward bending; $u_b > 0$ and $u_b < 0$ correspond to n- and p-type bulk materials. The condition in which v_s and u_b have the same sign corresponds to a surface accumulation layer; while the condition with v_s and u_b of opposite signs corresponds to an inversion or depletion layer. In Fig. 4, $u_b < 0$, $v_s > 0$ and we have an inversion layer.

The concentration of carriers is determined by the position of the Fermi level:

$$n = N_c \exp\left(\frac{E_F - E_c}{kT}\right), \quad p = N_v \exp\left(\frac{E_v - E_F}{kT}\right) \tag{4}$$

N_c and N_v being the effective numbers of states for the conduction and the valence band.

In terms of u and v the concentration of carriers becomes

$$n = n_i e^u = n_b e^v, \quad p = n_i e^{-u} = p_b e^{-v} \tag{5}$$

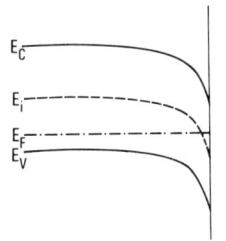

FIG.4. Schematic representation of the band bending at the surface of a semiconductor in the case of an inversion layer.

n_i, n_b, p_b are the intrinsic carrier concentrations and the concentrations of electrons and holes in the bulk. Under the hypothesis that the bulk donors and acceptors (of concentration N_D and N_A) are all ionized, the charge density is

$$\rho(x) = q(N_D - N_A - n + p) = q(n_b - p_b - n + p) \qquad (6)$$
$$= q[n_b(1 - e^v) - p_b(1 - e^{-v})]$$

The dependence on x, the co-ordinate normal to the surface, is, of course, through $v(x)$.

Neglecting for the moment what is going on at the surface plane (where the surface states are localized) the Poisson equation reads, for $x < 0$:

$$\frac{d^2\Phi}{dx^2} = \frac{d^2 V}{dx^2} = -\frac{q}{\epsilon\epsilon_0} \rho(x) \qquad (7)$$

If $v \ll 1$ (bending smaller than kT), the charge density reduces to

$$\rho(x) = -q(n_b + p_b) v \qquad (8)$$

and the solution of Eq. (7) is simply

$$v = v_s \exp\left(-\frac{x}{L}\right) \qquad (9)$$

L is the Debye effective length and is given by

$$L = \sqrt{\frac{\epsilon\epsilon_0 kT}{q^2(n_b + p_b)}} \qquad (10)$$

The space-charge region in such a simple case decays exponentially with a characteristic length L.

In the general case, one obtains (MGG, p.138)

$$x = L \int_{v_s}^{v} \frac{dv}{\pm F(u_b, v)} \qquad (11)$$

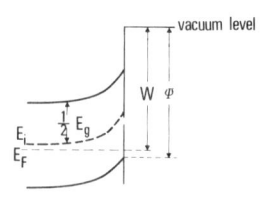

FIG.5. Work function W, photoelectric threshold ψ and their relation to Φ_s.

where the plus and minus signs hold when $v < 0$ and $v > 0$, respectively. The function F is defined by

$$F(u_b, v) = \sqrt{2} \left[\frac{\cosh(u_b + v)}{\cosh u_b} - v \tgh u_b - 1 \right]^{1/2} \quad (12)$$

The integral in (11) can be done numerically and it gives v as a function of x, i.e. the shape of the potential barrier. The results are shown in Fig. 4.6 on p.144 of MGG. It should be noted that the thickness of the space-charge region is proportional to L, which in turn decreases as $(n_b + p_b)^{-1/2}$, i.e. the space-charge region becomes thinner the greater the doping of the bulk. For intrinsic Ge the Debye length at room temperature is 0.67×10^{-4} cm.

The total space charge per unit surface can also be obtained from the value of F at the surface:

$$Q_{sc} = \pm qL(n_b + p_b) F_s(u_b, v_s) \quad (13)$$

If surface states are present they carry a charge Q_{ss}. The total charge is then

$$Q_T = Q_{ss} + Q_{sc} \quad (14)$$

When the external applied field is zero, $Q_T = 0$ and

$$Q_{ss} = -Q_{sc} \quad (15)$$

the charge in the surface states just balances the space charge, making the surface of the crystal neutral. Equation (15) (or 14)) is enough to determine the bending at the surface, provided the distribution of surface states and their acceptor or donor character is known.

Let us assume, for example, that we have a distribution of surface states at energies E_j^d (donors) and E_i^a (acceptors), with a density (per unit surface) N_j^d and N_i^a (i, j = 1, 2...). Then

$$Q_{ss} = q\Sigma_j N_j^d [1 - f_s(E_j^d)] - q\Sigma_i N_i^a f_s(E_i^a) \quad (16)$$

$f_s(E)$ being the Fermi function at the surface, which is a function of u_s. Since Q_{sc} and Q_{ss} are both functions of u_s (or v_s), Eq. (15) ((or (14)) can be solved for u_s. It is then seen that u_s can be determined once the distribution

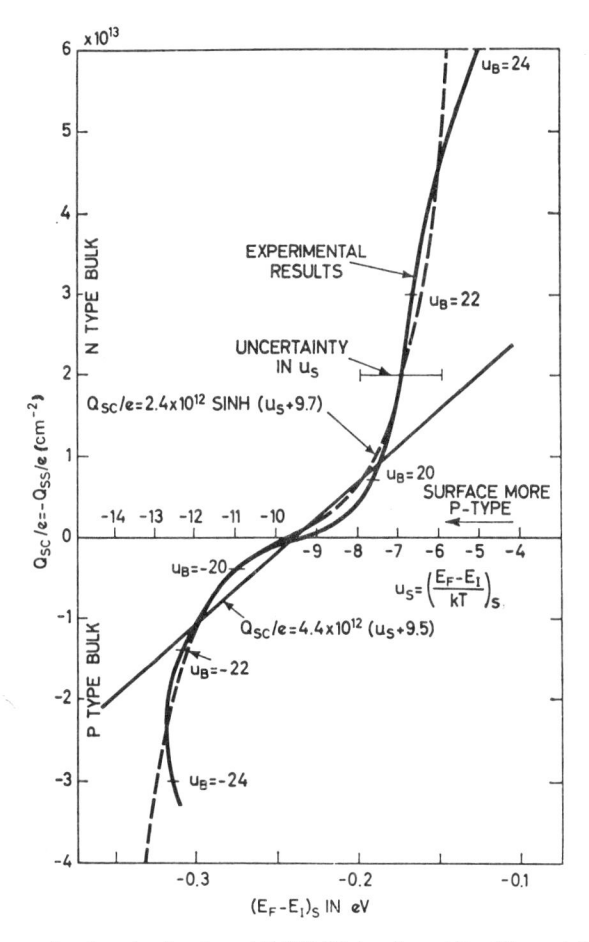

FIG.6. Q_{ss} as a function of u_s for cleaved Si(111) 2 × 1 surfaces (after Allen and Gobeli [6]).

of states is known. The converse, however, is not true since Eq. (15) has not in general a unique solution. The determination of the distribution of states through a measurement of u_s is generally obtained by best-fitting techniques.

One of the earliest experiments that gave information on the distribution of states on clean Si surfaces was in fact based upon the measurement of Q_{ss} as a function of u_s and was done by Allen and Gobeli in 1962 [6].

The experiments of Allen and Gobeli consisted in determining the photoelectric threshold Ψ and the work function W in cleaved Si samples of various doping. Ψ and W are related to Φ_s by

$$\Phi_s = (E_F - E_i)_s = (\Psi - W) - \tfrac{1}{2}E_g \tag{17}$$

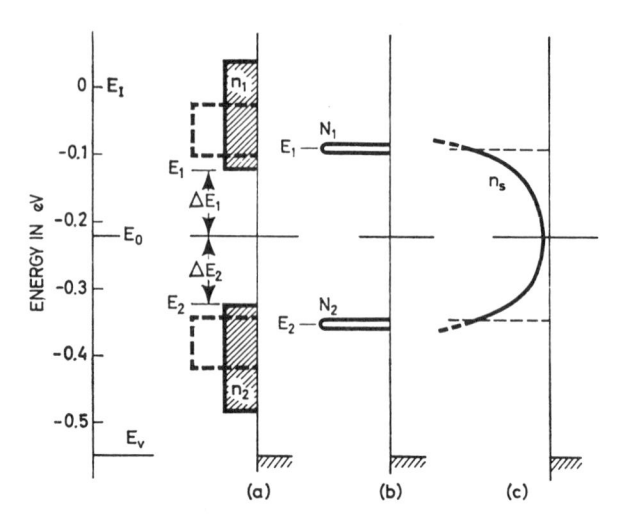

FIG. 7. Various distributions of surface states that fit the data of Fig. 6 (after Allen and Gobeli [6]).

which is evident from Fig. 5. For each sample of known u_b, $Q_{ss} = -Q_{sc}$ can be determined by means of Eq. (13) and gives rise to a point on the curve of Fig. 6.

Within a reasonable approximation, the experimental results can be fitted by the expression:

$$Q_{ss} = -2.4 \times 10^{12} \sinh(u_s + 9.7) \tag{18}$$

shown as a broken line in Fig. 6.

Any of the distributions of Fig. 7 fit the experimental data. They are all symmetric with respect to the neutrality point E_0 characterized by $E_0 - E_i = -9.7\,kT$. It can be easily demonstrated, however, that any distribution symmetric with respect to E_0 gives rise to a $Q_{ss} \propto -\sinh(u_s + (E_i - E_0)/kT)$ (at least for non-degenerate conditions) provided the states below E_0 have donor character and those above acceptor character (the general case is discussed in Ref. [7]).

Though the results of Allen and Gobeli do not show conclusive evidence, a two-band model is generally accepted for cleaved Si on the grounds that the states do not show direct conductivity [8], that the value of field-effect mobility (defined as $\mu_{FE} \equiv -d\Delta\sigma/dQ_T$, $\Delta\sigma$ being the change in conductivity) does not seem to show a large density of states at $E_F \simeq E_0$ [9] (see also Ref. [6] for a discussion on the interpretation), and by other considerations [10].

4. OPTICAL METHODS

A great deal of information on the spectrum of surface states in semiconductors has been obtained recently by optical methods: (1) optical absorption; (2) photoconductivity, and (3) photoemission spectroscopy.

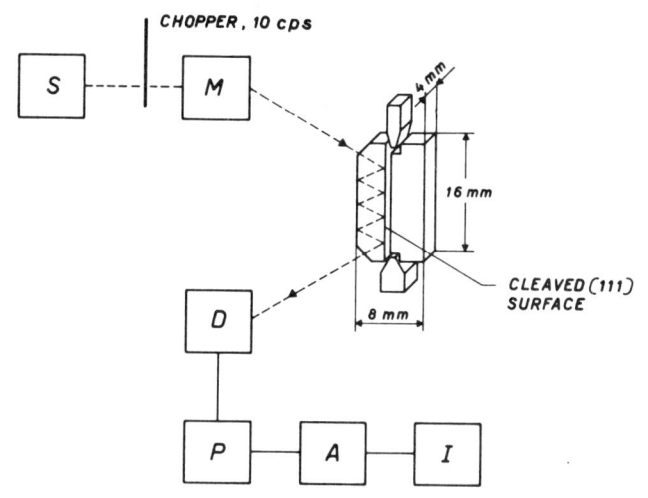

FIG.8. Block diagram of experimental apparatus and sketch of sample section; S is the electronically stabilized source, M the monochromator, D the PbSe detector, P the precision potentiometer, A the selective amplifier, and I the integrating digital voltmeter. The path of light is shown by the dashed line. (After Chiarotti et al.[12].)

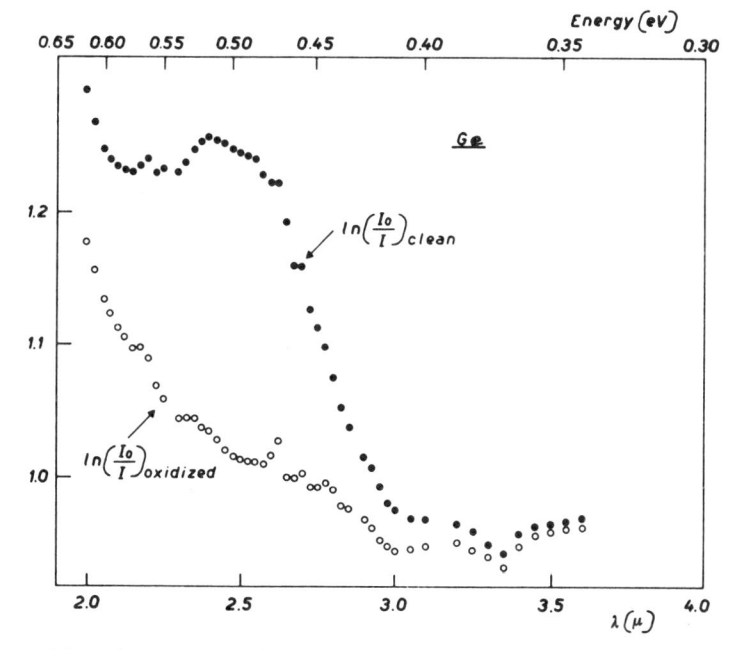

FIG.9. Natural logarithm of the ratio I_0/I as a function of wavelength for a cleaved surface of Ge and for the same surface after oxidation (after Chiarotti et al. [12]).

FIG. 11. Absorption constant α_s versus photon energy for an ultrahigh vacuum cleaved Si surface (after Chiarotti et al. [12].)

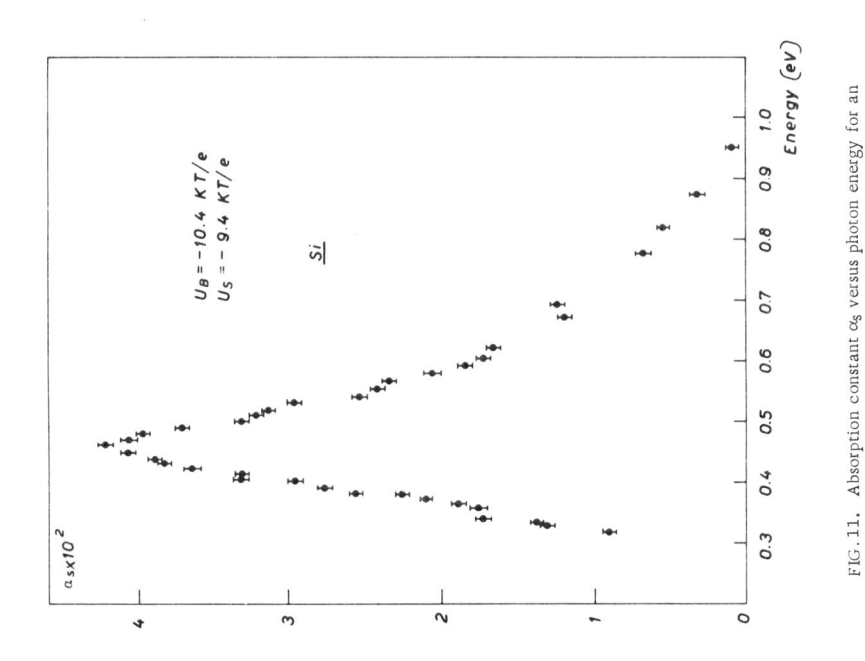

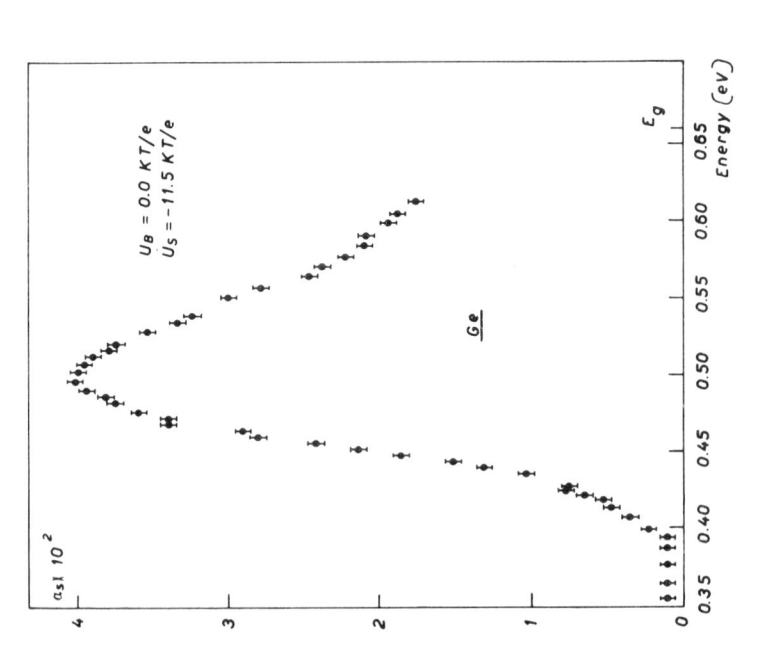

FIG. 10. Absorption constant α_s versus photon energy for a surface of Ge cleaved in ultrahigh vacuum. The almost constant absorption in the low-energy range is of the same order as the expected change of intraband hole absorption caused by the change of hole concentration during the oxidation. (After Chiarotti et al. [12].)

4.1. Optical absorption

In principle, light can induce transitions between the surface-state levels in semiconductors, allowing a measurement of optical absorption at energies lower than the gap. For a long time it was thought that the states at the surface (at most one per surface atom) were not enough to give an observable effect. However, a rough estimate shows that this is not true, at least for the intrinsic surface states: a density of 10^{15} states per cm^2 with an optical cross-section of 10^{-16} cm^2 would give rise to a decrease of 10% of the intensity of a beam of light transmitted by the surface layer.

In fact, the optical absorption of intrinsic surface states on cleaved (111) 2×1 surfaces of Si and Ge has been observed with the technique of multiple internal total reflection [11,12]. A beam of monochromatic light, of energy smaller than the gap, is totally reflected several times on the (111) surface of a crystal cleaved in ultrahigh vacuum ($\sim 10^{-10}$ Torr) (Fig. 8). The intensity of the light traversing the sample at the various wavelengths is recorded immediately after cleavage and after complete oxidation of the surface.

The results are shown in Fig. 9 for Ge and show that the oxidized surface transmits more light than the clean one, as if absorbing states had disappeared from the gap upon oxidation. The difference between the two curves of Fig. 9 gives then the contribution of the surface states on the clean surface. In Figs 10 and 11 the value

$$\alpha_s = \frac{1}{5}\left[\log\left(\frac{I_0}{I}\right)_{\text{clean}} - \log\left(\frac{I_0}{I}\right)_{\text{ox}}\right] \tag{19}$$

is plotted for Ge and Si, I_0 and I being the intensities of the light impinging on the sample and emerging from it. The factor 1/5 is introduced to take into account the multiple reflections.

Various processes could in principle explain the results of Figs 10 and 11:

(a) Optical transitions between two bands of surface states localized in the gap, one below and one above the Fermi level at the surface;
(b) Optical transitions from the valence band to an empty band of surface states localized in the gap;
(c) Optical transitions between a full band of surface states in the gap and the empty levels of the conduction band.

Although it is not easy to distinguish between these three processes, hypothesis (a) seems the most likely. The reason is that it appears difficult to explain peaked curves like those of Figs 10 and 11 under hypotheses (b) and (c). In fact, for transitions between the top of the valence band and a localized state, the joint density of states is expected to increase monotonically, at least as far as a parabolic expansion of the bulk three-dimensional band is valid. Moreover, hypothesis (c) in both Si and Ge is ruled out by the consideration that $(E_c - E_F)$ at the surface (~ 0.8 eV in Si and ~ 0.65 eV in Ge) is larger than the threshold for optical absorption (~ 0.3 and ~ 0.4 eV respectively).

On the other hand, a logarithmic singularity of the joint density of states (which bears some resemblance to the peaked curves of Figs 10 and 11) is

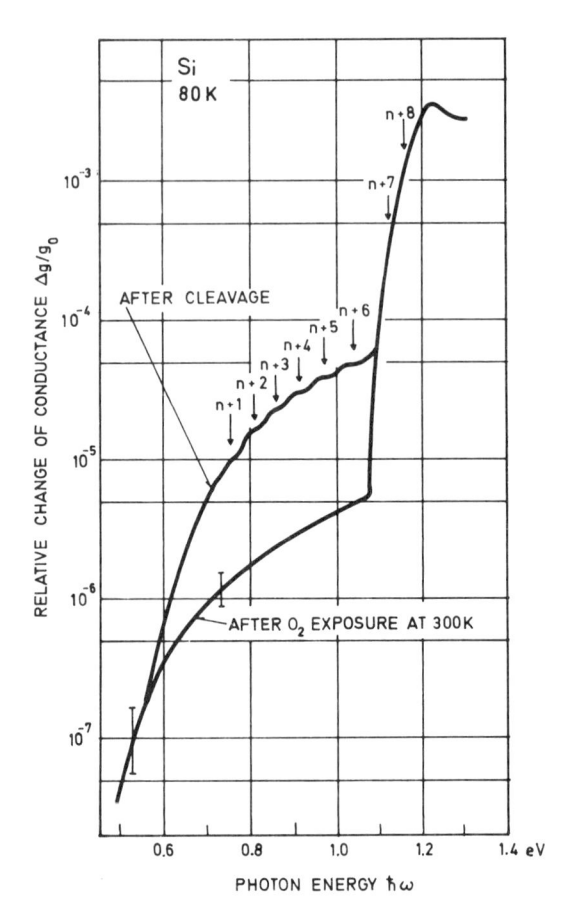

FIG.12. Si crystal (p-type, 3000 ohm · cm at r.t.). (111) surface cleaved in ultrahigh vacuum. Spectral dependence of relative photoconductance $\Delta g/g_0$ measured with chopped light (11 s^{-1}); g_0: dark conductance; Δg: photoconductance due to a photon flux of 2×10^{13} cm^{-2} · s^{-1}. (After Müller and Mönch [19]).

expected for transitions between the levels of a two-dimensional structure [13,14]. In fact the absorption coefficient for direct transitions in a two-dimensional Brillouin zone is given by [14]

$$\alpha(\omega) \sim \frac{1}{\omega} \int_{BZ} |Q_{if}|^2 \, \delta(E_f - E_i - \hbar\omega) \, d^2k \tag{20}$$

Q_{if} being the dipole matrix element for the i → f transition. The integral (20) has singular points when $\nabla_k(E_f) - \nabla_k(E_i) = 0$. Under the hypothesis that Q_{if} is a slowly varying function of k, the integral (20) can be easily done [14] by expanding the energy around a singular point E_0 (within a circle of radius R) and by using the property of the δ function:

$$\int f(x) \, \delta[g(x)] \, dx = f(x_0) \left|\frac{dg}{dx}\right|_{x=x_0}^{-1} \tag{21}$$

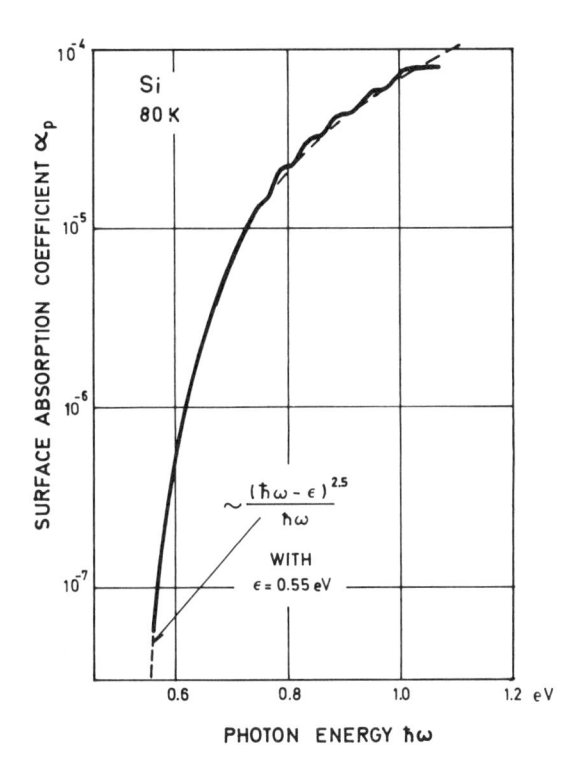

FIG.13. Spectral dependence of the surface absorption coefficient α_p for clean cleaved Si(111) surfaces as deduced from photoconductivity measurements. ϵ is threshold energy. (After Müller and Mönch [20]).

x_0 being the root of the equation $g(x) = 0$. The result is

$$\alpha(\omega) \sim \log \left[\frac{R + \sqrt{R^2 + |\hbar\omega - E_0|}}{\sqrt{\hbar\omega - E_0}} \right] \quad (22)$$

for a saddlepoint and a step function for a minimum or a maximum. Equation (22) shows a logarithmic singularity for $\hbar\omega = E_0$.

The case of Si appears most interesting since two bands of surface states symmetrically located with respect to the neutrality level and separated by a gap of ~0.30 eV would fit both the optical data and the results of Allen and Gobeli mentioned in Section 3. This two-band structure is also consistent with the experiments on photoconductivity to be discussed.

On real surfaces of Si and Ge, surface levels have also been localized by the method of modulated optical absorption [15-17]. By means of a field-effect electrode, the position of the Fermi level at the surface is periodically displaced with respect to the intrinsic level. The population of the surface states is then changed periodically and a modulation of the optical transitions originating from (or ending at) the surface levels is obtained. The method presents some difficulty when applied to clean surfaces because of the high density of surface states that 'lock' the Fermi level. It has, however, been

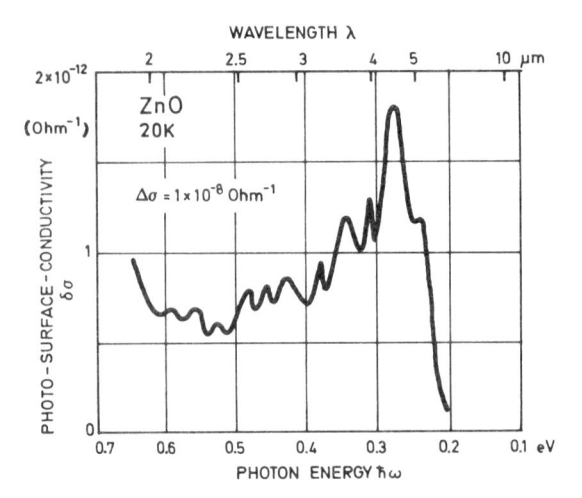

FIG.14. ZnO crystal (undoped, $\sigma = 10^{-2}$ ohm$^{-1} \cdot$ cm^{-1} at r.t.). Prism surface cleaved by heating to 400°C in ultrahigh vacuum. Spectral distribution of photoconductivity $\delta\sigma$, related to an incident photon flux of 6×10^{14} cm$^{-2} \cdot$ s^{-1}. (After Lüth and Heiland [24].)

applied to cleaved (111) Ge surfaces [18] with results that support the two-band model already discussed.

4.2. Surface photoconductivity

Photoconductivity due to surface states has been observed by Müller and Mönch in cleaved Si crystals [19] (Fig. 12).

At energies well below the gap, the photoconductivity spectrum shows a shoulder with a sharp threshold at approximately 0.55 eV. Since at the clean Si surface the Fermi level is ~0.80 eV below the bottom of the conduction band, the photoexcited carriers are holes in the valence band. The optical transition that causes the observed photoconductivity is then from the bulk valence band to empty surface states above the Fermi level.

In the two-band model described in Section 4.1, the threshold for photoconductive transitions is expected at ~0.5 eV. The agreement between optical and photoconductivity data is then very good. The slight discrepancy (0.05 eV) is probably related to the structure of the surface bands in the two-dimensional Brillouin zone.

Müller and Mönch [20, 21] have calculated the value of the absorption constant α_p for the optical transitions that give rise to photoconductivity, by measuring the lifetime τ of the excited electrons. The experimental value of τ is ~40 ms, a rather long time, which in part explains the great sensitivity of the photoconductivity method.

The surface photoconductivity per incident photon is $\Delta\sigma/N = q\mu_h \tau \alpha_p$, μ_h being the mobility of the holes in the valence band. The results for α_p are plotted in Fig. 13 as a function of the energy of the photons. It is seen that, experimentally, α_p has the expression:

$$\alpha_p \sim \frac{(\hbar\omega - E_g)^{5/2}}{\hbar\omega} \tag{23}$$

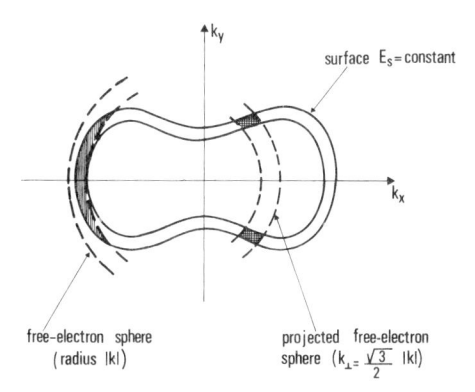

FIG.15. Curves of constant energy in the two-dimensional Brillouin zone.

as if the optical transitions were indirect forbidden transitions from a three-dimensional to a two-dimensional Brillouin zone.

The low value of α_p (in comparison with α_s) explains why the transition is not observed in the optical absorption spectrum of Fig. 11. On the other hand, the reason why the transitions responsible for the optical absorption of Fig. 11 do not give rise to photoconductivity is probably that the mobility of the electrons in the surface states is comparatively low and/or the lifetime for recombination is short.

A very interesting effect also shown in Fig. 12 is the oscillatory character of the surface photoconductivity.

Oscillations in bulk photoconductivity of various materials have been known for a long time and have been explained as due to the interaction of the excited electrons with LO phonons [22]. When the electrons are raised to an energy (with respect to the bottom of the conduction band) just below that necessary to excite a LO phonon and are then accelerated by the field, they are inelastically scattered. This extra acceleration is, however, limited to electrons that move against the field (electrons moving in the direction of the field being decelerated). The fastest electrons which move against the field (and which give a positive contribution to the current) are then removed from the distribution, causing a decrease of current at energies $n\hbar\omega_{LO}$, n being a small integer. The photocurrent then shows periodic oscillations as in Fig. 12. The value of ω_{LO} obtained from Fig. 12 is ~58 meV, to be compared with 65 meV for the bulk phonons and 55 meV for the surface phonons of Si [23].

An even more interesting effect is that found by Lüth and Heiland [24, 25] in the surface photoconductivity of ZnO (Fig. 14), where two series of oscillations are found, corresponding to the interaction with a bulk phonon of 72 meV and a surface phonon of 48 meV, in good agreement with energy loss experiments [26].

4.3. Surface photoemission spectroscopy

In recent years photoemission spectroscopy has proved a powerful tool for the study of the electronic structure of solids. Since the escape depth

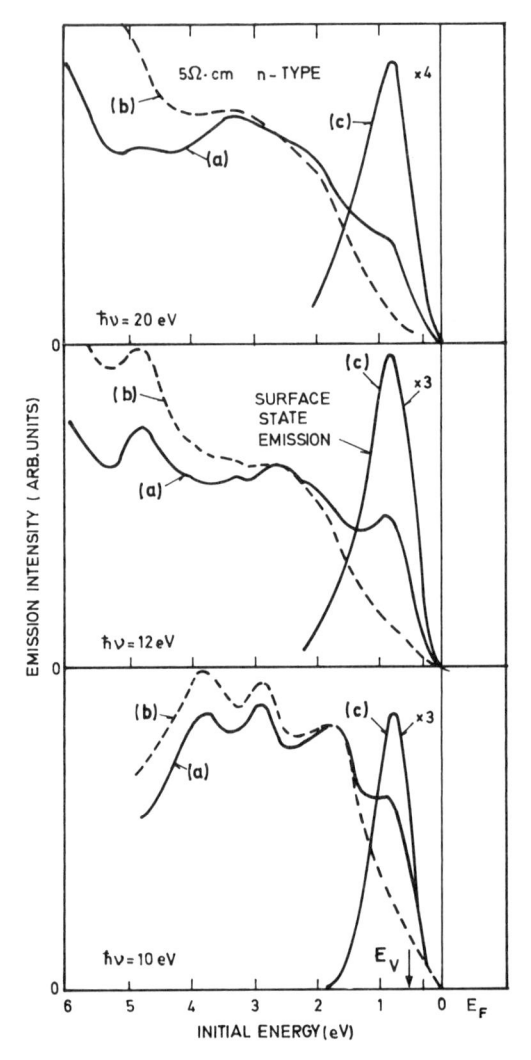

FIG.16. Energy distributions for Si. Curves (a) were measured within ~30 minutes of cleaving in ~3 × 10^{-10} Torr and show bulk plus surface-state emission. Curves (b) were measured about 7 hours later, by which time the surface-state emission has largely vanished. Curves (c) are the differences of (a) and (b) and depict the optical density of intrinsic surface states. (After Eastman and Grobman [28].)

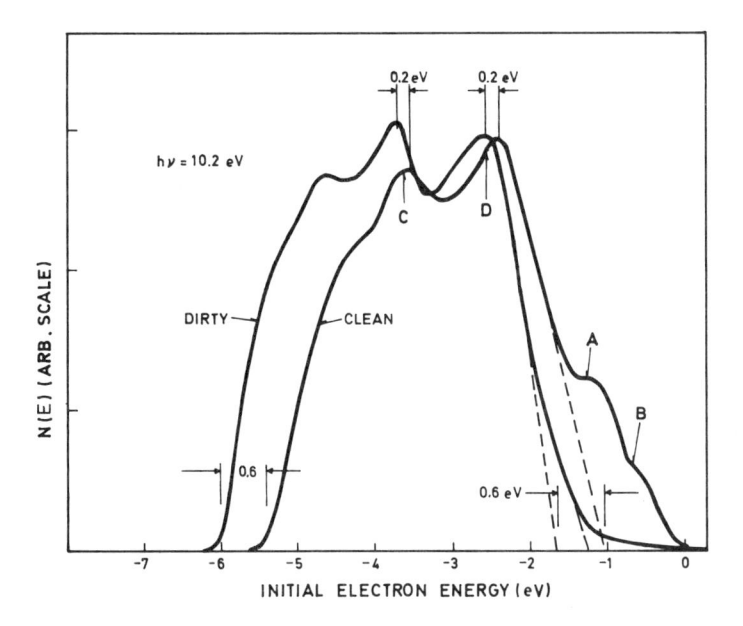

FIG. 17. Comparison of energy distribution curves of Si before and after vacuum contamination at 10^{-10} Torr. Note that band bending causes the work function (left edge) and the extrapolated high-energy edge of the valence-band structure to shift 0.6 eV while the positions of the bulk peaks, C and D, shift only 0.2 eV. (After Wagner and Spicer [29].)

of photoexcited electrons is of the order of 10 to 40 Å [27], photoemission spectroscopy is very sensitive to the state of the surface and to the presence of surface states [28, 29].

Photoemission in the bulk is generally treated as a three-step process consisting of: (a) optical excitation of an electron to levels higher than E_{vacuum}; (b) propagation of the electron to the surface; and (c) escape of the electron through some kind of potential barrier at the surface.

Description of process (b) requires knowledge of the scattering probability of a hot electron by phonons, valence band electrons (through e-h pair production or Auger processes) and (in metals) by other (nearly free) electrons. The theory for semiconductors has been worked out by Kane [27] and gives rise to an energy-dependent escaping length of the order of 10-40 Å, as already mentioned.

Process (a) is a bulk process and is governed by the optical joint density of states for direct transitions. Since, however, we now measure the energy E of the emitted electron, we are limited to those final states whose energy is E. The distribution of the energy of internal electrons (which have not yet undergone steps (b) and (c)) is then proportional to

$$N(\hbar\omega, E) \sim \int_{BZ} \delta(E_f - E_i - \hbar\omega) \, \delta(E_f - E) \, d^3k \tag{24}$$

The integral can be evaluated by using the properties of the δ function (see e.g. Ref.[30]) or by geometrical considerations [31]. The result is

$$N(\hbar\omega, E) \sim \int \frac{d\ell}{|\nabla_k(E_i) \times \nabla_k(E_f)|} \quad (25)$$

where the integration is carried out along the intersection between the two surfaces $E_f(k) = E$ and $E_f(k) - E_i(k) = \hbar\omega$.

$N(\hbar\omega, E)$ is large when the denominator in (25) is zero, i.e. when $\nabla_k(E_i) = 0$, $\nabla_k(E_f) = 0$ or $\nabla_k(E_i) \mathbin{/\mkern-5mu/} \nabla_k(E_f)$.

$\nabla_k(E_i) = 0$ defines the singularities of the distribution of the initial states which is the quantity of physical interest.

Some of the photoexcited electrons go through processes (b) and (c) without being scattered. It is then possible to extract $N(\hbar\omega, E)$ from the external energy distribution curve. However, $N(\hbar\omega, E)$ is by no means the density of the initial states and great care should be taken in interpreting the photoemission data. Only when the energy of the photon is large enough does the final-state electron behave as a free electron and the external energy distribution resemble the initial density of states. This is, however, rigorously true for surface states.

Let us consider (Fig.15) a two-dimensional Brillouin zone (k_x, k_y) and let $E_s = E_s^0$ and $E_s = E_s^0 + \Delta E_s$ be two curves of constant surface-states energy. Consider also a free-electron sphere corresponding to the energy $E_f = E - \hbar\omega$ and its intersection with the k_x, k_y plane. States in the hatched portion between the two sets of curves have $(k_f)_\| = (k_i)_\|$ and $(k_f)_\perp = 0$ so that they can be connected by a vertical optical transition. However, in the case of transitions from a surface state band, conservation of $k_\perp$ is no longer required, so that other parts of the free-electron sphere contribute as well. For example, in Fig.15 a projection on the k_x, k_y plane of the free-electron sphere is shown for $k_\perp = (\sqrt{3}/2)|k|$. Transitions which conserve $k_\|$ are also allowed from the cross-hatched areas. By taking into account all possible values of $k_\perp$, the area between the curves $E_s = E_s^0$ and $E_s = E_s^0 + \Delta E_s$ is completely covered and the energy distribution of emitted electrons gives simply the density of surface states with energy E_s.

The energy spectra of electrons photoemitted from ultrahigh vacuum cleaved surfaces of Si, Ge and GaAs, and their changes when the surface becomes oxidized, have recently been investigated by Eastman and Grobman [28], Wagner and Spicer [29], Erbudak and Fisher [32]; the results are shown in Figs 16 and 17. Energies of the emitted electrons are referred to the Fermi level and must be corrected for band bending. Eastman and Grobman use the values of V_s given by Allen and Gobeli, while Spicer and Wagner obtain it from the consideration that the work-function edge of their spectra (Fig.17, left-hand side) shifts by 0.6 eV (towards higher work-function values) when the Si crystal is oxidized. On the assumption that bands are flat in the oxidized samples, they obtain $qV_s = -0.6$ eV for the clean surface.

If band bending is taken into account, the results of Figs 16 and 17 give essentially the same distribution of surface states, with a main peak approximately 0.5 eV below the top of the valence band. However, Wagner and Spicer find a second smaller peak slightly above the top of the valence band which is apparently not present in the curves of Fig.16.

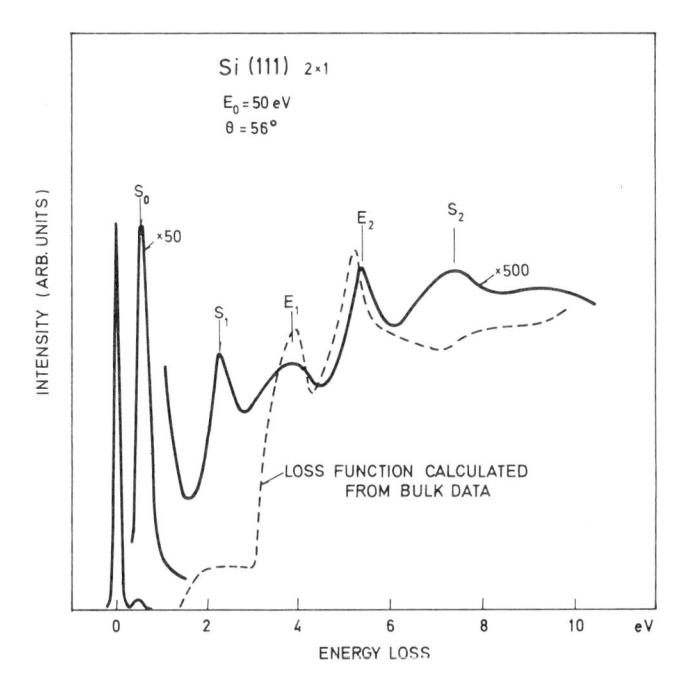

FIG.18. High-resolution electron energy-loss spectrum for cleaved clean Si(111) 2 × 1 at a primary energy of 50 eV and angle of incidence of 56°. The surface state transitions are labelled S_0, S_1, S_2 while the bulk transitions are labelled E_1 and E_2. The optical loss function calculated from bulk data is shown as a broken line. (After Rowe et al. [34].)

This second peak (B in Fig. 17) has the same energy as the full band responsible for the optical absorption shown in Fig. 11.

The main result of photoemission, however, is the presence of a large density of states well within the valence band which extends to the Fermi energy with no apparent gap in contrast with optical measurements. Probably a comparison of optical and photoemission measurements could be satisfactorily made only if the band structure of the surface states in the two-dimensional Brillouin zone were better understood (see Section 6 for more comments).

5. ELECTRON ENERGY LOSSES

An electron entering a solid displays a charge $q\delta(x-vt)$ and produces an electric field (on account of $\text{div}\,\vec{D} - \rho$) that interacts with the other electrons giving rise to an absorbed power:

$$\frac{1}{2}\frac{\epsilon_2}{\epsilon_1^2 + \epsilon_2^2}\,\omega\,\frac{DD^*}{\epsilon_0}$$

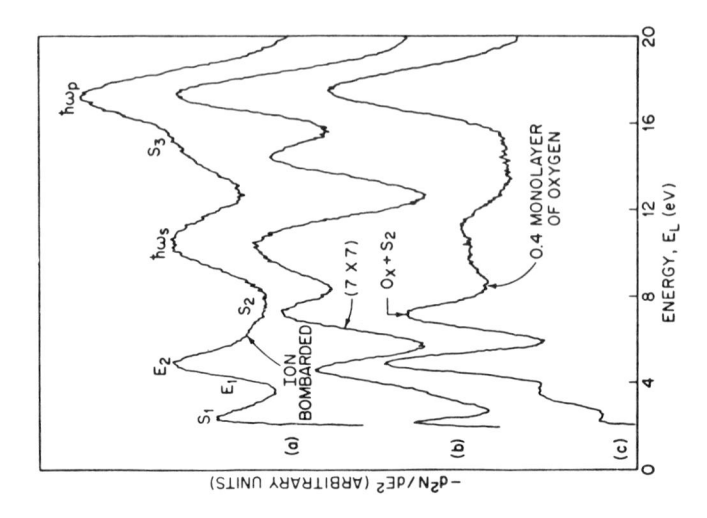

FIG. 20. Second-derivative electron energy-loss spectra at $E_p = 100$ eV for Si(111) surfaces after: (trace (a)) Ar^+-ion bombardment; (trace (b)) annealed to produce a sharp 7×7 LEED pattern; and (trace (c)) at $\theta = 0.4$ monolayers of a surface oxide. (After Rowe and Ibach [35].)

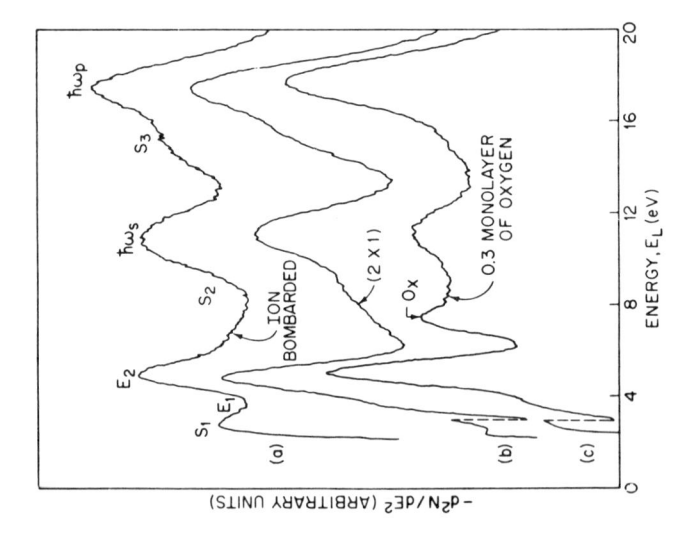

FIG. 19. Second-derivative electron energy-loss spectra at $E_p = 100$ eV for Si(100) surfaces which are: (trace (a)) clean but disordered following Ar^+-ion bombardment; (trace (b)) clean and ordered resulting in sharp 2×1 LEED pattern; and (trace (c)) after partial formation of a disordered surface oxide at $\theta = 0.3$ monolayers of oxygen coverage. (After Rowe and Ibach [35].)

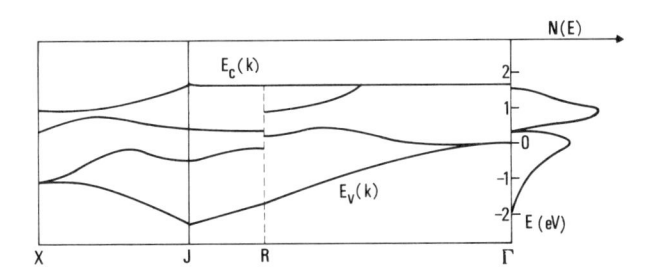

FIG.21. Energy bands of surface states of Si in the two-dimensional Brillouin zone. On the right, the density of states N(E). (After Rubio and Flores [36].)

ω being the Fourier component of the field and $\epsilon = \epsilon_1 + i\epsilon_2$ the complex dielectric function of the solid. For a review of energy-loss spectroscopy see Ref. [33].

Energy-loss spectroscopy measures excitations characterized by the bulk loss function:

$$\frac{\epsilon_2}{\epsilon_1^2 + \epsilon_2^2} = -\mathrm{Im}\,\frac{1}{\epsilon} \tag{26}$$

(or by the surface loss function $-\mathrm{Im}\,1/(\epsilon + 1)$) and can be compared with optical absorption that measures $\mathrm{Im}\,\epsilon$.

Energy loss experiments carried out in reflection are very sensitive to the structure of the surface, essentially for the same reasons mentioned in Section 4.3 on photoemission spectroscopy. This technique has been widely used by Rowe and Ibach, who have obtained several interesting results on surface states on Si and Ge that supplement in various ways the optical and photoelectric data.

Figure 18 shows a high-resolution energy-loss spectrum of cleaved Si(111) 2×1 [34]. The main peak S_0 at 0.52 eV is the surface state to surface state transition already observed in optical absorption (Section 4.1), the slight difference in energy (~0.05 eV) being due to the energy dependence of the denominator of (26).

Energy loss experiments also give evidence of surface states with energy well into the valence band, as shown in Figs 19 and 20 for the (100) and (111) Si surfaces [35].

Like optical absorption, however, energy-loss experiments give only energy differences, so that the actual position of the levels depends in general on the model assumed.

6. CONCLUSIONS

From the results of the previous sections a scheme of levels can be deduced for the (111) 2×1 surface of clean Si. The spectrum consists of:

(a) Two bands of surface states in the gap, symmetric with respect to the neutrality level (0.35 eV above the top of the valence band) centred 0.65 eV and 0.15 eV above the valence band and separated by a gap of 0.3 eV. Evidence for this comes from optical absorption, photoconductivity and energy losses;
(b) A band of states approximately 0.5 eV below the top of the valence band with a width of at least 0.5 eV as observed in photoemission;
(c) Two deep levels (or bands) 5.4 eV and 12.5 eV below the top of the valence band, as evidenced by energy loss experiments.

Whether such levels are true separated bands or simply critical points in the density or joint-density of states cannot be decided without a detailed knowledge of the band structure in the surface Brillouin zone.

Most recently, Rubio and Flores [36] have attempted a calculation of the surface bands on (111) surfaces of Si, taking into account, in a simplified way, the reconstruction of the surface. Their results are shown in Fig. 21. It is seen that the two bands of surface states run almost parallel for part of the ΓJ and JX lines in the surface Brillouin zone, giving a very high joint density of states, with a saddlepoint at R, that explains very well the optical results of Fig. 11. On the other hand, the density of states in the two bands, plotted in the right-hand side of Fig. 21, shows a lower band partially overlapping the valence band, in agreement with the results on photoemission. In this scheme the transitions giving rise to photoconductivity are indirect transitions from the top of the bulk valence band at Γ to the minimum of the (empty) upper band of surface states at point J in the surface Brillouin zone.

A convincing model of the bands at the Si(111) 2×1 surface has been given by Tosatti [37]. He assumes a single half-filled band at the unreconstructed surface, degenerate along the borders of the Brillouin zone of the 2×1 structure. Reconstruction removes this degeneracy, opening a gap that lowers the occupied levels, thus stabilizing the 2×1 structure.

A detailed calculation (done, however, for the (100) surface) by Betteridge and Heine [38] also shows good agreement with the optical data of Fig. 11.

REFERENCES

[1] MANY, A., these Proceedings.
[2] JONES, R.O., these Proceedings.
[3] FORSTMANN, F., "Quantum theory at surfaces and interfaces", Theory of Imperfect Crystalline Solids (Proc. Int. Course, Trieste, 1970), IAEA, Vienna (1971) 511.
[4] FRANKL, D.R., Electrical Properties of Semiconductor Surfaces, Pergamon, Oxford (1967).
[5] MANY, A., GOLDSTEIN, Y., GROVER, N.B., Semiconductor Surfaces, North-Holland, Amsterdam (1965), hereafter referred to as MGG.
[6] ALLEN, F.C., GOBELI, G.W., Phys. Rev. $\underline{127}$ (1962) 150.
[7] DAVISON, S.G., LEVINE, J.D., in Solid State Physics $\underline{25}$ (EHRENREICH, H., SEITZ, F., Turnbull, D., Eds), Academic Press (1970) 1.
[8] ASPNES, D.E., HANDLER, P., Surf. Sci. $\underline{4}$ (1966) 353.
[9] HENZLER, M., Phys. Status Sol. $\underline{19}$ (1967) 833.
[10] MÖNCH, W., Phys. Status Sol. $\underline{40}$ (1970) 257.
[11] CHIAROTTI, G., DEL SIGNORE, G., NANNARONE, S., Phys. Rev. Lett. $\underline{21}$ (1968) 1170.
[12] CHIAROTTI, G., NANNARONE, S., PASTORE, R., CHIARADIA, P., Phys. Rev. $\underline{4B}$ (1971) 3398.
[13] VAN HOVE, L., Phys. Rev. $\underline{89}$ (1953) 1189.

[14] BASSANI, F., Optical Properties of Solids (Proc. Int. School of Physics Enrico Fermi, Course 34, Varenna, 1965), Academic Press, New York (1966) 33.
[15] HARRICK, N.J., Phys. Rev. 125 (1962) 1165.
[16] CHIAROTTI, G., DEL SIGNORE, G., FROVA, A., SAMOGGIA, G., Nuovo Cim. 26 (1962) 403.
[17] SAMOGGIA, G., NICCIOTTI, A., CHIAROTTI, G., Phys. Rev. 144 (1966) 749.
[18] CHIARADIA, P., NANNARONE, S., Int. Rep. No. 446, Inst. Phys., University of Rome (1973).
[19] MÜLLER, W., MÖNCH, W., Phys. Rev. Lett. 27 (1971) 250.
[20] MÜLLER, W., MÖNCH, W., Advances in Solid State Physics, Pergamon, Oxford (1973) 241.
[21] HEILAND, G., MÖNCH, W., Surf. Sci. 37 (1973) 30.
[22] STOCKER, H.J., LEVINSTEIN, H., STANNARD, C.R., Phys. Rev. 150 (1966) 613.
[23] IBACH, H., Phys. Rev. Lett. 27 (1971) 253.
[24] LÜTH, H., HEILAND, G., Phys. Status Sol. (a) 14 (1972) 573.
[25] LÜTH, H., Phys. Rev. Lett. 29 (1972) 1377.
[26] IBACH, H., Phys. Rev. Lett. 24 (1970) 1416.
[27] KANE, E.O., Phys. Rev. 159 (1967) 624.
[28] EASTMAN, D.E., GROBMAN, W.D., Phys. Rev. Lett. 28 (1972) 1378.
[29] WAGNER, L.F., SPICER, W.E., Phys. Rev. Lett. 28 (1972) 1381.
[30] HEDIN, L., "Optical, X-ray and electron-emission phenomena", Electrons in Crystalline Solids (Proc. Int. Course, Trieste, 1972), IAEA, Vienna (1973) 665.
[31] EASTMAN, D.E., in Techniques of Metals Research 6 (PASSAGLIA, E., Ed.), Interscience, New York (1972) 411.
[32] ERBUDAK, M., FISHER, T.E., Phys. Rev. Lett. 29 (1972) 732.
[33] RAETHER, H., Springer Tracts in Modern Physics 38, Springer Verlag (1965) 84.
[34] ROWE, J.E., IBACH, H., FROIDZHEIM, H., Photoemission and energy-loss spectroscopy on semiconductor surfaces (to be published).
[35] ROWE, J.E., IBACH, H., Phys. Rev. Lett. 31 (1973) 102.
[36] RUBIO, J., FLORES, F., Reconstruction and surface states of (111) surfaces of Si (to be published).
[37] TOSATTI, E., "Instabilities and superlattices on semiconductor surfaces", Seminar at the Winter College on Surface Science, Trieste, 1974 (not published in these Proceedings).
[38] BETTERIDGE, G.P., HEINE, V., Calculations of the optical absorption between surface states of silicon, Rep. TCM/53/1973, Dept. Phys., Cambridge University (1973).

ELECTRICAL TRANSPORT IN THE SPACE-CHARGE REGION

A. MANY
Racah Institute of Physics,
Hebrew University of Jerusalem,
Jerusalem, Israel

Abstract

ELECTRICAL TRANSPORT IN THE SPACE-CHARGE REGION.
1. The surface space-charge region: Band bending in the bulk semiconductor; Origin of the surface space-charge layer; Space-charge density and shape of the potential barrier; Excess surface-carrier densities ΔN and ΔP; Space-charge region under non-equilibrium conditions; Degenerate surface conditions; Quantization effects in very strong accumulation and inversion layers.
2. Measurements on the space-charge layer: Surface conductance; Surface capacitance; Field effect; Contact potential and surface photovoltage; Galvanomagnetic measurements.
3. Surface transport processes: Concepts and definitions; Simple considerations; Calculations of average and surface mobilities for diffuse scattering; Partially diffuse scattering — the Fuchs boundary condition; Quantized space-charge layers.

1. THE SURFACE SPACE-CHARGE REGION

1.1. Band bending in the bulk semiconductor

The concept of band bending is of primary importance in understanding the space-charge region at the semiconductor surface. The simplest way to clarify this concept is to consider a non-homogeneous semiconductor, i.e. a semiconductor in which the carrier densities (in the bulk) are functions of position. The discussions that follow will be confined to non-degenerate semiconductors.

When the carrier densities are not uniform, carriers tend to diffuse from regions of high to regions of low concentrations. Thus, in addition to the drift current due to the external field, a diffusion current is also established. If the latter is assumed to be directly proportional to the concentration gradient, an assumption amply borne out by experiment, we may write the hole and electron current densities as

$$\left. \begin{array}{l} \vec{J}_p = q\mu_p p \vec{\mathscr{E}} - q D_p \operatorname{grad} p \\ \vec{J}_n = q\mu_n n \vec{\mathscr{E}} + q D_n \operatorname{grad} n \\ \vec{J} = \vec{J}_p + \vec{J}_n \end{array} \right\} \quad (1.1)$$

Here q is the electronic charge, μ_p and μ_n the hole and electron mobilities, and $\vec{\mathscr{E}}$ is the electric field; p and n are the position-dependent hole and electron concentrations, and D_p and D_n are the so-called diffusion constants for holes and electrons; $\vec{J}$ is the total current density. The signs in front of the diffusion terms are in accordance with the convention regarding the current polarity.

The relation between the diffusion constant and the mobility of either carrier type is obtained by considering an inhomogeneous semiconductor under thermal equilibrium and in the absence of external fields. According to the principle of detailed balance, the hole and electron currents must then vanish separately. Let us first consider the electrons and assume that a concentration gradient exists only in the x direction. Using Eq.(1.1), we have

$$\mu_n n \mathscr{E}_x = -D_n (dn/dx) \tag{1.2}$$

The field $\mathscr{E}_x$ is established internally to counteract the diffusion flow. If V denotes the electrostatic potential set up in this manner, then

$$\vec{\mathscr{E}} = -\text{grad } V \; ; \; \mathscr{E}_x = -dV/dx \tag{1.3}$$

Combining Eqs (1.2) and (1.3), we obtain

$$n = \text{const } \exp(\mu_n V/D_n) \tag{1.4}$$

Since -qV is the potential energy of an electron, we may express the Maxwell-Boltzmann distribution law in the form

$$n = A \exp(qV/kT) \tag{1.5}$$

where A is a normalizing constant, k is Boltzmann's constant and T the temperature. Thus it follows from Eqs (1.4) and (1.5) that

$$D_n = (kT/q)\mu_n \tag{1.6}$$

In a similar manner one obtains for holes

$$D_p = (kT/q)\mu_p \tag{1.7}$$

These are the well-known Einstein relations between mobility and diffusion constant. The units of D_n and D_p are m^2/s if μ is in $m^2/V \cdot s$ and cm^2/s if μ is in $cm^2/V \cdot s$.

Due to the dependence on position of the potential V in an inhomogeneous semiconductor, an electron in, say, the lowest state of the conduction band will have different energies in different parts of the crystal. On the other hand, at thermal equilibrium the Fermi level E_F must be constant throughout the crystal. This is because the Fermi level is closely related to the thermodynamic or chemical potential, which is a constant (at a given temperature) throughout a system in thermal equilibrium, even when the system is composed of different phases (such as occur in inhomogeneous semiconductors or in different materials in contact). Combining Eq.(1.5) with the well-known expression for the electron concentration in a non-degenerate semiconductor, one has

$$n = A \exp(qV/kT) = N_c \exp[(E_F - E_c)/kT]$$

Here N_c is the effective density of states in the conduction band, E_c the bottom of the conduction band and E_F the Fermi level. Thus the edge of the conduction band must be related to the potential by

$$E_c = \text{const} - qV \tag{1.8}$$

The edge of the valence band E_v, as well as all other levels in the crystal, follows the variation of the potential just as E_c does. This behaviour is a consequence of the fact that the separations between these various energy levels are determined by the large, short-range crystal fields and therefore remain practically unaffected by the much weaker fields associated with the inhomogeneity of the crystal. The energy band diagram of an inhomogeneous semiconductor is schematically represented in Fig.1 and should be contrasted with that of a homogeneous semiconductor (Fig.2). Also shown in Fig.1 is the corresponding variation of the electrostatic potential V. The bending of the bands shown in Fig.1 expresses the establishment of space charge and a position-dependent electrostatic potential. A well-known situation of this sort exists in a p-n junction.

It should be noted that in the foregoing discussion, thermal equilibrium conditions apply only to the carrier densities. The impurities and other imperfection centres giving rise to the inhomogeneity of the crystal do not of course have an equilibrium distribution — they maintain their position in the lattice only because of their vanishingly small mobility at the temperatures considered.

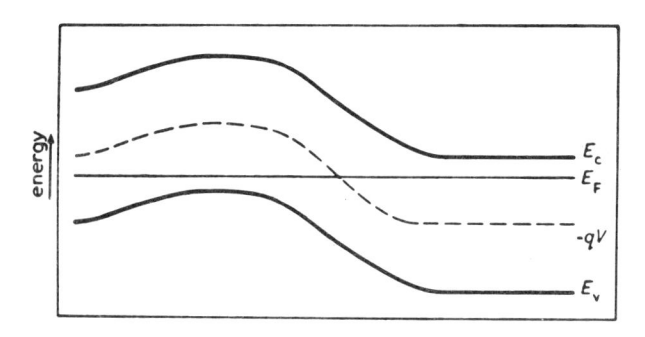

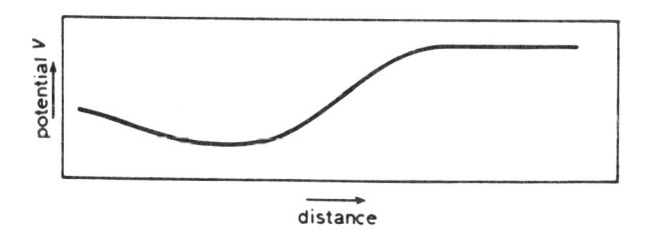

FIG.1. Energy-level and potential diagrams in an inhomogeneous semiconductor.

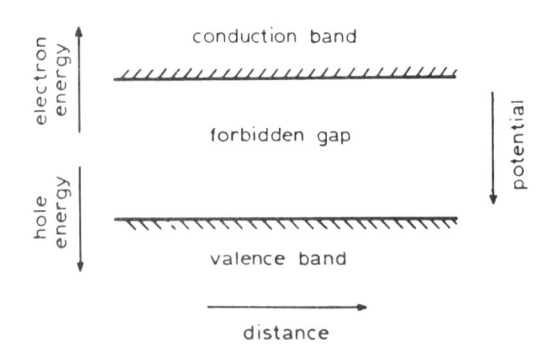

FIG.2. Energy-level diagram for a homogeneous semiconductor. Electron energy increases upwards, hole energy and potential increase downwards.

1.2. Origin of the surface space-charge layer

Obviously, in an infinite, homogeneous semiconductor under thermal equilibrium there is no space charge and hence no band bending. Close to the surface of a finite solid, however, there is often a space-charge layer and the bands which are horizontal throughout the bulk may bend upwards or downwards as the surface is approached. Such a space-charge layer may be produced by an electric field outside the solid, such field being established by an external source or by the proximity of another solid with a different work function. Alternatively, the space-charge region may result from the presence of a localized charged layer at the surface proper, due usually to surface states. These possibilities will now be discussed.

1.2.1. External field

Consider a parallel-plate capacitor having one electrode a metal and the other a semiconductor. The application of a voltage across the capacitor results in the establishment of an electric field $\mathscr{E}$ between the plates. A displacement of mobile charge-carriers near the surface of each plate takes place, thus giving rise to two space-charge regions. The density Q_s of the induced charge on each plate is given by Gauss' law: $Q_s = \epsilon_0 \mathscr{E}$. Since the free carrier density in a semiconductor is much smaller than that in a metal, the space-charge region will extend much farther into the bulk of the former (see below). This situation is illustrated schematically in Fig.3 for the case of an n-type semiconductor devoid of surface states, for various values of the bulk mobile carrier density n_b. The polarity of the applied voltage V_0 has been chosen so as to make the metal plate negative. Note that because the electrons have a negative charge, the electron energy increases in the direction of decreasing potential. To simplify the presentation in Fig.3, the discontinuity in potential caused by the electron affinity has been omitted (see below). Also, in order to avoid the discontinuity of the field at the surface, the position co-ordinate z has been changed to z/κ, where κ is the dielectric constant in the medium concerned. The penetration of the field into the semiconductor produces a potential barrier beneath the surface. To obtain an idea of the

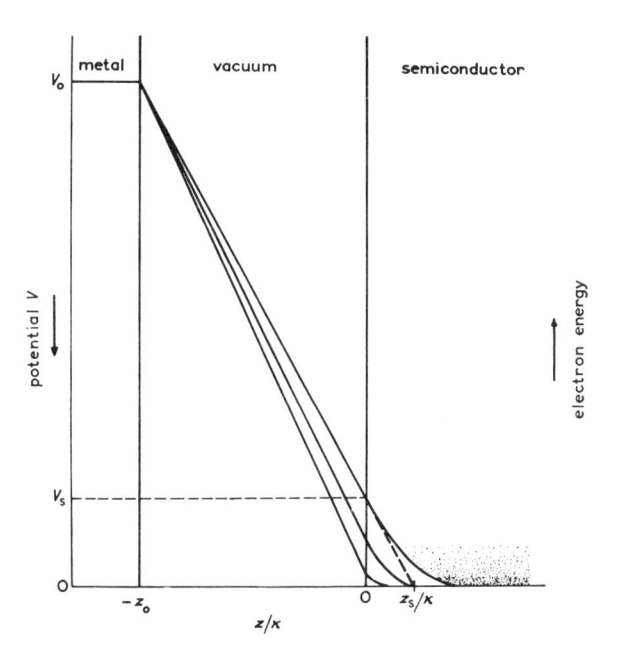

FIG.3. Space-charge region in an n-type semiconductor devoid of surface states as produced by an applied voltage between a metal and the semiconductor. The potential and electron energy are shown as functions of the normalized distance z/κ. The three curves are for different values of the bulk electron density n_b, the uppermost curve corresponding to the smallest n_b. The dots illustrate the electron energy distribution in the semiconductor.

magnitude of the potential drop V_s between the surface and the underlying bulk, we approximate the potential in the space-charge region by a linear extension (dashed) of that in the vacuum. From Fig.3 we see that

$$\frac{V_s}{V_0 - V_s} = \frac{z_s/\kappa}{z_0} \tag{1.9}$$

where z_0 is the separation of the capacitor plates and z_s the approximate distance (within the semiconductor) over which the potential drops to zero. An estimate of Q_s is readily obtained by assuming that all the electrons in the space-charge region up to the distance z_s are removed by the field, thus leaving behind an equal static positive charge. By Gauss' law,

$$-\frac{\epsilon_0(V_0 - V_s)}{z_0} \approx q n_b z_s \tag{1.10}$$

Combining Eqs (1.9) and (1.10), we have

$$z_s \approx \left(\frac{\kappa \epsilon_0 |V_s|}{q n_b}\right)^{1/2} \; ; \quad \frac{V_s}{V_0 - V_s} \approx -\frac{\epsilon_0(V_0 - V_s)}{\kappa z_0^2 q n_b} \tag{1.11}$$

TABLE I. VALUES OF V_s AND z_s FOR DIFFERENT VALUES OF n_b IN GERMANIUM ($\kappa = 16$)

n_b (cm^{-3})	V_s (V)	z_s (cm)
10^{13}	-3.5×10^{-1}	5.6×10^{-4}
10^{15}	-3.5×10^{-3}	5.6×10^{-6}
10^{17}	-3.5×10^{-5}	5.6×10^{-8}

$V_0 = -100$ V; $z_0 = 10^{-2}$ cm

Despite the fact that the approximations used in deriving Eq.(1.11) appear rather crude, nevertheless the results yield the correct order of magnitude (see below). It is seen from Eq.(1.11) that as n_b increases, $|V_s|$ and z_s decrease. This is illustrated in Table I, where values of V_s and z_s are given for various values of n_b in germanium ($\kappa = 16$). The applied voltage V_0 is -100 V and the plate separation z_0 is 10^{-2} cm. Obviously, although a space-charge region exists in metals as well, the corresponding values of V_s and z_s are so small as to be negligible for most purposes.

In the construction of Fig.3 and the derivation of Eqs (1.9) – (1.11), the effect of surface states has been omitted. As will be discussed, such states can give rise to a charged layer at the surface. A space-charge region will then be formed below the surface to compensate this surface charge. If now an external field is applied, charge will be displaced from both the surface states and the space-charge region already present. For a sufficiently large density of occupied surface states, the external field will be terminated almost entirely at the surface proper, leaving the space-charge region essentially undisturbed. Thus surface states can screen the underlying region from external effects.

We now consider in more detail the metal-semiconductor capacitor just discussed. Although conduction electrons in a solid are free to move about, they are not able to leave the crystal surface easily. This implies that their energy is lower inside the solid than outside. The work required to remove an electron at the Fermi level to a point in free space just outside the solid is defined as the work function W_0 of the solid. In semiconductors it is found useful to define a second quantity, the electron affinity χ, as being the work required to remove an electron from the bottom edge of the conduction band at the surface to a point in free space just outside the semiconductor. The corresponding changes in potential as one moves from the solid into free space are very steep since they rise from the short-range binding forces exerted by the crystal lattice on the electron (typically, 10^8 V/cm over an interatomic spacing). Such potential steps are sometimes referred to as micropotentials, to stress their microscopic character. The work function and affinity are illustrated in Fig.4(a), which represents an energy level diagram for an electron in a system composed of a metal and an n-type semi-conductor separated by free space (vacuum). For simplicity, we assume that the work functions of the metal and the semiconductor are equal. Under these conditions the free-space potential is horizontal (zero field), which

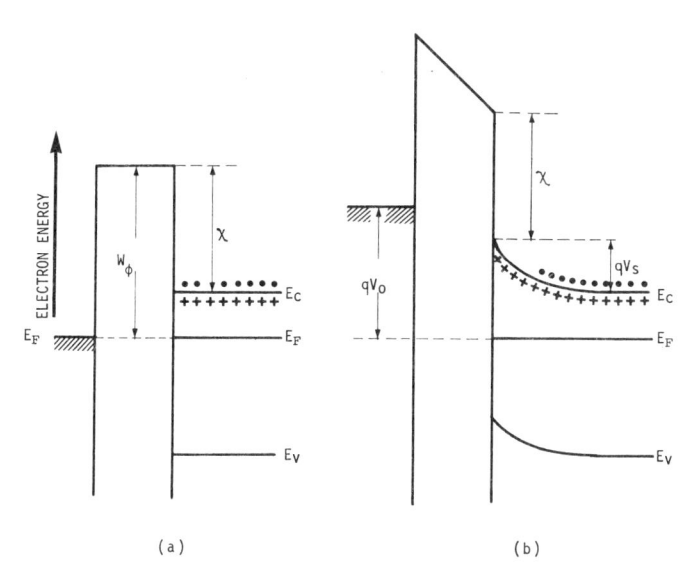

FIG.4. Energy level diagram in system composed of metal and n-type semiconductor separated by free space (vacuum). Work functions of two solids assumed equal. (a) No applied voltage; (b) With voltage V_0 applied across metal-semiconductor capacitor.

means that an electron in vacuum has the same energy everywhere. By the same token, the energy bands are flat all the way to the surface, as shown in Fig.4(a).

The electrostatic field established by the applied voltage (Fig.4(b)) is usually much weaker than the crystal field, so that the macroscopic potential from which it is derived leaves the micropotential associated with W_0 and χ practically unaffected. Thus if it is borne in mind that the macropotential is superimposed on the short-range micropotential, it is seen that practically the entire applied voltage distributes itself between the vacuum and the semiconductor. In the latter region the energy-band edges assume the shape of the macropotential. This follows from the fact that the carrier density at any point can be expressed in equivalent forms in terms of either the macropotential or the band edges, as discussed in Section 1.1. For these expressions to yield the same value for the carrier densities, the band edges at any point must differ from the electrostatic energy $-qV$ by at most a constant. Since the zero of electrostatic potential is arbitrary, it is possible to identify the macropotential energy with either of the band edges (or with any other energy level at a fixed distance from them). The procedure for estimating the barrier height V_s established by the applied voltage and the effective barrier width z_s is the same as discussed above. These estimates are based on the assumption that $|V_s|$ is not too high (which obtains if the applied voltage V_0 is not too large). Under these conditions the space-charge is composed almost entirely of positively charged donor centres, most of the compensating free electrons having been repelled from the space charge region by the external field. When $|V_s|$ becomes too large, minority carriers (holes) are produced, and the situation is drastically altered, as will be discussed below.

1.2.2. Surface states

The presence of surface states can give rise to band bending similar to that produced by an external field. The surface states may become sites of surface charge which forms and is balanced by a space-charge layer beneath the surface. Consider first the effect of acceptor-like surface states. (An acceptor state is one that is able to capture an electron, thereby becoming negatively charged.) When such surface states are introduced below the Fermi level, they will not be in equilibrium with the energy bands as long as they remain unoccupied. This situation is illustrated in Fig.5(a), where the surface states have been introduced at an energy level E_t. Since the states are empty and below the Fermi level, some of the electrons in the conduction band fall into them. In this process, the surface becomes negatively charged while a positive space-charge layer forms below it. Consequently, the energy bands at the surface bend upwards with respect to the Fermi level. Since the energy position of the surface states in the forbidden gap is determined by short-range atomic forces and is thus unaffected by the presence of the macropotential, the surface levels rise together with the band edges. The process of charge transfer continues until equilibrium is reached (Fig.5(b)). In the case shown, the surface states at thermal equilibrium are somewhat above the Fermi level and so are only partially filled. Actually, the final position of the surface states with respect to the Fermi level is determined by the condition that the positive charge in the space-charge region just balance the negative equilibrium charge in the surface states. This is made necessary by the requirement that overall neutrality should prevail. Thus the larger the surface-state density, the higher the bending of the bands at the surface. The situation for donor-like surface states placed above the Fermi level is completely analogous (Figs 5(c) and 5(d)). The introduction of acceptor-like states above the Fermi level or of donor-like states below the Fermi level has of course no influence on the shape of the energy bands.

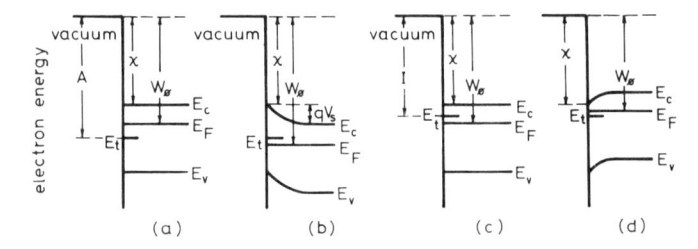

FIG.5. Energy-level diagram for an n-type semiconductor in the presence of acceptor-like ((a) and (b)) and donor-like ((c) and (d)) surface states at an energy level E_t. (a) and (c) represent conditions immediately following the introduction of the surface states; (b) and (d) represent conditions after thermal equilibrium has been reached.

1.3. Space-charge density and shape of the potential barrier

In the previous section we discussed possible origins of the space-charge region near the surface. Once such a region exists in a given semiconductor, its detailed characteristics are uniquely determined by the height of the

potential barrier at the surface proper. We now calculate the magnitude of
the space-charge density and the shape of the potential in the space-charge
region as a function of barrier height.

1.3.1. Concepts and definitions

It will be useful at this stage to define the function and parameters
associated with the space-charge region near the surface. We consider a
semi-infinite, homogeneous crystal under thermal equilibrium. The surface
is represented by the plane z = 0 and the bulk by positive values of z. The
potential V at any given point is then a function of z only. It is defined by

$$q V(z) = E_c(\infty) - E_c(z) = E_v(\infty) - E_v(z) \tag{1.12}$$

where $E_c(\infty)$, $E_v(\infty)$ are the energy positions of the band edges in the bulk, well
removed from the surface, and $E_c(z)$, $E_v(z)$ are the positions of the band edges
at point z. Thus V(z) represents the potential at any point in the space-
charge region with respect to its value in the bulk. In particular, the barrier
height V_s ($\equiv V(0)$) is the total potential difference between the surface and the
bulk (Fig.6). Another useful function is the so-called potential $\phi(z)$ defined by

$$q \phi(z) = E_F - E_i(z) \tag{1.13}$$

where $E_i(z)$ is parallel to the band edges and in the bulk ($z = \infty$) coincides with
the intrinsic Fermi level E_i. The intrinsic Fermi level is the position of the
Fermi level in an intrinsic sample, i.e. in a sample in which the electron and
hole concentrations n_b and p_b are equal. E_i is usually close to the mid-gap.
The value of ϕ in the bulk is called the bulk potential ϕ_b; its value at the
surface is called the surface potential ϕ_s. (For an n-type semiconductor
$\phi_b > 0$, while for a p-type sample $\phi_b < 0$.) Another quantity defined in Fig.6

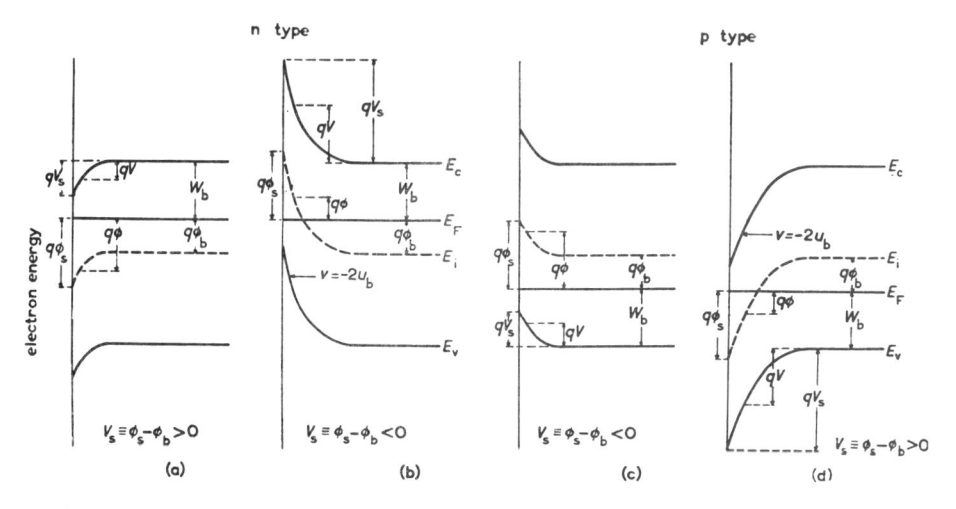

FIG.6. Energy-level diagram indicating the various energy parameters used to characterize the space-charge
region. (a) n-type semiconductor, accumulation layer; (b) Inversion layer; (c) p-type semiconductor,
accumulation layer; (d) Inversion layer.

is W_b, the energetic separation between E_F and the majority-carrier band edge in the bulk (E_c for n type; E_v for p type). Obviously,

$$V(z) = \phi(z) - \phi_b$$

$$V_s = \phi_s - \phi_b$$

It should be noted that a positive value of V_s means band bending downwards (Figs 6(a) and 6(d)); a negative value means band bending upwards (Figs 6(b) and 6(c)).

As we shall see, it would be convenient to define dimensionless potentials by the equations:

$$v(z) = q V(z)/kT, \qquad v_s = q V_s/kT$$

$$u(z) = q \phi(z)/kT, \qquad u_s = q \phi_s/kT \tag{1.14}$$

By means of these definitions, the electron and hole densities $n(z)$, $p(z)$ at any point of a non-degenerate semiconductor are given by

$$n(z) = n_b \exp[v(z)] = n_i \exp[u(z)]$$

$$p(z) = p_b \exp[-v(z)] = n_i \exp[-u(z)] \tag{1.15}$$

where n_i is the intrinsic carrier concentration. In particular, at the surface,

$$n_s = n_b \exp(v_s) = n_i \exp(u_s)$$

$$p_s = p_b \exp(-v_s) = n_i \exp(-u_s) \tag{1.16}$$

while in the bulk we have the well-known relations:

$$n_b = n_i \exp(u_b)$$

$$p_b = n_i \exp(-u_b) \tag{1.17}$$

The condition $u > 0$ signifies that at that point $n > p$, and conversely; $v > 0$ implies $n > n_b$ and $p < p_b$, and conversely. At the point where $u = 0$ the carrier densities are intrinsic; in particular $u_b = 0$ corresponds to an intrinsic bulk, $u_s = 0$ to an intrinsic surface. When $v_s = 0$ there is no bending of the energy bands and they continue straight from the bulk to the surface. This is known as the flat-band condition.

When the majority-carrier density in the space-charge region is greater than that in the bulk, the space-charge region is termed an accumulation layer. This condition obtains when the sign of v_s is the same as that of u_b (positive for n type and negative for p type). When the sign of v_s is opposite to that of u_b, we have either a depletion or an inversion layer. The space-charge region up to the point where the minority-carrier density equals the majority-carrier bulk density ($v = -2u_b$) is called the depletion region. Beween this point and the surface the minority-carrier density exceeds the majority-carrier bulk density, and this region is called the inversion region. (Of course the situation can exist where the depletion region continues right up to the surface.) Figure 6 illustrates the above definitions for an n-type and a p-type semiconductor.

1.3.2. Poisson's equation

The potential V(z) is determined by Poisson's equation which, for the planar geometry assumed, has the form:

$$\frac{d^2V}{dz^2} = -\frac{\rho}{\kappa\epsilon_0} \tag{1.18}$$

where ρ is charge density (Coulomb/m³) and is composed of positive and negative static charge (ionized donor and acceptor impurities) and the charge due to mobile electrons and holes. The appropriate boundary conditions at the surface (z = 0) and in the bulk (z → ∞) are $V = V_s$ and $V = 0$, respectively. We shall assume that the donor and acceptor impurities are completely ionized. Under these conditions the static charge density is the difference between the donor and acceptor concentrations ($N_D - N_A$) and, from the condition of charge neutrality in the bulk, is equal to the difference in electron and hole bulk densities ($n_b - p_b$). The mobile electron and hole densities in the case of non-degenerate statistics are given by Eq.(1.15). With the help of the reduced potential v, Poisson's equation can be rewritten as

$$\frac{d^2v}{dz^2} = -\frac{q^2}{\kappa\epsilon_0 kT}[n_b - p_b + p_b \exp(-v) - n_b \exp(v)] \tag{1.19}$$

To obtain a feeling for the problem, we shall first consider the case of small disturbances, corresponding to $|v| \lesssim \frac{1}{2}$ (or, at room temperature, to $|V_s| \lesssim 0.01$ V). Under these conditions it is permissible to neglect the non-linear terms in the series expansion of the exponential functions. We then have

$$\frac{d^2v}{dz^2} \approx \frac{v}{L^2} \tag{1.20}$$

where

$$L \equiv \left(\frac{\kappa\epsilon_0 kT}{q^2(n_b + p_b)}\right)^{1/2} \tag{1.21}$$

Equation (1.20) can be integrated directly, and yields the result

$$v = v_s \exp(-z/L) \tag{1.22}$$

It thus follows that in terms of z/L the potential barrier for small disturbances is independent of the bulk potential u_b. The value of $|v|$ decreases exponentially with characteristic length L, the *effective* Debye length (Eq.(1.21)).

It can be seen from the approximate solution (Eq.(1.22)) that the derivative dv/dz vanishes in the bulk and that the boundary condition v = 0 for z → ∞ is equivalent to v = 0 for z ≫ L. In the case of thin slabs, where the condition z ≫ L no longer holds, the situation is more complicated and will not be dealt with here.

With the help of Eqs (1.15) and (1.21), Poisson's equation (Eq.(1.19)) can be written as

$$\frac{d^2 v}{dz^2} = \frac{1}{L^2} \left[\frac{\sinh(u_b + v)}{\cosh u_b} - \tanh u_b \right] \qquad (1.23)$$

Multiplying both sides by $2 dv/dz$, we can integrate once and, using the condition $dv/dz = 0$ for $v = 0$, we obtain

$$\frac{dv}{dz} = \mp \frac{F(u_b, v)}{L} \quad ; \quad \frac{z}{L} = \int_{v_s}^{v} \frac{dv}{\mp F(u_b, v)} \qquad (1.24)$$

where

$$F(u_b, v) = \sqrt{2} \left[\frac{\cosh(u_b + v)}{\cosh u_b} - v \tanh u_b - 1 \right]^{1/2} \qquad (1.25)$$

The upper sign (minus) refers to v > 0, the lower sign (plus) to v < 0; this convention will be adhered to throughout. Equation (1.24) cannot be solved in closed form for the general case, and numerical solutions are given in the literature [1]. Approximate solutions are available for limiting cases, as will be discussed below.

The space-charge density Q_{sc} is defined as the total net charge in the space-charge region per unit surface area. As such, Q_{sc} satisfies Gauss' law:

$$Q_{sc} = \kappa \epsilon_0 \mathscr{E}_s \qquad (1.26)$$

where $\mathscr{E}_s$ is the electrostatic field inside the semiconductor just below the surface and is positive when directed from the surface outwards. The value and sign of $\mathscr{E}_s$ are given by the derivative of the potential V at the surface. With the help of Eqs (1.14), (1.21) and (1.24), we can express Q_{sc} in the form

$$Q_{sc} = \mp q(n_b + p_b) L F_s \qquad (1.27)$$

where F_s is the value of F at the surface ($v = v_s$).

The space-charge density Q_{sc} arises from the charge that is unevenly distributed throughout the space-charge region. We shall find it useful to define an effective charge distance L_c as the position (measured from the surface) of the centre of the space charge:

$$L_c \equiv \frac{\int_0^\infty \rho z \, dz}{\int_0^\infty \rho \, dz} \equiv \frac{\int_0^\infty \rho z \, dz}{Q_{sc}} \qquad (1.28)$$

Integrating by parts and using Eqs (1.18), (1.24) and (1.27), we obtain

$$L_c = -\frac{[zF]_0^\infty}{F_s} + \frac{\int_0^\infty F\,dz}{F_s} \tag{1.29}$$

The first term vanishes at the upper limit as well as at the lower, because $F \to 0$ exponentially as $z \to \infty$ (Eqs (1.22), (1.24)). The numerator of the second term is just $|v_s|L$, as can be seen from Eq.(1.24). Thus

$$L_c = \frac{|v_s|}{F_s} L \tag{1.30}$$

This equation has a very simple physical interpretation. If we define the space-charge capacitance C_{sc} as the ratio of the space-charge density Q_{sc} to the barrier height V_s, we obtain (using Eqs (1.14), (1.21) and (1.27)),

$$C_{sc} \equiv \left|\frac{Q_{sc}}{V_s}\right| = \frac{\kappa \epsilon_c}{L_c} \tag{1.31}$$

The space-charge capacitance is thus seen to be the capacitance of a parallel-plate capacitor of unit area having one plate situated at the surface and the other at the centre of the space charge. It is easily seen (Eqs (1.22), (1.24) and (1.30)) that, for small values of $|v_s|$, L_c approaches L.

It should be noted that two basic simplifications were made in solving Poisson's equation. First, the donor and acceptor ions were represented by a uniform charge distribution. In reality the impurity ions are distributed randomly and, since the spacing between neighbours can be of the same order of magnitude as the thickness of the space-charge layer, the potential will fluctuate along the surface on that scale. This effect would be particularly pronounced in depletion layers, where most of the space charge consists of ionized impurities. Experimentally measured quantities usually involve large surface areas, however, so that the fluctuations about the average value may be neglected for most purposes. A second assumption concerns the use of classical statistics in the expression for the free carrier densities in terms of the electrostatic potential (Eq.(1.15)). This is permissible as long as the variation in potential over a distance equal to the electron wavelength is small compared to kT/q. For steeper barriers (such as are encountered in very strong inversion or accumulation layers), a modification of the treatment becomes necessary in order to include quantization effects. Such quantization effects are discussed in detail in Section 1.7.

1.3.3. Approximate solutions for extrinsic semiconductors

The space-charge density Q_{sc} is determined by both the electrons and the holes that are either accumulated or depleted in the space-charge region. But when one type of carrier becomes dominant, it is this one that will effectively determine Q_{sc}. Simple approximations can then be derived for the function F and the shape of the potential barrier $v(z)$.

The majority-carrier bulk density exceeds that of the minority carrier by the factor $\exp(|2u_b|)$, so that for an extrinsic semiconductor ($|u_b| \gtrsim 2$) minority carriers in the bulk constitute at most 2% of the total number of carriers (see Eq.(1.17)). This leads to simple approximations for three limiting cases. These cases are of greatest interest and will now be discussed. For definiteness we consider an n-type semiconductor ($u_b > 0$). The case of a p-type semiconductor is completely analogous.

(a) <u>Accumulation layers</u> ($v_s > 0$): In this case minority carriers (holes) are negligible both in the bulk and in the space-charge region. Thus the function F can be approximated by

$$F(u_b, v) \approx \sqrt{2}\, [\exp(v) - v - 1]^{1/2} \tag{1.32}$$

and is independent of bulk potential u_b. For strong accumulation layers ($v_s \gtrsim 3$), an approximate solution of Eq.(1.24) yields

$$z/L \approx \sqrt{2}\left[\exp\left(-\tfrac{1}{2} v\right) - \exp\left(-\tfrac{1}{2} v_s\right)\right] \tag{1.33}$$

The function F_s, the space-charge density Q_{sc} and the effective charge distance L_c are given by (see Eqs (1.21), (1.27), (1.31) and (1.32))

$$F_s \approx \sqrt{2}\, \exp\left(\tfrac{1}{2} v_s\right) \tag{1.32'}$$

$$Q_{sc} \approx -(2\kappa \epsilon_0 kT n_b)^{1/2} \exp\left(\tfrac{1}{2} v_s\right) \tag{1.34}$$

$$L_c/L \approx (1/\sqrt{2})\, v_s \exp\left(-\tfrac{1}{2} v_s\right) \tag{1.35}$$

Thus Q_{sc} increases and L_c decreases very rapidly (exponentially) with v_s, i.e. with increasing strength of the accumulation layer.

(b) <u>Depletion layers</u>: This corresponds to the situation in which the majority carriers are depleted but as yet the minority carrier concentration remains small. Such a requirement is satisfied if $0 > v_s > -2u_b$. The space-charge density is then determined by the <u>static</u> space charge, composed (in the case of an n-type semiconductor being considered) of the completely ionized donor impurities. The function F can be approximated by

$$F(u_b, v) \approx \sqrt{2}\, [\exp(-|v|) + |v| - 1]^{1/2} \tag{1.36}$$

and is seen to increase slowly with $|v|$, approximately parabolically. The reason for this is that the space charge consists of static charge only, which remains uncompensated by the depleted majority carriers. When $|v_s|$ increases, the space charge can become larger only to the extent that it penetrates deeper into the bulk, and the centre of the space charge (Eq.(1.30)) recedes from the surface approximately as the square root of $|v_s|$. For

$|v| \gtrsim 3$, the exponential term can be neglected, and Eq.(1.24) can be integrated directly:

$$|v| \approx \tfrac{1}{2}[(2|v_s|-2)^{1/2} - z/L]^2 + 1 \tag{1.37}$$

This corresponds to the so-called Schottky barrier, which is characterized by a quadratic dependence of the potential on z.

For strong depletion layers ($v \lesssim -5$), F_s, Q_{sc} and L_c can be approximated by (see Eqs (1.21), (1.27), (1.31) and (1.36))

$$F_s \approx \sqrt{2}\,|v_s|^{1/2} \tag{1.36'}$$

$$Q_{sc} \approx +(2\kappa\epsilon_0\, kT\, n_b)^{1/2}\,|v_s|^{1/2} \tag{1.37'}$$

$$L_c/L \approx (1/\sqrt{2})\,|v_s|^{1/2} \tag{1.38}$$

Thus both Q_{sc} and L_c increase approximately as the square root of $|v_s|$.

(c) <u>Inversion layers</u>: For weak inversion layers, Q_{sc} is determined by both the static space charge and the minority carriers, and F can be approximated by

$$F(u_b, v) \approx \sqrt{2}\,[\exp(|v| - 2u_b) + |v| - 1]^{1/2} \tag{1.39}$$

The parabolic region continues up to where the exponential term can be neglected. Thus for large $|u_b|$, the parabolic dependence extends slightly into the inversion region. For larger values of $|v|$ (strong inversion region) the linear term can be neglected, and near the surface it is the minority carriers which are dominant. (Even though, farther removed from the surface, the space-charge region is still determined by the static charge.) Here F can be written approximately as

$$F(u_b, v) \approx \sqrt{2}\,\exp\left[\tfrac{1}{2}(|v| - 2u_b)\right] \tag{1.40}$$

which is equivalent to neglecting both the static space charge and the majority carriers in Poisson's equation. The function F increases exponentially with $\tfrac{1}{2}|v|$, as in the accumulation-layer case, but now its absolute magnitude is determined by $\exp[\tfrac{1}{2}(|v|-2u_b)]$. Equation (1.24) can be integrated directly to yield

$$z/L \approx \sqrt{2}\left\{\exp\left[\tfrac{1}{2}(2u_b - |v|)\right] - \exp\left[\tfrac{1}{2}(2u_b - |v_s|)\right]\right\} \tag{1.41}$$

This result becomes identical with that for accumulation layers if v is measured from the point where $v = -2u_b$, i.e. from the point where the minority-carrier density is equal to that of the bulk majority carriers. Physically, this follows from the fact that in both cases the slope of the potential barrier is determined only by the carriers dominant at the surface. In an inversion layer, the value of L_c decreases with increasing $|v_s|$ after having reached its maximum value at $v_s = -2u_b$.

For strong inversion layers, Q_{sc} and L_c are given by

$$Q_{sc} \approx (2\kappa\epsilon_0 kT n_b)^{1/2} \exp\left[\tfrac{1}{2}(|v_s| - 2u_b)\right] \tag{1.42}$$

$$L_c/L \approx (1/\sqrt{2}) |v_s| \exp\left[-\tfrac{1}{2}(|v_s| - 2u_b)\right] \tag{1.43}$$

The dependence of these functions on $|v_s|$ is similar to that for the case of strong accumulation layers.

For near-intrinsic semiconductors ($|u_b| < 2$), the situation is qualitatively the same, except for the absence of the parabolic region between the inversion region and the bulk.

Throughout the foregoing discussions we have considered a semiconductor in which the state of occupation of the various localized levels in the space-charge region is independent of the potential barrier. Such a situation applies quite well to germanium and silicon samples. Here the only significant localized levels arise from shallow donor and acceptor impurities, and these are always fully ionized except at very low temperatures. Only under these conditions can the charge density ρ in Poisson's equation be expressed in a simple form (Eq.(1.19)). In most wide-gap semiconductors (or insulators), however, deep-lying levels are the rule rather than the exception, and it is such traps that usually dominate the characteristics of the space-charge region. A general solution of Poisson's equation under these conditions along the lines discussed above is not possible, and each distribution of traps has to be treated individually. It should be noted, however, that in strong accumulation and inversion layers, the static charge due to impurities is small compared to the free-carrier charge, so that the various expressions obtained above remain valid even in the presence of deep traps (incomplete ionization of impurities).

1.4. Excess surface-carrier densities ΔN and ΔP

The present work is concerned mainly with the transport properties of carriers within the space-charge region in a direction parallel to the surface. In this context one is not interested so much in the space-charge density Q_{sc}, but rather in the individual contributions of the holes and electrons to Q_{sc}. The number (per unit surface area) of mobile electrons ΔN and of holes ΔP in the space-charge layer with respect to their numbers at flat bands ($v_s = 0$) are referred to as the excess surface-carrier densities. These densities can be expressed as

$$\Delta N = \int_0^\infty (n-n_b)\, dz \; ; \quad \Delta P = \int_0^\infty (p - p_b)\, dz \tag{1.44}$$

In terms of these quantities, the space-charge density Q_{sc} is given by

$$Q_{sc} = q(\Delta P - \Delta N) \tag{1.45}$$

It should be noted that if ΔN is positive then ΔP is negative, and conversely.

By changing the variable of integration in Eq.(1.44) from z to v and using Eqs (1.15) and (1.24), one obtains for the non-degenerate case being considered:

$$\Delta N = n_b L \int_{v_s}^{0} \frac{\exp(v) - 1}{\mp F(u_b, v)} dv \tag{1.46}$$

$$\Delta P = p_b L \int_{v_s}^{0} \frac{\exp(-v) - 1}{\mp F(u_b, v)} dv \tag{1.47}$$

Plots of these functions, obtained by numerical integration, are available in the literature [1]. However, in those ranges where the excess surface density of one type of carrier can be neglected with respect to that of the other, the dominant carrier density is obtainable from Q_{sc} (see Eq.(1.45)). Such is the case in extrinsic semiconductors ($|u_b| \gtrsim 2$) for majority carriers in accumulation and depletion layers, where the situation is essentially that of a single-carrier system. For the case of an n-type semiconductor, for example, we neglect $\exp(-2u_b)$ with respect to unity in the expression for $F(u_b, v)$, and $q\Delta N$ is then very nearly given by Eqs (1.34) and (1.37) in the accumulation and depletion ranges, respectively. These approximations are better than 1%. In the transition between depletion and strong inversion layers, Q_{sc} is determined by both majority and minority carriers, so that there is no convenient approximation to ΔN or ΔP. However, in strong inversion layers (and we consider again an n-type semiconductor) the hole concentration near the surface is very large, and every increment in $|v_s|$ falls across that portion of the space-charge region just below the surface. The space-charge region does not penetrate farther into the bulk, as in the case of depletion layers, so that ΔN tends to a (negative) saturation level. To a good approximation, which becomes better the stronger the inversion layers, $q\Delta P$ is then given by Eq.(1.42).

1.5. Space-charge region under non-equilibrium conditions

In the preceding sections the semiconductor was assumed to be at thermal equilibrium. Any disturbance that results in carrier excitation generally modifies the potential barrier and charge distribution at the surface. Such modifications give rise to a surface photovoltage, one of the phenomena used in studying the space-charge layer. Before we consider these effects we require some background on non-equilibrium situations. In the presence of a steady, external excitation, the equilibrium electron and hole concentrations n and p appearing in the expressions for the current densities (Eq.(1.11)) should be replaced by the corresponding non-equilibrium, steady-state values n* and p*. The combined action of carrier diffusion and drift can be conveniently expressed by means of the so-called quasi-Fermi levels F_n and F_p

for holes and electrons. These are defined in terms of the carrier densities p^*, n^* by

$$n^* = N_c \exp[(F_n - E_c)/kT]$$
$$p^* = N_v \exp[(E_v - F_p)/kT] \qquad (1.48)$$

where N_c and N_v are the effective densities of states in the conduction and valence bands [1], respectively. In the general case, n^*, F_n, E_c, p^*, E_v and F_p are functions of position. The quasi-Fermi levels are not to be confused with the Fermi level, which is defined only under thermal-equilibrium conditions. At thermal equilibrium, of course, F_p and F_n reduce to E_F and Eqs (1.48) reduce to the well-known expressions for the carrier concentrations in a non-degenerate semiconductor under thermal equilibrium.

Using Eqs (1.48), the hole and electron current densities (Eq.(1.1)) may be written as

$$\vec{J}_p = q\mu_p p^* \vec{\mathscr{E}} - (qD_p/kT)p^* \,\mathrm{grad}\,(F_n - E_c)$$
$$\vec{J}_n = q\mu_n n^* \vec{\mathscr{E}} + (qD_n/kT)n^* \,\mathrm{grad}\,(E_v - F_p)$$

Recalling that $\vec{\mathscr{E}} = -\mathrm{grad}\,E_c$ (see Section 1.1) and using the Einstein relations (Eqs (1.6), (1.17)), we obtain

$$\vec{J}_p = \mu_p p^* \,\mathrm{grad}\,F_p$$
$$\vec{J}_n = \mu_n n^* \,\mathrm{grad}\,F_n \qquad (1.49)$$

To obtain the characteristics of the space-charge region under such non-equilibrium conditions, we consider the case of uniform excitation of carriers throughout the semiconductor sample (produced, for example, by illumination in the absorption edge). If no recombination takes place at the surface, there can be no significant diffusion current of either carrier type towards the surface. Under these conditions $\vec{J}_p = \vec{J}_n = 0$, so that the quasi-Fermi levels for electrons F_n and for holes F_p are position-independent and continue in straight horizontal lines up to the surface (see Eqs (1.49)). Actually, to a good approximation, the quasi-Fermi levels are horizontal in the space-charge region even in the presence of weak surface recombination. Precise conditions for the range of validity of this approximation are given in Chap.5, Section 7, of Ref.[1].

The effect of illumination on the space-charge layer is illustrated in Fig.7 for an n-type semiconductor, V_s being the barrier height (negative for the case of a depletion layer shown) and E_t the energy of a given set of surface states. The dashed curves represent conditions under uniform illumination, with F_n and F_p depicting the quasi-Fermi levels for electrons and holes, respectively. The change δV_s in barrier height in the presence of photogenerated hole-electron pairs comes about by a re-adjustment of the potential barrier to maintain a constant space charge (equal to the fixed surface-state charge).

The re-adjustment in the potential barrier can be calculated easily so long as F_n and F_p are horizontal. Let n_b^* and p_b^* denote the (uniform) steady-

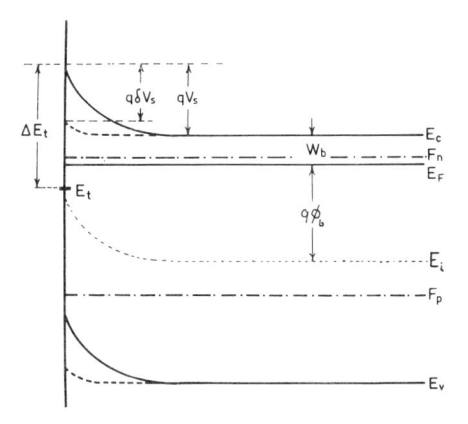

FIG. 7. Energy-level diagram at the surface in the dark (solid lines) and under illumination (dashed).

state electron and hole concentrations in the bulk and v*(z) the modified potential — both in the presence of illumination. Using Eqs (1.12), (1.14) and (1.48), we can then express the steady-state carrier densities n* and p* at any point in the space-charge layer as

$$n^* = n_b^* \exp(v^*), \quad p^* = p_b^* \exp(-v^*) \tag{1.50}$$

In analogy with Eq.(1.17), we can express the steady-state bulk concentrations as

$$n_b^* = n_i^* \exp(u_b^*), \quad p_b^* = n_i^* \exp(-u_b^*) \tag{1.51}$$

where the 'steady-state bulk potential' u_b^* and the 'steady-state intrinsic carrier density' n_i^* are defined by

$$u_b^* \equiv \left[\tfrac{1}{2}(F_n + F_p) - E_i \right]/kT$$

$$n_i^* \equiv (n_b^* p_b^*)^{1/2} = n_i \exp[(F_n - F_p)/2kT] \tag{1.52}$$

Keeping in mind that neutrality prevails in the bulk and neglecting bulk trapping, we have $N_D - N_A = n_b^* - p_b^*$. Poisson's equation (Eq.(1.23)) can then be written in terms of the new variables as

$$\frac{d^2 v^*}{dz^2} = \left(\frac{1}{L^*}\right)^2 \left[\frac{\sinh(u_b^* + v^*)}{\cosh u_b^*} - \tanh u_b^*\right] \tag{1.53}$$

where L* is the steady-state effective Debye length and is given by

$$L^* = \left(\frac{\kappa \epsilon_0 kT}{q^2 (n_b^* + p_b^*)}\right)^{1/2} \tag{1.54}$$

The boundary conditions are as before: $v^* = v_s^*$ at $z = 0$, and $v^* \to 0$ in the bulk ($z \to \infty$). It is thus seen that the whole problem has been reduced to that at thermal equilibrium, and the considerations of this section are therefore also valid for non-equilibrium steady-state conditions. Hence all the solutions (numerical and approximate) obtained for thermal equilibrium can be used for the present case provided that the parameters are replaced by their parallel, steady-state values (starred).

As a demonstration of the usefulness of the above treatment, we shall consider the case of surface photovoltage. This is defined as the change in barrier height due to illumination: $(v_s^* - v_s)(kT/q)$ (denoted by δV_s in Fig.7). The situation in the general case, when surface states are involved, is quite complicated and will not be discussed here. We shall consider here the simpler case in which surface states are absent or inoperative due to poor communication with the space-charge region. It then follows that the space-charge density Q_{sc} is unchanged upon illumination. Hence, by the use of Eqs (1.21), (1.27) and (1.54),

$$F(u_b^*, v_s^*) = \left(\frac{n_b + p_b}{n_b^* + p_b^*}\right)^{1/2} F(u_b, v_s) \tag{1.55}$$

Thus v_s^* can be calculated from the values of u_b and v_s in the dark and the values of the bulk carrier densities n_b^*, p_b^* under illumination. Since F_s is a monotonic function of v_s^*, it is seen that $|v_s^*|$ decreases as the steady-state bulk densities increase (increasing illumination). In the limit of very strong illumination, v_s^* approaches zero. As we shall see below, this behaviour provides a method for determining experimentally the barrier height.

The conclusion that the effect of illumination is always such as to decrease the absolute magnitude of the barrier height generally holds even when surface states are present. Such behaviour is just what is to be expected from physical considerations. If the quasi-Fermi level is to remain flat up to the surface, the only way a constant space charge can be maintained is by a re-adjustment of the barrier, resulting in a decrease in the magnitude of the barrier height.

As an example we shall calculate the surface photovoltage for strong depletion layers and strong accumulation layers. The light intensity will be assumed to be not too high so that the function F_s might be approximated by Eqs (1.36') and (1.32'), respectively, both in the dark and under illumination. For strong depletion layers (Fig.7) the use of Eqs (1.36') and (1.55) yields

$$\frac{\delta V_s}{V_s} = \frac{|v_s^*| - |v_s|}{|v_s|} = -\frac{\delta p_b + \delta n_b}{p_b^* + n_b^*} \tag{1.56}$$

where $\delta p_b \equiv p_b^* - p_b$ and $\delta n_b \equiv n_b^* - n_b$. For strong accumulation layers, Eqs (1.32') and (1.55) lead to

$$(q/kT)\,\delta V_s = v_s^* - v_s = -\ln[1 + (\delta p_b + \delta n_b)/(p_b + n_b)] \tag{1.57}$$

1.6. Degenerate surface conditions

So far, it has been assumed that the semiconductor is non-degenerate both in the bulk and in the space-charge region. This permitted the use of Boltzmann's statistics in calculating the carrier concentrations (see e.g. Eqs (1.15) - (1.17)), which in turn leads to considerable simplification in the solution of Poisson's equation. In sufficiently strong accumulation or inversion layers, however, one or the other band edge would cross the Fermi level at some point in the space-charge region and the electron or hole gas becomes degenerate. Under these conditions the Fermi-Dirac statistics must be used in the calculation of the carrier concentration, and the solution of Poisson's equation becomes more involved [2]. Degenerate inversion and accumulation layers have recently become of particular interest. We shall accordingly discuss such layers in some detail.

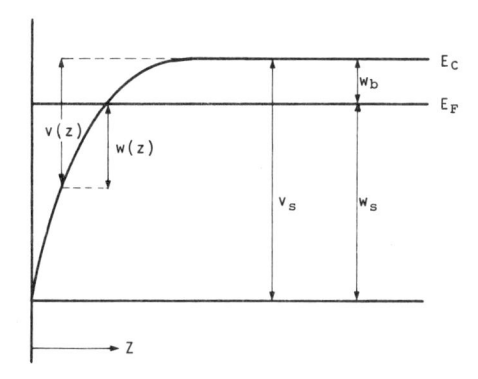

FIG. 8. Shape of conduction-band edge for degenerate accumulation layer on n-type semiconductor.

The case of interest is that of a non-degenerate extrinsic semiconductor. When the space-charge layer becomes degenerate, only one carrier is dominant at the surface, and similar approximations can be made [2] for the F function as in the non-degenerate case (Eqs (1.32), (1.40)). We first consider accumulation layers, and take again an n-type semiconductor (Fig.8). The space-charge density can then be very well approximated by $\rho(z) = -q[n(z) - n_b]$ so that Poisson's equation (Eq.(1.18)) reduces to

$$\frac{d^2v}{dz^2} = \frac{1}{L^2}\left(\frac{n}{n_b} - 1\right) \tag{1.58}$$

where the effective Debye length L is as defined by Eq.(1.21). The electron concentration n(z) is no longer given by Eq.(1.15) wherever $E_c(z)$ lies below or close to E_F. In the general case,

$$n(z) = \int_{E_c(z)}^{\infty} N(E) f_n(E) \, dE \tag{1.59}$$

where $f_n(E)$ is the Fermi-Dirac distribution function:

$$f_n(E) = \{1 + \exp[(E - E_F)/kT]\}^{-1} \tag{1.60}$$

and $N(E)$ is the density of states in the conduction band. For spherical energy bands (isotropic effective mass m_n), $N(E)$ is given by [1]

$$N(E) = 4\pi(2m_n/h^2)^{3/2} (E - E_c)^{1/2} \tag{1.61}$$

If $E_c > E_F$, $f_n \approx \exp[-(E-E_F)/kT]$ (Boltzmann's statistics) and

$$n(z) = N_c \exp[-(E_c(z) - E_F)/kT]$$

which is just the expression given in Eq.(1.15). Here

$$N_c \equiv 2(2\pi m_n kT/h^2)^{3/2} \tag{1.62}$$

is the effective density of states in the conduction band [1]. For $E_c < E_F$ the integration in Eq.(1.59) is more involved. Rather than use the (dimensionless) potential $v(z)$, it would be more convenient to introduce the function $w(z)$ defined as

$$w(z) \equiv [E_F - E_c(z)]/kT = v(z) - w_b \tag{1.63}$$

where w_b (positive for the case of an n-type semiconductor being considered) is the dimensionless energy separation between the conduction-band edge in the bulk and E_F. At the surface ($z = 0$)

$$w_s = v_s - w_b \tag{1.64}$$

It should be noted that with the convention used, $w > 0$ for $E_c < E_F$ and $w < 0$ for $E_c > E_F$. Let $x \equiv [E - E_c(z)]/kT$ so that $(E - E_F)/kT = x - w$ (see Fig.8). Using Eqs (1.60) and (1.61), Eq.(1.59) can be rewritten as

$$n(z) = 2\pi^{-1/2} N_c F_{1/2}(w) \tag{1.65}$$

where

$$F_{1/2}(w) \equiv \int_0^\infty \frac{x^{1/2} dx}{1 + \exp(x - w)} \tag{1.66}$$

and has been tabulated [2].

Using Eqs (1.17), (1.63) and (1.65), Eq. (1.58) can be rewritten as

$$\frac{d^2 w}{dz^2} = \frac{1}{L^2} [2\pi^{-1/2} \exp(w_b) F_{1/2}(w) - 1] \tag{1.67}$$

To integrate Eq.(1.67) once, we multiply both sides by 2(dw/dz), as we did in the non-degenerate case (Section 1.3). Here we use the relation [2]:

$$\int F_{1/2}(w)\,dw = \frac{2}{3} F_{3/2}(w) + \text{const}$$

where

$$F_{3/2}(w) \equiv \int_0^\infty \frac{x^{3/2}\,dx}{1+\exp(x-w)} \quad (1.68)$$

which is also tabulated [2]. Finally, by using the boundary condition dw/dz = 0 for $w = -w_b$ (v = 0), we obtain, similarly to Eq.(1.24),

$$\frac{dw}{dz} = -\frac{F(u_b,w)}{L} \quad (1.69)$$

where now

$$F(u_b, w) \equiv \sqrt{2}\left[\frac{4}{3}\pi^{-1/2}\exp(w_b)F_{3/2}(w) - w - w_b - 1\right]^{1/2} \quad (1.70)$$

The constant of integration in Eq.(1.70) is obtained from the assumption that the bulk is non-degenerate, so that (see Eq.(1.68))

$$F_{3/2}(-w_b) \approx \exp(-w_b)\int_0^\infty x^{3/2}\exp(-x)\,dx$$

The integral is a γ function, equal to $\frac{3}{4}\pi^{1/2}$.

To calculate the excess surface-electron density ΔN and the effective charge distance L_c we make use of Eqs (1.27) and (1.31), where now F_s is obtained from Eq.(1.70):

$$F_s \equiv F(u_b, w_s) = \sqrt{2}\left[\frac{4}{3}\pi^{-1/2}\exp(w_b)F_{3/2}(w_s) - w_s - w_b - 1\right]^{1/2} \quad (1.71)$$

For very strong accumulation layers ($w_s \gg 1$), $F_{1/2}(w_s)$ and $F_{3/2}(w_s)$ can be approximated by

$$F_{1/2}(w_s) \approx \int_0^{w_s} x^{1/2}\,dx = \frac{2}{3} w_s^{3/2}$$

$$\quad (1.72)$$

$$F_{3/2}(w_s) \approx \int_0^{w_s} x^{3/2}\,dx = \frac{2}{5} w_s^{5/2}$$

This is equivalent to neglecting in the integration of Eq.(1.59), the electron concentration above E_F compared to that below E_F.

A similar calculation for degenerate inversion layers of an n-type semiconductor yields [2] for F_s

$$F(u_b, e_g - w_b, v_s) \approx \sqrt{2}\left[\frac{4}{3}\pi^{-1/2}\exp(e_g - w_b - 2u_b)F_{3/2}(|v_s| - e_g + w_b) + |v_s| - 1\right]^{1/2} \quad (1.73)$$

where $e_g = (E_c - E_v)/kT$ is the dimensionless energy gap.

For practical purposes one is often interested in w_s (or v_s) and L_c as functions of ΔN, the latter being the only easily measurable variable (see further on). To illustrate the above calculations, these functions, as well as the surface electron concentration n_s (as obtained from Eq.(1.65)), are plotted in Fig.9 for accumulation layers on ZnO. Zinc oxide can possess the strongest accumulation layers ever attained and is therefore of particular interest [3]. Actually, however, the only parameters used in the calculations were the dielectric constant κ and the effective density of states N_c; the scales of Fig.9 can readily be modified to suit any other semiconductor (different κ and N_c). The calculations were carried out for an insulating sample ($w_b = 50$) and a semiconducting sample ($w_b = 10$). As one moves to the right along the abscissa, stronger and stronger accumulation layers are obtained. The onset of degenerate conditions ($w_s = 0$, i.e. E_c at the surface crossing E_F) occurs around $\Delta N = 10^{12}$ cm^{-2}, when w_s changes sign from negative to positive.

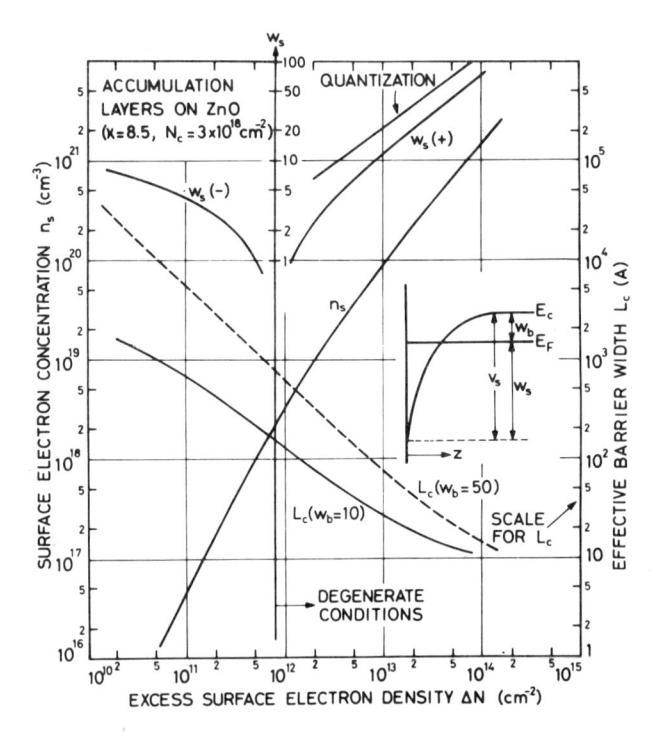

FIG.9. Surface electron concentration n_s, effective barrier width L_c and dimensionless energy separation w_s as functions of excess surface electron density ΔN in accumulation layers of ZnO.

It should be noted that for $\Delta N > 10^{13}$ cm^{-2} the volume concentration n_s of electrons at the surface ($z = 0$) assumes values characteristic of metallic concentrations ($n_s > 10^{20}$ cm^{-3}). With the choice of w_s rather than v_s for characterizing the potential barrier, the w_s and n_s curves are practically independent of the parameter w_b. Only the L_c curve depends on w_b.

1.7. Quantization effects in very strong accumulation and inversion layers

As can be seen from Fig.9, the effective charge distance L_c becomes very small (tens of angstroms) in very strong accumulation layers. An appreciable fraction of the electrons in the space-charge layer lies even closer to the surface. The distances involved are considerably smaller than the de Broglie wavelength of a conduction electron (~ 100 Å) so that quantization effects become very pronounced. Such quantization effects have received considerable attention in recent years. They are discussed in two review articles [4,5] and here we shall only touch on some of the consequences. The classical calculations depicted in Fig.9 are of course no longer valid; the more so the stronger the accumulation layer (larger ΔN and smaller L_c). One should then simultaneously solve Poisson's equation and the Schrödinger equation; a rather difficult task, performed so far only for a few limiting cases [4,5]. The upper curve at the top of Fig.9 represents a crude approximation of the variation of w_s with ΔN in the quantization regime. The actual variation lies in between this and the classical curve.

The changes introduced by quantization are illustrated schematically by Fig.10. The accumulation layer (a) corresponds to the classical approximation, with the conduction electrons being distributed more or less evenly throughout the potential well. The effect of quantization is two-fold. First, electrons can occupy only levels lying above a certain energy level E_0, and, second, they are concentrated mostly around the middle of the potential well (Fig.10(b)). An electron in the well is still free to move parallel to the surface but the component p_z of its momentum normal to the surface becomes quantized. Thus, the allowed energy levels for electrons become grouped in sub-bands, each being characterized by a fixed quantized value of p_z and consisting of a continuum of levels corresponding to motion parallel to the surface. The

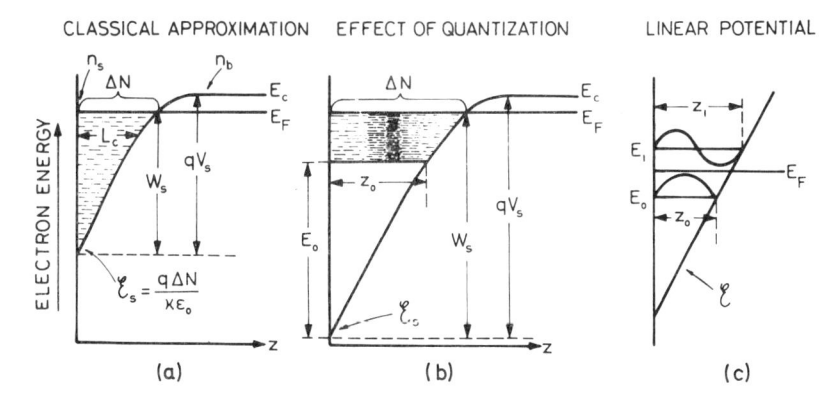

FIG.10. Schematic representation of strong accumulation layer in classical approximation and with quantization effects taken into account. (a) represents a triangular well (linear potential) with its first two sub-bands.

problem is essentially that of an electron confined in a box or rather in a two-dimensional trench parallel to the surface. One has, in effect, a two-dimensional electron gas.

An estimate for E_0 can be obtained in a very simple way. For this purpose we approximate the potential in the accumulation layer by a linear potential. In other words, we take the slope of the potential (the field $\mathscr{E}$) as constant throughout the space-charge region (Fig.10(c)). The electron energy can be expressed as

$$E = (p_x^2 + p_y^2 + p_z^2)/2m_n \tag{1.74}$$

where p_x and p_y are the momentum components parallel to the surface and m_n is the electron effective mass (assumed to be isotropic, as is very nearly the case in ZnO). The lowest allowed energy E_0 corresponds to no motion parallel to the surface ($p_x = p_y = 0$) so that

$$E_0 + p_z^2/2m_n \tag{1.75}$$

It remains to estimate the minimum value p_z can assume. Let z_0 denote the width of the potential barrier at energy E_0. Since the electron wave function must vanish at both ends, z_0 should be set equal to half the electron wavelength λ. But $\lambda = h/p_z$, where h is Planck's constant. Thus

$$z_0 = \tfrac{1}{2}\lambda = \tfrac{1}{2}(h/p_z) \tag{1.76}$$

Combining Eqs (1.75) and (1.76), one obtains

$$E_0 = h^2/8m_n z_0^2 \tag{1.77}$$

At the same time, since the electric field $\mathscr{E}$ in the potential well is assumed constant, E_0 must also be equal to

$$E_0 = q\mathscr{E} z_0 \tag{1.78}$$

Elimination of z_0 from Eqs (1.77) and (1.78) yields

$$E_0 = [qh\mathscr{E}/2(2m_n)^{1/2}]^{2/3} \tag{1.79}$$

This differs from the result of the accurate calculation for a linear potential [4] by less than 10%.

E_0 is the bottom of the first sub-band. The continuum of the sub-band lying above E_0 is given by

$$E = E_0 + (p_x^2 + p_y^2)/2m_n \tag{1.80}$$

where p_x and p_y vary continuously, corresponding to motion parallel to the surface (with the same p_z). The spatial distribution of the electrons occupying the first sub-band is represented by the half-sine shape of the wave function (Fig.10). The distribution is thus peaked somewhere in the middle of the well. The bottom E_1 of the second sub-band is determined by the requirement $z = \lambda$, and so on for the higher sub-bands. Figure 10(c) corresponds to the so-called quantum limit, wherein only the first sub-band is populated by electrons (E_F lying between E_0 and E_1).

The actual potential is of course not linear, but if one takes the surface field $\mathscr{E}_s$ as representing some average of the field in the potential well, a fairly good idea is obtained of the energy position of the first sub-band. By Gauss' theorem, $\mathscr{E}_s = q\Delta N/\kappa\epsilon_0$, so that

$$E_0 \approx \left[\frac{q^2 h}{2(2 m_n)^{1/2}\kappa\epsilon_0}\right]^{2/3} (\Delta N)^{2/3} \qquad (1.81)$$

For $\Delta N = 5 \times 10^{13}$ cm^{-2}, which is attainable in ZnO, $E_0 \approx 1$ eV, a very large value indeed. This demonstrates the drastic effect of quantization on the characteristics of strong accumulation layers.

Equation (1.81) should not be taken too literally since several assumptions on which it is based are rather poor, especially at very strong accumulation conditions.

The effect of quantization in strong inversion layers is similar [4,5].

2. MEASUREMENTS ON THE SPACE-CHARGE LAYER

In the previous section the equations and expressions governing the behaviour of the space-charge region were developed. Particular attention was given to processes that can be directly correlated with measurable quantities so as to yield information on the surface parameters. In this section some of the experimental methods employed in studying the space-charge layer are reviewed. One of these methods, surface conductance, will receive particular attention since it leads to the most direct information on the transport characteristics of the carriers within the space-charge layer. We shall be concerned not so much with experimental detail as with the principles and theory of measurement.

The electrical properties of the surface are usually derived from measurements on regularly shaped homogeneous samples having as large a surface-to-volume ratio as is practical. Some of the electronic phenomena, such as carrier transport, can be detected only for the sample as a whole. If the bulk properties are known, however, the contribution of the bulk can be accounted for so as to yield information on the surface. One way of obtaining such a separation of bulk and surface effects is to vary the potential barrier height at the surface and to follow the resulting changes in the electrical properties of the sample. Another method makes use of inversion layers on p-n-p (or n-p-n) transistor structures or on metal-oxide-semiconductor (MOS) structures (see below). In either case the contribution of the bulk to the sample's conductance is suppressed so that the measurement yields the surface conductance directly.

The equilibrium or quiescent value of the barrier height at the surface of a given semiconductor sample is determined by the type, density and distribution of surface states. These in turn are functions of surface preparation and of the surrounding ambient. Both effects can therefore be used as means of varying the barrier height.

Another method of varying the potential barrier is based on the effect of a capacitively applied field normal to the surface, as discussed in Section 1.2.

The charge induced by the field is distributed between the surface states and the space-charge region, and a change in the latter is accompanied by a corresponding change in barrier height. The electric field can be produced in a number of ways. The semiconductor sample is usually shaped in the form of a thin rectangular filament with one cross-sectional dimension large compared to the other, most of the surface area thus being confined to the two larger faces. A metal electrode, placed parallel to one of these faces and insulated from it by a suitable spacer, forms one plate of a plane-parallel capacitor, the semiconductor sample constituting the other. Thin Mylar sheets (~ 10 μm thick) are often used as spacers. By applying a voltage across such a capacitor, fields as high as 2×10^6 V/cm can in practice be attained at the semiconductor surface. Higher fields can be attained in some cases by using a thin (~ 1000 Å) thermally grown oxide as the spacer between the metal and the semiconductor (MOS structure). The highest surface fields, and hence the largest swing in barrier height, can be obtained in some cases by using an electrolyte as the field electrode.

For any work of a quantitative nature it is essential to determine the potential barrier height throughout its variation. The measurements to be described, singly or in suitable combination, can serve for this purpose. At the same time, some of these measurements are directly applicable in the study of surface transport. We begin by a general discussion of surface conductance and surface capacitance, both of which are intimately associated with the space-charge layer. Next we consider the field-effect technique, one of the most powerful tools in studying surface phenomena. The section concludes with a discussion of surface-photovoltage and galvanomagnetic measurements. The former can be very useful in determining the quiescent barrier height while the latter can yield the most direct information on surface transport.

2.1. Surface conductance

The presence of a potential barrier results in a surface layer — the space-charge region — whose conductance (parallel to the surface) is different from that of a parallel layer of comparable thickness in the underlying bulk. Thus a given semiconductor filament consists in effect of two conductors in parallel, one associated with the fixed bulk carrier densities and the other with the barrier-dependent surface densities. Quite often, only changes in the conductance of the space-charge layer corresponding to different barrier heights can be measured directly. This experimental limitation suggests the feasibility of expressing the surface conductance in terms of these changes rather than in absolute terms. Conditions at flat bands ($V_s = 0$), where the excess surface-carrier densities are zero, are very convenient as a reference point. Thus, the surface conductance $\Delta\sigma$ at any barrier height V_s is defined as the change in filament conductance per square area of its surface resulting from a change in barrier height from 0 to V_s. For a homogeneous filament of uniform cross-section, we have

$$\Delta\sigma = (\ell^2/A)((1/R) - (1/R_0)) \tag{2.1}$$

Here A is the total surface area parallel to the direction of current flow, ℓ is the filament length, R is the filament resistance, and R_0 is the value of R at

$V_s = 0$. If we assume that the carrier mobilities in the space-charge layer are the same as those in the bulk, then evidently,

$$\Delta\sigma = q(\mu_n \Delta N + \mu_p \Delta P) \tag{2.2}$$

where ΔN and ΔP are the excess surface-carrier densities as defined in Section 1.4. Note that the dimensions of $\Delta\sigma$ are those of conductance (m.h.o.s.) and do not involve length or area, the conductance associated with a square area of surface being independent of the size of the square.

The dependence of surface conductance on barrier height can be obtained from that of ΔN and ΔP calculated in Section 1.4. Conversely, if the surface conductance can be measured, it may be used to derive the corresponding value of barrier height. It is apparent that both accumulation and inversion layers are characterized by high conductances. In accumulation layers this is due to the large number of majority carriers, in inversion layers to the large number of minority carriers. The surface conductance is less in depletion layers and passes through a minimum value $\Delta\sigma_{min}$ where very few mobile carriers are present in the space-charge region. The exact value V_{sm} of the barrier height at which the minimum occurs can be evaluated by differentiating Eq.(2.2). Substituting for ΔN, ΔP from Eqs (1.46) and (1.47), we obtain

$$v_{sm} \equiv qV_{sm}/kT \approx -2u_b - \ln(\mu_n/\mu_p) \tag{2.3}$$

In Fig.11, $\Delta\sigma$ is plotted as a function of the dimensionless barrier height v_s for an n-type and a p-type germanium sample at 300 K. The impurity concentrations were chosen to yield bulk potentials of $u_b = +2$ and $u_b = -2$, respectively. The dashed curves have been obtained from Eq.(2.2) by the use of the normal (bulk) mobilities. The solid curves have been calculated employing surface mobility values, as explained further on. In the right-hand branches of the curves the surface conductances arise mainly from electrons, and in the left-hand branches from holes. Only in the vicinity of

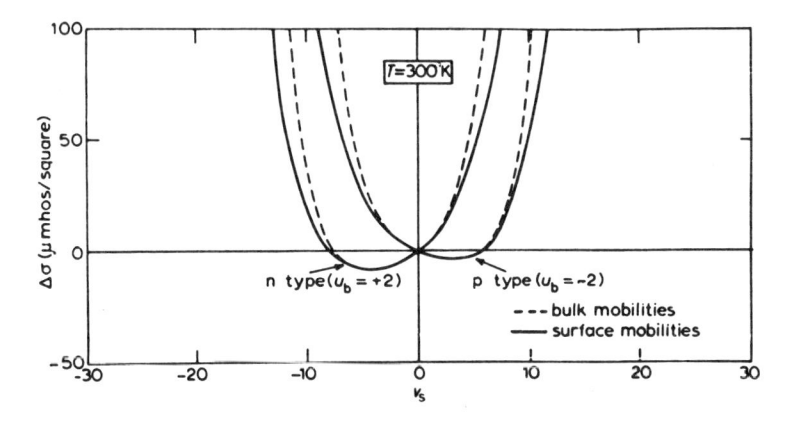

FIG.11. Surface conductance $\Delta\sigma$ as a function of barrier height v_s for two extrinsic germanium samples ($N_D - N_A \approx \pm 2 \times 10^{14}$ cm^{-3}) at 300 K. The dashed curves correspond to bulk mobilities, the solid curves to surface mobilities.

the minimum does $\Delta\sigma$ consist of contributions from both electrons (whose density increases with v_s) and holes (whose density decreases with increasing v_s). The value $\Delta\sigma = 0$ is obtained twice: once for the case of flat bands ($v_s = 0$) where $\Delta\sigma = 0$ by definition, and again on the other side of the minimum where the minority carrier conductance just cancels the negative contribution of the depleted majority carriers.

Actually, the bulk mobilities μ_n, μ_p in Eq.(2.2) should be replaced by lower values because of the scattering by the surface of carriers moving in its vicinity. Theoretical estimates are available for the appropriate surface mobilities to be used in this case, as will be discussed in the next section. There is some uncertainty, however, about the exact nature of the surface scattering mechanism. The solid curves in Fig.11 have been calculated on the basis of the diffuse scattering model, which assumes that each carrier scattered by the surface emerges with a completely random velocity. As such, these curves represent the lowest limit to the surface conductance, the actual values lying somewhere in between the solid and dashed curves. It is seen that the surface mobilities depart significantly from the corresponding bulk values only in strong accumulation and inversion layers (large $|v_s|$), where the carriers in the space-charge region are constrained to move in deep potential wells. In these regions, however, the curves are steep and no appreciable error in the derivation of v_s from $\Delta\sigma$ is introduced by the uncertainty in surface mobility.

To determine the surface conductance, the filament is provided with two non-rectifying end contacts and its resistance is measured while the potential barrier is varied by external means. Obviously, the maximum in filament resistance corresponds to the minimum in surface conductance $\Delta\sigma_{min}$. As can be seen from Eq.(2.3), the value v_{sm} for which this occurs is a unique (and known) function of temperature and impurity concentration. The maximum in resistance can therefore be employed, in conjunction with the appropriate plot of $\Delta\sigma$ versus v_s, to correlate the filament resistance with the corresponding values of barrier height. Using Eq.(2.1), one obtains

$$\Delta\sigma - \Delta\sigma_{min} = (\ell/w)((1/R) - (1/R_M)) \qquad (2.4)$$

where ℓ is the length of the filament, w the width of the surface being monitored, and R_M the maximum value of the filament resistance R.

A convenient procedure for determining R_M consists in varying v_s monotonically (by an external field or by a gaseous ambient cycle) while the filament resistance is continuously monitored. The resistance first increases, as R_M is approached, and then decreases as v_s is changed further.

It should be stressed that the usefulness of surface conductance as a tool for determining the barrier height hinges on the possibility of attaining $\Delta\sigma_{min}$. In germanium the use of either an external field or an ambient cycle is generally sufficient by itself to provide the necessary range of variation in v_s. In some cases, particularly for silicon, the combined action of both gases and fields may be required, while in others even this procedure is of no avail. Other methods, such as surface photovoltage, must then be used.

We have tacitly assumed that the measurement of filament conductance represents the true parallel combination of bulk and surface conductances. For this to be the case, at least one of two requirements must be satisfied:

either the metal contacts are non-rectifying towards the surface layer as well, or the conductance (perpendicular to the surface) between the surface and the underlying bulk is sufficiently large. While both requirements are evidently met when an accumulation layer exists at the surface, they may well be violated under inversion-layer conditions. Because the contacts were chosen to be non-rectifying to the bulk and thus capable of handling the majority-carrier flow, they will generally be unable to supply the minority-carrier flow in the inversion layer. Thus the method described may be inadequate for measuring surface conductance in inversion layers. Other more suitable methods for this range employ p-n-p or MOS structures, as will be discussed below.

2.2. Surface capacitance

The surface capacitance is a convenient quantity by means of which the change in the overall surface charge can be correlated with the corresponding change in potential barrier height. Consider, for example, a semiconductor where no potential barrier exists at the surface. (Such a condition may arise from a special distribution of surface states or, more commonly, from an external biasing field of the proper magnitude.) If now a potential barrier is set up by varying the external field, a space-charge layer will form and, in addition, a change may take place in the charge stored in surface states. In Section 1.3 the space-charge capacitance C_{sc} was defined as the ratio of the space-charge density (per unit surface area) Q_{sc} to the barrier height V_s. In an analogous manner we can introduce the surface-state capacitance C_{ss} as

$$C_{ss} \equiv \left| \Delta Q_{ss} / V_s \right| \tag{2.5}$$

where ΔQ_{ss} represents the change in surface-state charge density brought about by changing the barrier height from 0 to V_s. The surface capacitance is defined as the ratio of the total change in surface-charge density ΔQ_s to the barrier height V_s, and is given by

$$C_s \equiv \left| \frac{\Delta Q_s}{V_s} \right| = \left| \frac{Q_{sc} + \Delta Q_{ss}}{V_s} \right| = C_{sc} + C_{ss} \tag{2.6}$$

The space-charge capacitance C_{sc} is a unique function of V_s for a given impurity concentration and temperature; it can be calculated from Eqs (1.30) and and (1.31). The surface-state capacitance C_{ss}, on the other hand, depends on the particular distribution of surface states present on the surface.

To measure the surface capacitance C_s, use can be made of the field-effect configuration described above: a plane-parallel capacitor is formed between the semiconductor sample and a metal plate separated from the free surface by a thin insulating layer (Fig.12 (a)). The insulating medium ensures that no conduction current flows between the metal and the semiconductor. Suppose that a change V_0 in the applied voltage across the capacitor is necessary to change the barrier height from 0 to V_s. In this process the overall surface charge is changed by ΔQ_s (and the charge on the metal plate

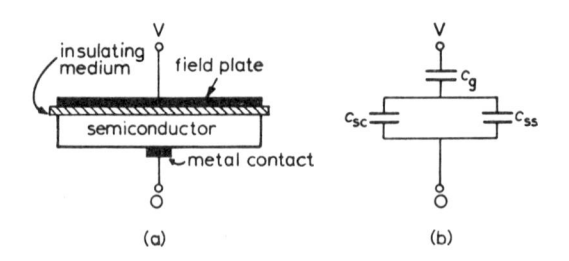

FIG.12. Field plate — semiconductor capacitance. (a) Schematic representation of experimental configuration; (b) Equivalent capacitance circuit.

by $-\Delta Q_s$). The change in potential drop across the insulating medium is evidently $V_0 - V_s$ and can be expressed by the relation:

$$V_0 - V_s = - \frac{\Delta Q_s}{C_g} \qquad (2.7)$$

where C_g is the geometric capacitance (per unit area) between the metal electrode and the surface proper — i.e. the capacitance that would be measured if the semiconductor were replaced by a metal. Combining Eqs (2.6) and (2.7), we obtain

$$\frac{1}{C_0} \equiv \left| \frac{V_0}{\Delta Q_s} \right| = \frac{1}{C_g} + \frac{1}{C_s} = \frac{1}{C_g} + \frac{1}{C_{sc} + C_{ss}} \qquad (2.8)$$

Thus the effective surface capacitance C_0, as measured by an external circuit, consists of the geometric capacitance in series with the parallel combination of the space-charge and surface-state capacitances, as illustrated in Fig.12(b).

In practice, it is generally easier to measure the differential capacitance $c_0 \equiv \left| dQ_s/dV_0 \right|$. A small ac signal is superimposed on the dc bias voltage and the capacitance measured as a function of bias voltage. Similarly to Eq.(2.8), we have

$$\frac{1}{c_0} = \frac{1}{C_g} + \frac{1}{c_s} = \frac{1}{C_g} + \frac{1}{c_{sc} + c_{ss}} \qquad (2.9)$$

where C_g is the geometric capacitance as before, and c_{sc} and c_{ss} are the differential space-charge and surface-state capacitances, respectively:

$$c_{sc} \equiv \left| \frac{dQ_{sc}}{dV_s} \right| = \frac{q}{kT} \left| \frac{dQ_{sc}}{dv_s} \right|$$

$$c_{ss} \equiv \left| \frac{dQ_{ss}}{dV_s} \right| = \frac{q}{kT} \left| \frac{dQ_{ss}}{dv_s} \right| \qquad (2.10)$$

For the case of fully ionized impurities, we obtain from Eqs (1.27) and (2.10)

$$c_{sc} = \frac{\kappa\epsilon_0}{L} \left| \frac{dF_s}{dv_s} \right| \qquad (2.11)$$

A study of the characteristics of the function F_s reveals that c_{sc} attains a minimum value in depletion layers and increases rapidly in both accumulation and inversion layers.

Equations (2.9) and (2.10) assume that the frequency of the ac signal is sufficiently small for the surface states to have enough time to assume their steady-state occupation. The higher the frequency, the less efficient becomes the charge exchange between the surface states and the space-charge region, until in the limit of very high frequencies the contribution of the surface states to the measured capacitance C_0 is completely eliminated. Thus, at least in principle, measurements of surface capacitance can yield direct information on the space-charge capacitance without interference from surface states.

The principal difficulty in such measurements is the determination of c_s with sufficient accuracy. The surface capacitance is usually of the order of tenths of a $\mu F/cm^2$, and to achieve the necessary sensitivity the geometric capacitance should not be too small by comparison (see Eq.(2.9)). The ordinary field-effect configuration, which employs a Mylar spacer, is inadequate since the geometric capacitance in such structures cannot in practice exceed a few hundred pF/cm^2, a value that is much smaller than c_s. On the other hand, it is sometimes possible to increase the geometric capacitance considerably by evaporating a very thin insulating layer or by growing an oxide film on the surface and then evaporating a metal electrode on top (see Section 2.3).

The semiconductor-electrolyte system is perhaps the most suitable one for surface capacitance measurements. The electronic processes taking place at the interface of such a system [6,7] will not be discussed here. Suffice it to say that in certain cases a potential barrier at the electrolyte side of the interface limits the flow of majority carriers (and sometimes of minority carriers as well) into the semiconductor, such that under the proper polarity the semiconductor-electrolyte contact is blocking to a good approximation. At the same time, for not too dilute electrolytes the 'geometric capacitance' associated with the blocking layer is very large (10-100 $\mu F/cm^2$), so that the measured capacitance is in effect equal to the semiconductor surface capacitance $c_{sc} + c_{ss}$. This in turn results in almost the entire voltage drop between the semiconductor and the electrolyte occurring across the space-charge layer of the former. Thus the changes in barrier height are given directly by the values of the applied voltage. In practice, one measures the potential difference between the semiconductor and a standard electrode immersed in the electrolyte while the barrier height is being varied by passing a small current between the semiconductor and another electrode. The voltage changes brought about by varying the current yield in most cases the changes in barrier height V_s, except for the small ohmic drop across the electrolyte and the semiconductor bulk. The differential surface capacitance c_s can be conveniently measured by using a third electrode connected to a small ac or pulsed voltage source. To determine the absolute magnitude of the barrier height, a reference point is necessary. This is provided by the

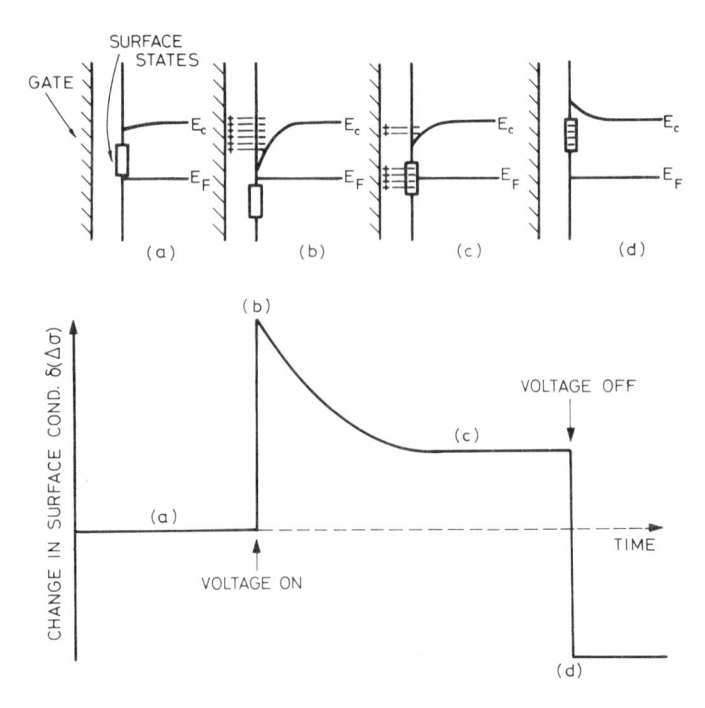

FIG.13. Shape of conduction-band edge in n-type semiconductor under voltage pulse applied across the metal-semiconductor capacitor. The lower diagram demonstrates the accompanying change $\delta(\Delta\sigma)$ in surface conductance (the field effect).

minimum value of the measured capacitance (neglecting the surface-state contribution to c_s), by surface-photovoltage measurements (see Section 2.4), or by surface conductance data.

2.3. Field effect

The term 'field effect' describes the change in sample conductance taking place as a result of a capacitively applied field normal to the surface. This effect has proved to be a powerful tool in studying surface phenomena, especially those involving the surface states. Its use in studies of surface transport is limited to those cases in which surface states are either absent or else their effect can be eliminated. The field-effect technique is illustrated in Fig.13. By applying a voltage pulse across the metal-semiconductor capacitor, charge is induced on the semiconductor surface, the magnitude of which can easily be calculated from the known capacitance and the applied voltage. We consider an n-type semiconductor in which minority carriers (holes) are absent both in the bulk and at the surface. We have in effect a one-carrier system and in Fig.13 only the conduction-band edge (E_c) need be shown. We assume that under equilibrium conditions (a) there is a set of empty surface states (lying above the Fermi level E_F). Suppose that the voltage pulse is applied at $t = 0$. If the metal plate is made positive then the field, which momentarily penetrates into the semiconductor, will cause a flow of electrons from the contacts to the surface until the metal-semiconductor capacitor is charged up. The time constant associated with this process in semiconductors is, under normal experimental conditions, very small (of the order of 10^{-8} seconds). Accordingly, a very short time (t = 0 +) after the application of the field, the induced charge in the conduction band attains an equilibrium distribution, the Fermi level being well defined in the space-charge region as well as in the bulk. The barrier height becomes more positive (bands bending downwards) to permit the induced charge to be accommodated in the conduction band. This situation is shown schematically in (b) by the termination of the lines of force at free electrons in the accumulation layer. At this stage the surface states are not as yet in equilibrium with the induced charge. Excess free electrons tend to drop into the states and a relaxation in the barrier height takes place. When equilibrium conditions are reached, part of the induced charge resides in the space-charge region (and is thus mobile) while the rest is trapped in the surface state. The lower part of Fig.13 illustrates schematically the changes in surface conductance. Regions (a), (b) and (c) correspond to the three stages described above. Also shown (d) is the behaviour of the surface conductance after the voltage is switched off. At this point of time the trapped charge in surface states repels free electrons from the space-charge region and the conductance decreases abruptly to a value below the initial level. A second type of relaxation now takes place, associated with the thermal release of electrons from surface states back into the conduction band. If the surface states are energetically deep, the time constant characterizing this relaxation can be very large and it would take a long time before the surface conductance decays to its initial, equilibrium level. A similar behaviour obtains for a voltage pulse of the opposite polarity, except that now the relaxation in surface conductance would be expected to be slower following the pulse onset than following the pulse termination. This polarity permits a more convenient and accurate measurement of the _initial_ change in surface conductance, before electrons trapped in any surface states present can be emitted into the conduction band. In other words, the initial change in $\Delta\sigma$ corresponds to the entire induced charge being mobile, and is hence a direct measure of the surface mobility associated with the (known) induced charge.

Surface states can sometimes be eliminated also by applying a high-frequency ac voltage instead of a voltage pulse across the metal-semiconductor capacitor of Fig.13. In this experiment one measures the so-called field-effect mobility defined as

$$\mu_{fe} \equiv - \frac{d(\Delta\sigma)}{dQ_s} \qquad (2.12)$$

where, as before, Q_s is the charge on the semiconductor surface. If indeed the effect of surface states has been eliminated, μ_{fe} can readily be expressed in terms of the surface mobility μ_s. For accumulation or depletion layers of an n-type semiconductor, for example, $Q_{sc} \approx -q\Delta N$ and one has (see Eq.(2.2))

$$\mu_{fe} = \mu_s + \Delta N \frac{d\mu_s}{d\Delta N} \qquad (2.13)$$

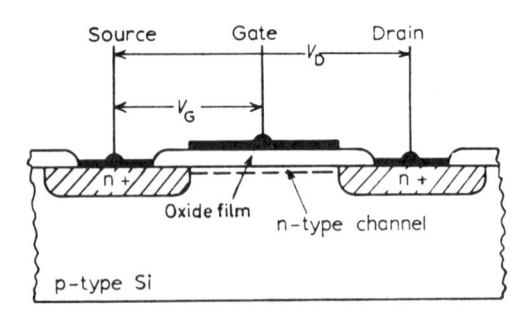

FIG. 14. MOS field-effect transistor.

The metal-oxide-semiconductor (MOS) transistor has proved extremely useful in the study of transport processes on silicon surfaces. A schematic diagram of the MOS transistor is shown in Fig.14. The structure here is based on a p-type silicon slab. The surface is thermally oxidized, the oxide thickness being typically several hundred Å. A metal electrode (referred to as the gate) is then evaporated on top of the oxide layer. The n-type inversion layer is either formed during the oxidation process or it can be induced by biasing the gate positively. The diffused n+ regions on both sides of the gate ensure that the current flowing between the two ohmic contacts ('source' and 'drain') be confined to the inversion layer. This follows from the fact that for any polarity of the voltage applied across the source and drain, one of the n+/p junctions would be reversely biased and hence rectifying. Thus the bulk is effectively eliminated and the measured conductance gives directly the surface conductance of the inversion layer. There are two additional and very valuable advantages of the MOS structure. First, the oxide films can be made very thin and yet with high uniformity and dielectric strength. As a consequence the geometric capacitance C_g is very large, comparable to the space-charge capacitance C_{sc}. This in turn permits accurate measurement of C_{sc} (see Section 2.2) as well as large variation in the barrier height. Second, suitable oxidation procedures result in a marked reduction in the density of surface states which, as we have seen, is a valuable asset in surface transport measurements. These advantages are of course essential in the operation of the MOS transistor as a device.

2.4. Contact potential and surface photovoltage

Measurements of contact potential are widely used for determining the electron affinity and other surface properties of semiconductors. The contact potential between two conducting solids is defined as the difference in their work functions:

$$V_{cp} = \frac{W_\emptyset - W_\emptyset'}{q} \tag{2.14}$$

Considering, for example, an n-type semiconductor, we may express the work functions as (see Figs 4 and 5)

$$W_\emptyset = \chi + W_b - qV_s \tag{2.15}$$

If both χ and W_b are known it is possible, at least in principle, to determine V_s from the measurement of $W_\emptyset$. In practice, however, $q|V_s|$ is usually small compared to $W_\emptyset$ so that a small relative error in the absolute magnitude of the latter may introduce a large error in the former. Accordingly, it is more feasible to derive the surface characteristics from the observed changes in work function under varying conditions at the surface. Such changes can occur by varying the ambient gas or by illumination. While in the former case the changes in work function may be due to a variation in both χ and V_s, illumination can alter only V_s.

In practice one measures the contact potential between the semiconductor and some reference electrode whose work function is not expected to alter in the course of varying the conditions at the semiconductor surface. The most common technique is based on the classical Kelvin method. A field-effect configuration (Figs 12-14) can be used, with a suitable detector connected in series with the plate-semiconductor capacitor. The field plate (gate) is mechanically coupled to a vibrating reed driven by an electromagnet or by other means. The vibrations modulate the plate-semiconductor capacitance C_g sinusoidally, and give rise to an electric signal (of the same frequency) across the detector when any potential drop (contact or other) is present between the two media. The circuit includes a series dc voltage source that can be adjusted to balance out the contact potential, a high-input-impedance amplifier serving as a null indicator. At balance, the voltage of the external source is clearly equal and opposite to the contact potential difference.

Of the various measurements of contact potential, that of its change δV_{cp} with light can be most readily correlated with the characteristics of the space-charge region. In the Kelvin method, δV_{cp} is simply given by the change in the compensating voltage required to restore balance under illumination. The periodic variation of the plate-semiconductor spacing can be dispensed with, however, by the use of chopped or flash illumination rather than steady light. The signal due to the modulation in contact potential is amplified and displayed on a CRO. The magnitude of δV_{cp} can be easily derived from the constants of the circuit.

Several processes may contribute to the measured change in V_{cp}. First and foremost is the surface photovoltage discussed in Section 1.5. The change δV_s in barrier height brought about by the light (see Fig.7) is picked up by the reference electrode, and if no other voltage drops occur in the semiconductor bulk, δV_{cp} must be equal to δV_s. Moreover, because of the large separation between the metal plate and the semiconductor compared to the width of the space-charge layer ($C_g \ll C_{sc}$), the change in surface charge is negligible and the value of δV_s is essentially unaffected by the proximity of the reference electrode. As was pointed out in Section 1.5, δV_s should vary monotonically with light intensity, reaching a saturation level at high intensities. The saturation level corresponds to almost complete flattening of the energy bands by the light and therefore it very nearly represents the magnitude V_s of the quiescent barrier height ($\delta V_s \approx -V_s$). Thus, such measurements can yield one of the most fundamental parameters of the space-charge region, the barrier height.

Other contributions to the measured change in V_{cp} may originate from photovoltages generated at the metal-semiconductor contacts, from inhomogeneities in the bulk and from the Dember potential [1]. These contributions can often be eliminated by suitable experimental conditions.

2.5. Galvanomagnetic measurements

Measurements of Hall effect and magnetoresistance have been widely used to estimate the surface mobility associated with carriers constrained to move in a potential well at the surface. Unless inversion layers on n-p-n (or p-n-p) transistors and MOS structures are used, the results reflect the contribution from both bulk and surface carriers. A simplified representation of these contributions is shown in Fig.15. The sample is considered as consisting of two parallel conductors, one corresponding to the space-charge layer (thickness d_s, conductance G_s), the other to the underlying bulk region (thickness d_b, conductance G_b). The two conductors will usually differ in both carrier density and mobility. The Hall voltage produced by a magnetic field applied normal to the surface is measured across the two side arms shown protruding from the sample. It can be calculated from elementary circuit theory if one assumes that the two conductors are insulated from each other, except at the upper and lower ends where they are shorted by the Hall probes. The equivalent circuit applicable under these conditions is shown in Fig.15(b). The individual currents and Hall voltages are given by (see Eq.(2.159) of Ref.[1])

$$I_b = \frac{\sigma_b d_b I}{\sigma_b d_b + \sigma_s d_s} \quad ; \quad I_s = \frac{\sigma_s d_s I}{\sigma_b d_b + \sigma_s d_s}$$

$$V_{Hb} = \frac{R_{Hb} I_b B}{d_b} \quad ; \quad V_{Hs} = \frac{R_{Hs} I_s B}{d_s} \quad \quad (2.16)$$

$$I = I_b + I_s \quad ; \quad d = d_b + d_s$$

The subscripts b and s refer to the bulk and the surface respectively; σ_b and σ_s are the corresponding conductivities; R_{Hb} and R_{Hs} are the Hall coefficients; I_b and I_s are the currents; and B is the magnetic induction. An overall Hall

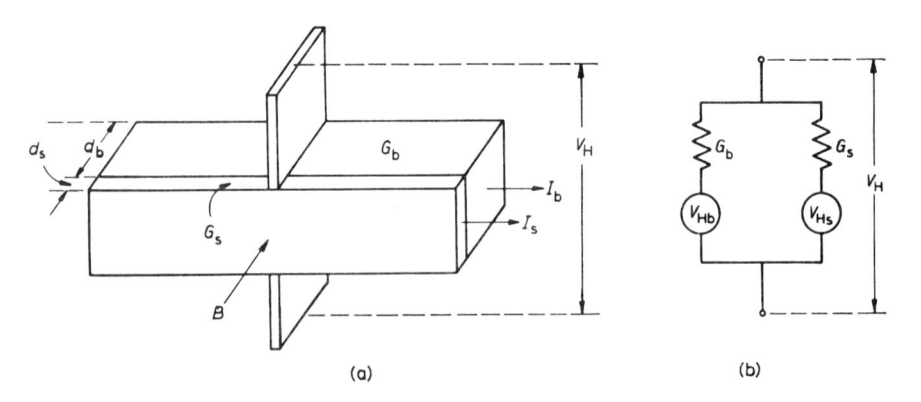

FIG.15. A simplified representation of the bulk and surface contribution to the Hall effect. (a) Hall configuration; (b) equivalent circuit used in calculating the resultant Hall coefficient.

coefficient R_H can be defined in terms of the measured open-circuit Hall voltage V_H:

$$R_H = \frac{V_H d}{IB} \qquad (2.17)$$

By reference to the equivalent circuit of Fig.15(b), it can readily be shown that R_H is given by

$$R_H = \frac{R_{Hb}\sigma_b^2 d_b + R_{Hs}\sigma_s^2 d_s}{\bar{\sigma}^2 d} \qquad (2.18)$$

where $\bar{\sigma} \equiv (\sigma_b d_b + \sigma_s d_s)/d$ is the average sample conductivity.

This approach is approximate in that it does not take into account the variation of carrier density with depth in the space-charge layer, nor does it include the possibility of circulating currents due to communication between the surface layer and the bulk. It does illustrate, however, how the surface contributes to the measured Hall effect. A more rigorous analysis of bulk and surface effects [1] yields much the same expression for the Hall coefficient as Eq.(2.18).

To separate bulk and surface effects so as to determine μ_s as a function of V_s, one can vary the barrier height and measure the resulting changes in sample conductance and Hall coefficient. Here again the conductance minimum can serve as a reference point or, if unfeasible, surface-photovoltage measurements can be used to determine the barrier height at one point of its variation.

To maximize the surface contribution to the Hall coefficient, the sample thickness should be made as small as possible (see Eq.(2.18)). Furthermore, the higher the bulk resistivity, the larger the relative contribution of the surface. In insulators, as well as in p-n-p or MOS structures, the bulk contribution is of course essentially absent.

3. SURFACE TRANSPORT PROCESSES

Free carriers drifting along a filament of finite cross-section are subject to scattering by the boundary surfaces in addition to the normal bulk scattering. This additional scattering will involve predominantly the carriers moving close to the surface and will generally reduce their mobility below that in the bulk. A study of the transport properties of such carriers can yield valuable information on the structure of the surface, and in particular on the surface scattering processes. At the same time it is of considerable practical importance because of the vital role surface conductance plays in many of the electrical measurements on semiconductor surfaces. For these reasons the problem of surface transport has received a considerable amount of attention [1,8]. With the growing interest in quantization effects in degenerate surface layers [4,5], the emphasis has shifted in recent years to transport processes in such layers. The scope of this paper does not permit a review of the rather extensive theoretical and experimental work on the subject [8,9]. Instead, we shall attempt to convey some of the important concepts involved and indicate the basis underlying the theoretical treatments. Rather than present the rigorous analysis we shall apply simple physical considerations in obtaining estimates for the surface mobility.

3.1. Concepts and definitions

The two extreme ways in which the surface can behave towards free carriers striking it are as a completely diffuse (random) scatterer and as a specular (perfect) reflector. The former implies that the carriers emerge from the surface with a Maxwell-Boltzmann distribution, having lost all memory of their velocities prior to the collision. This type of scattering clearly leads to a reduction in mobility for the carriers drifting within a mean free path from the surface. Specular reflection, on the other hand, requires that only the momentum component normal to the surface change, the parallel components and the energy remaining constant. In the case of spherical constant-energy surfaces, the electrons (and holes) can be treated as free particles with a scalar effective mass. Reflection from a specular surface then results only in a reversal of the sign of the velocity component normal to the surface, the velocities parallel to the surface remaining unchanged. There can obviously not be any mobility reduction under these conditions. Once we depart from the assumption of spherical energy surfaces, however, the effective mass becomes a tensor quantity and the velocity and momentum are in general no longer parallel. In this case there may be a mobility reduction even for specular scattering, such reduction being determined by the crystallographic orientation.

Specular reflection is the type of scattering one expects from an ideal surface. The disorder present on a non-ideal surface, on the other hand, will result in a measure of diffusivity, the exact extent of which should depend on the density and scattering cross-sections of the defects that go to make up the disorder.

Any surface scattering that is not completely specular will result in a decrease in the conductance of a given sample. This reduction is appreciable only when the thickness of the sample is not too large compared to the mean free path of the carriers, thereby affording access to the surface to a significant fraction of the carriers. In thin samples it will be found useful to consider the average effect of surface scattering on the contribution of <u>all</u> the carriers to the current parallel to the surface. For this purpose, we define average electron and hole mobilities $\bar{\mu}_n$ and $\bar{\mu}_p$ as

$$\bar{\mu}_n = \frac{I_{nx}}{q\bar{n}\mathscr{E}_x 2d} \quad ; \quad \bar{\mu}_p = \frac{I_{px}}{q\bar{p}\mathscr{E}_x 2d} \tag{3.1}$$

where I_{nx} and I_{px} are the electron and hole currents (per unit width of a rectangular filament) set up by an electric field $\mathscr{E}_x$, 2d is the thickness, and $\bar{n}$ and $\bar{p}$ are the average electron and hole densities (given by the total number of carriers in the sample, divided by the volume).

While the average mobility is a useful concept for describing transport phenomena in thin films, it becomes inadequate in the case of thick samples. As the thickness increases, less and less of the bulk carriers are able to reach the surface before suffering a collision in the bulk. Thus surface scattering becomes increasingly less significant in the expression for the total sample conductance, and $\bar{\mu} \to \mu_b$. Here the effect of surface scattering is best seen by considering the contribution to the conductance of only those carriers that are in the space-charge region adjacent to the surface. This leads to a definition of the electron and hole surface mobilities μ_{ns} and μ_{ps} in terms of $\Delta\sigma$, the change in conductance (per square area) caused by the excess surface-carrier densities ΔN and ΔP (cf. Eq.(2.2), which is applicable in the case of no surface scattering):

$$\Delta\sigma = q(\mu_{ns} \Delta N + \mu_{ps} \Delta P) \qquad (3.2)$$

The surface mobilities so defined can also be expressed as

$$\mu_{ns} = \frac{\Delta I_{nx}}{q \Delta N \, \mathcal{E}_x} \; ; \; \mu_{ps} = \frac{\Delta I_{px}}{q \Delta P \, \mathcal{E}_x} \qquad (3.3)$$

where ΔI_{nx} and ΔI_{px} are the increments (positive or negative) in the electron and hole currents for each surface (per unit of its width) with respect to their values at flat bands. The definition of the surface mobilities in Eq.(3.2) or (3.3) has the advantage of being directly related to $\Delta\sigma$, which is a measurable quantity.

Since the contributions of the two carrier types to the total current are always additive, it will be sufficient to consider only one of them. Accordingly, all the expressions appearing subsequently for the current and mobility will involve only one of the carrier types. Wherever there is no danger of ambiguity, we shall use the symbols I_x, ΔI_x, $\bar{\mu}$ and μ_s (with the subscripts n and p omitted) for either electrons or holes as the case may be.

3.2. Simple considerations

To gain some insight into the main physical processes controlling the surface mobility, a simplified treatment of carrier transport is presented in this section. A one-carrier system, corresponding to an extrinsic n-type sample, will be assumed throughout. We first estimate the average mobility in thin slabs and then extend the treatment to the calculation of surface mobility in accumulation and depletion layers of thick samples. Both completely diffuse and partially specular scattering are considered. Comparison of the surface mobility values derived by these simple considerations with those obtained numerically from more rigorous treatments shows the agreement to be quite good. The former have the particular advantage that they are given in closed form and are thus very convenient to use in practice.

3.2.1. Thin slabs

We consider a semiconductor sample bounded by the surfaces $z = 0$, $z = 2d$ and calculate the current flowing parallel to the surfaces as a result of an applied external field. The band edges are assumed to continue flat up to the surfaces ($V_s = 0$), the electron density n being uniform throughout the sample and equal to n_b. The effect of surface scattering will be introduced in the form of some average collision time τ_s, just as bulk scattering is characterized by the relaxation time τ_b. It is further assumed that these two processes, involving the bulk and the surface, act in parallel and independently in determining the average electron mobility, so that

$$\frac{1}{\bar{\tau}} = \frac{1}{\tau_s} + \frac{1}{\tau_b} \qquad (3.4)$$

where $\bar{\tau}$ is the average collision time of the electrons in the sample. Another way of expressing Eq.(3.4) is to say that the probability per unit time that an electron be scattered, and thus lose all memory of the previous action of the field, is given by the sum of the scattering probabilities for the bulk and the surface taken separately.

For diffuse surface scattering, τ_s represents the average time an electron requires to collide with the surface towards which it is moving and thus lose all memory of its energy. As an estimate of τ_s we take the mean distance d of a carrier from the surface divided by the unilateral mean velocity $\bar{c}_z$:

$$\tau_s \approx \frac{d}{\bar{c}_z} = \frac{d}{\lambda} \tau_b \qquad (3.5)$$

The unilateral mean velocity is defined, analogously to the root mean square velocity ($\langle c^2 \rangle^{1/2} = (3kT/m_n)^{1/2}$), as the average over the positive (or negative) velocity components c_z of all electrons; it is easily seen to be given by

$$\bar{c}_z = \left(\frac{kT}{2\pi m_n}\right)^{1/2} \qquad (3.6)$$

where m_n is the scalar effective mass. The unilateral mean free path λ is related to $\bar{c}_z$ by means of the expression:

$$\lambda = \tau_b \bar{c}_z = \tau_b \left(\frac{kT}{2\pi m_n}\right)^{1/2} = \mu_b \left(\frac{m_n kT}{2\pi q^2}\right)^{1/2} \qquad (3.7)$$

The average electron mobility $\bar{\mu}$ is taken as $\bar{\mu} = q\bar{\tau}/m_n$, in analogy with the corresponding expression for the bulk mobility $\mu_b = q\tau_b/m_n$. By using Eqs (3.4) and (3.5) we obtain

$$\frac{\bar{\mu}}{\mu_b} = \frac{1}{1 + \lambda/d} \qquad (3.8)$$

It is seen that the average mobility decreases with decreasing sample thickness, as expected, while for $d \gg \lambda$ it approaches its value in the bulk. In the latter case, Eq.(3.8) can be written in the form $\bar{\mu}/\mu_b \approx (d-\lambda)/d$. In other words, instead of considering all the carriers as moving with an average mobility $\bar{\mu}$, one can look upon the sample as though the carriers within a distance λ of each of the surfaces have zero mobility while the carriers in the remaining part of the sample (of thickness $2d - 2\lambda$) move with their bulk mobility μ_b.

For very thin slabs such that d is small compared to the effective Debye length L (Eq.(1.21)), the potential in the sample will be essentially constant regardless of whether $V_s = 0$ (as assumed above) or not. In both cases the electron density will be uniform, although for $V_s \neq 0$ its value will be different from n_b. This does not affect at all the derivation of Eq.(3.8), which is thus always valid for sufficiently thin slabs. For thick slabs, on the other hand, Eq.(3.8) is just a special case (flat bands), and a more general expression will be derived below.

3.2.2. Thick samples

The potential barriers at the two surfaces will be assumed symmetrical about the plane $z = d$. Consider first an accumulation layer. The electrons at the surface can be divided into two categories, bounded and unbounded. The former are constrained to move inside the potential well while the latter have energies above the well. The current increment ΔI_x in Eq.(3.3) consists not only of the contribution of the bounded electrons but also of that arising from the change in scattering conditions for the unbounded electrons close enough to the surface to be affected by it. The latter contribution is clearly negligible for strong accumulation layers or when the mean free path λ is small compared to the width of the well. Under these conditions the surface mobility μ_s associated with the excess surface-carrier density ΔN is given to a good approximation by the average mobility $\bar{\mu}$ defined in Eq.(3.1) but now corresponding to a thin slab whose thickness is no longer that of the sample but rather of the order of the width of the space-charge region. To estimate μ_s, we approximate the potential in the accumulation layer by a square-well potential and take for its width the effective charge distance L_c (Eq.1.28). The carriers in the accumulation layer can then be looked upon as moving in a thin slab with one surface (the actual surface, $z = 0$) a diffuse scatterer and the other ($z = L_c$) a specular reflector. The mean distance of the carriers from the scattering surface is now L_c (and not half the sample thickness, as in the case of a slab in which both surfaces are diffuse scatterers). The discussion in the preceding subsection must be modified in another, much more important, respect. The electrons in the accumulation layer are now accelerated towards the surface and their kinetic energy is increased due to the potential V through which they drop. As a result, the unilaterial mean velocity $\bar{c}_z$ will be given by

$$\bar{c}_z = (\pi m_n)^{-1/2} (\tfrac{1}{2} kT + q\bar{V})^{1/2} \tag{3.9}$$

where $\bar{V}$ is some average value of the potential barrier V. As an estimate of $\bar{V}$ we take $\tfrac{1}{2} V_s$, half the value of the barrier height. Substitution from Eqs (3.5) and (3.7) then yields the following expression for τ_s:

$$\tau_s \approx \frac{L_c \tau_b}{\lambda(1 + v_s)^{1/2}} \tag{3.10}$$

The average collision time $\bar{\tau}$ of the bound electrons is again given by Eq.(3.4), but now the value of τ_s is as in Eq.(3.10) rather than in (3.5). The surface mobility is thus given by

$$\frac{\mu_s}{\mu_b} = \frac{1}{1 + (\lambda/L_c)(1 + v_s)^{1/2}} \tag{3.11}$$

Substituting Lv_s/F_s for L_c from Eq.(1.30), we have

$$\frac{\mu_s}{\mu_b} = \frac{1}{1+(rF_s/v_s)(1+v_s)^{1/2}} \tag{3.12}$$

where

$$r \equiv \frac{\lambda}{L} = \mu_b\left(\frac{m_n}{2\pi\kappa\epsilon_0}\right)^{1/2}(n_b+p_b)^{1/2} \tag{3.13}$$

A glance at Eq.(1.32') shows that F_s increases rapidly with v_s, so that μ_s is a decreasing function of barrier height. This is to be expected, since the carriers are constrained to move in ever narrower and deeper wells as the accumulation layer becomes stronger. Note that $\mu_s/\mu_b \to (1+r)^{-1}$ as $v_s \to 0$, implying that, even for very slight curvature of the bands, the surface mobility can differ from its bulk value.

In the derivation of Eq.(3.12) we have neglected the change in scattering conditions for the unbounded electrons. The circumstances under which this procedure is valid can be obtained from the following considerations. As pointed out above, under flat-band conditions the effect of surface scattering can be expressed by looking upon the carriers situated within a distance λ of each of the surfaces as though moving with zero mobility. In the presence of the potential barrier, the distance λ should be replaced by $\lambda(1+v_s)^{1/2}$ to allow for the increased unilateral mean velocity $\bar{c}_z$ (see Eq.(3.9)). (We implicitly require that $\lambda < L$.) Our assumption amounts to neglecting $\mu_b n_b \lambda[(1+v_s)^{1/2}-1]$ with respect to $\mu_s \Delta N$. Recalling that $\Delta N \approx \sqrt{2}\,n_b L\,\exp(\frac{1}{2}v_s)$ (see Eqs (1.27), (1.32'), (1.44)), we see that this condition is equivalent to

$$r\exp(-\tfrac{1}{2}v_s)[(1+v_s)^{1/2}-1] \ll \frac{\mu_s}{\mu_b} \tag{3.14}$$

For strong accumulation layers, (3.14) is always satisfied, since $\exp(-\tfrac{1}{2}v_s)\ll 1$ and is much smaller than μ_s/μ_b. Such is no longer the case for weak accumulation layers, however, and only for r sufficiently small will the inequality (3.14) be maintained.

We shall now derive the surface mobility for depletion layers. Here only those electrons having sufficient energy to surmount the potential barrier are able to reach the surface. Such carriers, which are present with a density of approximately $n_b\exp v_s$, will be assumed to move with an average mobility $\bar{\mu}$ given by Eq.(3.8). All other carriers are specularly reflected at the potential barrier and are not expected to be affected by surface scattering. Under these assumptions we obtain for the current

$$I_x = 2[n_b d(1-\exp v_s) + \Delta N]q\,\mathscr{E}_x\mu_b + 2n_b d\,q\,\mathscr{E}_x\bar{\mu}\,\exp v_s \tag{3.15}$$

The square brackets represent the electrons moving with their bulk mobility. (Note that both ΔN and v_s are negative in depletion layers.) By subtracting from I_x its value at flat bands (obtained by equating v_s to zero in Eq.(3.15)) and substituting for $\bar{\mu}$ from Eq.(3.8), we have (for each surface)

$$\Delta I_x = [n_b d(1 - \exp v_s) + \Delta N] \, q \, \mathscr{E}_x \mu_b - n_b d(1 - \exp v_s) q \, \mathscr{E}_x \mu_b (1 + \lambda/d)^{-1}$$

$$= \Delta N q \, \mathscr{E}_x \mu_b \left[1 + \frac{\lambda n_b}{\Delta N(1 + \lambda/d)} (1 - \exp v_s) \right] \quad (3.16)$$

For thick samples ($\lambda/d \ll 1$), the electron surface mobility can be expressed with the help of Eqs (1.27), (1.45), (3.3) and (3.13) in the form

$$\frac{\mu_s}{\mu_b} = \left(\frac{r}{F_s} \right) [1 - \exp(-|v_s|)] \quad (3.17)$$

For strong depletion layers, $F_s = \sqrt{2} |v_s|^{1/2}$ (see Eq.(1.36')). Thus μ_s is seen to approach μ_b for large (negative) values of v_s, as expected. At flat bands ($v_s \to 0$), $\mu_s/\mu_b \to 1 - r$, as can easily be verified from Eqs (3.17) and (1.36). For small values of r, this result is the same as that obtained when flat-band conditions are approached from accumulation layers (Eq.(3.12)).

To obtain the hole surface mobility in accumulation and depletion layers it is necessary only to change the sign of v_s in Eqs (3.12) and (3.17).

3.2.3. Partially specular scattering

The simple arguments presented above can readily be extended to the case where the scattering surface is partially diffuse, partially specular. We define ω as the probability that an electron reaching the surface be specularly reflected. Thus ω = 0 corresponds to completely diffuse scattering, ω = 1 to totally specular reflection.

The reciprocal of τ_s (see Eq.(3.4)) expresses the probability that an electron be scattered by the surface in unit time. Thus for diffuse scattering it also represents the probability that an electron loses all memory of its energy. For partially specular reflection, however, this probability must be modified to include the fact that of all the reflections taking place during unit time, only the fraction (1 - ω) will lead to a loss of memory, so that $1/\tau_s$ must be replaced by $(1 - \omega)/\tau_s$. This is equivalent to replacing d in Eq.(3.5) by d/(1-ω). The average mobility for thin slabs (Eq.(3.3)) then becomes

$$\frac{\bar{\mu}}{\mu_b} = \frac{1}{1 + (1 - \omega)\lambda/d} \quad (3.18)$$

Identical reasoning for the case of thick samples leads to the replacing of L_c in Eq.(3.10) by $L_c/(1 - \omega)$ so that the surface mobility for majority carriers in accumulation layers (Eq.(3.12)) becomes

$$\frac{\mu_s}{\mu_b} = \frac{1}{1 + (1 - \omega)(rF_s/v_s)(1 + v_s)^{1/2}} \quad (3.19)$$

A similar result is obtained for depletion layers (Eq.(3.17)):

$$\frac{\mu_s}{\mu_b} = 1 - \left[(1-\omega)\frac{r}{F_s}\right][1 - \exp(-|v_s|)] \quad (3.20)$$

3.3. Calculations of average and surface mobilities for diffuse scattering

The formal treatment of surface transport is based, as in the case of the bulk, on the Boltzmann transport equation. In this section we formulate the Boltzmann equation and indicate how it can be applied in the analysis of surface transport processes. We consider a non-degenerate n-type semiconductor having spherical energy surfaces (isotropic effective mass m_n). The sample will be taken again in the form of a slab bounded by the surfaces $z = 0$ and $z = 2d$, and we shall calculate the current set up by a small electric field $\mathscr{E}_x$ parallel to the surface.

3.3.1. The Boltzmann transport equation

Let $f(\vec{c}, \vec{r}, t)$ represent the carrier distribution function in velocity and co-ordinate space in the presence of the electric field. The distribution at thermal equilibrium will be denoted by $f_0(\vec{c},\vec{r})$. When expressed as a function of the electron energy $E (\equiv \frac{1}{2} m_n c^2)$, then f_0 is just the Fermi-Dirac distribution function, Eq.(1.60).

The steady-state distribution established under the applied field is obtained by equating to zero the total rate of change of f with time. The total rate consists of the rates of change due to the external field and to scattering:

$$\frac{df}{dt} = \left(\frac{\partial f}{\partial t}\right)_{field} + \left(\frac{\partial f}{\partial t}\right)_{scattering} \quad (3.21)$$

Each group of electrons in an element of $(\vec{c},\vec{r})$ space moves as an incompressible fluid according to the relations:

$$d\vec{c} = \vec{a}\,dt \quad ; \quad d\vec{r} = \vec{c}\,dt \quad (3.22)$$

where $\vec{a}$ is the acceleration of an electron due to the field. Thus

$$f(\vec{c}+\vec{a}\,dt, \vec{r}+\vec{c}\,dt, t+dt) = f(\vec{c},\vec{r},t)$$

Expansion of the left-hand side yields

$$\vec{a}\cdot\text{grad}_c f + \vec{c}\cdot\text{grad}_r f + \left(\frac{\partial f}{\partial t}\right)_{field} = 0$$

and Eq.(3.21) becomes

$$\vec{a}\cdot\text{grad}_c f + \vec{c}\cdot\text{grad}_r f = \left(\frac{\partial f}{\partial t}\right)_{scattering} \quad (3.23)$$

This is the general form of the Boltzmann transport equation.

For sufficiently weak external forces, the deviations from thermal equilibrium are small and one can usually characterize the scattering processes by a relaxation time $\tau(\vec{c},\vec{r})$ such that

$$\left(\frac{\partial f}{\partial t}\right)_{scattering} = - \frac{f(\vec{c},\vec{r},t) - f_0(\vec{c},\vec{r})}{\tau(\vec{c},\vec{r})} \quad (3.24)$$

In the general case τ is a function of $\vec{c}$ (or E) and $\vec{r}$. For simplicity, however, one often assumes τ to be a constant (denoted by τ_b). Combining Eqs (3.23) and (3.24), we can express the Boltzmann equation as

$$\vec{a} \cdot \text{grad}_c f + \vec{c} \cdot \text{grad}_r f = - \frac{(f - f_0)}{\tau_b} \quad (3.25)$$

Obviously f is determined by both bulk and surface scattering. The former is represented by the (constant) bulk relaxation time τ_b in the right-hand side of Eq.(3.25), while the latter is introduced by appropriate boundary conditions imposed on f at the scattering surfaces. Equation (3.25) is solved by substituting $f = f_0 + \delta f$, where $\delta f = \delta f(c,z)$ is a perturbing function (assumed small) which, in addition to the dependence on c as in the case of bulk scattering, is also a function of z.

3.3.2. The average mobility

Consider first the case of flat bands ($v_s = 0$). Here $a = (a_x,0,0)$, the acceleration component normal to the surface being zero, and f_0 is independent of z. The assumption of diffuse scattering requires that the carriers leaving the surface be randomly distributed. Thus the boundary conditions are

$$\delta f(\vec{c},0) = 0 \quad \text{for} \quad c_z \geq 0$$
$$\delta f(\vec{c},2d) = 0 \quad \text{for} \quad c_z \leq 0 \quad (3.26)$$

It can be verified by direct substitution that, by neglecting $a_x \partial(\delta f)/\partial c_x$ with respect to $c_z \partial(\delta f)/\partial z$ (this is justified since $a_x = -(q/m_n)\mathscr{E}_x$, and $\mathscr{E}_x$ is taken sufficiently small), an approximate solution to the Boltzmann equation with these boundary conditions is obtained in the form:

$$\delta f = \delta f_b \left[1 - \exp\left(-\frac{z}{c_z \tau_b}\right)\right] \quad \text{for} \quad c_z \geq 0$$

$$\delta f = \delta f_b \left[1 - \exp\left(-\frac{(z-2d)}{c_z \tau_b}\right)\right] \quad \text{for} \quad c_z \leq 0 \quad (3.27)$$

where δf_b is the deviation from the thermal equilibrium distribution in the bulk. As we are dealing with the case of flat bands (c_z independent of z), the value of z/c_z or of $(z-2d)/c_z$ is just the time elapsed since the electron left the relevant surface. Thus the factors in square brackets in Eq.(3.27) are the probability that an electron having been scattered by the surface $z = 0$ (or $z = 2d$) suffer a bulk collision within the time z/c_z (or $(z-2d)/c_z$).

The electron current (per unit width) parallel to the surface is

$$I_x = -q \int_0^{2d} \int_{-\infty}^{\infty} \int_{-\infty}^{\infty} \int_{-\infty}^{\infty} c_x N(c) \delta f \, dc_x \, dc_y \, dc_z \, dz \qquad (3.28)$$

where $N(c)$ is the density of states in velocity space and is given by [1] $2(m_n/h)^3$. Substituting for δf, we can integrate directly over all the variables but c_z. The introduction of the dimensionless energy parameter $\epsilon \equiv m_n c_z^2/2kT$ then yields

$$I_x = 2n_b q d \, \mathscr{E}_x \mu_b \left[1 - \frac{\lambda}{d} + \frac{\lambda}{d} \Gamma_1\left(\frac{\lambda}{d}\right) \right] \qquad (3.29)$$

where

$$\Gamma_1\left(\frac{\lambda}{d}\right) \equiv \int_0^{\infty} \exp\left[-\epsilon - \frac{d}{\lambda} (\pi\epsilon)^{-1/2} \right] d\epsilon \qquad (3.30)$$

and λ is the unilateral mean free path defined by Eq.(3.7). Using Eq.(3.1), we obtain for the average mobility $\bar{\mu}$

$$\frac{\bar{\mu}}{\mu_b} = 1 - \frac{\lambda}{d} + \frac{\lambda}{d} \Gamma_1\left(\frac{\lambda}{d}\right) \qquad (3.31)$$

This function has been evaluated numerically [10] and is plotted in Fig.16 against λ/d. The crosses represent values calculated on the basis of the simplified analysis of the previous section (Eq.(3.8)). They are seen to agree quite well with the results of the more rigorous treatment.

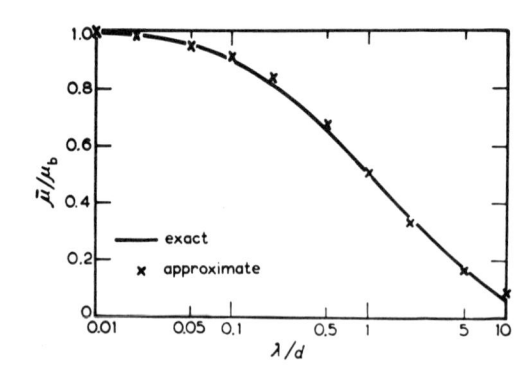

FIG.16. Calculated average mobility for the case of flat bands (thin slabs) as a function of λ/d. (After Flietner, Ref.[10]). The crosses represent approximate values obtained from Eq.(3.8).

3.3.3. Surface mobility

To calculate the surface mobility for the case of thick samples, it is necessary to extend the above treatment to include the influence of the potential barrier v. Two of the terms appearing in the Boltzmann transport equation (Eq.(3.25)) are modified. First, the electron acceleration now has an additional component a_z due to the electric field $\mathscr{E}_z = -(kT/q)(dv/dz)$ associated with the potential barrier. Second, the expression for the equilibrium distribution function f_0 in the case of flat bands must be multiplied by the factor $\exp v$.

Consider first accumulation layers in n-type samples. The boundary condition for all electrons leaving the surface $z = 0$ is the same as before:

$$\delta f(c,0) = 0 \quad \text{for} \quad c_z \geq 0 \tag{3.32}$$

This condition determines δf uniquely at any point in the space-charge region for all bounded electrons ($m_n c_z^2/2kT \leq v$), the potential barrier acting as a specular reflector towards such electrons. For the unbounded electrons ($m_n c_z^2/2kT \geq v$) approaching the surface ($c_z \leq 0$), however, an additional condition is required. It is justified to assume that far removed from the surface ($z \to d$) these electrons have their bulk distribution — i.e. $\delta f = \delta f_b$ for $z \to d$. It is again easily verified by direct substitution that, by neglecting non-linear terms, the solution to the transport equation subject to these boundary conditions is

$$\delta f = \delta f_b \left[1 - \exp\left(-\frac{1}{\tau_b} \int_0^z \frac{dz}{c_z} \right) \right] \tag{3.33}$$

for the bounded electrons and

$$\delta f = \delta f_b \quad , \quad c_z \leq 0$$

$$\delta f = \delta f_b \left[1 - \exp\left(-\frac{1}{\tau_b} \int_0^z \frac{dz}{c_z} \right) \right] \quad , \quad c_z \geq 0 \tag{3.34}$$

for the unbounded electrons. The above integrals are to be evaluated with ϵ constant, where ϵ is a dimensionless energy parameter defined by $\epsilon \equiv (m_n c_z^2/2kT) - v$. The solutions (3.33) and (3.34) are similar to those for flat bands (Eq.(3.27)). But as c_z is no longer independent of position, the time elapsed since the electron left the relevant surface must now be expressed in the form of an integral over z.

The electron current (per unit width) I_x is again given by Eq.(3.28). The surface mobility can then be calculated similarly to the case of flat bands. The expressions obtained in this manner are, however, rather cumbersome and will not be given here [1]. Results of numerical integration for majority carriers in accumulation layers are illustrated in Fig.17. Here μ_s/μ_b is plotted against $|v_s|$ for various values of the parameter $r = \lambda/L$ (see Eq.(3.13)). Results for an intrinsic semiconductor ($u_b = 0$) are also included (dashed curves), and are seen not to differ much from those for extrinsic samples.

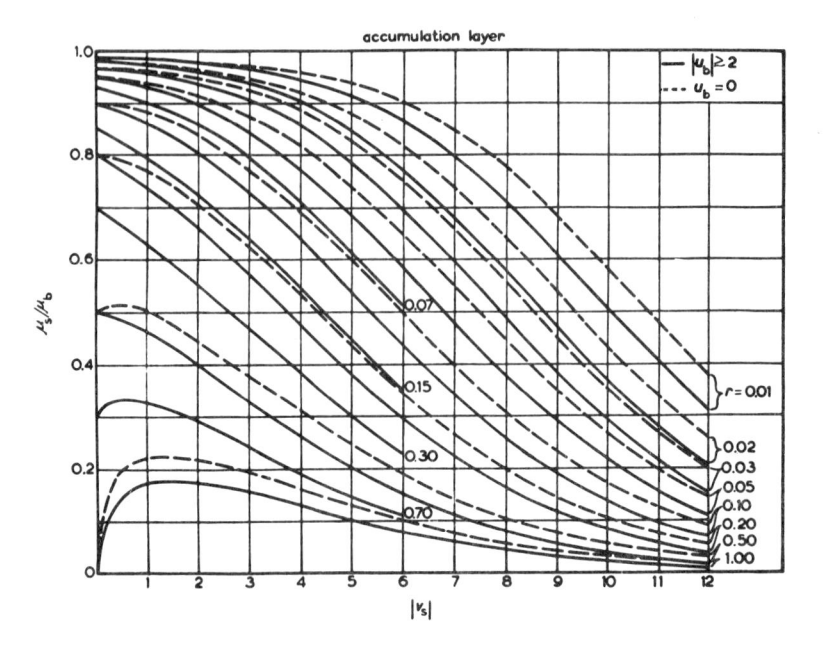

FIG.17. Calculated surface mobility for majority carriers in accumulation layers as a function of $|v_s|$ for various values of the parameter r. The solid curves correspond to extrinsic semiconductors ($|u_b| \gtrsim 2$), the dashed curves to intrinsic semiconductors ($u_b = 0$). (After Ref.[1]).

When r is close to unity, we see that for small $|v_s|$ the function μ_s/μ_b decreases with decreasing $|v_s|$, which is somewhat surprising at first glance. Such behaviour is associated with the contribution of the unbounded electrons which, under these conditions, is not negligible compared to the contribution of the bounded electrons.

As discussed in Section 3.2, the relative contribution of the unbounded carriers is negligible near flat bands for $r \ll 1$, and in strong accumulation layers for r not necessarily small. Under these conditions the simple calculations of the surface mobilities (Eq.(3.12)) should constitute a good approximation, and indeed the values so obtained are found to agree well (to within a few hundredths) with those calculated numerically (Fig.17).

The calculations above have been derived for accumulation layers. As long as the excess electrons are constrained to move in a potential well, however, it is immaterial whether they are majority or minority carriers. And indeed, similar results are obtained for inversion layers.

3.4. Partially diffuse scattering — the Fuchs boundary condition

Evidence accumulated from a large number of experimental studies on semiconductor surfaces [1, 8, 9, 11] shows that these surfaces are always strongly scattering. In most cases, however, the scattering is not completely diffuse, especially when the carriers are confined to deep potential wells, and some degree of specularity must be involved. In fact, in at least one case, that of PbSe, the scattering was found to be almost completely specular [12].

Thus the boundary condition, Eq.(3.26), can by no means be assumed to hold uniformly. A simple and plausible boundary condition for describing a partially specular, partially diffuse surface scatterer has been introduced by Fuchs [13] as

$$\delta f(\vec{c}_i, \vec{r}_s) = \omega \cdot \delta f(\vec{c}_r, \vec{r}_s) \qquad (3.35)$$

where $\vec{r}_s$ indicates the position at the surface, and $\vec{c}_i$ and $\vec{c}_r$ are the incident and reflected velocities, respectively. For the semiconductor slab considered above, the Fuchs boundary condition can be expressed as

$$\delta f(c_x, c_y, c_z, 0) = \omega \delta f(c_x, c_y, -c_z, 0)$$
$$\delta f(c_x, c_y, c_z, 2d) = \omega \delta f(c_x, c_y, -c_z, 2d) \qquad (3.36)$$

The parameter ω, referred to as the Fuchs reflectivity, has already been introduced in Section 3.2. It represents the probability that an electron reaching the surface be specularly reflected.

The Fuchs reflectivity has usually been chosen so as to give the best fit between theory and experiment. Since $\omega = 1$ corresponds to totally specular reflection of electrons and $\omega = 0$ to completely diffuse scattering, it seemed reasonable to interpret intermediate values of ω as the probability of specular reflection. It should be noted, however, that in the Boltzmann transport equation (Eq.(3.25)) there is no explicit reference to the nature of the surface; all the relevant properties of the scattering surface enter the transport theory through the Fuchs boundary condition, i.e. through the single parameter ω. It would not be surprising, therefore, if this boundary condition were too simple for reality.

And indeed, both experimental and theoretical work [8] has indicated that ω may have a strong angular dependence. It appears that the scattering is nearly specular for electrons hitting the surface at grazing angles, and becomes increasingly diffuse as normal-incidence angles are approached. Theoretical treatments for simple scattering models, such as those involving surface charges and surface roughness [8, 11] show this sort of angular dependence. More generally, ω is expected to depend on the specific form of the incident distribution (angular and velocitywise).

The theoretical effort in recent years has been directed at deriving a transport boundary condition from basic considerations involved in the reflection and scattering of electrons at a crystal surface. The ultimate objective of such studies is the determination of the physical origin of the Fuchs boundary condition and its relation to the atomic structure of the surface. In particular, one has to derive the specific form of the reflectivity function ω for the various scattering mechanisms. A Boltzmann transport calculation of surface mobility using the calculated Fuchs reflectivity then becomes possible. This in turn will permit the scattering model assumed to be tested against surface-mobility measurements. The reader is referred to the literature [8, 9, 11] for recent advances in the theory and experiment of surface transport. Here only the three main mechanisms of surface scattering will be mentioned.

The simplest kind of imperfection is that of charged centres randomly arrayed on the crystal surface. The theory of this effect is similar to the

Conwell-Weisskopf treatment of charged impurity scattering in the bulk semiconductor. The charged centres can consist of occupied surface states or of surface charges localized at chemisorbed atoms. The centres are viewed as a partially ordered array of scatterers and use is made of the statistical properties of the scattering potential. In other words, the surface distribution of point charges is treated as a partially correlated stochastic function of position [11]. Thermal vibrations constitute another source of scattering. This scattering originates from an electron-phonon coupling localized at the surface proper. It can be described in terms of the phonon-produced distortion of the surface shape, a distortion resembling moving corrugations which act as reflection gratings. The phonons can be associated with either bulk waves or surface (Rayleigh) waves. Finally, one has to consider scattering originating from surface roughness of a scale larger than atomic dimensions. The scattered electron intensity depends on the details of the surface shape, which is probably complicated enough for the roughness parameter to be considered as a random variable.

3.5. Quantized space-charge layers

As we have seen in Section 1.7, in strong accumulation and inversion channels the potential well associated with the space-charge layer can be narrow enough for quantization effects to become important. Such quantization alters the carrier wave functions and energy levels which in turn should be expected to modify the surface transport properties. One aspect of the altered conditions becomes immediately obvious if it is recalled that quantization broadens the potential well above its classical width. In other words, for a given excess surface-carrier density ΔN (or ΔP) the effective charge distance L_c is larger than the classically derived value. The effect of this broadening on surface transport can be seen by looking at the simple approximate expression for μ_s/μ_b in accumulation layers, Eq.(3.11). With L_c in quantized channels decreasing not as rapidly with increasing v_s, one should expect higher μ_s/μ_b values than those given by the classical transport theory we have discussed.

Actually, however, the entire approach to the problem of surface scattering in a two-dimensional electron gas is different conceptually from that used for ordinary surface transport. One can no longer make the separation into the Boltzmann equation in (c_x, c_y, c_z, z) space plus a Fuchs boundary condition characterizing the degree of specularity or diffusiveness of the surface $(z = 0)$. In the quantum limit all carriers in the surface channel have their motion perpendicular to the surface quantized, with wave functions that go to zero at the channel boundaries (see Fig.10). It is only in a direction parallel to the surface that the carriers have a free-like (two-dimensional) character. In other words, the carriers have zero velocity component c_z normal to the surface so that the conventional picture of carriers impinging on and being reflected (diffusely or otherwise) at the surface is not appropriate. The Boltzmann equation for the quantum limit must now be formulated in the (c_x, c_y) space only, taking into account the various scattering mechanisms present. It turns out that the two-dimensional character of the quantized channel drastically changes the scattering properties of the carriers, usually producing a considerable curtailment in scattering.

In recent years there has been a surge of activity in the study of transport processes in quantized channels [5, 8, 9, 14]. In this paper neither the theoretical nor the experimental work will be covered; only some of the main points will be briefly discussed.

On the experimental side, n-type inversion channels of Si on MOS structures constitute the system most comprehensively studied. The main features of the results obtained [14] with this system are: (1) At low temperatures the surface Hall mobility increases with ΔN, reaches a peak value and then drops with further increase in ΔN. The peak becomes less pronounced at higher temperatures until it disappears completely at room temperature, when the mobility drops monotonically with increasing ΔN throughout the range studied (10^{11} to 10^{13} cm^{-2}). (2) The peak mobility decreases with increasing temperature. (3) In weaker inversion layers ($\Delta N = 2 \times 10^{11}$ to 8×10^{11} cm^{-2}), the surface mobility is thermally activated (increasing exponentially with temperature). The activation energy drops with increasing ΔN (taken as a parameter) until it falls off to zero when $\Delta N \approx 10^{12}$ cm^{-2}.

The drop in mobility in strong inversion layers, as well as the reduction in peak mobility with increasing temperature, has been attributed to phonon and surface-roughness scattering. The increasing mobility with increasing ΔN observed at low temperatures over some range of ΔN is opposite to the behaviour encountered in non-degenerate layers and appears to be a unique feature of quantized channels. This feature is more pronounced in accumulation layers on ZnO, where even at room temperature a nearly tenfold enhancement in surface mobility has been observed [3] when ΔN was increased from 5×10^{12} to 5×10^{13} cm^{-2}.

It has been suggested [5, 15] that such behaviour, as well as the activated nature of the mobility in weak inversion layers, may be due to scattering by charged centres located at or near the Si/SiO$_2$ interface. The potential fluctuations associated with the charged centres either scatter or localize predominantly carriers in states of low energy. The lowest-energy states correspond to large potential wells, which are relatively scarce and therefore widely spaced along the channel. Carrier transport in these states takes place by hopping processes and is therefore characterized by low mobilities. At higher energy states, which become populated as ΔN increases, the barriers between adjacent wells are reduced and the mobility increases. A similar process may account for the thermal activation of the mobility. With increasing temperature, more and more carriers are excited from low-lying states into higher, more mobile energy states. Another possible explanation is that the activation energy corresponds to excitation of carriers from bound states associated with the surface charges, and that the reduction of the activation energy as carriers are added is a screening effect.

It should be pointed out in conclusion that while considerable progress has been achieved in the theory of surface transport in quantized channels, and a fair amount of experimental data on several two-dimensional systems has been accumulated, our understanding of surface transport processes is far from complete. In fact, the same statement applies also to transport in non-degenerate layers, but future work will most probably be centred on the quantized channel, the more interesting system by far.

REFERENCES

[1] MANY, A., GOLDSTEIN, Y., GROVER, N.B., Semiconductor Surfaces, North-Holland, Amsterdam (1965).
[2] SEIWATZ, R., GREEN, M., J. Appl. Phys. 29 (1958) 1034.
[3] MANY, A., CRC Crit. Rev. Solid State Sci. 4 (1974) 515.

[4] DORDA, G., in Advances in Solid State Physics (QUEISSER, H.J., Ed.), Vol.13, p.215, Pergamon Press, Oxford (1973), (references).
[5] STERN, F., CRC Crit.Rev.Solid State Sci. 4 (1974) 499 (references).
[6] HARTEN, H.U., in Festkörperprobleme 3 (SAUTER, F., Ed.), Braunschweig.
[7] ROSE, A., Concepts in Photoconductivity and Allied Problems, Interscience, New York (1963).
[8] GREENE, R.F., in Solid State Surface Science 1 (GREEN, M., Ed.), M. Dekker, New York (1973) 87 (references).
[9] GREENE, R.F., CRC Crit.Rev.Solid State Sci. 4 (1974) 477 (references).
[10] FLIETNER, H., Phys.Status Solidi 1 (1961) 483.
[11] GREENE, R.F., Phys.Rev. B7 (1973) 1384 (references).
[12] BRODSKY, M.H., ZEMEL, J.N., Phys.Rev. 155 (1967) 780.
[13] FUCHS, K., Proc.Camb.Philos.Soc. 34 (1938) 100.
[14] FANG, F.F., FOWLER, A.B., Phys.Rev. 169 (1968) 619.
[15] STERN, F., Phys.Rev. B9 (1974) 2762 (references).

FACULTY

An asterisk indicates that a lecturer's contribution is not published in these Proceedings

DIRECTORS

V. Celli
International Centre for Theoretical Physics,
Trieste, Italy
(Present address: Department of Physics,
University of Virginia,
Charlottesville,
Va. 22901, United States of America)

G. Chiarotti
Istituto di Fisica,
Università degli Studi di Roma,
Piazzale delle Scienze 5,
I-00100 Rome, Italy

F. García-Moliner
Departamento de Física,
Facultad de Ciencias, C-XII, 6,
Universidad Autónoma de Madrid,
Canto Blanco, Madrid 34, Spain

S. Lundqvist
Institute of Theoretical Physics,
Chalmers University of Technology,
Fack,
S-402 20 Göteborg, Sweden

N. H. March*
Department of Physics,
Imperial College of Science & Technology,
Prince Consort Road,
London SW7 2AZ, United Kingdom

J. M. Ziman*
H. H. Wills Physics Laboratory,
University of Bristol,
Royal Fort, Tyndall Avenue,
Bristol BS8 1TL, United Kingdom

LECTURERS

S. Andersson
Department of Physics,
Chalmers University of Technology,
Fack,
S-402 20 Göteborg 5, Sweden

M. V. Berry
H. H. Wills Physics Laboratory,
University of Bristol,
Royal Fort,
Tyndall Avenue,
Bristol BS81TL, United Kingdom

W. Brenig
Physik-Abteilung der Technischen Universität München,
James-Franck-Strasse,
8040 Garching b. München,
Federal Republic of Germany

FACULTY

F. Clementi* Department of Pharmacology,
University of Milan,
Via Festa del Perdono 7,
Milan, Italy

G. Dearnaley Nuclear Physics Division, H.8,
AERE Harwell,
Didcot, Oxfordshire,
United Kingdom

D. Dowden Imperial Chemical Industries Ltd.,
Agricultural Division,
PO Box 6,
Billingham,
Cleveland TS23 1LE,
United Kingdom

K. Dransfeld* Max-Planck Institut,
Grenoble, France

A.J. Forty Department of Physics,
University of Warwick,
Coventry, War., United Kingdom

R. Gomer James Franck Institute,
University of Chicago,
5640 Ellis Avenue,
Chicago, Ill. 60637,
United States of America

R. Jones Institut für Festkörperforschung,
Postfach 365,
D 517 Jülich 1,
Federal Republic of Germany

B. Makin Department of Electrical Engineering,
University of Southampton,
United Kingdom
(Present address: Department of Electrical Engineering
and Electronics,
University of Dundee,
Dundee DD1 4HN, United Kingdom)

A. Many Racah Institute of Physics,
Hebrew University of Jerusalem,
Jerusalem 91000, Israel

H. Nahr Physics Institute IV,
Erwin-Rommel-Strasse 1,
Erlangen,
Federal Republic of Germany

C.A. Neugebauer* General Electric Company,
Research and Development Center,
PO Box 8,
Schenectady, N.Y. 12301,
United States of America

R. Parsons	Department of Physical Chemistry, University of Bristol, Cantock's Close, Bristol BS8 1TS, United Kingdom
J. Schnakenberg	Institut für Theoretische Physik der Technischen Universität, Templer Graben 64, Aachen, Federal Republic of Germany
J.R. Schrieffer*[1]	Department of Physics, University of Pennsylvania, Philadelphia, Pa. 19174, United States of America
G. Scoles*	Chemistry Department, University of Waterloo, Waterloo, Ont. N2L 3G1, Canada
G.A. Somorjai	Department of Chemistry, University of California, Berkeley, Calif. 94720, United States of America
D. Tabor	Cavendish Laboratory, Cambridge University, Madingley Road, Cambridge CB3 0HE, United Kingdom
M. Tomášek	J. Heyrovský Institute of Physical Chemistry and Electrochemistry, Máchova 7, 121 38 Prague 2, Czechoslovakia
E. Tosatti*	Institut für Theoretische Physik, Universität Stuttgart, Azenbergstrasse 12, 7 Stuttgart 1, Federal Republic of Germany

EDITOR

Miriam Lewis	Division of Publications, IAEA, Vienna, Austria

---

[1] Professor Schrieffer's lectures on Electron Theory of Chemisorption and Catalysis are published in the Proceedings of Course LVIII of the Enrico Fermi Summer School in Physics, Varenna, 1973.

The following conversion table is provided for the convenience of readers and to encourage the use of SI units.

FACTORS FOR CONVERTING UNITS TO SI SYSTEM EQUIVALENTS*

SI base units are the metre (m), kilogram (kg), second (s), ampere (A), kelvin (K), candela (cd) and mole (mol).
[For further information, see International Standards ISO 1000 (1973), and ISO 31/0 (1974) and its several parts]

Multiply		by	to obtain
Mass			
pound mass (avoirdupois)	1 lbm	= 4.536×10^{-1}	kg
ounce mass (avoirdupois)	1 ozm	= 2.835×10^{1}	g
ton (long) (= 2240 lbm)	1 ton	= 1.016×10^{3}	kg
ton (short) (= 2000 lbm)	1 short ton	= 9.072×10^{2}	kg
tonne (= metric ton)	1 t	= 1.00×10^{3}	kg
Length			
statute mile	1 mile	= 1.609×10^{0}	km
yard	1 yd	= 9.144×10^{-1}	m
foot	1 ft	= 3.048×10^{-1}	m
inch	1 in	= 2.54×10^{-2}	m
mil (= 10^{-3} in)	1 mil	= 2.54×10^{-2}	mm
Area			
hectare	1 ha	= 1.00×10^{4}	m^2
(statute mile)^2	1 mile^2	= 2.590×10^{0}	km^2
acre	1 acre	= 4.047×10^{3}	m^2
yard^2	1 yd^2	= 8.361×10^{-1}	m^2
foot^2	1 ft^2	= 9.290×10^{-2}	m^2
inch^2	1 in^2	= 6.452×10^{2}	mm^2
Volume			
yard^3	1 yd^3	= 7.646×10^{-1}	m^3
foot^3	1 ft^3	= 2.832×10^{-2}	m^3
inch^3	1 in^3	= 1.639×10^{4}	mm^3
gallon (Brit. or Imp.)	1 gal (Brit)	= 4.546×10^{-3}	m^3
gallon (US liquid)	1 gal (US)	= 3.785×10^{-3}	m^3
litre	1 l	= 1.00×10^{-3}	m^3
Force			
dyne	1 dyn	= 1.00×10^{-5}	N
kilogram force	1 kgf	= 9.807×10^{0}	N
poundal	1 pdl	= 1.383×10^{-1}	N
pound force (avoirdupois)	1 lbf	= 4.448×10^{0}	N
ounce force (avoirdupois)	1 ozf	= 2.780×10^{-1}	N
Power			
British thermal unit/second	1 Btu/s	= 1.054×10^{3}	W
calorie/second	1 cal/s	= 4.184×10^{0}	W
foot-pound force/second	1 ft·lbf/s	= 1.356×10^{0}	W
horsepower (electric)	1 hp	= 7.46×10^{2}	W
horsepower (metric) (= ps)	1 ps	= 7.355×10^{2}	W
horsepower (550 ft·lbf/s)	1 hp	= 7.457×10^{2}	W

* Factors are given exactly or to a maximum of 4 significant figures

Multiply	by	to obtain

Density

pound mass/inch³	1 lbm/in³ = 2.768 × 10⁴	kg/m³
pound mass/foot³	1 lbm/ft³ = 1.602 × 10¹	kg/m³

Energy

British thermal unit	1 Btu = 1.054 × 10³	J
calorie	1 cal = 4.184 × 10⁰	J
electron-volt	1 eV ≃ 1.602 × 10⁻¹⁹	J
erg	1 erg = 1.00 × 10⁻⁷	J
foot-pound force	1 ft·lbf = 1.356 × 10⁰	J
kilowatt-hour	1 kW·h = 3.60 × 10⁶	J

Pressure

newtons/metre²	1 N/m² = 1.00	Pa
atmosphere[a]	1 atm = 1.013 × 10⁵	Pa
bar	1 bar = 1.00 × 10⁵	Pa
centimetres of mercury (0°C)	1 cmHg = 1.333 × 10³	Pa
dyne/centimetre²	1 dyn/cm² = 1.00 × 10⁻¹	Pa
feet of water (4°C)	1 ftH₂O = 2.989 × 10³	Pa
inches of mercury (0°C)	1 inHg = 3.386 × 10³	Pa
inches of water (4°C)	1 inH₂O = 2.491 × 10²	Pa
kilogram force/centimetre²	1 kgf/cm² = 9.807 × 10⁴	Pa
pound force/foot²	1 lbf/ft² = 4.788 × 10¹	Pa
pound force/inch² (= psi)[b]	1 lbf/in² = 6.895 × 10³	Pa
torr (0°C) (= mmHg)	1 torr = 1.333 × 10²	Pa

Velocity, acceleration

inch/second	1 in/s = 2.54 × 10¹	mm/s
foot/second (= fps)	1 ft/s = 3.048 × 10⁻¹	m/s
foot/minute	1 ft/min = 5.08 × 10⁻³	m/s
mile/hour (= mph)	1 mile/h = 4.470 × 10⁻¹	m/s
	1 mile/h = 1.609 × 10⁰	km/h
knot	1 knot = 1.852 × 10⁰	km/h
free fall, standard (= g)	= 9.807 × 10⁰	m/s²
foot/second²	1 ft/s² = 3.048 × 10⁻¹	m/s²

Temperature, thermal conductivity, energy/area·time

Fahrenheit, degrees −32	°F − 32 × 5/9	°C
Rankine	°R × 5/9	K
1 Btu·in/ft²·s·°F	= 5.189 × 10²	W/m·K
1 Btu/ft·s·°F	= 6.226 × 10¹	W/m·K
1 cal/cm·s·°C	= 4.184 × 10²	W/m·K
1 Btu/ft²·s	= 1.135 × 10⁴	W/m²
1 cal/cm²·min	= 6.973 × 10²	W/m²

Miscellaneous

foot³/second	1 ft³/s = 2.832 × 10⁻²	m³/s
foot³/minute	1 ft³/min = 4.719 × 10⁻⁴	m³/s
rad	rad = 1.00 × 10⁻²	J/kg
roentgen	R = 2.580 × 10⁻⁴	C/kg
curie	Ci = 3.70 × 10¹⁰	disintegration/s

[a] atm abs: atmospheres absolute; atm (g): atmospheres gauge.

[b] lbf/in² (g) (= psig): gauge pressure; lbf/in² abs (= psia): absolute pressure.

HOW TO ORDER IAEA PUBLICATIONS

Exclusive sales agents for IAEA publications, to whom all orders
and inquiries should be addressed, have been appointed
in the following countries:

UNITED KINGDOM	Her Majesty's Stationery Office, P.O. Box 569, London SE 1 9NH
UNITED STATES OF AMERICA	UNIPUB, Inc., P.O. Box 433, Murray Hill Station, New York, N.Y. 10016

In the following countries IAEA publications may be purchased from the
sales agents or booksellers listed or through your
major local booksellers. Payment can be made in local
currency or with UNESCO coupons.

ARGENTINA	Comisión Nacional de Energía Atómica, Avenida del Libertador 8250, Buenos Aires
AUSTRALIA	Hunter Publications, 58 A Gipps Street, Collingwood, Victoria 3066
BELGIUM	Service du Courrier de l'UNESCO, 112, Rue du Trône, B-1050 Brussels
CANADA	Information Canada, 171 Slater Street, Ottawa, Ont. K 1 A OS 9
C.S.S.R.	S.N.T.L., Spálená 51, CS-11000 Prague
	Alfa, Publishers, Hurbanovo námestie 6, CS-80000 Bratislava
FRANCE	Office International de Documentation et Librairie, 48, rue Gay-Lussac, F-75005 Paris
HUNGARY	Kultura, Hungarian Trading Company for Books and Newspapers, P.O. Box 149, H-1011 Budapest 62
INDIA	Oxford Book and Stationery Comp., 17, Park Street, Calcutta 16
ISRAEL	Heiliger and Co., 3, Nathan Strauss Str., Jerusalem
ITALY	Libreria Scientifica, Dott. de Biasio Lucio "aeiou", Via Meravigli 16, I-20123 Milan
JAPAN	Maruzen Company, Ltd., P.O.Box 5050, 100-31 Tokyo International
NETHERLANDS	Marinus Nijhoff N.V., Lange Voorhout 9-11, P.O. Box 269, The Hague
PAKISTAN	Mirza Book Agency, 65, The Mall, P.O.Box 729, Lahore-3
POLAND	Ars Polona, Centrala Handlu Zagranicznego, Krakowskie Przedmiescie 7, Warsaw
ROMANIA	Cartimex, 3-5 13 Decembrie Street, P.O.Box 134-135, Bucarest
SOUTH AFRICA	Van Schaik's Bookstore, P.O.Box 724, Pretoria
	Universitas Books (Pty) Ltd., P.O.Box 1557, Pretoria
SPAIN	Nautrónica, S.A., Pérez Ayuso 16, Madrid-2
SWEDEN	C.E. Fritzes Kungl. Hovbokhandel, Fredsgatan 2, S-10307 Stockholm
U.S.S.R.	Mezhdunarodnaya Kniga, Smolenskaya-Sennaya 32-34, Moscow G-200
YUGOSLAVIA	Jugoslovenska Knjiga, Terazije 27, YU-11000 Belgrade

Orders from countries where sales agents have not yet been appointed and
requests for information should be addressed directly to:

Publishing Section,
International Atomic Energy Agency,
Kärntner Ring 11, P.O.Box 590, A-1011 Vienna, Austria